Advances in Behavioral Biology

ADVANCES IN BEHAVIORAL BIOLOGY

Recent Volumes in this series

Volume 46   BIOLOGY AND PHYSIOLOGY OF THE BLOOD-BRAIN BARRIER: Transport, Cellular Interactions, and Brain Pathologies
Edited by Pierre-Olivier Couraud and Daniel Scherman

Volume 47   THE BASAL GANGLIA V
Edited by Chihoto Ohye, Minoru Kimura and John S. Mckenzie

Volume 48   KINDLING 5
Edited by Michael E. Corcoran and Solomon Moshé

Volume 49   PROGRESS IN ALZHEIMER'S AND PARKINSON'S DISEASES
Edited by Abraham Fisher, Israel Hanin and Mitsuo Yoshida

Volume 50   NEUROPSYCHOLOGY OF CHILDHOOD EPILEPSY
Edited by Isabelle Jambaqué, Maryse Lassonde and Olivier Dulac

Volume 51   MAPPING THE PROGRESS OF ALZHEIMER'S AND PARKINSON'S DISEASE
Edited by Yoshikuni Mizuno, Abraham Fisher and Israel Hanin

Volume 52   THE BASAL GANGLIA VI
Edited by Louise F.B. Nicholson and Richard L.M. Faull

Volume 53   CATECHOLAMINE RESEARCH: From Molecular Insights to Clinical Medicine
Edited by Toshiharu Nagatsu, Toshitaka Nabeshima, Richard McCarty and David S. Goldstein

Volume 54   THE BASAL GANGLIA VII
Edited by Ann M. Graybiel, Mahlon R. Delong and Stephen T. Kitai

Volume 55   KINDLING 6
Edited by Michael E. Corcoran and Solomon L. Moshé

Volume 56   THE BASAL GANGLIA VIII
Edited by J. Paul Bolam, Cali A. Ingham and Peter J. Magill

Volume 57   ADVANCES IN ALZHEIMER'S AND PARKINSON'S DISEASE
Edited by Abraham Fisher, Maurizio Memo, Fabrizio Stocchi and Israel Hanin

Abraham Fisher · Maurizio Memo ·
Fabrizio Stocchi · Israel Hanin

Editors

Advances in Alzheimer's and Parkinson's Disease

Insights, Progress, and Perspectives

 Springer

Editors

Abraham Fisher
Israel Institute for Biological Research
IIBR, Ness-Ziona
Israel

Maurizio Memo
Universitá degli Studi di Brescia
25124 Brescia
Italy

Fabrizio Stocchi
Universitá La Sapienze Roma
86077 Pozilli, Isernia
Italy

Israel Hanin
Loyola University Stritch
School of Medicine
Department of Pharmacology
Maywood, IL 60153-5589
USA

Library of Congress Control Number: 2007930197

ISBN 978-0-387-72074-6   ISBN 978-0-387-72076-0

Printed on acid-free paper

9 8 7 6 5 4 3 2 1

springer.com

Preface

The 7th International Congress on Alzheimer's and Parkinson's Disease took place during March 9–13, 2005 high in the hills of Sorrento, Italy, overlooking the splendid blue Mediterranean Sea with Mt. Vesuvius in the foreground.

This event continues to earn its growing reputation of success. With approximately 1500 scientific participants in attendance and the plenary, oral presentation, and poster sessions, the four and a half day meeting provided much scientific stimulation and many opportunities for discussion, exchange of ideas, and development of new collaborations. This book includes chapters contributed by selected invited speakers at the congress.

The generosity of several past sponsors as well as an impressive number of new corporate sponsors enabled us to ensure a meaningful and comprehensive congress. Our appreciation goes to Astra Zeneca, Axonyx, Bracco, Elan, Wyeth, and GE Healthcare (major sponsors) and to Amgen, Biogen Idec, Boehringer-Ingelheim, Merck Research Laboratories, Neurochem, Neuropharma, Newron, Novartis, and Voyager (sponsors). In addition we acknowledge with thanks the contributions of GSK, Teva, and Zapaq.

We are particularly grateful to members of the Scientific Board, whose scientific stature and contributions to the conference program contributed to the high level and excellent quality of the scientific content of ADPD 2005. The highly efficient and professional organizational skills of Kenes International were evident throughout the conference and contributed to the smooth functioning of the entire event. Professor Umberto Cornelli's provision of four competitive travel awards to junior investigators started a new trend for these conferences, encouraging and assisting young and upcoming scientists in the field of Alzheimer's disease and Parkinson's disease research to attend these meetings.

Last but not least, the success of a conference of this caliber depends on the participants themselves. The diverse international clinical and preclinical AD and PD research and treatment community was well represented at this conference. We hope that the conference served the purpose of bringing together old friends, helped create new acquaintanceships, and resulted in new, fruitful collaborative ventures.

Abraham Fisher PhD, Maurizio Memo PhD,
Fabrizio Stocchi MD, PhD, and Israel Hanin PhD

v

Contents

1. **Clinical Features and Treatment of Dementia Associated With Parkinson's Disease** 1
   Murat Emre

2. **Neurogenetics in Parkinson's Disease** 9
   Yoshikuni Mizuno and Nobutaka Hattori

3. **Impaired DNA Repair Systems: Generation of Mitochondrial Dysfunction and Increased Sensitivity to Excitotoxicity** 17
   Simona Francisconi, Mara Codenotti, Erminia Poli,
   Daniela Uberti, Giulia Ferrari-Toninelli, and
   Maurizio Memo

4. **Current Problems and Strategies in Motor Rehabilitation for Parkinson's Disease** 23
   Giovanni Abbruzzese, Elisa Pelosin, and Roberta Marchese

5. **Critical Acute Akinesia in Parkinson's Disease Patients Concomitant with Additional Precipitating Factors Other Than Treatment Withdrawal: A Possible Lethal Complication** 31
   Marco Onofrj and Astrid Thomas

6. **Biomarkers for Early Detection of Parkinson's Disease: An Essential Challenge** 35
   Florian Tribl and Peter Riederer

7. **Neurotransmitter and Neurotrophic Factor-Secreting Cell Line Grafting for the Treatment of Parkinson's Disease** 51
   Isao Date, Tetsuro Shingo, and Takao Yasuhara

8. **Novel Therapeutic Target in PD: Experimental Models** 57
   Francesco Fornai

9. *LRRK2* (Leucine-Rich Repeat Kinase 2) Gene on PARK8 Locus in
    Families with Parkinsonism 75
    Zbigniew K. Wszolek, Alexander Zimprich, Saskia Biskup,
    Ryan J. Uitti, Donald B. Calne, A. Jon Stoessl, Akiko Imamura,
    Matthew J. Farrer, Judith Miklossy, Thomas Meitinger,
    Thomas Gasser, Dennis W. Dickson, and Patrick L. McGeer

10. Microglia Activation and Gene Expression of Cytokines in
    Parkinson's Disease 91
    Makoto Sawada, Kazuhiro Imamura, and Toshiharu Nagatsu

11. Dissecting the Biochemical Pathways Mediated by Genes
    Implicated in Parkinson's Disease: Induction of DJ-1 Expression in
    A30P α-Synuclein Mice 97
    Mark Frasier, Shanti Frausto, Daniel Lewicki, Lawrence Golbe,
    and Benjamin Wolozin

12. Neuroinflammation in Early Stages of Alzheimer's Disease and
    Parkinson's Disease 113
    Piet Eikelenboom, A. Crosswell, C. van Engen, M. Limper,
    J.J.M. Hoozemans, R. Veerhuis, W.A. van Gool, and
    J.M. Rozemuller

13. Alzheimer's Disease, Parkinson's Disease, and Frontotemporal
    Dementias: Different Manifestations of Protein Misfolding 123
    John Q. Trojanowski, Mark S. Forman, and Virginia M-Y. Lee

14. Role of Oxidative Insult and Neuronal Survival in Alzheimer's and
    Parkinson's Diseases 133
    Akihiko Nunomura, Paula I. Moreira, Xiongwei Zhu,
    Adam D. Cash, Mark A. Smith, and George Perry

15. Redox Proteomics Identification of Oxidatively Modified Proteins
    in Alzheimer's Disease Brain and in Brain from a Rodent Model of
    Familial Parkinson's Disease: Insights into Potential Mechanisms
    of Neurodegeneration 149
    Rukhsana Sultana, H. Fai Poon, and D. Allan Butterfield

16. Biomarkers for Alzheimer's Disease and Parkinson's Disease 169
    John H. Growdon, Michael C. Irizarry, and Clemens Scherzer

17. Bioluminescent Imaging of Excitotoxic and Endotoxic Brain Injury
    in Living Mice 175
    Jian Luo, Amy H. Lin, and Tony Wyss-Coray

18. **Alzheimer's Disease Neuroimaging Initiative** 183
Susanne G. Mueller, Michael W. Weiner, Leon J. Thal, Ronald C.
Petersen, Clifford Jack, William Jagust, John Q. Trojanowski,
Arthur W. Toga, and Laurel Beckett

19. **Two Hits and You're Out? A Novel Mechanistic Hypothesis of
Alzheimer Disease** 191
Xiongwei Zhu, George Perry, and Mark A. Smith

20. **Vascular Risk Factors and Risk for Alzheimer's Disease and Mild
Cognitive Impairment: Population Based Studies** 205
Hilkka Soininen, Miia Kivipelto, and Aulikki Nissinen

21. **Cholesterol Transport and Production in Alzheimer's Disease** 211
Judes Poirier, Louise Lamarre-Théroux, Doris Dea, Nicole
Aumont, and Jean Francois Blain

22. **Cholesterol and Aβ Production: Methods for Analysis of Altered
Cholesterol De Novo Synthesis** 221
Jakob A. Tschäpe, Marcus O.W. Grimm, Heike S. Grimm, and
Tobias Hartmann

23. **Glycosaminoglycans and Analogs in Neurodegenerative
Disorders** 231
Lucilla Parnetti and Umberto Cornelli

24. **Prospective Role of Glycosaminoglycans in Apoptosis Associated
with Neurodegenerative Disorders** 247
Bertalan Dudas, Amira Lemes, Umberto Cornelli, and
Israel Hanin

25. **Stem Cell Therapy in Alzheimer's Disease** 255
Kiminobu Sugaya, Young-Don Kwak, and Angel Alvarez

26. **Oral Aβ Vaccine Using a Recombinant Adeno-Associated Virus
Vector in an Alzheimer's Disease Mouse Model** 265
Takeshi Tabira and Hideo Hara

27. **In Vivo Targeting of Amyloid Plaques Via Intranasal
Administration of Phage Anti-β-Amyloid Antibodies** 273
Beka Solomon

28. **Decreased ProBDNF: The Cause of Alzheimer's-Associated
Neurodegeneration and Cognitive Decline?** 279
Margaret Fahnestock, S. Peng, D.J. Garzon, and Elliott J. Mufson

29. **Shift in the Balance of TRKA and ProNGF in Prodromal Alzheimer's Disease** 285
Elliott J. Mufson, Scott E. Counts, S. Peng, and Margaret Fahnestock

30. **Neuroprotective Effects of Trophic Factors and Natural Products: Involvement of Multiple Extracellular Kinases** 291
Stéphane Bastianetto, Wen-Hua Zheng, Yingshan Han, Lixia Gan, and Rémi Quirion

31. **Intraneuronal Aβ and Alzheimer's Disease** 297
Lauren M. Billings and Frank M. LaFerla

32. **Physiological Processing of the Cellular Prion Protein and βAPP: Enzymes and Regulation** 305
Bruno Vincent, Moustapha Alfa Cisse, and Frédéric Checler

33. **Expression of Wnt Receptors, Frizzled, in Rat Neuronal Cells** 317
Marcelo A. Chacón, Marcela Columbres, and Nibaldo C. Inestrosa

34. **Cell Models of Tauopathy** 325
J. Biernat, I. Khlistunova, Y-P. Wang, M. Pickhardt, M. von Bergen, Z. Gazova, Eckhart Mandelkow, and Eva-Marie Mandelkow

35. **Co-expression of FTDP-17 Human Tau and GSK-3β (or APP^{SW}) in Transgenic Mice: Induction of Tau Polymerization and Neurodegeneration** 337
Jesús Avila, Tobias Engel, José J. Lucas, Mar Pérez, Alicia Rubio, and Félix Hernández

36. **Enhanced Activation of NF-κB Signaling by Apolipoprotein E4** 343
Gal Ophir, Liza Mizrahi, and Daniel M. Michaelson

37. **Pleiotropic Effects of Apolipoprotein E in Dementia: Influence on Functional Genomics and Pharmacogenetics** 355
Ramón Cacabelos

38. **Up-regulation of the α-Secretase Pathway** 369
Falk Fahrenholz, Claudia Prinzen, Rolf Postina, and Elżbieta Kojro

39. **Frequency and Relation of Argyrophilic Grain Disease and Thorn-Shaped Astrocytes in Alzheimer's Disease** 375
Hirotake Uchikado, Yasuhiro Fujino, Wenlang Lin, and Dennis Dickson

40. **Cellular Membranes as Targets in Amyloid Oligomer Disease Pathogenesis** 381
Erene W. Mina and Charles G. Glabe

41. Ganglioside-Dependent Generation of a Seed for Alzheimer's Disease Amyloid 387
Katsuhiko Yanagisawa

42. Evidence That Amyloid Pathology Progresses in a Neurotransmitter-Specific Manner 393
Karen F.S. Bell and A. Claudio Cuello

43. Rationale for Glutamatergic and Cholinergic Approaches for the Treatment of Alzheimer's Disease 403
Paul T. Francis and Sara L. Kirvell

44. Cortical Cholinergic Lesion Causes Aβ Deposition: Cholinergic-Amyloid Fusion Hypothesis 411
Thomas Beach, Pamela Potter, Lucia Sue, Amanda Newell,
Marissa Poston, Raquel Cisneros, Yoga Pandya, Abraham
Fisher, Alex Roher, Lih-Fen Lue, and Douglas Walker

45. Expression of Acetylcholinesterase in Alzheimer's Disease Brain: Role in Neuritic Dystrophy and Synaptic Scaling 429
David H. Small, Steven Petratos, Sharon Unabia, and Danuta Maksel

46. Role of β-Amyloid in the Pathophysiology of Alzheimer's Disease and Cholinesterase Inhibition: Facing the Biological Complexity to Treat the Disease 439
Stefano Govoni, Michela Mazzucchelli, S. Carolina Lenzken,
Emanuela Porrello, Cristina Lanni, and Marco Racchi

47. Dissociation Between the Potent β-Amyloid Protein Pathway Inhibition and Cholinergic Actions of the Alzheimer Drug Candidates Phenserine and Cymserine 445
Nigel H. Greig, Tada Utsuki, Qian-sheng Yu, Harold W.
Holloway, Tracyann Perry, David Tweedie, Tony Giordano,
George M. Alley, De-Mao Chen, Mohammad A. Kamal,
Jack T. Rogers, Kumar Sambamurti, and Debomoy K. Lahiri

48. Allosteric Potentiators of Neuronal Nicotinic Cholinergic Receptors: Potential Treatments for Neurodegenerative Disorders 463
Emanuele Sher, Giovanna De Filippi, T. Baldwinson, Ruud
Zwart, Kathy H. Pearson, Martin Lee, Louise Wallace, Gordon
I. McPhie, Martine Keenan, Renee Emkey, Sean P. Hollinshead,
Colin P. Dell, S. Richard Baker, Michael J. O'Neil, and Lisa M. Broad

Index 473

Contributors

Giovanni Abbruzzese
Department of Neurosciences, Ophthalmology and Genetics, University of
Genoa, Genova, Italy

George M. Alley
Department of Psychiatry, Indiana University School of Medicine,
Indianapolis, IN 46202, USA

Angel Alvarez
Biomolecular Sciences Center, Burnett College of Biomedical Sciences,
University of Central Florida, Orlando, FL 32816-2364, USA

Nicole Aumont
McGill Center for studies in Aging, McGill University, Montreal, Quebec,
Canada

Jesús Avila
Centro de Biología Molecular "Severo Ochoa" (CSIC-UAM). Facultad de
Ciencias. Campus de Cantoblanco. Universidad Autónoma de Madrid. 28049-
Madrid, Spain.

S. Richard Baker
Eli Lilly and Company, Lilly Research Centre, Erl Wood Manor, Windlesham,
Surrey GU20 6PH, UK

Tristan Baldwinson
Eli Lilly and Company, Lilly Research Centre, Erl Wood Manor, Windlesham,
Surrey GU20 6PH, UK

Stéphane Bastianetto
Douglas Hospital Research Centre, Department of Psychiatry, McGill
University, Montréal, Québec, H4H 1R3, Canada

Thomas Beach
Civin Laboratory for Neuropathology, Sun Health Research Institute, Sun City, AZ, USA

Laurel Beckett
University of California, Davis, California, USA

Karen F. S. Bell
Department of Pharmacology and Therapeutics, McGill University, Montreal, QC, Canada

M. von Bergen
Max-Planck-Unit for Structural Molecular Biology, c/o DESY, Hamburg, Germany

J. Biernat
Max-Planck-Unit for Structural Molecular Biology, c/o DESY, Hamburg, Germany

Lauren M. Billings
Department of Neurobiology and Behavior, University of California, Irvine, CA 92612, USA; Allergan Inc., Irvine, CA 92612, USA

Saskia Biskup
Institute of Human Genetics GSF National Research Center for Environment and Health, Neuherberg 85764, Germany

Jean Francois Blain
McGill Center for studies in Aging, McGill University, Montreal, Quebec, Canada

Lisa M. Broad
Eli Lilly and Company, Lilly Research Centre, Erl Wood Manor, Windlesham, Surrey GU20 6PH, UK

D. Allan Butterfield
Department of Chemistry; Sanders-Brown Center on Aging; Center of Membrane Sciences; University of Kentucky, Lexington, KY 40506-0055, USA.

Ramón Cacabelos
EuroEspes Biomedical Research Center, Institute for CNS Disorders, 15166-Bergondo, Coruña, Spain

Donald B. Calne
Pacific Parkinson's Research Centre University of British Columbia,
Vancouver, British Columbia V6T 2B5 Canada

Adam D. Cash
Department of Pathology, Case Western Reserve University, Cleveland, OH
44106, USA

Marcelo A. Chacón
Centro de Regulación Celular y Patología "Joaquín V. Luco" (CRCP), MIFAB,
Facultad de Ciencias Biológicas, Pontificia Universidad Católica de Chile

Frédéric Checler
Institut de Pharmacologie Moléculaire et Cellulaire, CNRS UMR6097,
Valbonne 06560, France

De-Mao Chen
Department of Psychiatry, Indiana University School of Medicine,
Indianapolis, IN 46202, USA

Raquel Cisneros
Civin Laboratory for Neuropathology, Sun Health Research Institute, Sun
City, AZ, USA

Moustapha Alfa Cisse
Department of Neurology and Neurological Sciences, Stanford University
School of Medicine, Stanford, CA 94305; and GRECC, VA Palo Alto Health
Care System, Palo Alto, CA 94304, USA

Mara Codenotti
Department of Biomedical Sciences and Biotechnologies, University of Brescia,
Brescia, Italy

Marcela Colombres
Centro de Regulación Celular y Patología "Joaquín V. Luco" (CRCP), MIFAB,
Facultad de Ciencias Biológicas, Pontificia Universidad Católica de Chile, Chile

Umberto Cornelli
Department of Pharmacology, Loyola University Chicago Stritch School of
Medicine, Maywood, IL 60153, USA

Scott E. Counts
Department of Neurological Sciences, Rush University Medical School, 1735
W. Harrison St., Chicago, IL, USA; Department of Psychiatry and Behavioural
Neurosciences, McMaster University, Hamilton, Canada

A. Crosswell
Department of Neuropathology, Academic Medical Center, University of
Amsterdam, Amsterdam, The Netherlands

A. Claudio Cuello
Departments of Pharmacology and Therapeutics, Anatomy and Cell Biology,
and Neurology and Neurosurgery, McGill University, Montreal, QC, Canada

Isao Date
Department of Neurological Surgery, Okayama University, Okayama
700-8558, Japan

Doris Dea
McGill Center for studies in Aging, McGill University, Montreal, Quebec,
Canada

Colin P. Dell
Eli Lilly and Company, Lilly Research Centre, Erl Wood Manor, Windlesham,
Surrey GU20 6PH, UK

Dennis W. Dickson
Department of Neuroscience, Mayo Clinic Jacksonville, Jacksonville, FL
32224, USA

Bertalan Dudas
Neuroendocrine Organization Laboratory (NEO), Lake Erie College of
Osteopathic Medicine, Erie, PA 16509, USA

Piet Eikelenboom
Department of Neurology, Academic Medical Center, University of
Amsterdam, Amsterdam, The Netherlands and Department of Psychiatry,
Vrije Universiteit Medical Center, Amsterdam, The Netherlands

Renee Emkey
Eli Lilly and Company, Lilly Research Centre, Erl Wood Manor, Windlesham,
Surrey GU20 6PH, UK

Murat Emre
Professor of Neurology, Istanbul University, Istanbul Faculty of Medicine,
Department of Neurology, Behavioral Neurology and Movement Disorders
Unit, 34390 Çapa Istanbul Turkey

Tobias Engel
Centro de Biología Molecular "Severo Ochoa" (CSIC-UAM). Facultad de
Ciencias. Campus de Cantoblanco. Universidad Autónoma de Madrid. 28049-
Madrid, Spain.

C. Van Engen
Department of Neuropathology, Academic Medical Center, University of Amsterdam, Amsterdam, The Netherlands

Margaret Fahnestock
Department of Psychiatry and Behavioural Neurosciences, McMaster University, Hamilton, Canada

Falk Fahrenholz
Institute of Biochemistry, University of Mainz, 55099 Mainz, Germany

Matthew J. Farrer
Department of Neurology Mayo Clinic Jacksonville, Jacksonville, FL 32224, USA

Giulia Ferrari-Toninelli
Department of Biomedical Sciences and Biotechnologies, University of Brescia, Brescia, Italy

Giovanna De Filippi
Eli Lilly and Company, Lilly Research Centre, Erl Wood Manor, Windlesham, Surrey GU20 6PH, UK

Abraham Fisher
Israel Institute for Biological Research, Ness Ziona, Israel

Mark S. Forman
The Center for Neurodegenerative Disease Research, Department of Pathology and Laboratory Medicine, and Institute on Aging, The University of Pennsylvania School of Medicine, Philadelphia, PA 19104, USA

Francesco Fornai
Department of Human Morphology and Applied Biology, University of Pisa, 56126 Pisa, Italy,

Paul T. Francis
Wolfson Centre for Age-Related Diseases, King's College London, Guy's Campus, St Thomas Street, London SE11UL, UK

Simona Francisconi
Department of Biomedical Sciences and Biotechnologies, University of Brescia, Brescia, Italy

Mark Frasier
Michael J. Fox Foundation, NewYork, NY 10163, USA

Shanti Frausto
Northwestern University, Chicago, IL 60611, USA

Yasuhiro Fujino
Kawanami Hospital, Fukuoka 814-0171, Japan

Lixia Gan
Douglas Hospital Research Centre, Department of Psychiatry, McGill
University, Montréal, Québec, H4H 1R3 Canada

D.J. Garzon
Department of Psychiatry and Behavioural Neurosciences, McMaster
University, Hamilton, Canada

Thomas Gasser
Hertie Institute for Clinical Brain Research, University of Tübingen, 72076
Tübingen, Germany

Z. Gazova
Max-Planck-Unit for Structural Molecular Biology, c/o DESY, Hamburg,
Germany

Tony Giordano
Department of Biochemistry & Molecular Biology, Louisiana State University,
Shreveport, LO 71130, USA

Charles G. Glabe
Department of Molecular Biology and Biochemistry, University of California,
Irvine, CA 92696-3900, USA

Lawrence Golbe
Department of Pharmacology, Boston University School of Medicine, Boston,
MA 02118-2526

W.A. Van Gool
Department of Neurology, Academic Medical Center, University of
Amsterdam, Amsterdam, The Netherlands

Stefano Govoni
CEBA – Centre of Excellence in Applied Biology and Department of
Experimental and Applied Pharmacology, Pavia 27100, Italy

Nigel H. Greig
Drug Design & Development Section, Laboratory of Neurosciences, Intramural
Research Program, National Institute on Aging, Baltimore, MD 21224, USA

Heike S. Grimm
Center for Molecular Biology Heidelberg (ZMBH), University of Heidelberg,
Heidelberg D-69120, Germany

Marcus O.W. Grimm
Center for Molecular Biology Heidelberg (ZMBH), University of Heidelberg,
Heidelberg D-69120, Germany

John H. Growdon
Department of Neurology, Partners HealthCare System, Inc. Boston,
MA. USA

Yingshan Han
Douglas Hospital Research Centre, Department of Psychiatry, McGill
University, Montréal, Québec, H4H 1R3 Canada

Israel Hanin
Department of Pharmacology, Loyola University Chicago Stritch School of
Medicine, Maywood, IL 60153, USA

Hideo Hara
National Institute for Longevity Sciences, NCGG, 36-3 Gengo, Morioka, Obu
City, Aichi 474-8522, Japan

Tobias Hartmann
Center for Molecular Biology Heidelberg (ZMBH), University of Heidelberg,
Heidelberg D-69120, Germany

Nobutaka Hattori
Department of Neurology, Juntendo University School of Medicine, Tokyo
113-8421, Japan

Félix Hernández
Centro de Biología Molecular "Severo Ochoa" (CSIC-UAM). Facultad de
Ciencias. Campus de Cantoblanco. Universidad Autónoma de Madrid. 28049-
Madrid, Spain

Sean P. Hollinshead
Eli Lilly and Company, Lilly Research Centre, Erl Wood Manor, Windlesham,
Surrey GU20 6PH, UK

Harold W. Holloway
Drug Design & Development Section, Laboratory of Neurosciences,
Intramural Research Program, National Institute on Aging, Baltimore, MD
21224, USA

J.J. M. Hoozemans
Department of Neuropathology, Academic Medical Center, University of
Amsterdam, Amsterdam, The Netherlands

Akiko Imamura
Department of Neurology, Mayo Clinic Jacksonville, Jacksonville, FL 32224, USA

Kazuhiro Imamura
Department of Neurology, Okazaki City Hospital, Okazaki, Aichi 444-8553,
Japan

Nibaldo C. Inestrosa
Centro de Regulación Celular y Patología "Joaquín V. Luco" (CRCP), MIFAB,
Facultad de Ciencias Biológicas, Pontificia Universidad Católica de Chile

Michael C. Irizarry
Department of Neurology, Partners HealthCare System Inc., Boston,
MA, USA

Clifford Jack
Mayo Clinic, Rochester, N.Y., USA

William Jagust
University of California, Berkeley, California, USA

Mohammad A. Kamal
Enzymoics, Hebersham, NSW 2770, Australia

Martine Keenan
Eli Lilly and Company, Lilly Research Centre, Erl Wood Manor, Windlesham,
Surrey, GU20 6PH, UK

I. Khlistunova
Max-Planck-Unit for Structural Molecular Biology, c/o DESY, Hamburg
22607, Germany

Sara L. Kirvell
Wolfson Centre for Age-Related Diseases, King's College London, Guy's
Campus, London SE11UL, UK

Miia Kivipelto
Neurotec; Karolinska University Hospital, Huddinge, Stockholm, Sweden

Elżbieta Kojro
Institute of Biochemistry, University of Mainz, Mainz, Germany

Young-Don Kwak
Biomolecular Sciences Center, Burnett College of Biomedical Sciences,
University of Central Florida, Orlando, FL 32816-2364, USA

Debomoy K. Lahiri
Department of Psychiatry, Indiana University School of Medicine,,
Indianapolis, IN 46202, USA.

Frank M. LaFerla
Department of Neurobiology and Behavior, University of California, Irvine,
CA 92612, USA

Louise Lamarre-Théroux
McGill Center for studies in Aging, McGill University, Montreal, Quebec,
Canada

Cristina Lanni
CEBA – Centre of Excellence in Applied Biology and Department of
Experimental and Applied Pharmacology, Pavia 27100, Italy

Martin Lee
Eli Lilly and Company, Lilly Research Centre, Erl Wood Manor, Windlesham,
Surrey GU20 6PH, UK

Virginia M-Y. Lee
The Center for Neurodegenerative Disease Research, Department of Pathology
and Laboratory Medicine, and Institute on Aging, The University of
Pennsylvania School of Medicine, Philadelphia, PA 19104, USA

Amira Lemes
Department of Pharmacology, Loyola University Chicago Stritch School of
Medicine, Maywood, IL 60153, USA

S. Carolina Lenzken
CEBA – Centre of Excellence in Applied Biology and Department of
Experimental and Applied Pharmacology, Pavia 27100, Italy

Danielle Lewicki
Argonne Laboratories, Argonne, IL 60439, USA

M. Limper
Department of Neuropathology, Academic Medical Center, University of
Amsterdam, Amsterdam, The Netherlands

Amy H. Lin
Department of Neurology and Neurological Sciences, Stanford University
School of Medicine, Stanford, CA 94305, USA

Wenlang Lin
Mayo Clinic College of Medicine, Jacksonville, FL 32224, USA

José J. Lucas
Centro de Biología Molecular "Severo Ochoa" (CSIC-UAM). Facultad de
Ciencias. Campus de Cantoblanco. Universidad Autónoma de Madrid. 28049-
Madrid, Spain

Lih-Fen Lue
Civin Laboratory for Neuropathology, Sun Health Research Institute, Sun
City, AZ, USA

Jian Luo
Department of Neurology and Neurological Sciences, Stanford University
School of Medicine, Stanford, CA, USA

Danuta Maksel
Department of Biochemistry and Molecular Biology, Monash University,
Victoria 3800, Australia

Eckhart Mandelkow
Max-Planck-Unit for Structural Molecular Biology, c/o DESY, Hamburg
22607, Germany

Eva-Marie Mandelkow
Max-Planck-Unit for Structural Molecular Biology, c/o DESY, Hamburg
22607, Germany

Roberta Marchese
Department of Neurosciences, Ophthalmology and Genetics, University of
Genoa, Via De Toni 5, Genova, Italy

Michela Mazzucchelli
CEBA – Centre of Excellence in Applied Biology and Department of
Experimental and Applied Pharmacology, Pavia 27100, Italy

Patrick L. McGeer
Pacific Parkinson's Research Centre University of British Columbia,
Vancouver, British Columbia V6T 2B5 Canada

Gordon I. McPhie
Eli Lilly and Company, Lilly Research Centre, Erl Wood Manor, Windlesham, Surrey GU20 6PH, UK

Thomas Meitinger
Institute of Human Genetics GSF National Research Center for Environment and Health, Neuherberg 85764, Germany

Maurizio Memo
Department of Biomedical Sciences and Biotechnologies, University of Brescia, Italy

Daniel M. Michaelson
Department of Neurobiochemistry, George S. Wise Faculty of Life Sciences, Tel Aviv University, Tel Aviv 69978, Israel

Judith Miklossy
Pacific Parkinson's Research Centre University of British Columbia, Vancouver, British Columbia V6T 2B5 Canada

Erene W. Mina
Department of Molecular Biology and Biochemistry, University of California, Irvine, CA 92696-3900, USA

Liza Mizrahi
Department of Neurobiochemistry, George S. Wise Faculty of Life Sciences, Tel Aviv University, Tel Aviv 69978, Israel

Yoshikuni Mizuno
Department of Neurology, Juntendo University School of Medicine, Tokyo 113-8421, Japan

Paula I. Moreira
Center for Neuroscience and Cell Biology of Coimbra, University of Coimbra, Coimbra 3004-517, Portugal

Susanne G. Mueller
University of California, San Francisco, California, USA

Elliott J. Mufson
Rush University Medical School, Department of Neurological Sciences, Chicago, USA

Toshiharu Nagatsu
Department of Brain Life Science, Research Institute of Environmental
Medicine, Nagoya University, Nagoya, Aichi 464-8601, Japan; Department of
Pharmacology, School of Medicine, Fujita Health University, Toyoake, Aichi
470-1192, Japan

Amanda Newell
Civin Laboratory for Neuropathology, Sun Health Research Institute, Sun
City, AZ, USA

Aulikki Nissinen
National Institute of Public Health, Helsinki, Finland

Akihiko Nunomura
Department of Psychiatry and Neurology, Asahikawa Medical College,
Asahikawa 078-8510, Japan

Michael J. O'Neil
Eli Lilly and Company, Lilly Research Centre, Erl Wood Manor, Windlesham,
Surrey GU20 6PH, UK

Marco Onofrj
Neurophysiopathology, Movement Disorders Center, Department of
Oncology and Neuroscience, University "G.D'Annunzio" Chieti-Pescara,
and Aging Research Center, Ce.S.I., "Gabriele D'Annunzio" University
Foundation, Chieti-Pescara, Italy, Ospedale Civile di Pescara, Italy

Gal Ophir
Department of Neurobiochemistry, George S. Wise Faculty of Life Sciences,
Tel Aviv University, Tel Aviv 69978, Israel

Yoga Pandya
Civin Laboratory for Neuropathology, Sun Health Research Institute, Sun City,
AZ, USA

Lucilla Parnetti
Neurology Section, University of Perugia, Italy

Kathy H. Pearson
Eli Lilly and Company, Lilly Research Centre, Erl Wood Manor, Windlesham,
Surrey GU20 6PH, UK

Elisa Pelosin
Department of Neurosciences, Ophthalmology and Genetics, University of
Genova, Genova, Italy

S. Peng
Department of Psychiatry and Behavioural Neurosciences, McMaster
University, Hamilton, Canada

Mar Pérez
Centro de Biología Molecular "Severo Ochoa" (CSIC-UAM). Facultad de
Ciencias. Campus de Cantoblanco. Universidad Autónoma de Madrid. 28049-
Madrid, Spain

George Perry
Department of Pathology, Case Western Reserve University, Cleveland
44106, OH, USA; College of Sciences, University of Texas at San Antonio,
San Antonio, Texas 78249, USA

Tracyann Perry
Drug Design & Development Section, Laboratory of Neurosciences,
Intramural Research Program, National Institute on Aging, Baltimore,
MD 21224, USA

Ronald C. Petersen
Mayo Clinic, Rochester, N.Y., USA

Steven Petratos
Department of Biochemistry and Molecular Biology, Monash University,
Victoria 3800, Australia

M. Pickhardt
Max-Planck-Unit for Structural Molecular Biology, c/o DESY, Hamburg
22607, Germany

Judes Poirier
McGill Centre for Studies in Aging, McGill University, Montreal, Quebec,
Canada

Erminia Poli
Department of Biomedical Sciences and Biotechnologies, University of Brescia,
Brescia, Italy

H. Fai Poon
Department of Chemistry; Sanders-Brown Center on Aging; University of
Kentucky, Lexington, KY 40506-0055, USA

Emanuela Porrello
CEBA – Centre of Excellence in Applied Biology and Department of
Experimental and Applied Pharmacology, Pavia 27100, Italy

Rolf Postina
Institute of Biochemistry, University of Mainz, Mainz, Germany

Marissa Poston
Civin Laboratory for Neuropathology, Sun Health Research Institute, Sun City, AZ, USA

Pamela Potter
Midwestern University, Glendale, AZ, USA

Claudia Prinzen
Institute of Biochemistry, University of Mainz, Mainz, Germany

Rémi Quirion
Douglas Hospital Research Centre, Department of Psychiatry, McGill University, Montréal, Québec, H4H 1R3 Canada

Marco Racchi
CEBA – Centre of Excellence in Applied Biology and Department of Experimental and Applied Pharmacology, Pavia 27100, Italy

Peter Riederer
Department of Clinical Neurochemistry, Clinic and Polyclinic for Psychiatry and Psychotherapy, University Würzburg, Würzburg 97080, Germany

Jack T. Rogers
Genetics and Aging Research Unit, Department of Psychiatry, Harvard Medical School, Charlestown, MA 02129, USA

Alex Roher
Civin Laboratory for Neuropathology, Sun Health Research Institute, Sun City, AZ, USA

J.M. Rozemuller
Department of Neuropathology, Academic Medical Center, University of Amsterdam, Amsterdam, The Netherlands

Alicia Rubio
Centro de Biología Molecular "Severo Ochoa" (CSIC-UAM). Facultad de Ciencias. Campus de Cantoblanco. Universidad Autónoma de Madrid. 28049-Madrid, Spain

Kumar Sambamurti
Department of Physiology & Neuroscience, Medical University S. Carolina, Charleston, SC 29425, USA

Makoto Sawada
Department of Brain Life Science, Research Institute of Environmental
Medicine, Nagoya University, Nagoya, Aichi 464-8601, Japan

Clemens Scherzer
Department of Neurology, Partners HealthCare System Inc., Boston,
MA, USA

Emanuele Sher
Eli Lilly and Company, Lilly Research Centre, Erl Wood Manor, Windlesham,
Surrey GU20 6PH, UK

Tetsuro Shingo
Department of Neurological Surgery, Okayama University, Okayama
700-8558, Japan

David H. Small
Department of Biochemistry and Molecular Biology, Monash University,
Victoria 3800, Australia

Mark A. Smith
Department of Pathology, Case Western Reserve University, Cleveland, OH
44106, USA

Hilkka Soininen
Department of Neurology, University of Kuopio, Kuopio 70211, Finland

Beka Solomon
Department of Molecular Microbiology and Biotechnology, George S. Wise
Faculty of Life Sciences, Tel Aviv University, Ramat Aviv, Tel Aviv 69978,
Israel

A. Jon Stoessl
Pacific Parkinson's Research Centre University of British Columbia,
Vancouver, British Columbia V6T 2B5 Canada

Lucia Sue
Civin Laboratory for Neuropathology, Sun Health Research Institute, Sun
City, AZ, USA

Kiminobu Sugaya
Biomolecular Sciences Center, Burnett College of Biomedical Sciences,
University of Central Florida, Orlando, FL 32816-2364, USA

Rukhsana Sultana
Department of Chemistry; Sanders-Brown Center on Aging; University of
Kentucky, Lexington, KY 40506-0055, USA

Takeshi Tabira
National Institute for Longevity Sciences, NCGG, Morioka, Obu City, Aichi
474-8522, Japan

Leon J. Thal
University of California, San Diego, USA

Astrid Thomas
Neurophysiopathology, Movement Disorders Center, Department of
Oncology and Neuroscience, University "G.D'Annunzio" Chieti-Pescara, and
Aging Research Center, Ce.S.I., "Gabriele D'Annunzio" University
Foundation, Chieti-Pescara, Italy

Arthur W Toga
University of California, Los Angeles, California, USA

Florian Tribl
The National Parkinson Foundation (NPF) Research Laboratories, Miami,
FL, USA

John Q. Trojanowski
The Center for Neurodegenerative Disease Research, Department of Pathology
and Laboratory Medicine, and Institute on Aging, The University of
Pennsylvania School of Medicine, Philadelphia, PA 19104, USA

Jakob-A. Tschäpe
Center for Molecular Biology Heidelberg (ZMBH), University of Heidelberg,
Heidelberg D-69120, Germany

David Tweedie
Drug Design & Development Section, Laboratory of Neurosciences,
Intramural Research Program, National Institute on Aging, Baltimore,
MD 21224, USA

Daniela Uberti
Department of Biomedical Sciences and Biotechnologies, University of Brescia,
Brescia, Italy

Hirotake Uchikado
Mayo Clinic College of Medicine, Jacksonville, FL 32224, USA

Ryan J. Uitti
Department of Neurology, Mayo Clinic Jacksonville, Jacksonville,
FL 32224, USA

Sharon Unabia
Department of Biochemistry and Molecular Biology, Monash University,
Victoria 3800, Australia

Tada Utsuki
Pennington Biomedical Research Center, Louisiana State University, Baton
Rouge, LA 70808, USA; Drung Design & Development Section, Laboratory
of Neurosciences, Intramural Research Program, National Institute on Aging,
Baltimore, MD 21224, USA

R, Veerhuis
Department of Psychiatry, Vrije Universiteit Medical Center, Amsterdam, The
Netherlands

Bruno Vincent
Institut de Pharmacologie Moléculaire et Cellulaire, CNRS
UMR6097,Valbonne 06560, France

Douglas Walker
Civin Laboratory for Neuropathology, Sun Health Research Institute, Sun
City, AZ, USA

Louise Wallace
Eli Lilly and Company, Lilly Research Centre, Erl Wood Manor, Windlesham,
Surrey GU20 6PH, UK

Y.-P. Wang
Max-Planck-Unit for Structural Molecular Biology, c/o DESY, Hamburg
22607, Germany

Michael W. Weiner
Center for Imaging of Neurodegenerative Disease, NIH, USA

Benjamin Wolozin
Department of Pharmacology, Boston University School of Medicine, Boston,
MA 02118-2526

Zbigniew K. Wszolek
Department of Neurology, Mayo Clinic Jacksonville, Jacksonville,
FL 32224, USA

Tony Wyss-Coray
Department of Neurology and Neurological Sciences, Stanford University
School of Medicine, Stanford, CA, USA

Katsuhiko Yanagisawa
National Institute for Longevity Sciences, 36-3, Gengo, Morioka, Obu, Aichi
474-8522, Japan

Takao Yasuhara
Department of Neurological Surgery, Okayama University, Okayama
700-8558, Japan

Qian-sheng Yu
Drug Design & Development Section, Laboratory of Neurosciences,
Intramural Research Program, National Institute on Aging, Baltimore, MD
21224, USA

Xiongwei Zhu
Department of Pathology, Case Western Reserve University, Cleveland,
OH 44106, USA

Alexander Zimprich
Medical University of Vienna, 1090 Vienna, Austria

Ruud Zwart
Eli Lilly and Company, Lilly Research Centre, Erl Wood Manor, Windlesham,
Surrey GU20 6PH, UK

Chapter 1
Clinical Features and Treatment of Dementia Associated With Parkinson's Disease

Murat Emre

Introduction

In contrast to the traditional perception of Parkinson's disease (PD) as a disorder of motor functions, dementia has only much later been increasingly recognized as an associated feature of PD. This is probably because patients with PD are now surviving longer thanks to modern treatment. Although cognitive deficits can be found even in newly diagnosed patients [1], dementia itself is strongly associated with advanced age and severe disease [2]. In population-based cross-sectional studies, the prevalence of dementia has been reported to be 28% to 41% [3–5]. Its incidence has increased up to sixfold [6]; and in a recent study, 78% of patients with PD were reported to have developed dementia over 8 years of follow-up [7]. The main risk factors are old age, severity of motor symptoms, the akinetic-rigid form of the disease, and subtle deficits in verbal fluency, executive functions, and memory performance at baseline [8].

Clinical Features of Dementia Associated With PD

Many patients with PD are elderly individuals, and as such they are prone to develop disorders of old age as much as their peers without PD. Therefore, they can theoretically be affected by all of the etiologies that can cause dementia in an elderly population at large, including symptomatic forms such as vascular dementia or other degenerative dementias such as Alzheimer disease (AD). The clinical profile of dementia in such cases is compatible with the underlying etiology. Dementia is, however, highly prevalent in PD, suggesting that the disease process

M. Emre
Professor of Neurology, Istanbul University, Istanbul Faculty of Medicine, Department of Neurology, Behavioral Neurology and Movement Disorders Unit, 34390 Çapa, Istanbul, Turkey

A. Fisher et al. (eds.), *Advances in Alzheimer's and Parkinson's Disease*,
© Springer 2008

itself is responsible for the dementia syndrome encountered with this disease. The prototypical dementia syndrome associated with PD (PDD) has characteristic clinical features that can be best summarized as a dysexecutive syndrome with prominently impaired attention, visuospatial dysfunction, moderately impaired memory, and accompanying behavioral symptoms such as apathy and psychosis. The cognitive and behavioral features of PDD have been described in a number of studies and reviewed in a few recent articles as summarized below [8,9].

Cognitive Features

Impaired attention is an early, prominent feature of patients with PDD. Impairments in reaction time and vigilance as well as fluctuating attention are present to an extent comparable to that seen in patients with a similar condition, dementia with Lewy bodies (DLB). Compared to patients with AD, PDD patients are more apathetic and have more prominent cognitive slowing. The magnitude of cognitive slowing is usually disproportionate to the general level of cognitive performance.

Impaired executive function is the core feature of PDD. Deficits include poor performance in tasks involving rule finding, problem solving, planning, set elaboration, set shifting, and set maintenance. Patients have more difficulty when they have to develop their own strategies, and their performance improves substantially when external cues are provided. Abnormalities in executive functions occur early in the course of PDD and are prominent throughout the course of the illness.

Memory functions are impaired in PDD, including working memory, explicit memory, and implicit memory such as procedural learning. Deficits in working memory can be found early in the disease course. The relative severity of amnesia compared to other cognitive deficits and the profile of impairment in different memory functions differ from those seen in AD. Memory impairment is less severe than that in patients with AD, and it is characterized by a deficit in free recall with preserved recognition, indicating that information is stored but not adequately accessed. Memory performance usually improves when semantic cues or multiple choices are provided. In fact, memory impairment in PDD may, rather, be due to executive dysfunction (i.e., difficulty accessing memory traces because of a deficiency in developing appropriate search strategies). This is in contrast to the "limbic" or "hippocampal" type of amnesia, with abnormalities of storage and consequent deficits of recall as well as recognition, which are the key neuropsychological features of AD.

Another characteristic feature of PDD is an early, prominent deficit in visuospatial functions that is disproportionately severe compared to the overall severity of dementia. Impaired visuospatial function is also more severe than that seen in AD patients with a similar severity of dementia; PDD patients perform worse in all perceptual scores, and those with visual hallucinations tend to have worse visual perception than those without these hallucinations [10]. Visuospatial abstraction and reasoning were found to be more impaired in patients with PDD than in those with AD, whereas visuospatial memory tasks were worse in patients with AD. Impairment becomes especially evident with more complex tasks that require planning and sequencing, so deficits in visuomotor tasks may be partly due to deficits in executive functions [11].

In contrast to prominent visuospatial deficits, another lateralized function, language, seems to be largely preserved. Core language functions are usually normal in PDD; the language deficits consist of mild anomia, rather than the prominent aphasia that develops early in patients with AD. Impaired verbal fluency, the main language impairment found in PDD, is more severe than that seen in AD patients. Other deficits include decreased content of spontaneous speech and impaired comprehension of complex sentences, but they are to a significantly lesser extent than in patients with AD. Ideomotor apraxia is also not a common feature of PDD. As for impairments in other domains, it was suggested that most of the language deficits, such as impaired verbal fluency and word-finding difficulty, may not reflect a true involvement of language functions but, rather, may be related to the dysexecutive syndrome [9].

Behavioral Features

Patients with PDD frequently have behavioral symptoms. and almost all of them demonstrate changes in personality, usually in the form of social withdrawal and apathy. The most common neuropsychiatric symptoms are depression, hallucinations, delusions, apathy, and anxiety. Hallucinations and delusions can follow treatment with dopaminergic agents, but they occur disproportionately often in patients with dementia [12]. Altogether, 83% of those with PDD were found to have at least one psychiatric symptom, in contrast to 95% of those with AD; hallucinations were more severe in PDD, whereas increased psychomotor activity (including aberrant motor behavior, agitation, disinhibition, and irritability) were more common in AD. In PDD, apathy was more common in the mild stages, whereas delusions increased with more severe motor and cognitive dysfunction [12]. Rapid-eye-movement (REM) sleep behavior disorder is also frequent in PDD, similar to other synucleinopathies such as DLB and multiple system atrophy (MSA) but which is not the case in other degenerative diseases such as AD or progressive supranuclear palsy (PSP).

Treatment of PDD

During the last two decades the neuropathology and neurochemical deficits accompanying PDD have become increasingly better understood. Morphological and biochemical studies have revealed prominent cholinergic deficits associated with PDD. These deficits include loss of cholinergic neurons in the nucleus basalis of Meynert (nbM) [13], which is observed to a greater extent than in patients with AD [14]. This cellular loss is associated with cholinergic deficits [reduced choline acetyltransferase (ChAT) and acetylcholinesterase (AChE) activity] in the nbM and cerebral cortex and correlates with the presence and severity of dementia [15,16]. Among the various pathological and chemical indices studied, only presynaptic cholinergic markers (including the number of neurons in the nbM) were found to be related to dementia in PD [17]. In a comparative study of patients with AD, DLB, and PD, the mean midfrontal ChAT activity was found to be markedly reduced in PD and DLB patients compared to normal controls and patients with AD [18]. Similar findings were revealed in imaging studies of cortical cholinergic function using positron emission tomography (PET): Compared with controls, the mean cortical AChE activity was lowest in patients with PDD (-20%), followed by patients with PD without dementia (-13%) and AD patients with equal severity of dementia (-9%) [19]. Furthermore, PDD is associated with neuronal loss also in the pedinculopontine cholinergic pathways that project to structures such as the thalamus [20].

Based on these substantial cholinergic deficits, cholinergic enhancement strategies using cholinesterase inhibitors (ChE-I) have been investigated in patients with PDD. Following an inital study with tacrine that described a rather dramatic improvement in cognition and no worsening of motor functions, a number of studies have been reported with commercially available ChE-I. They included two open studies with tacrine, seven studies (one case series and three open and three double-blind, placebo-controlled studies) with donepezil, one open study with galantamine, and three open and one large randomized, placebo-controlled study with rivastigmine. These studies, except for the last one, were summarized in a recent review [21]. Despite their methodological limitations with regard to study designs and small sample sizes, the review of these studies suggested that there was an improvement in cognition in almost all studies; behavioral symptoms diminished in most of them where they were measured, and motor symptoms did not worsen in most of the patients.

Recently, the first large, randomized, placebo-controlled study assessing the efficacy and safety of a ChE-I (rivastigmine) in patients with PDD was published [22]. This study, abbreviated EXPRESS, involved 68 centers from 12 countries and included 541 patients with a diagnosis of PD, according to the UK Brain Bank criteria, and dementia due to PD, according to DSM-IV. Patients were randomized to rivastigmine or placebo at a ratio of 2:1. To differentiate from those fullfilling the current criteria for DLB, patients were required to have an interval of at least 2 years between the onset of their cognitive symptoms and the PD diagnosis. Those with exposure to ChE-I or anticholinergics within the last 3 months before entry into the study and those with evidence of other neurodegenerative disorders or any unstable systemic disease were excluded. Primary efficacy parameters included the Alzheimer Disease Assessment Scale–Cognitive Section (ADAS-cog) for cognitive functions, and the Alzheimer Disease Collaborative Study–Clinical Global Impression of Change (ADCS-CGIC) for clinical assessment of overall change from baseline. Secondary clinical efficacy parameters included the Mini-Mental Status Examination (MMSE) for screening and staging, Activities of Daily Living Scale (ADCS-ADL) to assess activities of daily living, Neuropsychiatric Inventory (NPI) to assess neuropsychiatric symptoms, a computerized test battery to assess attention (CDR power of attention tests), and two tests for the assessment of executive functions including verbal fluency from the D-KEFS Test Battery and the Ten Point Clock-Drawing test. Safety parameters included recording adverse events, laboratory evaluations, pulse, blood pressure, body weight, electrocardiography (ECG) results, and Unified Parkinson Disease Rating Scale (UPDRS) Part III for the assessment of motor functions.

Of the 541 patients entered in the study, 410 completed it and 131 discontinued it prematurely. Discontinuation was seen more often in the rivastigmine group (27.3% vs. 17.9% in the placebo group); this was also the case for discontinuation because of adverse events (17.1% vs. 7.8%, respectively). Both primary efficacy endpoints showed statistically significant improvement in favor of rivastigmine. On ADAS-cog evaluation, patients on rivastigmine showed a 2.1-point improvement at 26 weeks from a baseline value of 23.8, whereas patients on placebo deteriorated by 0.7 point (from a baseline score of 24.3), yielding a 2.9-point treatment difference from baseline ($p < 0.001$). The mean scores for the ADCS-CGIC at week 24 were 3.8 in the rivastigmine group and 4.3 in the placebo group (score of 4 indicated no change, lower scores indicated improvement, and higher scores indicating worsening from baseline). A comparison of the outcomes across all response categories revealed a statistically significant difference in favor of rivastigmine ($p = 0.007$). More patients on rivastigmine showed an improvement (40.8% vs. 29.7%), and more patients on placebo deteriorated (42.5% on placebo vs. 33.7% on rivastigmine). There were statistically significant differences in favor of rivastigmine on all secondary efficacy parameters: As measured with the NPI, neuropsychiatric symptoms diminished on rivastigmine but did not change from baseline on placebo; attention improved on rivastigmine and worsened on placebo; improvement from baseline was also seen on the Ten Point Clock Drawing test, verbal fluency, and MMSE in patients on rivastigmine, whereas patients on placebo worsened compared to baseline scores. According to the ADCS-ADL, patients on rivastigmine showed minimal worsening, whereas those on placebo had significantly more deterioration.

Adverse events occurred significantly more often on rivastigmine. The main adverse events were those related to the gastrointestinal system, with nausea and vomiting occurring most frequently (29.0% vs. 11.2% for nausea and 16.6.% vs. 1.7% for vomiting on rivastigmine and placebo, respectively). Worsening of parkinsonian symptoms was reported as an adverse event more frequently in patients on rivastigmine than in those on placebo (27.3% vs. 15.6%, respectively) mainly driven by worsening of tremor (10.2% on rivastigmine vs. 3.9% on placebo). The objective measures of motor symptoms, however, as assessed by UPDRS Part III did not reveal any signicant differences or trends between the two treatments. There were no clinically relevant changes in the vital signs, body weight, ECG, or laboratory parameters.

Conclusions

Parkinson's disease is frequently associated with dementia. In cross-sectional studies, up to 40% of patients with PD have been estimated to be affected, and the incidence has increased sixfold. Advanced age, the akinetic-rigid form of the disease, and severe motor symptoms are particular risk factors. Prominent cholinergic deficits accompany the dementia associated with PD. A number of small, mostly open studies with ChE-I and and one large, randomized, controlled study with rivastigmine revealed the benefical effects of ChE-I in these patients. The large, randomized, placebo-controlled EXPRESS study demonstrated that rivastigmine alleviated deficits in all symptom domains in patients with PDD, including cognitive dysfunction and behavioral symptoms. These improvements seemed to be reflected in the overall status and functioning of patients. Adverse events were mainly gastrointestinal; notably, there was no worsening of motor functions except for worsening of tremor in 10% of the patients. ChE-I thus constitutes a promising treatment approach in patients with PDD.

References

1. Foltynie T, Brayne CEG, Robbins TW, Barker RA. The cognitive ability of an incident cohort of Parkinson's patients in the UK: the CamPaIGN study. Brain 2004;127:1–11.
2. Levy G, Schupf N, Tang MX, et al. Combined effect of age and severity on the risk of dementia in Parkinson's disease. Ann Neurol 2002;51:722–729.
3. Marttila RJ, Rinne UK. Epidemiology of Parkinson's disease in Finland. Acta Neurol Scand 1976;53:81–102.
4. Aarsland D, Tandberg E, Larsen JP, Cummings JL. Frequency of dementia in Parkinson disease. Arch Neurol 1996;53:538–542.
5. Mayeux R, Denaro J, Hemenegildo N, et al. A population-based investigation of Parkinson's disease with and without dementia: relationship to age and gender. Arch Neurol 1992;49:492–497.
6. Aarsland D, Andersen K, Larsen JP, et al. Risk of dementia in Parkinson's disease: a community-based, prospective study. Neurology 2001;56:730–736.
7. Aarsland D, Andersen K, Larsen JP, et al. Prevalence and characteristics of dementia in Parkinson disease: an 8-year prospective study. Arch Neurol 2003;60:387–392.
8. Emre M. Dementia associated with Parkinson's disease. Lancet Neurol 2003;2:229–237
9. Pillon B, Boller F, Levy R, Dubois B. Cognitive deficits and dementia in Parkinson's disease. In: Boller F, Cappa S (eds) Handbook of neuropsychology. 2nd ed. Elsevier Sciences, Amsterdam, 2001, pp 311–371.
10. Mosimann UP, Mather G, Wesnes KA, et al. Visual perception in Parkinson disease dementia and dementia with Lewy bodies. Neurology 2004;63:2091–2096.

11. Crucian GP, Okun MS. Visual-spatial ability in Parkinson's disease. Front Biosci 2003;8:992–997.
12. Aarsland D, Cummings JL, Larsen JP. Neuropsychiatric differences between Parkinson's disease with dementia and Alzheimer's disease. Int J Geriatr Psychiatry 2001;16:184–191.
13. Whitehouse PJ, Hedreen JC, White CL 3rd, Price DL. Basal forebrain neurons in the dementia of Parkinson disease. Ann Neurol 1983;13:243–248.
14. Candy JM, Perry RH, Perry EK, et al. Pathological changes in the nucleus of Meynert in Alzheimer's and Parkinson's diseases. J Neurol Sci 1983;59:277–289.
15. Dubois B, Ruberg M, Javoy-Agid F, et al. A subcortico-cortical cholinergic system is affected in Parkinson's disease. Brain Res 1983;288:213–218.
16. Perry EK, Curtis M, Dick DJ, et al. Cholinergic correlates of cognitive impairment in Parkinson's disease: comparisons with Alzheimer's disease. J Neurol Neurosurg Psychiatry 1985;48:413–421.
17. Perry RH, Perry EK, Smith CJ, et al. Cortical neuropathological and neurochemical substrates of Alzheimer's and Parkinson's diseases. J Neural Transm Suppl 1987;24:131–136.
18. Tiraboschi P, Hansen LA, Alford M, et al. Cholinergic dysfunction in diseases with Lewy bodies. Neurology 2000;54:407–411.
19. Bohnen NI, Kaufer DI, Ivanco LS, et al. Cortical cholinergic function is more severely affected in parkinsonian dementia than in Alzheimer disease: an in vivo positron emission tomographic study. Arch Neurol 2003;60:1745–1748.
20. Rub U, Del Tredici K, Schultz C, et al. Parkinson's disease: the thalamic components of the limbic loop are severely impaired by alpha-synuclein immunopositive inclusion body pathology. Neurobiol Aging 2002;23:245–254.
21. Aarsland D, Mosimann UP, McKeith IG. Role of cholinesterase inhibitors in Parkinson's disease and dementia with Lewy bodies. J Geriatr Psychiatry Neurol 2004;17:164–171
22. Emre M, Aarsland D, Albanese A, et al. Rivastigmine for dementia associated with Parkinson's disease. N Engl J Med 2004;351:2509–2518.

Chapter 2
Neurogenetics in Parkinson's Disease

Yoshikuni Mizuno and Nobutaka Hattori

Introduction

To date, 11 forms of familial Parkinson's disease (PD) have been mapped to different loci on chromosomes. As a result, six causative genes have been identified: *alpha-synuclein* [1], *parkin* [2], *UCH-L1* [3], *PINK1* [4], *DJ-1* [5], and *LRRK2* [6,7]. UCHL-1 still needs additional families before it has full acceptance as a causative gene for PARK5 (only one family has been reported so far) [3]. In this chapter, we review recent progress in the genetics of PD and discuss the molecular mechanism of nigral neurodegeneration.

PARK1

PARK1 is an autosomal dominant familial PD caused by mutations of *alpha-synuclein (SNCA)*, which has been mapped to the long arm of chromosome 4 at 4q21–q23. To date, three missense mutations are known: A30P [8], E46K [9], and A53T [1]. Alpha-synuclein is a neuron-specific protein localized mainly in the presynaptic terminal membranes and synaptic vesicles. Aggregated alpha-synuclein accumulates in the nigral neurons in PD. In this sense, alpha-synuclein appears to be an important protein in the pathogenesis of PD. Interestingly, triplication [10] and duplication [11,12] of the alpha-synuclein locus were found in other autosomal dominant PD families. Thus, overexpression of alpha-synuclein per se appears to be responsible for nigral neurodegeneration. Moreover, triplication was associated with widespread neuropathology consisting of diffuse Lewy body disease with clinical dementia in addition to L-dopa-responsive parkinsonism [13]. On the other hand, duplication was associated with pure L-dopa-responsive parkinsonism without dementia.

Y. Mizuno
Department of Neurology, Juntendo University School of Medicine, Tokyo 113-8421, Japan

A. Fisher et al. (eds.), *Advances in Alzheimer's and Parkinson's Disease,*
© Springer 2008

Regarding the molecular mechanism of nigral neuronal death with missense mutations of *alpha-synuclein*, it has been shown that mutated alpha-synuclein proteins show an increased tendency for self-aggregation [14]. Particularly oligomers of alpha-synuclein have been shown to be toxic, inducing release of dopamine into the cytoplasm from synaptic vesicles [15] and impairing 26S proteasome [16] as well as mitochondrial functions [17]. Release of dopamine into the cytoplasm induces oxidative stress to nigral neurons. Both oxidative stress and mitochondrial impairment enhance alpha-synuclein aggregation. In addition, mitochondrial impairment results in reduced ATP synthesis: As 26S proteasome is an ATP-dependent protein-degrading enzyme, mitochondrial impairment reduces its catalytic activity. Thus, vicious cycles are formed in nigral neurons, leading them to progress slowly toward neuronal death. Furthermore, aggregated insoluble alpha-synuclein proteins are likely to impair transport of vital substances in nigral neurons.

Clinical features of PARK1 consist of typical L-dopa-responsive parkinsonism, with or without cognitive impairment. Clinical features of the Glu46Lys mutation are consistent with the clinical diagnosis of diffuse Lewy body disease. In fact, many cortical Lewy bodies were reported in the brain of an autopsied patient [9]. Thus, Glu46Lys missense mutation and triplication of *alpha-synuclein* cause dementia and parkinsonism, and other mutations of *alpha-synuclein* may be associated with variable degrees of cognitive impairment.

PARK2

PARK2 is an autosomal recessive familial PD caused by mutations of *parkin* [2], which has been mapped to the long arm of chromosome 6 at 6q25.2–q27. To date, more than 30 exon rearrangements (deletion, duplication, triplication), 30 missense mutations 8, nonsense mutations, and close to 20 small deletions or insertions have been reported [18–22].These numbers are expected to increase.

Regarding the functions of parkin protein, one of the most important functions is its enzymatic activity as a ubiquitin-protein ligase (E3) of the ubiquitin system [23]. The ubiquitin system consists of three enzymes: a ubiquitin-activating enzyme (E1), a ubiquitin-conjugating enzyme (E2), and a ubiquitin-protein ligase (E3). The ubiquitin-proteasome system (UPS) is an important intracellular proteolytic system responsible for a wide variety of biologically important cellular processes, such as cell cycle progression, signaling cascades, developmental programs, the protein quality control system, DNA repair, apoptosis, signal transduction, transcription, metabolism, immunity, and neurodegeneration. The E3 transfer ubiquitin molecules to target proteins, forming a polyubiquitin chain, which is recognized by 26S proteasome as the proteolytic signal [24].Therefore, in the presence of mutated parkin proteins, accumulation of parkin-substrate proteins is expected to be the major cause of nigral neuronal death. To date, however, there is no clear immunohistochemical evidence to indicate accumulation of parkin-substrates despite the fact that many parkin-interacting proteins have been reported, such as CDCrel-1 [25], glycosylated alpha-synuclein [26], PAEL receptor [27], and synphilin-1 [28].

Therefore, other mechanisms may be operating for nigral degeneration in PARK2. Polyubiquitin chains are formed by repeated reactions through which another ubiquitin links a lysine residue at position 48 of the ubiquitin protein. Ubiquitin has seven lysine residues. A polyubiquitin chain formed via lysine at position 48 mainly becomes a marker for proteolytic attack by the 26S proteasome. Lysine 63-linked polyubiquitylation and monoubiquitylation without the formation of a ubiquitin tree have many biological roles other than proteolysis, such as endocytosis, DNA repair, translation, IkB activation, DNA silencing, virus budding, protein sorting, and protein trafficking [24]. Parkin promotes polyubiquitylation not only at lysine 48 but also at lysine 63. Recently, Lim and co-investigators [29] reported that parkin enhanced Lys-63-mediated polyubiquitylation of synphilin-1. Recently, we found accumulation of dopa/dopaminequinone in parkin-knockdown cells [30]. Therefore, an absence of normal parkin would predispose nigral neurons to oxidative stress. Oxidative damage is an important mechanism for nigral neuronal death in sporadic PD as well.

Clinical features of PARK2 include early-onset (usually at less than 50 years of age) L-dopa-responsive parkinsonism with early development of motor fluctuations (wearing off and dyskinesia) once L-dopa is given. Since gene analysis became possible, however, an age of onset as late as 72 years has been reported [31]. Also, some atypical features such as dementia [32], psychosis and behavioral problems [33], cerebellar ataxia [34], peripheral neuropathy [33,35,36], hyperhydrosis [37], orthostatic hypotension [33], urinary urgency and impotence [33], and hemiparkinsonism-hemiatrophy [38] have been reported.

PARK3

PARK3 is an autosomal dominant familial PD linked to the short arm of chromosome 2 at 2p13 [39]. The causative gene has not yet been identified. What is interesting is that recently *HtrA2/Omi*, which has been mapped to the same locus (2p13), was reported as a possible causative gene in four sporadic PD patients [40].

HtrA2 is a serine protease that has extensive homology to bacterial heat shock endoprotease [41]. Knockout mice [42] showed striatal neuronal loss. Interestingly, this is a mitochondrial protein localized in the intermembrane space and is released from mitochondria into the cytoplasm upon apoptotic stimuli, inducing the apoptosis cascade by activating caspase 3 [43].Thus, it is a proapoptotic protein. Nonetheless, its mutation in its PDZ domain (carboxy-terminal side) was associated with familial PD. Furthermore, a mutation in the protease domain caused motor neuron degeneration type 2 in mice [44]. This protein thus appears to be an interesting new protein in some way related to familial PD.

PARK4

PARK4 is an autosomal dominant familial PD characterized by L-dopa-responsive parkinsonism associated with dementia. It is caused by triplication of the *alpha-synuclein* gene. This family was initially mapped to the short arm of chromosome 4 [13], but later triplication of *alpha-synuclein* was identified (see PARK1, above).

PARK5

PARK5 is an autosomal dominant familial PD linked to the short arm of chromosome 4 at 4p14–p15.1. To date, only one family has been reported [3], in which a I93M missense mutation was found.

UCH-L1 is a neuron-specific enzyme that cleaves the carboxy-terminal peptide bond of polyubiquitin chains. Thus, UCH-L1 is a ubiquitin-recycling enzyme. Catalytic activity of I93M-mutated UCH-L1 was reported to be half that of the wild enzyme [3]. Thus, it is expected that the supply of ubiquitin for proteins that have to be destroyed by 26S proteasome is reduced with this mutation. Interestingly, deletion of exons 7 and 8 in mouse UCH-L1 causes gracile axonal dystrophy (*gad*) in the mouse; this is an autosomal recessive condition characterized by axonal degeneration and formation of spheroid bodies in motor and sensory nerve terminals [45]. It is also interesting to note that recently discovered mutations of *HtrA2/Omi* cause PD in humans and a motor neuron disease in mice.

The clinical features of PARK5 are similar to those of late-onset sporadic PD.

PARK6

PARK6 is an autosomal recessive young-onset familial PD caused by mutations of *PINK1* (PTEN-induced kinase 1) [4], which has been mapped to the short arm of chromosome 1 at 1p35–p36. To date 17, missense mutations 3, nonsense mutations 1, insertion, and 1 exon deletion are known [4,46–49]. We recently found a novel missense mutation and an exonic deletion from exons 6 to 8; the latter was the first documented case with an exonic deletion mutation in PARK6 [49]. PARK6 appears to be the second most common autosomal recessive PD after PARK2.

Functions of PINK1 are not known. PINK1 stands for PTEN-induced kinase 1. PTEN (protein tyrosine phosphatase with homology to tensin) is a tumor-suppressor gene on chromosome 10, mutated in many human tumors [50]. PINK1 is a mitochondrial matrix protein and has protein kinase activity. Thus, carcinogenesis and neurodegeneration appear to have a relation similar to that of two sides of a coin. The exact functions of PINK1 have yet to be studied.

Clinical features of PARK6 are similar to those of PARK2, but its onset is at a somewhat older age. The age of onset of the original family studied by Valente and co-investigators [51] ranged from 32 to 48 years. Reflecting this somewhat older age of onset, dystonia and sleep benefit are usually not seen.

PARK7

PARK7 is an autosomal recessive familial PD caused by mutations of *DJ-1* [5], which has been mapped to the short arm of chromosome 1 at 1p36. To date, 6 missense mutations, 1 intronic mutation, 1 small deletion, and 2 exonic deletions (exons 1 to 5 and exons 5 to 7) have been identified [5,52–54]. *DJ-1* mutations are rare compared with *parkin* and *PINK1* mutations. We have not been able to find a *DJ-1* mutation among Japanese families so far tested.

DJ-1 was identified as a novel oncogene that transformed mouse NIH3T3 cells in cooperation with activated Ras, and mapped to 1p36 [55]. The function of DJ-1 protein is not yet elucidated. One of the PD-inducing missense mutations, L166P, interferes with dimer formation [56]; and dimer formation is considered to be important for normal activity. In addition, this mutant DJ-1 protein is mislocalized to mitochondria; on the other hand, wild-type DJ-1 protein is ubiquitously localized in cells [5]. Furthermore, parkin interacted with mutated DJ-1 (L166P) protein but not with wild DJ-1 [57], suggesting that parkin might be acting as a quality-control protein for DJ-1. In addition, DJ-1 is a potent antioxidative protein, and this characteristic depends on its 106-cysteine residue [58]; therefore, DJ-1 may be an important protein for the survival of nigral neurons. The oxidized DJ-1 protein is relocated to mitochondria [59].

Clinical features of PARK7 are essentially similar to those of PARK2, including the age of onset. Interestingly, three of four patients in the original family showed psychiatric disturbances (anxiety attacks) [60]. Atypical clinical features include short stature and brachydactyly, which were found in a Dutch kindred [61].

PARK8

PARK8 is an autosomal dominant PD mapped to the centromeric region of chromosome 12 [62] and is caused by mutations of *LRRK2/dardarin* [6,63]. LRRK2 stands for leucine-rich repeat kinase 2, and dardar means tremor in the Basque language where families with PARK8 are found. *LRRK2* is a huge gene encompassing 144 kb, and the open reading frame consists of 1449 basepairs in 51 exons. LRRK2 protein consists of 2527 amino acids, and it is ubiquitously expressed in the cytoplasm of many organs. To date, 7 missense mutations are known [6,7,63].

LRRK2 protein belongs to the ROCO protein family. ROCO proteins are a group of proteins that have ROC and COR domains [64]. ROC stands for Ras of complex proteins and COR stands for carboxy-terminal of ROC. In addition, many ROCO proteins have an LRR (leucine-rich repeat) domain, an MAPKKK (mitogen-induced protein kinase kinase kinase) domain, and a WD domain, which is rich in tryptophan and aspartate repeats. The function of LRRK2 is still unknown; but because it has a protein kinase domain, it is likely that its role is phosphorylation of proteins that are important for the survival of nigral neurons. It is interesting to note that alpha-synuclein aggregates in PD are highly phosphorylated in Ser-129 [65], but it is not known if LRRK2 is in some way related to phosphorylation of alpha-synuclein.

The clinical features of PARK8 are similar to those of sporadic PD, although early-onset cases are not uncommon. The patients show L-dopa-responsive parkinsonism. Cognitive impairment can be seen. Neuropathological changes are interesting in that Wszolek and co-investigators [66] reported four types of neuropathology in four members from the same family: One showed brain stem-type PD, the second diffuse Lewy body disuse, the third tau-positive inclusions in nigral neurons, and the fourth simple atrophy of the substantia nigra. In the family reported by Funayama and co-investigators [62], both Lewy body-positive and Lewy body-negative patients were present in the same family.

PARK9, PARK10, PARK11

PARK9 is an autosomal recessive familial PD linked to the short arm of chromosome 1 at 1p36 [67]. The causative gene has not been identified. Clinical features consist of levodopa-responsive parkinsonism, supranuclear gaze palsy, pyramidal signs, and dementia, called Kufor-Rakeb syndrome. The age of onset is 10 to 20 years. Neuropathologically, not only the substantia nigra but also the pyramidal tract, putamen, and pallidum show degeneration.

The PARK10 locus was found by genome-wide scanning on familial as well as sporadic cases of PD in Iceland [68]. This group included 117 PD patients and 168 of their unaffected relatives in 51 families using 781 microsatellite markers. They showed linkage to chromosome 1p32 with a lod score of 4.9. The disease gene has not yet been identified. The clinical features are essentially similar to those of sporadic PD, and the mean age of onset was 65.8 years.

PARK11 is an autosomal dominant familial PD linked to the long arm of chromosome 2 at 2q36 to q37 [69]. The causative gene has not yet been identified. The clinical features are essentially similar to those of sporadic PD, with a mean age of onset at 58 years.

Other Forms of Familial PD

There are many families in which analysis failed to show linkage to any one of the known loci associated with familial forms of PD. Such reports are increasing every year. According to our observations, to date we have analyzed 347 families with either autosomal dominant or recessive inheritance for known PD-causing genes. We found 116 families with *parkin* mutations, 8 families with *PINK1* mutations/no *DJ-1* mutation, 10 families with *LRRK2* mutations, and 2 families with *alpha-synuclein* duplication. The overall mutation rate was 136 positive families of 347 (39.2%). In other words, approximately 60% of patients with familial PD do not have known mutations. Progress in molecular cloning of new genes for familial PD and elucidation of the functions of the disease genes will definitely contribute to our understanding of the molecular mechanism of nigral neurodegeneration of sporadic PD. Such information can help in the development of disease-modifying new treatments for PD.

Acknowledgment This study was supported in part by a Grant-in-Aid for Priority Area Comprehensive Brain Research from the Ministry of Education, Culture, and Science, Japan; a Grant-in-Aid for Intractable Disorders, from the Ministry of Health and Labor, Japan; and a Center of Excellence Grant from the National Parkinson Foundation, USA.

References

1. Polymeropoulos MH, Lavedan C, Leroy E, et al. Science 1997;276:2045–2047
2. Kitada T, Asakawa S, Hattori N, et al. Nature 1998;392:605–608.
3. Leroy E, Boyer R, Auburger G, et al. Nature 1998;395:451–452.
4. Valente EM, Abou-Sleiman PM, Caputo V, et al. Science 2004;304:1158–1160.
5. Bonifati V, Rizzu P, van Baren MJ, et al. Science 2003;299:256–259.
6. Paisan-Ruiz C, Jain S, Evans EW, et al. Neuron 2004;44:595–600.
7. Zimprich A, Biskup S, Leitner P, et al. Neuron 2004;44:601–607.
8. Zarranz JJ, Alegre J, Gémez-Esteban J, et al. Ann Neurol 2004;55:164–173.
9. Singleton AB, Farrer M, Johnston J, et al. Science 2003;302:841.
10. Chartier-Harlin MC, Kachergus J, Roumier C, et al. Lancet. 2004;364:1167–1169.
11. Ibanez P, Bonnet AM, Debarges B, et al. Lancet 2004;364:1169–1171.
12. Farrer M, Gwinn-Hardy K, Muenter M, et al. Hum Mol Genet 1999;8:81–85.
13. El Agnaf OM, Jakes R, Curran MD, et al. FEBS Lett 1998;440:67–70.
14. Volles MJ, Lansbury PT Jr. Biochemistry 2002;41:4595–4602.
15. Snyder H, Mensah K, Theisler C, et al. J Biol Chem 2003;278:11753–11759.
16. Tanaka Y, Engelender S, Igarashi S, et al. Hum Mol Genet 2001;10:919–926.
17. Hattori N, Matsumine H, Kitada T, et al. Ann Neurol 1998;44:935–941.
18. Abbas N, Lücking CB, Ricard S, et al. Hum Mol Gen 1999;8:567–574.
19. Lücking CB, Dürr A, Bonifati V, et al. N Engl J Med 2000;342:1560–1567.
20. Oliviera SA, Scott WK, Martin ER, et al. Ann Neurol 2003;53:624–629.
21. Hedrich K, Eskelson C, Wilmot B, et al. Mov Disord 2004;19:1146–1157.
22. Shimura H, Hattori N, Kubo S, et al. Nat Genet 2000;25:302–305.
23. Tanaka K, Suzuki T, Hattori N, Mizuno Y. Biochim Biophys Acata 2004;1695:235–247.
24. Zhang Y, Gao J, Chung KK, et al. Proc Natl Acad Sci U S A 2000;21:13354–13359.

25. Shimura H, Schlossmacher MG, Hattori N, et al. Science 2001;293:263–269.
26. Imai Y, Soda M, Inoue H, et al. Cell 2001;105:891–902.
27. Chung KKK, Zhang Y, Lim KL, et al. Nat Med 2001;7:1144–1150.
28. Lim KL, Chew KC, Tan JM, et al. J Neurosci 2005;25:2002–2009.
29. Machida Y, Chiba T, Takayanagi A, et al. Biochem Biophys Res Commun 2005;332: 233–240.
30. Lincolon SJ, Maraganore DM, Lesnick TG, et al. Mov Disord 2003;18:1306–1311.
31. Benbunan BR, Korczyn AD, Giladi N. J Neural Transm 2004;111:47–57.
32. Khan NL, Graham E, Critchley P, et al. Brain 2003;126:1279–1292.
33. Kuroda Y, Mitsui T, Akaike M, et al. Nat Genet 1998;18:106–108.
34. Tassin J, Dürr A, de Brouchker T, et al. Am J Hum Genet 1998;63:88–94.
35. Okuma Y, Hattori N, Mizuno Y. Parkinsonism Rel Disord 2003;9:313–314.
36. Yamamura Y, Kuzuhara S, Kondo K, et al. Parkinsonism Rel Disord 1998;4:65–72.
37. Pramistaller PP, Künig G, Leenders K, et al. Neurology 2002;58:808–810.
38. Gasser T, Müller-Myhsok B, Wszolek ZK, et al. Nat Genet 1998;18:262–265.
39. Strauss KM, Martins M, Plun-Favreau H, et al. Hum Mol Genet (advance access published June 16, 2005).
40. Faccio R, Fusco C, Chen A, et al. Hum Mol Genet 1999;8:81–85.
41. Martins LM, Morrison A, Klupsch K, et al. Mol Cell Biol 2004;24:9848–9862.
42. Suzuki Y, Imai Y, Nakayama H, et al. Mol Cell 20018:613–621.
43. Jones JM, Datta P, Srinivasula SM, et al. Nature 2003;425:721–727.
44. Saigoh K, Wang YL, Suh JG, et al. Nat Genet 1999;23:47–51.
45. Hatano Y, Li Y, Sato K, et al. Ann Neurol 2004;56:424–427.
46. Healy DG, Abou-Sleiman PM, Gibson JM, et al. Neurology 2004;63:1486–1488.
47. Rohe CF, Montagna P, Breedveld G, et al. Ann Neurol 2004;56:427–431.
48. Li Y, Tomiyama H, Sato K, et al. Neurology 2005;64:1955–1957.
49. Steck PA, Pershouse MA, Jasser SA, et al. Nat Genet 1997;15:356–362.
50. Valente EM, Bentivolglio AR, Dixon PH, et al. Am J Hum Genet 2001;68:895–900.
51. Abou-Sleiman PM, Healy DG, Quinn N, et al. Ann Neurol 2003;54:283–286.
52. Hague S, Rogaeva E, Hernandez D, et al. Ann Neurol 2003;54:271–274.
53. Hering R, Strauss KM, Tao X, et al. Hum Mutat 2004;24:321–329.
54. Nagakubo D, Taira T, Kitaura H, et al. Biochem Biophys Res Commun 1997;231: 509–513.
55. Wilson MA, Collins JL, Hod Y, et al. Proc Natl Acad Sci U S A 2003;100:9256–9261.
56. Moore DJ, Zhang L, Troncoso J, et al. Hum Mol Genet 2005;14:71–84.
57. Taira T, Saito Y, Niki T, et al. EMBO Rep 2004;5:213–218.
58. Canet-Aviles RM, Wilson MA, Miller DW, et al. Proc Natl Acad Sci U S A 2004;101:9103–9108.
59. Dekker M, Bonifati V, van Swieten J, et al. Mov Disord 2003;18:751–757.
60. Dekker MC, Galjaard RJ, Snijders PJ, et al. Am J Med Genet A 2004;130:102–104.
61. Funayama M, Hasegawa K, Kowa H, et al. Ann Neurol 2002,51:296–301.
62. Nichols WC, Pankratz N, Hernandez D, et al. Lancet 2005;365: 410–412.
63. Bosgraaf L, Haastert PJMV. Biochim Biophys Acta 2003;1643:5–10.
64. Fujiwara H, Hasegawa M, Dohmae N, et al. Nat Cell Biol 2002;4:160–164.
65. Wszolek ZK, Pfeiffer RF, Tsuboi Y, et al. Neurology 2004;62:1619–1622.
66. Hampshire DJ, Roberts E, Crow Y, et al. J Med Genet 2001;38:690–692.
67. Hicks AA, Petursson H, Jonsson T, et al. Ann Neurol 2002;52:549–555.
68. Pankratz N, Nichols WC, Uniacke SK, et al. Am J Hum Genet 2003;72:1053–1057.

Chapter 3
Impaired DNA Repair Systems: Generation of Mitochondrial Dysfunction and Increased Sensitivity to Excitotoxicity

Simona Francisconi, Mara Codenotti, Erminia Poli, Daniela Uberti, Giulia Ferrari-Toninelli, and Maurizio Memo

Introduction

Preservation of genomic stability is an essential biological function, and cells engage efficient mechanisms to maintain DNA stability over time and prevent the generation and the persistence of impaired cells that may be detrimental to the organism. These mechanisms involve several DNA surveillance and repair proteins, such as the DNA mismatch repair system (MMR), nucleotide excision repair (NER), and base excision repair (BER) [1]. DNA damage may arise in cells as the result of endogenous and exogenous sources, such as free radical production during cellular metabolism, replication errors, ultraviolet ionizing radiation, and/or mutagenic agents.

MMR has been widely investigated because mutations in MMR genes in humans are associated with hereditary nonpolyposis colorectal cancer (HNPCC) and a wide variety of sporadic tumors [2,3]. In eukaryotic cells, mismatches in DNA are recognized by the MutS homologs MSH2, MSH3, and MSH6, which initiate an MMR event. They form heterodimeric complexes, MSH2-MSH6 (MutSα) and MSH2-MSH3 (MutSβ), which are required for the repair of base-base mispairs as well as single-base insertion/deletion loops (IDLs). An interaction between MutS and MutL homologs is required to activate subsequent repair events [4].

DNA repair is not the only function of the MMR system; the key proteins of MMR appear to play a dual role in (1) repairing DNA damage and (2) activating an apoptotic program in response to extensive DNA damage. MMR mutations allow improved survival of neoplastic cells, preventing the damage-induced apoptotic response [5].

Although the primary role of MMR takes place during replication, several independent observations indicate that MMR is active also in postmitotic cells, such as neurons. Little is known about the role of MMR in neurons, even if

S. Francisconi
Department of Biomedical Sciences and Biotechnologies, University of Brescia, Brescia, Italy

A. Fisher et al. (eds.), *Advances in Alzheimer's and Parkinson's Disease*,
© Springer 2008

several syndromes associated with defects in DNA damage response exhibit neurological symptoms as the most important feature of their clinical manifestation [6,7].

The possible role of the MMR system in the brain was investigated only recently. Independent studies demonstrated the expression of MutS complexes in rat brain tissues [8] and their activity in repairing DNA mismatches [9]. In this context, we investigated the expression and modulation of MMR proteins in neuronal cells as well as the effects of defects in the repair of DNA damage on brain functions.

Results

We investigated the expression of MMR proteins in neurons by in vivo and in vitro studies. The distribution of MSH2, one of the key proteins involved in recognition of damaged DNA, was evaluated by immunohistochemistry in the adult rat brain. The results from these experiments showed that MSH2 was expressed in several areas of the brain, including the hippocampus, cerebellum, cortex, striatum, substantia nigra, and spinal cord. MSH2 was detected only in neuronal cells with nuclear localization [10].

The role of MSH2 in neuron degeneration was studied in rats treated with kainic acid, a well known experimental paradigm of excitotoxicity characterized by specific cell loss in CA3/CA4 hippocampal subfields. Kainate injection resulted in a dose-dependent increase of MSH2 expression, specifically in pyramidal cells of the CA3/CA4 subfield [10], that could be interpreted as a component of proapoptotic signaling.

Involvement of MMR in the repair of base oxidation induced by H_2O_2 was explored in human neuronal cell lines [11,12]. Human SH-SY5Y neuroblastoma cells were differentiated to a neuron-like phenotype by treatment with retinoic acid and then treated with a pulse of H_2O_2. This lesion also caused a marked oxidative DNA damage, as detected by 8-OH-deoxyguanosine immunoreactivity. The calculated number of DNA-damaged cells reached 50% of the total 2 hours after the treatment and then declined. This observation suggested that a given number of neurons were able to recognize and repair the oxidized DNA while the others underwent apoptosis. H_2O_2 also induced MSH2 nuclear translocation of several members of the MMR system, including MSH2, MSH6, and MLH1, suggesting that neurons, like nonproliferating cells, are able to recognize and possibly repair DNA damage or activate an apoptotic program [13].

The effects of defects in DNA repair activity on brain functions were investigated in $Msh2+/-$ mice. Mice lacking MSH2 comprise a well characterized model for studies in cancer predisposition [14]. $Msh2$-null mice are viable and display no major abnormalities, except for reduced survival rate, due to the fact that they are prone to develop lymphomas at an early age. Mice heterozygous for $Msh2$ do not show reduced survival compared to wild-type control mice, nor are they predisposed to increased intestinal neoplasia. Heterozygosis for the *Msh2* gene leads to a constant, small increase in mutability that becomes evident at old age [15].

We chose to use cDNA microarray technology as a strategy to look simultaneously at the activity of thousand of genes. Gene expression profiles of hippocampus from wild-type and *Msh2+/−* mice were compared, and only genes with at least a twofold change in expression levels were considered as up- or down-regulated. Data analysis revealed about 110 up- or down-regulated genes in the hippocampus of heterozygous mice, including genes involved in various biochemical and cellular functions (Fig. 1).

Most of them are "unknown function" genes. *Msh2+/−* mice had reduced expression levels in genes encoding for transcriptional and splicing factors, DNA-binding protein, and genes involved in protein synthesis, modification, and transport. Array data strongly suggested mitochondrial function impairment, particularly involving the OXPHOS system. Moreover, these mice showed reduced expression levels in genes involved in mitochondrial protein transport (TOM complex), mitochondrial protein synthesis (ribosomal subunits), and both nuclear and mitochondrial encoded genes [16].

These alterations in mitochondrial functions were confirmed also by a corresponding reduction in protein levels, suggesting mitochondrial metabolism impairment in the hippocampus of heterozygous mice, perhaps due to

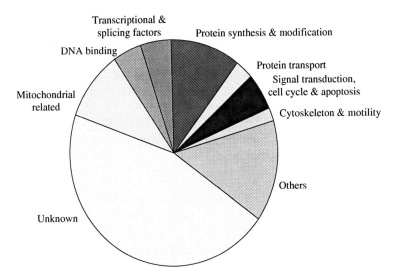

Fig. 1 Distribution of up-regulated and down-regulated genes in the hippocampus of *Msh2+/−* mice. Compared to age-related wild-type mice, *Msh2+/−* mice show about 110 up- or down-regulated genes involved in various cellular functions, including genes encoding for transcriptional and splicing factors, DNA-binding protein, mitochondrial proteins, and genes involved in protein synthesis, modification, and transport

a compensatory effect that could protect neurons from DNA damage by a reduction of reactive oxygen species (ROS) generation.

It is well known that mitochondrial cytopathies in the human central nervous system are characterized by seizure hypersensitivity, mental disorders, and myoclonu [17]. To investigate the effect of this brain mitochondrial impairment in $Msh2+/-$ mice, we evaluated whether a reduction in MMR activity affected neuronal sensitivity to an excitotoxic state in vivo by applying systemic kainic acid (KA) injections into the CA1 and CA3 areas of the hippocampus to evoke seizures and following neuronal death. Systemic KA injection caused a more severe convulsive response in heterozygous mice than in their age-matched controls. $Msh2+/-$ mice were also found to be more sensitive to KA-induced neurodegeneration in the hippocampal CA3 field. The response to KA-mediated excitotoxicity includes also the induction of cyclooxygenase-2 (COX-2) expression, which parallels the appearance of neuronal apoptotic features in the same cell types affected by KA treatment. Immunohistological studies in wild-type and $Msh2+/-$ mice treated with KA showed that COX-2 protein expression was enhanced in the CA3 neurons of heterozygous mice compared to those in wild-type mice.

Conclusions

Taken together these data suggest that the lack of efficient systems involved in recognizing and repairing DNA damage, in addition to an accompanying high risk for the development of colon cancer, modifies the gene expression profile in selected brain regions to reduce mitochondrial activity. This adaptive response, although preventing ROS generation and ROS-induced DNA damage, results in mitochondrial cytopathy characterized by a lower threshold to seizure, excitotoxicity, and neurodegeneration.

These data can be added to the growing list of evidence accumulated in recent years suggesting an involvement of cancer-related mechanisms in the development of neurodegenerative diseases.

References

1. Marti TM, Kunz C, Fleck O. DNA mismatch repair and mutation avoidance pathways. J Cell Physiol 2002;191:28–41.
2. Liu B, Parson R, Papadopoulos N. Analysis of mismatch repair genes in hereditary nonpolyposis colorectal cancer patients. Nat Med 1996;2:169–174.
3. Eshelman JR, Markowitz SD. Mismatch repair defects in human carcinogenesis. Hum Mol Genet 1996;5:1489–1494.
4. Acharya S, Foster PL, Brooks P, Fischel R. The coordinated functions of the E. coli MutS and MutL proteins in mismatch repair. Mol Cell 2003;12:233–246.

5. Bellacosa A. Functional interactions and signaling properties of mammalian DNA mismatch repair proteins, Cell Death Differ 2001;8:1076–1092.
6. Rolig RL, McKinnon PJ. Linking DNA damage and neurodegeneration. Trends Neurosci 2000;23:417–424.
7. Brooks PJ. DNA repair in neural cells: basic science and clinical implications. Mutat Res 2002;509:93–108.
8. Marietta C, Palombo F, Gallinari P, et al. Expression of long-patch and short-patch DNA mismatch repair proteins in the embryonic and adult mammalian brain. Mol Brain Res 1998;53:317–320.
9. Brooks PJ, Marietta C, Goldman D. DNA mismatch repair and DNA methylation in adult brain neurons. J Neurosci 1996;16:939–945.
10. Belloni M, Uberti D, Rizzini C, et al. Distribution and kainate-mediated induction of the DNA mismatch repair protein MSH2 in rat brain. Neuroscience 1999;94:1323–1331.
11. Belloni M, Uberti D, Rizzini C, et al. Induction of two DNA mismatch repair proteins, MSH2 and MSH6, in differentiated human neuroblastoma SH-SY5Y cells exposed to doxorubicin. J Neurochem 1999;72:974–979.
12. Uberti D, Ferrari Toninelli G, Memo M. Involvement of DNA damage and repair systems in neurodegenerative process. Toxicol Lett 2003;139:99–105.
13. Francisconi S, Codenotti M, Ferrari-Toninelli G, et al. Preservation of DNA integrity and neuronal degeneration. Brain Res Rev 2005;48:347–351.
14. Reitmair AH, Schmits R, Ewel A, et al. Msh2 deficient mice are viable and susceptible to lymphoid tumours. Nat Genet 1995;11:64–70.
15. De Wind N, Dekker M, van Rossum A, et al. Mouse models for hereditary nonpolyposis colorectal cancer. Cancer Res 1998;58:248–255.
16. Francisconi S, Codenotti M, Ferrari Toninelli G, et al. Mitochondrial dysfunction and increased sensitivity to excitotoxicity in mice deficient in DNA mismatch repair. J Neurochem 1998;98:223–233.
17. Schmiedel J, Jackson S, Schäfer J, Reichmann H. Mitochondrial cytopathies. J Neurol 2003;250:267–277.

Chapter 4
Current Problems and Strategies in Motor Rehabilitation for Parkinson's Disease

Giovanni Abbruzzese, Elisa Pelosin, and Roberta Marchese

Introduction

Idiopathic Parkinson's disease (PD) can be defined pathologically by the loss of dopaminergic neurons in the pars compacta of the substantia nigra (with the presence of intracytoplasmic neuronal inclusions, the Lewy bodies) and clinically by some combination of rest tremor, rigidity, bradykinesia, and loss of postural reflexes [1]. PD is a common neurodegenerative disorder affecting more than 250 patients per 100,000 population [2] and can be considered one of the major causes of neurological disability.

Parkinson's disease has a chronic progressive course, mainly affecting the motor system, and is responsible for severe motor impairment. The cause of the disease is still unknown, so no "cure" or neuroprotective treatment is yet available. On the other hand, symptomatic dopaminergic treatment is highly effective during the early phases but progressively loses its efficacy and induces motor and nonmotor complications. In addition, there are symptoms (e.g., postural instability, gait disturbances, cognitive impairment) that are not responsive to dopaminergic stimulation, and it has been demonstrated [3] that the most disabling long-term problems of PD are related to the occurrence of symptoms that are not alleviated by L-dopa. Eventually, autonomy in the daily living activities is severely reduced during the late phases, and there is no rescue treatment for mobility or postural problems.

There is therefore a solid rationale for motor rehabilitation in PD with the aim of improving quality of life. Different-level goals may be addressed: preventing or cutting down secondary complications due to reduced mobility, optimizing the residual functional capacities, and compensating for the defective abilities by means of new attitudinal strategies or environmental changes. Unfortunately, rehabilitation remains a relatively neglected aspect of PD, with a large variety of approaches and great uncertainty as to their effectiveness.

G. Abbruzzese
Department of Neurosciences, Ophthalmology and Genetics, University of Genoa, Via De Toni 5, Genova, Italy

A. Fisher et al. (eds.), *Advances in Alzheimer's and Parkinson's Disease*,
© Springer 2008

Clinical Research on Rehabilitation for PD

It is generally accepted that physical activity and exercise are good for patients with PD. This assumption is supported by experimental data in the 6-hydroxydopamine (6-OHDA) rat model of PD, which has demonstrated that preinjury forced limb use is able to reduce behavioral and neurochemical deficits, possibly because of the neuroprotective effects of glial cell line-derived neurotrophic factor (GDNF) [4]. Moreover, survival of people with PD has been associated with increased physical activity [5]; and recently a large prospective study [6] suggested that higher levels of physical activity may lower the risk of PD developing in men.

Although common experience suggests that parkinsonian patients may benefit from physical therapy added to their standard medication [7], it should be pointed out that referrals of PD patients to rehabilitative procedures are often arbitrary. It has been observed that approximately 40% of patients have no serious problems (posture, balance, gait, transfers) requiring referral to physiotherapy [8].

Also, the available clinical studies on rehabilitation for PD [9] are far from being conclusive and suffer frequent methodological flaws. Most of the studies have been observational (with few randomized controlled trials) and small scale (few patients included using uncertain diagnostic criteria) or suffered lack of control. A peculiar picture of the rehabilitative approach to PD stands out where highly heterogeneous treatment protocols (from karate training or stretching exercises to proprioceptive neuromuscular facilitation techniques or behavioral-occupational therapy) and largely variable outcome measures (from a four-point rating scale to instrumental motor analysis) have been adopted. Not surprisingly, therefore, systematic reviews of paramedical therapies for PD concluded that there is insufficient evidence "to support or refute the efficacy of paramedical therapies [10]." Such a conclusion reflects largely the difficulty of conducting rehabilitation research with a methodological procedure similar to that adopted for pharmaceutical trials, where blinded patients, blinded investigators, and placebo treatments are mandatory.

Indeed, there are some major problems in documenting the effectiveness of physical therapy in PD. There is certainly an absolute need for methodologically more robust research, including large, homogeneous patient samples, well defined inclusion criteria, placebo groups with adequate randomization and allocation, rigorous protocol designs (distinguishing specific and unspecific effects), meaningful outcome measures, prolonged follow-up periods, and evaluation by means of statistical analysis as well as clinical significance.

However, it should also be remembered that the rehabilitative approach in neurological patients is procedural in nature. It intrinsically is based on the mutual adaptation of patient and environment and involves learning new compensatory strategies or using repetitive activities that might modify the brain functional organization.

Current Problems in Rehabilitation for PD

There are some basic unresolved questions. When should we start the treatment? If rehabilitation is planned for the most advanced phases it may suffer several limitations, whereas earlier in the course of the disease hierarchically higher goals may be achieved. Which patients should we treat? Should we consider all the patients eligible for treatment? Definitive referral criteria for physiotherapy are not available. In particular, we do not have indications on which factors may be regarded as indicative of a poor therapeutic result. How should we treat the patients—with standard or individually tailored protocols? Various rehabilitative interventions have been tried in PD patients and have shown possible beneficial effects. Ultimately, there is little evidence to suggest which approach is the best.

Some possible answers to these questions can be derived from the model proposed by Morris [11]. The movement disorders leading to disability in PD are characterized by large variations across individuals and over time: resting tremor, slowness of movement, and gait hypokinesia are usually present beginning in the early stages, whereas severe akinesia (and possible dyskinesia) with postural instability and falls prevail in the later stages. In addition, the consequences of motor abnormalities are largely dependent on individual needs and environmental context. Finally, different pathophysiological mechanisms underlie the various PD-related movement disorders. Morris [11] therefore suggested some basic requirements for a rehabilitation approach to PD: (1) up-to-date knowledge of the pathophysiology of the motor impairment; (2) ability to implement a basic management plan according to the stage of disability; and (3) task-specific approaches within the context of functional tasks of everyday living and problem-solving skills that enable individually tailored treatment plans.

In keeping with the suggestion by Morris [11], we believe that a physiologically based rehabilitation program can effectively interact with some critical features of motor disability in PD.

Neurophysiological Approach to Motor Rehabilitation

Neurophysiological studies have elucidated several mechanisms underlying various aspects of motor impairment in PD [12]. They have shown that the basal ganglia are mainly involved in the maintenance of movement amplitude, velocity, and sequence and in the generation of automatic motor plans. In a certain way, the motor impairment in PD is characterized by a mismatch between the cortex (i.e., the desired or intended movement) and the basal ganglia (i.e., actual movement). Understanding these mechanisms may offer a unique opportunity to individualize possible substrates for rehabilitation by developing new strategies to overcome, or at least restrict, the motor deficits.

Parkinson disease patients have difficulty initiating movement and are globally slow in executing motor commands. Such abnormalities result from reduced motor excitability [13] and insufficient activation of the cortical areas (supplementary motor area, primary motor cortex) involved in planning and executing motor activities [12,14]. Parkinsonian patients therefore may benefit from all the procedures addressed to more efficiently build up motor excitability.

Some motor activities in PD are more impaired than others. Complex (sequential, simultaneous, or bimanual) movements are significantly more affected than simple movements. It has been shown that additional slowness can be observed in parkinsonian patients when they try to perform two rapid sequential movements [15]. The amplitude of individual movements progressively decreases as the execution of motor sequences progresses ("sequence effect"). Similarly, the concomitant execution of two motor (or cognitive) tasks is specifically impaired in PD [16,17]. This "dual task effect" reflects the difficulty of distributing the limited processing resources among several tasks and the greater resource consumption to compensate for the lack of automatic behavior [18]. The "dual task effect" is a critical factor for motor ability of parkinsonian patients. In a recent study analyzing the effect of cognitive loading on postural stability, we demonstrated that, at variance with normal controls, postural stability is significantly impaired in parkinsonian patients during dual performance of an additional motor or cognitive task [19]. They therefore should learn new strategies for motor planning using individual plans with well defined spatial and temporal features.

There is also considerable evidence that patients with PD have greater difficulty with self-initiated movements than with externally triggered movements [20,21]. Indeed, sensorimotor integration is impaired in PD [22], and parkinsonian patients show increased dependence on external sensory cues. Their motor performance is facilitated by visual, auditory, or proprioceptive information [23–26]. Various explanations have been proposed to interpret the effect of external cues; for example, possibly these cues facilitate motor processing through the intact lateral premotor cortex or simply favor focusing the attention, perhaps by using frontocortical mechanisms [24,27]. Finally, the ability to manipulate objects or perform skilled manual activities (reaching, grasping, writing, dressing) is severely compromised in parkinsonian patients. This impairment results from defective sensorimotor integration [28,29] as well as from bradykinesia and deficits in scaling motor activation [30] and grip force production [31]. Again, simplifying motor plans, adopting external cues, and adapting targets can facilitate manual activities.

The knowledge of defective physiological mechanisms in PD therefore supports using some basic strategies (although not yet specific protocols) in the rehabilitative treatment of parkinsonian patients. Some examples are reported in Table 1. In summary, the physiological approach to rehabilitation in PD is aimed at changing the skilled and automatic motor activities into new movement strategies that are not routinely processed through the faulty basal ganglia system.

Table 1 Examples of physiological strategies used to compensate for motor problems in Parkinson's disease

Motor problems	Physiological suggestions
Difficulty initiating movements	Prepare movement in advance.
	Increase cortical activation with attentional strategies (e.g., motor imagery, visualization)
Additional impairment during sequential movements ("sequence effect")	Break down complex motor sequences into smaller individual components
Specific impairment during simultaneous tasks ("dual task effect")	Avoid simultaneous motor (or motor and cognitive) activities
	Focus attention on voluntary control of a single task
Greater impairment in self-initiated versus externally triggered movements	Use external sensory cues (visual, auditory, proprioceptive) to start and maintain motor activities
Difficulty manipulating objects	Use larger targets to facilitate movement
Sudden inability to move ("freezing")	Modify environmental context or favor prior environmental recognition

Possible Limitations of Rehabilitation for PD

The positive effect of rehabilitation is enhanced if patients learn to select the correct movement strategy and are trained to use it in the appropriate environmental context. The basic assumption underlying the rehabilitative approach in parkinsonian patients is the possibility of "plastic" learning. The issue, therefore, of procedural motor learning is a critical one. Several studies document that patients with PD have the capability to learn new motor tasks and to improve their performance through practice [32–34]. What is relatively uncertain is the persistence over time of newly learned strategies. A recent study [35] demonstrated that after short-term practice (1–7 days) a benefit was observed that was similar in parkinsonian patients and normal controls; however, after prolonged (2 weeks) practice, healthy controls continued to improve but the patients did not. The authors suggested that the time spent in the rehabilitation of patients with PD should be optimized by training more tasks for a shorter time; they also suggested that frequently changing the training task may promote more efficient attentional strategies.

Although procedural motor learning is possible in PD it may be defective because of progressive basal ganglia pathology, or it may suffer important limitations. PD patients have perceptual difficulties, particularly with time and space estimation [36,37]. In addition, minor (mainly executive) cognitive deficits are frequently observed in PD, and dementia can develop with an average prevalence of 30% to 40% [38].

Cognitive rehabilitation in PD patients with mild cognitive deficits is able to improve verbal fluency and logic memory, thereby reinforcing cognitive strategies [39]. However, obviously rehabilitative procedures that rely on attentional strategies cannot be used in patients with severe cognitive impairment. Indeed, the mental status is probably the most critical issue when predicting the rehabilitative outcome [40], although several other factors should be considered, including disease severity, aging, mood, co-morbidity, drug regimen, and changes in the musculoskeletal and cardiovascular systems.

How long the benefits induced by rehabilitation can be maintained is still uncertain and represents another possible limitation of rehabilitation for PD. Improvements can usually be maintained at least 3 months [25,41]. Wade et al., [42] using a randomized controlled crossover approach, showed that group educational activities and individual multidisciplinary rehabilitation improved mobility in PD patients. However, 6 months after treatment, a significant increase in the disability and decline in the quality of life, as well as increased caregivers' strain, were observed. This suggests that follow-up treatments may be needed to maintain any benefit and raises the problem of cost-effectiveness of physical therapy programs for parkinsonian patients. In this regard, it should be pointed out that a self-supervised exercise program was found to be equally as effective as a physiotherapist-supervised exercise program in assuaging motor symptoms in patients with PD [43].

Conclusions

Definitive answers are still awaited with regard to the actual effectiveness of physical therapy for PD and the precise contribution of rehabilitation to the management of the disease. Possible suggestions depend on the results of double-blind, placebo-controlled, crossover clinical trials. Nevertheless, there is already considerable evidence to support the extensive use of physiological principles and strategies in the rehabilitation program of parkinsonian patients. Such a "physiological" approach is addressed mainly to transform some daily motor activities from an automatic to a voluntarily controlled behavior and to favor task-specific training in the natural environmental context.

References

1. Lang AE, Lozano AM. Parkinson's disease. N Engl J Med 1998;339:1130–1143.
2. Morgante L, Rocca WA, Di Rosa AE, et al. Prevalence of Parkinson's disease and other types of parkinsonism: a door-to-door survey in three Sicilian municipalities—the Sicilian Neuro-Epidemiologic Study (SNES) group. Neurology 1992;42:1901–1907.
3. Hely MA, Morris JGL, Reid WGJ, Trafficante R. Sydney Multicenter Study of Parkinson's Disease: non-L-dopa–responsive problems dominate at 15 years. Mov Disord 2004;20:190–199.
4. Cohen AD, Tillerson JL, Smith AD, et al. Neuroprotective effects of prior limb use in 6-hydroxydopamine-treated rats: possible role of GDNF. J Neurochem 2003;85: 299–305.
5. Kuroda K, Tatara K, Shinsho F. Effect of physical exercise on mortality in patients with Parkinson's disease. Acta Neurol Scand 1992;86:55–59.

6. Chen H, Zhang SM, Schwarzschild MA, et al. Physical activity and the risk of Parkinson disease. Neurology 2005;64:664–669.
7. De Goede CJ, Kreus SH, Kwakkel G, Wagenaar RC. The effects of physical therapy in Parkinson's disease: a research synthesis. Arch Phys Med Rehabil 2001;82:509–515.
8. Keus SH, Bloem BR, Verbaan D, et al. Physiotherapy in Parkinson's disease: utilization and patient satisfaction. J Neurol 2004;251:680–687.
9. Gage H, Storey L. Rehabilitation for Parkinson's disease: a systematic review of available evidence. Clin Rehabil 2004;118:463–482.
10. Deane KHO, Ellis-Hill C, Jones D, et al. Systematic review of paramedical therapies for Parkinson's disease. Mov Disord 2002;17:984–991.
11. Morris M. Movement disorders in people with Parkinson's disease: a model for physical therapy. Phys Ther 2000;80:578–597.
12. Berardelli A, Rothwell JC, Thompson PD, Hallett M. Pathophysiology of bradykinesia in Parkinson's disease. Brain 2001;124:2131–2146.
13. Pascual-Leone A, Valls-Solè J, Brasil-Neto JP, et al. Akinesia in Parkinson's disease. I. Shortening of simple reaction time with focal, single-pulse transcranial magnetic stimulation. Neurology 1994;44:884–891.
14. Gilio F, Currà A, Inghilleri M, et al. Repetitive magnetic stimulation of cortical motor areas in Parkinson's disease: implications for the pathophysiology of cortical function. Mov Disord 2002;17:467–473.
15. Benecke R, Rothwell JC, Dick JP, et al. Disturbance of sequential movements in patients with Parkinson's disease. Brain 1987;110:361–379.
16. Schwab RS, Chafetz ME, Walker S. Control of two simultaneous voluntary motor acts in normals and in parkinsonians. Arch Neurol Psychiatry 1954;72:591–598.
17. Benecke R, Rothwell JC, Dick JP, et al. Performance of simultaneous movements in patients with Parkinson's disease. Brain 1986;109:739–757.
18. Brown RG, Marsden CD. Dual task performance and processing resources in normal subjects and patients with Parkinson's disease. Brain 1991;114:215–231.
19. Marchese R, Bove M, Abbruzzese G. Effect of cognitive and motor tasks on postural stability in Parkinson's disease: a posturographic study. Mov Disord 2003;18:652–658.
20. Jahanshahi M, Jenkins IH, Brown RG, et al. Self-initiated versus externally triggered movements. I. An investigation using measurement of regional cerebral blood flow with PET and movement-related potentials in normal and Parkinson's disease subjects. Brain 1995;118:913–933.
21. Currà A, Berardelli A, Agostino R, et al. Performance of sequential arm movements with and without advance knowledge of motor pathways in Parkinson's disease. Mov Disord 1997;12:646–654.
22. Abbruzzese G, Berardelli A. Sensorimotor integration in movement disorders. Mov Disord 2003;18:231–240.
23. Cunnington R, Iansek R, Bradshaw JL, Philips JG. Movement-related potentials in Parkinson's disease: presence and predictability of temporal and spatial cues. Brain 1995;118:935–950.
24. Morris ME, Iansek R, Matyas TA, Summers JJ. Stride length regulation in Parkinson's disease: normalization strategies and underlying mechanisms. Brain 1996;119:551–568.
25. Marchese R, Diverio M, Zucchi F, et al. The role of sensory cues in the rehabilitation of parkinsonian patients: a comparison of two physical therapy protocols. Mov Disord 2000;15:879–883.
26. Rubinstein TC, Giladi N, Hausdorff JM. The power of cueing to circumvent dopamine deficits: a review of physical therapy treatment of gait disturbances in Parkinson's disease. Mov Disord 2002;17:1148–1160.
27. Soliveri P, Brown RG, Jahanshahi M, Marsden CD Effects of practice in performance of a skilled motor task in patients with Parkinson's disease. J Neurol Neurosurg Psychiatry 1992;55:454–460.

28. Klockgether T, Borutta M, Rapp H, et al. A defect of kinesthesia in Parkinson's disease. Mov Disord 1995;10:460–465.
29. Demirci M, Grill S, McShane L, Hallett M. A mismatch between kinesthetic and visual perception in Parkinson's disease. Ann Neurol 1997;41:781–788.
30. Berardelli A, Dick JP, Rothwell JC, et al. Scaling of the size of the first agonist EMG burst during rapid wrist movements in patients with Parkinson's disease. J Neurol Neurosurg Psychiatry 1986;49:1273–1279.
31. Fellows SJ, Noth J, Schwarz M. Precision grip and Parkinson's disease. Brain 1998;121:1771–1784.
32. Behrman AL, Cauraugh JH, Light KE. Practice as an intervention to improve speeded motor performance and motor learning in Parkinson's disease. J Neurol Sci 2000;174:127–136.
33. Nutt JG, Lea ES, van Houten L, et al. Determinants of tapping speed in normal control subjects and subjects with Parkinson's disease: differing effects of brief and continued practice. Mov Disord 2000;15:843–849.
34. Swinnen SP, Steyvers M, van Den Bergh L, Stelmach GE. Motor learning and Parkinson's disease: refinement of within-limb and between-limb coordination as a result of practice. Behav Brain Res 2000;111:45–59.
35. Agostino A, Currà A, Soldati G, et al. Prolonged practice is of scarce benefit in improving motor performance in Parkinson's disease. Mov Disord 2004;19:1285–1293.
36. Artieda J, Pastor MA, Lacruz F, Obeso JA. Temporal discrimination is abnormal in Parkinson's disease. Brain 1992;115:199–210.
37. Crucian GP, Barrett AM, Schwartz RL, et al. Cognitive and vestibule-proprioceptive components of spatial ability in Parkinson's disease. Neuropsychologia 2000;38:757–767.
38. Emre M. Dementia associated with Parkinson's disease. Lancet Neurol 2003;2:229–237.
39. Sinforiani E, Banchieri L, Zucchella C, et al. Cognitive rehabilitation in Parkinson's disease. Arch Gerontol Geriatr 2004;9:387–391.
40. Nieuwboer A, De Weerdt W, Dom R, Bogaerts K. Prediction of outcome of physiotherapy in advanced Parkinson's disease. Clin Rehabil 2002;16:886–893.
41. Pellecchia MT, Grasso A, Biancardi LG, et al. Physical therapy in Parkinson's disease: an open long-term rehabilitation trial. J Neurol 2004;251:595–598.
42. Wade DT, Gage H, Owen C, et al. Multidisciplinary rehabilitation for people with Parkinson's disease: a randomized controlled study. J Neurol Neurosurg Psychiatry 2003;74:158–162.
43. Lun V, Pullan N, Labelle NJ, et al. Comparison of the effects of a self-supervised home exercise program with a physiotherapist-supervised exercise program on the motor symptoms of Parkinson's disease. Mov Disord 2005;20:971–975.

Chapter 5
Critical Acute Akinesia in Parkinson's Disease Patients Concomitant with Additional Precipitating Factors Other Than Treatment Withdrawal: A Possible Lethal Complication

Marco Onofrj and Astrid Thomas

Introduction

Around 3% of Parkinson's disease (PD) patients per year suffer from a critical syndrome that consists of worsening acute motor symptoms with akinesia occurring concomitantly with infectious disease, surgery, trauma, or because of drug manipulation or withdrawal [1]. In the most severe cases, this syndrome is characterized by alterations of mental status, total akinesia with dysphagia, hyperthermia, dysautonomia, and incremental increases of muscle enzymes. It strictly resembles the neuroleptic malignant syndrome (NMS) induced by typical and some atypical neuroleptic drugs [2]. In less severe cases, the acute crisis may appear only with a sudden increment of parkinsonian motor symptoms. Symptomatic recovery may appear several days or weeks after onset and may be incomplete. In the international literature the akinetic crisis of PD patients was described solely as a consequence of antiparkinsonian treatment withdrawal and was defined as neuroleptic malignant-like syndrome (NMLS) [3], parkinsonian hyperpyrexia (PH) [4], or more recently malignant syndrome (MS) [5].

In a recent article we showed that this syndrome may definitely appear in vigorously treated patients because of precipitating factors other than treatment withdrawal. The occurrence of the acute crisis can explain the apparent acute onset of PD immediately after a surgical intervention, controversially attributed to anesthesia or PD onset after infectious diseases or trauma. These occurrences often incur unanswered questions in patient-to-doctor forums. Moreover, our study showed that during the akinetic crisis PD patients do not respond for days or weeks to nasogastric administration of current antiparkinsonian treatment, to increments of dopaminergic drugs, or to continuous subcutaneous apomorphine infusion. Moreover, these patients mostly have

M. Onofrj
Neurophysiopathology, Movement Disorders Center, Department of Oncology and Neuroscience, University "G.D' Annunzio" Chieti-Pescara, and Aging Research Center, Ce.S.I., "Gabriele D' Annunzio" University Foundation, Chieti-Pescara, Italy, Ospedale Civile di Pescara, Italy

A. Fisher et al. (eds.), *Advances in Alzheimer's and Parkinson's Disease*,
© Springer 2008

normal L-dopa kinetics. Thus, this critical condition can be differentiated from the well known continuum of motor hypokinesias observed in PD [6], such as "wearing-off" or other "off" states, because of refractoriness to dopaminergic rescue drugs [7].

In our study we defined the critical condition as acute akinesia because this term appeared, although not in the English-language literature [8], before other definitions of NMLS, PH, or MS. By showing that refractoriness to dopaminergic drugs and apomorphine is a key factor independent of the precipitating event (concomitant diseases or drug manipulation), we suggested that a common definition may encompass all critical sudden drug refractory worsening of parkinsonism in PD patients.

We suggest that this acute akinesia of PD patients should be recognized not only by movement disorder experts and PD clinicians but also by general practitioners, geriatricians, and general and orthopedic surgeons as the syndrome can unpredictably occur in PD patients during respiratory tract infections, with gastrointestinal diseases, and after bone fractures or surgery. Inadequate recognition might entail insufficient prevention of co-morbid factors (thrombosis, disseminated intravascular coagulation, cardiac arrhythmias) or inappropriate management strategies, including treatment withdrawal or administration of typical neuroleptics, leading to lethal consequences [1,5].

Misdiagnosis is common. Even though our study was conducted on a cohort of patients regularly followed in our PD clinic, in 61% (16/26) of the patients in the study and in 18 more patients observed during the same time period the diagnosis at admission to the hospital with sepsis, aphagia, or anarthria was laryngitis. Yet these diagnoses were not illogical as the akinetic crisis appeared 2 to 3 days after onset of flu, bronchopneumonia, bone fracture (17 patients), or jejunal volvolus, gastric stasis, and duodenal ulcer (3 patients). Moreover, 76% of the patients were febrile, and 84% had increased creatine phosphokinase blood levels.

However, the rapid evolution of the crisis with increments of parkinsonian motor symptoms (increased by 80% based on the Unified Parkinson Disease Rating Scale, or UPDRS) and the acute onset of severe invalidism clarified the diagnosis. Refractoriness to continuous subcutaneous apomorphine (50–200 mg/day) infusion lasted 4 to 26 days. We suggest that rapid deterioration and refractoriness to apomorphine are key factors for the diagnosis: A PD patient who presents with these two symptoms should be treated intensively and promptly with rehydration, treatment of hyperthermia, antithrombotic prophylaxis, and management of blood pressure instability and cardiac arrhythmias. This treatment strategy, together with continuation of antiparkinsonian treatment despite a lack of response, and with avoidance of neuroleptic sedative treatment favored a recovery in 84% of the patients (four fatalities among 26 patients), which was complete (i.e., to the same level of motor symptoms as prior to the crisis) in 61% of the patients.

During the same observation period we found that flu, bronchopneumonia (603 patients), surgery (7 patients), and even paralytic ileum with gastric stasis (1 patient) did not induce worsening of relevant motor symptoms or refractoriness in other PD patients. Our study therefore showed that a distinct clinical entity can appear in PD patients independently of treatment withdrawal or of antidopaminergic drug administration. This entity is not predictable on the basis of PD duration, stage, or treatment.

By administering apomorphine, treating intensively with L-dopa, and measuring L-dopa plasma levels, we showed that acute akinesia is unlikely to be related to failure of L-dopa uptake or conversion or to a reduction in dopamine storage. The persistence of acute akinesia after the disappearance of fever or its occurrence without fever suggested also that it is not dependent on energetic uncoupling. Furthermore, acute akinesia appeared in patients not treated with L-dopa or appeared because of amantadine withdrawal, suggesting that critical symptoms cannot be simply attributed to changes in dopamine pharmacodynamics [1].

This critical condition is historically known in the German literature [8–10] and is today listed in the treatment guidelines of the German Neurological Society [11] as akinetic syndrome or crisis. Acute akinesia, akinetic crisis, or akinetic syndrome appeared as a mention in a handbook on movement disorders [12] and in three review papers on amantadine [8–10,13].

A recent Japanese multicenter study [5], probably unaware of the previous literature, coined the term malignant syndrome (MS) to define patients affected by severe worsening of the parkinsonian symptom akinesia, with fever as a consequence of treatment withdrawal, thus abbreviating the NMLS definition. Yet the study states that the syndrome "may even develop without the interruption of antiparkinsonian drugs."

Responses to treatment were not investigated in the Japanese study, whereas the German guidelines indicated that the first treatment should be amantadine sulfate given via intravenous infusion, 500 mg b.i.d., and second-line apomorphine in continuous infusion or bolus. Despite these indications, evidence of the effects of amantadine are restricted to the few articles cited above, presenting heterogeneous descriptions lacking in details and including patients not affected by PD or patients under neuroleptic treatment.

In our study, we could not assess the effect of amantadine sulfate, as the drug is not available in our country, but we could assess apomorphine, which had been available since 1986 in military hospitals and since 1998 in a commercial formulation. We administered apomorphine together with intravenous ondansetron (8–16 mg/day) to avoid emesis, yet no changes in the UPDRS scores were observed. The observed refractoriness to apomorphine demonstrates that controlled trials are needed despite the strong incentive to intervene because of the possible lethal consequence.

A 15% death rate was observed among our patients (treated with apomorphine), and similar rates have been described for NMLS and NMS [2,14]. For all these critical conditions, there is no validated evidence that pharmacological treatments (e.g., bromocriptine, dantrolene, steroids, apomorphine, amantadine) modify the critical course.

Conclusion

Our report suggests the need for further studies on the comparative effects of treatments (assuming that double-blind treatment studies versus placebo are deemed unacceptable) and on the mechanism of the occurrence of this disease. The overlapping features of critical akinesia and NMS, the increase sensitivity of PD and Lewy body dementia [15] to neuroleptics, and the frequent finding of Lewy bodies in the hypothalamus of patients with PD and Lewy body dementia [16] suggest that critical conditions with severe and sudden motor deterioration and hyperthermia are a core feature of synucleopathies. More adequate knowledge of this feature could have relevant implications for the politics governing typical and atypical neuroleptic prescriptions for patients.

References

1. Onofrj M, Thomas A. Acute akinesia in Parkinson disease. Neurology 2005;64:1162–1169.
2. Kipps CM, Fung VSC, Grattan-Smith P, de Moore GM, Morris JGL. Movement disorders emergencies. Mov Disord doi: 10.1002/mds.20325.
3. Friedman JH, Feinberg SS, Feldman RC. Neuroleptic malignant-like syndrome due to l-dopa withdrawal. Ann Neurol 1984;16:126.
4. Granner MA, Wooten GF. Neuroleptic malignant syndrome or parkinsonism hyperpyrexia syndrome. Semin Neurol 1991;11:228–235.
5. Takubo H, Harada T, Hashimoto T, et al, A collaborative study on the malignant syndrome in Parkinson's disease and related disorders. Parkinsonism Relat Disord 2003;9:S31–S41.
6. Quinn NP. Classification of fluctuations in patients with Parkinson's disease. Neurology 1998;51(Suppl 2):S25–S29.
7. Stacy M, Factor S. Rapid treatment of "off" episodes: will this change Parkinson disease therapy? Neurology 2004;62(Suppl 4):S1–S2.
8. Danielczyk W. Die Behandlung von akinetischen Krisen. Med Welt 1973;24:1278–1282.
9. Muschard G, Völler GW. Wirksamkeit von Amantadinsulfat als Infusionslösung bei der Behandlung des Parkinsons Sindroms. Med Welt 1973;24:183–184.
10. Oppel F, Klaes H. Die Amantadin-Infusionen in der akuten Behandlung der Morbus Parkinson. Klinikarzt 1981;10:559–565.
11. Deutsche Gesellschaft für Neurologie: Leitlinien-Parkinson-Syndrome; 4,3,4,2 Akinetische Krise. http://www.dgn.org/168.0htlm#926.
12. Tanner CM, Goetz CG, Klawans HL. Autonomic nervous system disorders. In: Koller WC (ed) Handbook of Parkinson's Disease. Marcel Dekker, New York, 1987, pp 161–163.
13. Danielczyk W. Twenty-five years of amantadine therapy in Parkinson's disease. J Neural Transm Suppl 1995;46:399–409.
14. Shalev A, Hermesh H, Munitz H. Mortality from neuroleptic malignant syndrome. J Clin Psychiatry 1989;50:18–25.
15. McKeith I, Misumann U. Dementia with Lewy bodies and Parkinson's disease. Parkinsonism Relat Disord 2004;10(Suppl 1):S15–S18.
16. Dickson DW. Neuropathology of Parkinsonian disorders. In: Jankovic J, Tolosa E (eds) Parkinson's Disease and Movement Disorders, 4th ed. Lippincott Williams & Wilkins, Philadelphia, 2002, pp 256–269.

Chapter 6
Biomarkers for Early Detection of Parkinson's Disease: An Essential Challenge

Florian Tribl and Peter Riederer

Introduction

Today, Parkinson's disease (PD) is a clinical diagnosis that is available when the campaign has already been lost. There is an urgent demand for neuroprotective drugs to treat PD and eventually halt neurodegeneration at an early, preclinical stage. We therefore need new diagnostic criteria by which to detect the onset and the progression of the disease at a significantly earlier stage. A revision of the degeneration of the substantia nigra pars compacta as the current lead concept of PD may support the translation of recent pathological findings into new biomarkers to enhance a considerably earlier diagnosis.

Parkinson's disease is one of the most frequent neurological disorders of elderly or aged people. About 1% of people in the sixth decade of life are affected, with that figure increasing to 3% of people in the eighth decade. Approximately 3 million people in Europe are estimated to have PD.

Parkinson's disease is a chronically progressing neurodegenerative movement disorder characterized by akinesia, rigidity, resting tremor, the "freezing" phenomenon, a flexed posture, and loss of postural reflexes. The underlying molecular causes for PD and the exact etiology remain unknown. An increasing number of genes and relevant chromosomal regions have been found to be associated with parkinsonism. However, the early onset and the symptoms of these "familial" cases greatly differ from the more common (95% of cases) "sporadic" form of PD and render unclear the contribution of genetics to idiopathic Parkinson's syndrome (IPS) [1,2]. The causes of "sporadic, idiopathic" cases are therefore considered a complex network of genetic and environmental factors that result in a broad range of phenomena that obviously play significant roles in the parkinsonian brain, including defective protein degradation, the aggregation of proteins such as α-synuclein,

F. Tribl
The National Parkinson Foundation (NPF) Research Laboratories', Miami, FL, USA, and Department of Clinical Neurochemistry, Clinic and Polyclinic for Psychiatry and Psychotherapy, University Würzburg, Füchsleinstrasse 15, Würzburg 97080, Germany

A. Fisher et al. (eds.), *Advances in Alzheimer's and Parkinson's Disease*,
© Springer 2008

and the formation of Lewy bodies; inhibition and disruption of several mitochondrial respiratory chain complexes and increased formation of free radicals; severely disturbed iron metabolism; disturbed Ca2+ homeostasis; increased formation of nitrous oxide; glutamate excitotoxicity; apoptosis; inflammation; and reduced neurotrophic support [3–8].

The rationale of an environmental contribution to the pathogenesis of PD emanated from the evidence that parkinsonian symptoms resulted from 1-methyl-4-phenyl-1,2,3,6-tetrahydropyridine (MPTP) intoxication [9,10]. Several environmental neurotoxins, including paraquat, rotenone, 1,2,3,4-tetrahydro-isoquinolines, β-carbolines and their respective derivatives, have subsequently been identified to evoke symptoms similar to those of PD [11–15]. The *MDR1* gene encoding the blood-brain barrier (BBB) protein P-glycoprotein (P-gp) has recently been pinpointed as a link between environmental and genetic causes of PD because P-gp is engaged in the active transport of various molecules out of the brain across the BBB [16]. A functional C3435T polymorphism of the *MDR1* gene is suspected to aggravate the toxic effects of pesticides and xenobiotics due to their accumulation in the brain and thereby may account for PD. Several toxic compounds (e.g., MPP+, herbicides, pesticides) but also dopaminergic drugs accumulate in the neuromelanin granules in the substantia nigra (SN) [17–20].

Since the appearance of parkinsonism in postencephalitic cases of von Economo encephalitis lethargica [21,22], viral and bacterial infections—at least at an early, prenatal developmental stage of the brain—have been discussed as possible trigger processes that ultimately may end up in parkinsonism [23–26].

In principle, PD appears to be a heterogeneous syndrome in which multiple factors lead to various subtypes of idiopathic PD [27–29].

Current Diagnosis of Parkinson's Disease

Currently, PD is a pure clinical diagnosis and depends solely on the judgment of the clinician, as there is still no test system or biological marker available to give certainty to the diagnosis. Prerequisites for the current clinical diagnosis of PD are (1) occurrence of the cardinal motor symptoms (rigidity, resting tremor, akinesia) as the phenotypical correlates of dopamine depletion in the striatum due to the cell loss in the SN; and (2) a good response to L-dopa (levodopa) treatment [30,31]. After an initial diagnosis of PD, several diagnostic tools are available for grading the severity of the disease and evaluating disease progression, such as the Hoehn and Yahr scale [3] (the first clinical rating scale to judge PD) and the Unified Parkinson's Disease Rating Scale (UPDRS). These methods are complemented by imaging technologies such as positron emission tomography (PET) and single photon emission computed tomography (SPECT). Although the reliability of a clinical diagnosis of PD is rather high (79-92%) [32,33], postmortem histopathological confirmation of the diagnosis is still required.

The current diagnosis of PD, however, comes late. By the time the movement symptoms have manifested clinically, a dopaminergic cell loss of about 60% to 70% has taken place in the SN, as calculated form early postmortem studies [34,35]. This process of cellular decline develops over a considerably long prodromal stage, so the clinical motor symptoms become evident only 5 to 10 years after the actual onset of the disease [35–38]. Because the disease progresses undetected during this premotor symptomatic, preclinical period, valuable years have elapsed without therapeutic intervention. There is a great demand, especially for neuroprotective drugs, to eventually halt neurodegeneration at an early stage [39–42], but as yet no preclinical data on demonstrable neuroprotection could be translated into successful clinical applications [43,44]. This is mainly due to: (1) the focus mostly on the dopaminergic nigrostriatal system; and (2) the late arrival of putative neuroprotective therapies.

Thus, an essential challenge is delineation of the preclinical phase of PD. As most symptoms in early PD are rather unspecific and ambiguous, early disease biomarkers might be supportive, effective diagnostic tools. New concepts for the development of test systems, however, will largely depend on clear pathological guidelines. The more precisely the pathological "locating" of early PD correlates succeeds, the more specific the definition and the search for new biomarkers will be.

Translation of Pathological Findings into Biomarkers

From Neurochemical Findings...

The lead concept of PD pathology over the last decades has focused mainly on the degeneration of the SN. It entailed a variety of concepts to ameliorate the severe motor symptomatology by substituting for the reduced striatal dopamine (DA), ranging from administration of L-dopa and/or DA receptor agonists, tissue or cell transplantation, and gene therapeutic approaches. In addition to the nigrostriatal projections, however, apparently most dopaminergic systems are affected in PD.

Dopaminergic neurons in the mesocortical-mesolimbic system degenerate in the ventral tegmental area, which leads to considerably decreased DA in the gyrus cinguli, the hypothalamus, the amygdala, and the area olfactoria [45–47]. Considerably reduced olfactory sensitivity is reported to occur at an early stage of PD. Pathological changes in the gastrointestinal and cardiovascular systems account for autonomic disturbances preceding PD [48,49]. DA is also decreased in the retina, which correlates with impaired visual conception during the early stages of the disorder [50–53]. Neurochemical findings revealed significant dysfunction of the blood-brain barrier (BBB) in PD [54]. These findings of a dysfunctional BBB in PD have recently been substantiated by PET via functional imaging of the *MDR1* gene encoded P-glycoprotein [55], which mediates the efflux of various molecules, including pesticides, from the brain to the blood [56]. Moreover, peripheral regions are affected, as was highlighted by the decreased synthesis of DA in the adrenal medulla of PD patients [57,58].

Although excitatory neurotransmitter systems—γ-aminobutyric acid (GABA)—and glutamatergic transmission are obviously not affected in PD, additional systems apart from the dopaminergic system are equally comprised by partial degeneration in PD: the noradrenergic locus coeruleus—in conjunction with the affected serotonergic transmission from the raphe nuclei to the limbic system and the neocortex—might account for the occurrence of depression in PD patients [59]. Cholinergic neurons of the nucleus basalis of Meynert and the tegmental pedunculopontine nucleus might account for concomitant phenomena, such as dementiform cognitive impairments, hallucinations, and rapid-eye-movement (REM) sleep disturbances. Again, peripheral systems are considerably involved, such as the dorsal motor nucleus of the vagus (dmnX) and the sympathetic ganglia, which may explain vegetative complications.

...To Pathological Staging ...

(Neuro)pathological observations of a multiple-system neurodegeneration largely parallel these neurochemical and clinical observations [47]. Insoluble, eosinophilic protein inclusion bodies [e.g., Lewy bodies (LB)] may serve as a pathological hallmark of massively derailed metabolic networks of a cell in such a way as to facilitate a delineation of the neurodegenerative processes over the course of time. Tracing α-synuclein-immunopositive Lewy neurites and LBs in postmortem brains, the laboratories of Braak and of Del Tredici pioneered the succession of pathological developments in PD by suggesting a staging scheme that this disorder might follow [60,61]. Rather than a simultaneous evolving pathology, the authors suggested an onset obviously initiated outside the brain, most notably not in the SN, and an upwardly directed progression of the disease. Following a sequence of six stages, the pathology rises from the dmnX and the olfactory bulb (stage I) via the medulla oblongata, the pontine tegmentum, and the locus coeruleus (stage II) to the midbrain, involving the amygdala and the SN pars compacta (stage III), but continues via the basal proencephalon and the mesocortex (stage IV) to neocortical areas (stages V and VI).

Although the pathological findings have largely been confirmed [62], this stimulating concept is still hypothetical and needs further examination. This pathological staging has been suggested on condition that LBs are specific for PD; however, LBs are found in a variety of neurodegenerative disorders summarized as "synucleinopathies," including PD, dementia with Lewy bodies, pure autonomic failure, multiple system atrophy, or neurodegeneration with brain iron accumulation (NBIA, formerly known as Hallervorden-Spatz disease) [63]. Additionally, this concept provides no clues about the asymmetrical onset of the motor symptomatology in PD.

...*Toward Diagnostic Staging*...

En route to an earlier diagnosis of PD prior to manifestation of the motor symptoms, Przuntek et al. correlated the pathological findings of Braak and co-investigators with corresponding incidental "preclinical" features of PD [64]. Following the principle of structural substrates, which account for the clinical features, the concept of Przuntek et al. aspires to bridge the pathological findings classified as Braak stages I to III and preclinical nonmotor dysfunctions, including reduced gastrointestinal motility or olfactory dysfunction (stage I); REM sleep disturbances; and changes in the emotional and personal behavior, such as anxiety, a reduction in motivation, or tolerance to stress; but also pain syndromes (stage II). This stage is followed by deficits in visual perception, depression, unspecific dull pain syndromes of the locomotor system, and finally a sporadic onset of motor complications (stage III). As these nonmotor symptoms are rarely specific and thus mostly unrecognized or difficult to diagnose [65], an anchoring of preclinical symptoms with predetermined pathological changes is an essential step toward translating new neuropathological findings into future biomarkers for PD.

... *En Route to Translation into Biomarkers*

The need of early correlates, which are useful for diagnostics, is obvious. As a tool to facilitate an earlier diagnosis of PD, there is an urgent demand to identify disease-specific biomarkers. According to the Biomarkers Definitions Working Group, a biomarker constitutes "a characteristic that is objectively measured and evaluated as an indicator of normal biological processes, pathogenic processes, or pharmacological responses to a therapeutic intervention" [66,67]. Future approaches for early diagnosis of PD, especially a valid test system to monitor biomarkers, should be (1) sufficiently objective to prospectively differentiate frank PD from PD-related disorders; (2) sensitive enough to allow reliable discrimination of persons at risk; and (3) should not rely on symptoms, as they may fluctuate over time and are, to a considerable extent, subjectively assessed by the patients. (4) A given test system should be cost-efficient in order to be applicable to screen for predisposed individuals in large cohorts. (5) With respect to the development of future neuroprotective strategies, longitudinal clinical studies should be practical. The biomarker should thus be unaffected by pharmacological intervention in order to monitor the disease progression and demonstrate treatment efficacy.

Potential Biomarkers for PD

At the moment, there is no suitable biomarker available for PD. Several approaches have been elaborated, which range from highly sophisticated imaging techniques, classic biochemical testing, clinical tests on olfaction and vision, and genetic markers. Their potential as biomarkers, however, is not yet fully characterized.

Functional Imaging

The diagnostic application of PET and SPECT comes relatively late; and a preventive, longitudinal screen of large cohorts is not applicable for economic reasons. Moreover, functional imaging provides insight into dopamine transporter activities and postsynaptic actions of dopamine on receptors rather than directly measuring the decline of neuronal cells in the SN [68]. Because functional imaging monitors protein levels (transports, receptors) and protein expression is highly influenced by pharmacological treatment, the interpretation of such data is difficult and the conclusions mostly questionable [69].

It remains elusive whether the aggregation of α-synuclein and the concomitant formation of LBs might equally be imaged as the accumulation of β-amyloid in Alzheimer's disease [70,71]. PET or SPECT, however, might be valuable for clinically differentiating atypical parkinsonian syndromes, such as multiple system atrophy (MSA), progressive supranuclear palsy (PSP), dementia with Lewy bodies (DLB), and corticobasal degeneration (CBD) [72,73].

Transcranial Ultrasonography

Based on the transmissibility of the acoustic window for ultrasound located at the temporal bone, transcranial ultrasonography is a new technology for noninvasive structural imaging and determines the echogenicity of the SN [74,75]. The pathological correlate of increased echogenicity of the SN in PD seems to be an increase of iron, [76] which is selectively located in neuromelanin granules [77,78]. Some evidence suggests hyperechogenicity of the SN as a susceptibility marker for increased SN damage: a hyperechogenic SN was observed in some healthy, asymptomatic adults, who also exhibited reduced uptake of [18] F-dopa into the putamen [76]. Especially elderly people with SN hyperechogenicity, but without a diagnosis of PD, developed extrapyramidal symptoms of motor slowing and akinesia more often than age-matched subjects with a normal echogenic pattern [79].

Tests on Olfaction

Hyposmia—the loss of smell detection, identification or discrimination—seems to be an early event in PD and precedes manifestation of the cardinal movement symptoms. Olfactory dysfunction obviously correlates with degeneration occurring in the olfactory bulb [61,64]. Impaired olfaction, however, appears not to be specific for PD but, rather, is a symptom of several degenerative disorders, including Alzheimer's disease, progressive SPS, the parkinsonism-dementia complex of Guam [80], restless legs syndrome [81], and diffuse Lewy body disease [82]. In a refined application of the University of Pennsylvania's 12-item Brief Smell Identification Test (B-SIT; Sensonics, Haddon Heights, NJ, USA), only five specific odors appeared to be primarily affected in patients with PD [83]. Selective patterns of hyposmia, as observed in PD, might thus assist an early differential diagnosis of several neurodegenerative disorders on the basis of olfaction.

Tests on Vision

In PD vision might be compromised in various ways [52,84]. The most prominent dysfunctions of the visual system include abnormalities of color vision and chromatic contrast sensitivity, as highlighted by luminance pattern electroretinograms [85,86]. These abnormalities result from impaired dopaminergic transmission in the retinal pathways and dysfunctions of different subpopulations of retinal ganglion cells. It remains unclear to what extent tests on visual dysfunction may be used to uncover early manifestations of PD.

Biochemical Markers

Most biochemical markers are monitored in the periphery. Following the concept of oxidative stress in PD, several markers have been tested in the blood or cerebrospinal fluid, such as moderately elevated levels of malondialdehyde and reactive oxygen species [87], 8-hydroxy-2'-deoxyguanosine, and 8-hydroxyguanosine from oxidized DNA or RNA, respectively [88–90]. Platelets have been used to mirror serotonergic and dopaminergic neuronal functions [91]. For example, the uptake of ^{14}C-DA was reduced in platelets of PD patients [92] and the inhibitory action of selegiline [(–)-deprenyl] on monoamine oxidase B (MAO-B) could be monitored in the periphery [93]. An altered redox state of coenzyme Q10 was measured in platelets from PD patients, and although it did not correlate with disease severity it was suggested as an early disease marker [94]. In analogy to a reduction of mitochondrial complex I in the SN in PD [95–99], similar results were obtained from platelets [100–103],which proved to be more accessible than mitochondria from muscle biopsies [104]. Studies on postmortem tissue did not reveal pathogenic mitochondrial mutations in the mitochondrial genome [105]. Although some point mutations and polymorphisms have been detected, these associations could not be consistently reproduced [106–108]. Nevertheless, mitochondrial complex I defects can be transmitted from PD patients' mtDNA in cell cybrid experiments [109,110], although the reasons remain largely unknown. Interestingly, similar defects may be found in the accumulation of homoplasmic point mutations in the human brain during aging [111].

Several biochemical test systems could clearly shed light on pathogenic mechanisms of PD, but most factors either exhibit poor sensitivity to be applicable as biomarkers, or are influenced by pharmacological treatment.

Neuromelanin Blood Test

Neuromelanin (NM) is a polyphenolic pigment mainly found in the SN of primates [112]. In the SN, NM is localized in NM granules, a new type of lysosome-related organelles [113]. In PD either a protective or a cell-toxic action of NM in the pigmented neurons of the SN has been suggested [4,114–118]. Based on the observation of NM release from SN neurons during neurodegeneration, a test system is under way to quantify markers in the serum that mirror the neuronal decay of the SN. The underlying working hypothesis assumes a humoral response of the immune system following the release of NM, which could be measured by a novel enzyme-linked immunosorbent assay [119,120]. A pilot study on an Australian population and a German population demonstrated a significantly higher immune response in the sera of patients with diagnosed PD compared to healthy controls and patients with diagnosed depression. The dopaminergic system is affected during depression but exhibits no neurodegeneration. The immune response did not correlate with age and exhibited no difference between the sexes. In regard to disease severity, the immune response initially was high during the early clinical stages (Hoehn and Yahr stage I–II), which is in accordance with reinforced degeneration of the pigmented SN neurons at the clinical manifestation of PD.

This test system might thus prove useful either to confirm the clinical diagnosis or to facilitate the differential diagnosis of PD from related disorders. Moreover, it might have the potential to facilitate detection of nigral neurodegeneration when PD is still in the preclinical stage.

Genetic Markers

In a number of cases, parkinsonism is inherited in a mendelian autosomal dominant or recessive way. Most mutations of these familial cases do not occur with the most common sporadic PD [2]. Although investigation of the currently known genes has contributed significantly to the elucidation of neuro-degenerative processes that ultimately lead to parkinsonian syndromes, their benefit as biological markers remains elusive [121].

Gene Expression in Blood—A Future Diagnostic Tool?

In contrast to the focus on single nucleotide polymorphisms of the "classic" genes associated with familial parkinsonism (PARK1 to PARK10), gene expression patterns may alternatively provide insights into derailed metabolic, signaling, and stress response networks in PD. A recent global screen of gene expression of human SN of PD patients, covering 20,000 genes, revealed 137 genes to be misregulated in this disorder [121], most of which are involved in (down-regulated) signal transduction, protein modification and degradation, ion transport and metabolism, or in (up-regulated) cell adhesion, transcription, inflammation, and stress.

Several mRNA levels in the periphery, most conveniently in the blood, may reflect gene expression changes in the brain. Thus, a sophisticated panel of several genes that cover various species known to change dramatically in PD (e.g., genes involved in immune response, metal ion transport, apoptosis, or protein metabolism) may together serve as a powerful tool to monitor disease progression or to evaluate the efficacy of pharmacological treatments. Such a test panel may not only increase the sensitivity of the diagnosis but also may facilitate the discrimination of various subtypes of parkinsonian syndromes for the differential diagnosis of PD. It remains to be evaluated which gene expression alterations in the blood reflect the best PD-related autonomic changes as well as changes in the brain.

Conclusion

Progress in the treatment of PD largely depends on the definition of valid biomarkers, which not only should facilitate earlier detection of PD than is possible today but also longitudinal monitoring of disease progression under the influence of pharmacological intervention. Clinical test systems may comprise tests on REM sleep, visual, olfactory dysfunctions, and the echogenicity of the SN, which may be complemented by additional laboratory test systems, including assays monitoring the NM-release response, and gene-cluster analyses in the blood.

Different biomarkers may be useful to cover the spectrum of the multiple neurotransmitter systems affected in the autonomic and the central nervous system in PD, so that a combination of various biomarkers might ultimately provide an earlier diagnosis and better treatment of this disorder.

References

1. Gasser T. Genetics of Parkinson's disease. J Neu rol 2001;248:833–840.
2. Huang Y, Cheung L, Rowe D, Halliday G. Genetic contributions to Parkinson's disease. Brain Res Brain Res Rev 2004;46:44–70.
3. Hoehn MM, Yahr MD. Parkinsonism: onset, progression and mortality. Neurology 1967;17:427–442.
4. Hirsch E, Graybiel AM, Agid YA. Melanized dopaminergic neurons are differentially susceptible to degeneration in Parkinson's disease. Nature 1988;334:345–348.
5. Jenner P, Olanow CW. Oxidative stress and the pathogenesis of Parkinson's disease. Neurology 1996;47:S161–S170.
6. Gerlach M, Riederer P, Youdim MB. Molecular mechanisms for neurodegeneration: synergism between reactive oxygen species, calcium, and excitotoxic amino acids. Adv Neurol 1996;69:177–194.
7. Jellinger KA. Recent developments in the pathology of Parkinson's disease. J Neural Transm Suppl 2002:347–376.
8. Koutsilieri E, Scheller C, Tribl F, Riederer P. Degeneration of neuronal cells due to oxidative stress–microglial contribution. Parkinsonism Relat Disord 2002;8:401–406.
9. Davis GC, Williams AC, Markey SP, et al. Chronic parkinsonism secondary to intravenous injection of meperidine analogs. Psychiatry Res 1979;1:249–254.
10. Langston JW, Ballard P, Tetrud JW, Irwin I. Chronic Parkinsonism in humans due to a product of meperidine-analog synthesis. Science 1983;219:979–980.
11. Collins MA, Neafsey EJ. Beta-carboline analogs of N-methyl-4-phenyl-1,2,5,6-tetrahydropyridine (MPTP): endogenous factors underlying idiopathic parkinsonism? Neurosci Lett 1985;55:179–184.
12. Tanner CM. The role of environmental toxins in the etiology of Parkinson's disease. Trends Neurosci 1989;12:49–54.
13. Bringmann G, God R, Feineis D, et al. The TaClo concept: 1-trichloromethyl-1,2,3,4-tetrahydro-beta-carboline (TaClo), a new toxin for dopaminergic neurons. J Neural Transm Suppl 1995;46:235–244.
14. Bringmann G, Feineis D, Bruckner R, et al. Bromal-derived tetrahydro-beta-carbolines as neurotoxic agents: chemistry, impairment of the dopamine metabolism, and inhibitory effects on mitochondrial respiration. Bioorg Med Chem 2000;8:1467–1478.
15. Bringmann G, Feineis D, God R, et al. 1-Trichloromethyl-1,2,3,4-tetrahydro-beta-carboline (TaClo) and related derivatives: chemistry and biochemical effects on catecholamine biosynthesis. Bioorg Med Chem 2002;10:2207–2214.
16. Drozdzik M, Bialecka M, Mysliwiec K, et al. Polymorphism in the P-glycoprotein drug transporter MDR1 gene: a possible link between environmental and genetic factors in Parkinson's disease. Pharmacogenetics 2003;13:259–263.
17. Salazar M, Sokoloski TD, Patil PN. Binding of dopaminergic drugs by the neuromelanin of the substantia nigra, synthetic melanins and melanin granules. Fed Proc 1978;37: 2403–2407.
18. D'Amato RJ, Lipman ZP, Snyder SH. Selectivity of the parkinsonian neurotoxin MPTP: toxic metabolite MPP+ binds to neuromelanin. Science 1986;231:987–989.
19. Lindquist NG, Larsson BS, Lyden-Sokolowski A. Autoradiography of [^{14}C]paraquat or [^{14}C]diquat in frogs and mice: accumulation in neuromelanin. Neurosci Lett 1988;93:1–6.
20. Lyden-Sokolowski A, Larsson BS, Lindquist NG. Disposition of 1-methyl-4-phenyl-1,2,3,6-tetrahydropyridine (MPTP) in mice before and after monoamine oxidase and catecholamine reuptake inhibition. Pharmacol Toxicol 1988;63:75–80.
21. Hayase Y, Tobita K. Influenza virus and neurological diseases. Psychiatry Clin Neurosci 1997;51:181–184.
22. Dale RC, Church AJ, Surtees RA, et al. Encephalitis lethargica syndrome: 20 new cases and evidence of basal ganglia autoimmunity. Brain 2004;127:21–33.

23. Mattock C, Marmot M, Stern G. Could Parkinson's disease follow intra-uterine influenza? A speculative hypothesis. J Neurol Neurosurg Psychiatry 1988;51:753–756.
24. Koutsilieri E, Sopper S, Scheller C, et al. Parkinsonism in HIV dementia. J Neural Transm 2002;109:767–775.
25. Ling Z, Gayle DA, Ma SY, et al. In utero bacterial endotoxin exposure causes loss of tyrosine hydroxylase neurons in the postnatal rat midbrain. Mov Disord 2002;17:116–124.
26. Riederer P. Is there a subtype of developmental Parkinson's disease? Neurotox Res 2003;5:27–34.
27. Birkmayer W, Riederer P, Youdim BH. Distinction between benign and malignant type of Parkinson's disease. Clin Neurol Neurosurg 1979;81:158–164.
28. Graham JM, Sagar HJ. A data-driven approach to the study of heterogeneity in idiopathic Parkinson's disease: identification of three distinct subtypes. Mov Disord 1999;14:10–20.
29. Riederer P, Foley P. Mini-review: multiple developmental forms of parkinsonism: the basis for further research as to the pathogenesis of parkinsonism. J Neural Transm 2002;109:1469–1475.
30. Koller WC. How accurately can Parkinson's disease be diagnosed? Neurology 1992;42:6–16; discussion 57–60.
31. Fahn S. Description of Parkinson's disease as a clinical syndrome. Ann N Y Acad Sci 2003;991:1–14.
32. Hughes AJ, Daniel SE, Ben-Shlomo Y, Lees AJ. The accuracy of diagnosis of parkinsonian syndromes in a specialist movement disorder service. Brain 2002;125:861–870.
33. Jellinger KA. How valid is the clinical diagnosis of Parkinson's disease in the community? J Neurol Neurosurg Psychiatry 2003;74:1005–1006.
34. Bernheimer H, Birkmayer W, Hornykiewicz O, et al. Brain dopamine and the syndromes of Parkinson. Arch Neurol 1973;59:999–1005.
35. Riederer P, Wuketich S. Time course of nigrostriatal degeneration in parkinson's disease: a detailed study of influential actors in human brain amine analysis. J Neural Transm 1976;38:277–301.
36. McGeer PL, Itagaki S, Akiyama H, McGeer EG. Rate of cell death in parkinsonism indicates active neuropathological process. Ann Neurol 1988;24:574–576.
37. Fearnley JM, Lees AJ. Aging and Parkinson's disease: substantia nigra regional selectivity. Brain 1991;114(Pt 5):2283–2301.
38. Marek K, Innis R, van Dyck C, et al. [^{123}I]β-CIT SPECT imaging assessment of the rate of Parkinson's disease progression. Neurology 2001;57:2089–2094.
39. Jenner P. Presymptomatic detection of Parkinson's disease. J Neural Transm Suppl 1993;40:23–36.
40. Gerlach M, Riederer P, Youdim MB. Neuroprotective therapeutic strategies: comparison of experimental and clinical results. Biochem Pharmacol 1995;50:1–16.
41. Riederer P, Sian J, Gerlach M. Is there neuroprotection in Parkinson syndrome? J Neurol 2000;247(Suppl 4):8–11.
42. Mandel S, Grünblatt E, Riederer P, et al. Neuroprotective strategies in Parkinson's disease: an update on progress. CNS Drugs 2003;17:729–762.
43. Riederer P, Gille G, Muller T, et al. Practical importance of neuroprotection in Parkinson's disease. J Neurol 2002;249(Suppl 3):53–56.
44. Stocchi F, Olanow CW. Neuroprotection in Parkinson's disease: clinical trials. Ann Neurol 2003;53(Suppl 3):S87-S97; discussion S97-S89.
45. Javoy-Agid F, Ruberg M, Pique L, et al. Biochemistry of the hypothalamus in Parkinson's disease. Neurology 1984;34:672–675.
46. Javoy-Agid F, Ruberg M, Taquet H, et al. Biochemical neuropathology of Parkinson's disease. Adv Neurol 1984;40:189–198.
47. Jellinger KA. Pathology of Parkinson's disease: changes other than the nigrostriatal pathway. Mol Chem Neuropathol 1991;14:153–197.

48. Przuntek H, Muller T, Riederer P. Diagnostic staging of Parkinson's disease: conceptual aspects. J Neural Transm 2004;111:201–216.
49. Gurevich T, Korczyn AD. Autonomic disturbances in Parkinson's disease. In: Gálvez-Jiménez N (ed) Scientific Basis of the Treatment of Parkinson's Disease. Taylor & Francis, London, 2005, pp 333–347.
50. Hadjiconstantinou M, Neff NH. Catecholamine systems of retina: a model for studying synaptic mechanisms. Life Sci 1984;5:1135–1147.
51. Harnois C, Di Paolo T. Decreased dopamine in the retinas of patients with Parkinson's disease. Invest Ophthalmol Vis Sci 1990;31:2473–2475.
52. Bodis-Wollner I, Tagliati M. The visual system in Parkinson's disease. Adv Neurol 1993;60:390–394.
53. Djamgoz MB, Hankins MW, Hirano J, Archer SN. Neurobiology of retinal dopamine in relation to degenerative states of the tissue. Vision Res 1997;37:3509–3529.
54. Oestreicher E, Sengstock GJ, Riederer P, et al. Degeneration of nigrostriatal dopaminergic neurons increases iron within the substantia nigra: a histochemical and neurochemical study. Brain Res 1994;660:8–18.
55. Kortekaas R, Leenders KL, van Oostrom JC, et al. Blood-brain barrier dysfunction in parkinsonian midbrain in vivo. Ann Neurol 2005;57:176–179.
56. Drozdzik M, Bialecka M, Mysliwiec K, et al. Polymorphism in the P-glycoprotein drug transporter MDR1 gene: a possible link between environmental and genetic factors in Parkinson's disease. Pharmacogenetics 2003;13:259–263.
57. Riederer P, Rausch WD, Birkmayer W, et al. CNS modulation of adrenal tyrosine hydroxylase in Parkinson's disease and metabolic encephalopathies. J Neural Transm Suppl 1978;(14):121–131.
58. Stoddard SL, Ahlskog JE, Kelly PJ, et al. Decreased adrenal medullary catecholamines in adrenal transplanted parkinsonian patients compared to nephrectomy patients. Exp Neurol 1989;104:218–222.
59. Mossner R, Henneberg A, Schmitt A, et al. Allelic variation of serotonin transporter expression is associated with depression in Parkinson's disease. Mol Psychiatry 2001;6:350–352.
60. Del Tredici K, Rub U, De Vos RA, et al. Where does Parkinson disease pathology begin in the brain? J Neuropathol Exp Neurol 2002;61:413–426.
61. Braak H, Del Tredici K, Rub U, et al. Staging of brain pathology related to sporadic Parkinson's disease. Neurobiol Aging 2003;24:197–211.
62. Jellinger KA. Alpha-synuclein pathology in Parkinson's and Alzheimer's disease brain: incidence and topographic distribution—a pilot study. Acta Neuropathol (Berl) 2003;106:191–201.
63. Jellinger KA. The pathology of Parkinson's disease: recent advances. In: Gálvez-Jiménez N (ed) Scientific Basis for the Treatment of Parkinson's Disease. Taylor & Francis, London, 2005, pp 53–114.
64. Przuntek H, Muller T, Riederer P. Diagnostic staging of Parkinson's disease: conceptual aspects. J Neural Transm 2004;111:201–216.
65. Pfeiffer RF, Bodis-Wollner I. Parkinson's Disease and Nonmotor Dysfunction. Humana-Springer, Totowa, NJ, 2005.
66. Biomarkers Definitions Working Group. Biomarkers and surrogate endpoints: preferred definitions and conceptual framework. Clin Pharmacol Ther 2001;69:89–95.
67. Michell AW, Lewis SJ, Foltynie T, Barker RA. Biomarkers and Parkinson's disease. Brain 2004;127:1693–1705.
68. Leenders KL. Neuroimaging methods applied in Parkinson's disease. J Neurol 2004;251(Suppl 6):7–12.
69. Brooks DJ, Frey KA, Marek KL, et al. Assessment of neuroimaging techniques as biomarkers of the progression of Parkinson's disease. Exp Neurol 2003; 184(Suppl 1):S68–S79.

70. Kung HF, Lee CW, Zhuang ZP, et al. Novel stilbenes as probes for amyloid plaques. J Am Chem Soc 2001;123:12740–12741.
71. Zhuang ZP, Kung MP, Hou C, et al. Radioiodinated styrylbenzenes and thioflavins as probes for amyloid aggregates. J Med Chem 2001;44:1905–1914.
72. Parkinson Study Group. A multicenter assessment of dopamine transporter imaging with DOPASCAN/SPECT in parkinsonism: Parkinson Study Group. Neurology 2000;55:1540–1547.
73. Antonini A, DeNotaris R. PET and SPECT functional imaging in Parkinson's disease. Sleep Med 2004;5:201–206.
74. Becker G, Seufert J, Bogdahn U, et al. Degeneration of substantia nigra in chronic Parkinson's disease visualized by transcranial color-coded real-time sonography. Neurology 1995;45:182–184.
75. Becker G, Berg D. Neuroimaging in basal ganglia disorders: perspectives for transcranial ultrasound. Mov Disord 2001;16:23–32.
76. Berg D, Roggendorf W, Schroder U, et al. Echogenicity of the substantia nigra: association with increased iron content and marker for susceptibility to nigrostriatal injury and Huntington: clinical, morphological and neurochemical correlations. J Neurol Sci 2002;20:415–455.
77. Ben-Shachar D, Riederer P, Youdim MB. Iron-melanin interaction and lipid peroxidation: implications for Parkinson's disease. J Neurochem 1991;57:1609–1614.
78. Jellinger K, Kienzl E, Rumpelmair G, et al. Iron-melanin complex in substantia nigra of parkinsonian brains: an x-ray microanalysis. J Neurochem 1992;59:1168–1171.
79. Berg D, Siefker C, Ruprecht-Dorfler P, Becker G. Relationship of substantia nigra echogenicity and motor function in elderly subjects. Neurology 2001;56:13–17.
80. Doty RL, Perl DP, Steele JC, et al. Olfactory dysfunction in three neurodegenerative diseases. Geriatrics 1991;46(Suppl 1):47–51.
81. Adler CH, Gwinn KA, Newman S. Olfactory function in restless legs syndrome. Mov Disord 1998;13:563–565.
82. Westervelt HJ, Stern RA, Tremont G. Odor identification deficits in diffuse Lewy body disease. Cogn Behav Neurol 2003;16:93–99.
83. Double KL, Rowe DB, Hayes M, et al. Identifying the pattern of olfactory deficits in Parkinson disease using the brief smell identification test. Arch Neurol 2003;60:545–549.
84. Bodis-Wollner I, Onofrj M. The visual system in Parkinson's disease. Adv Neurol 1987;45:323–327.
85. Price MJ, Feldman RG, Adelberg D, Kayne H. Abnormalities in color vision and contrast sensitivity in Parkinson's disease. Neurology 1992;42:887–890.
86. Sartucci F, Orlandi G, Lucetti C, et al. Changes in pattern electroretinograms to equiluminant red-green and blue-yellow gratings in patients with early Parkinson's disease. J Clin Neurophysiol 2003;20:375–381.
87. Ilic TV, Jovanovic M, Jovicic A, Tomovic M. Oxidative stress indicators are elevated in de novo Parkinson's disease patients. Funct Neurol 1999;14:141–147.
88. Bogdanov MB, Beal MF, McCabe DR, et al. A carbon column-based liquid chromatography electrochemical approach to routine 8-hydroxy-2'-deoxyguanosine measurements in urine and other biologic matrices: a one-year evaluation of methods. Free Radic Biol Med 1999;27:647–666.
89. Kikuchi A, Takeda A, Onodera H, et al. Systemic increase of oxidative nucleic acid damage in Parkinson's disease and multiple system atrophy. Neurobiol Dis 2002;9:244–248.
90. Abe T, Isobe C, Murata T, et al. Alteration of 8-hydroxyguanosine concentrations in the cerebrospinal fluid and serum from patients with Parkinson's disease. Neurosci Lett 2003;336:105–108.
91. Stahl SM. The human platelet: a diagnostic and research tool for the study of biogenic amines in psychiatric and neurologic disorders. Arch Gen Psychiatry 1977;34:509–516.

92. Barbeau A, Campanella G, Butterworth RF, Yamada K. Uptake and efflux of ^{14}C-dopamine in platelets: evidence for a generalized defect in Parkinson's disease. Neurology 1975;25:1–9.

93. Elsworth JD, Glover V, Reynolds GP, et al. Deprenyl administration in man: a selective monoamine oxidase B inhibitor without the 'cheese effect.' Psychopharmacology (Berl) 1978;57:33–38.

94. Götz ME, Gerstner A, Harth R, et al. Altered redox state of platelet coenzyme Q10 in Parkinson's disease. J Neural Transm 2000;107:41–48.

95. Suzuki K, Mizuno Y, Yoshida M. Selective inhibition of complex I of the brain electron transport system by tetrahydroisoquinoline. Biochem Biophys Res Commun 1989;162:1541–1545.

96. Reichmann H, Riederer P. Biochemical analyses of respiratory chain enzymes in different brain regions of patients with Parkinson's disease. Presented at the BMBF Symposium "Morbus Parkinson und andere Basalganglienerkrankungen," Bad Kissingen, abstract 1989, p 44.

97. Mizuno Y, Ohta S, Tanaka M, et al. Deficiencies in complex I subunits of the respiratory chain in Parkinson's disease. Biochem Biophys Res Commun 1989;163:1450–1455.

98. Schapira AH, Cooper JM, Dexter D, et al. Mitochondrial complex I deficiency in Parkinson's disease. Lancet 1989;1:1269.

99. Schapira AH, Cooper JM, Dexter D, et al. Mitochondrial complex I deficiency in Parkinson's disease. J Neurochem 1990;54:823–827.

100. Parker WD Jr, Boyson SJ, Parks JK. Abnormalities of the electron transport chain in idiopathic Parkinson's disease. Ann Neurol 1989;26:719–723.

101. Yoshino H, Nakagawa-Hattori Y, Kondo T, Mizuno Y. Mitochondrial complex I and II activities of lymphocytes and platelets in Parkinson's disease. J Neural Transm Park Dis Dement Sect 1992;4:27–34.

102. Jenner P. Presymptomatic detection of Parkinson's disease. J Neural Transm Suppl 1993;40:23–36.

103. Blandini F, Nappi G, Greenamyre JT. Quantitative study of mitochondrial complex I in platelets of parkinsonian patients. Mov Disord 1998;13:11–15.

104. Nakagawa-Hattori Y, Yoshino H, Kondo T, et al. Is Parkinson's disease a mitochondrial disorder? J Neurol Sci 1992;107:29-33.

105. Lestienne P, Nelson J, Riederer P, et al. Normal mitochondrial genome in brain from patients with Parkinson's disease and complex I defect. J Neurochem 1990;55:1810–1812.

106. Lestienne P, Nelson I, Riederer P, et al. Mitochondrial DNA in postmortem brain from patients with Parkinson's disease. J Neurochem 1991;56:1819.

107. Simon DK, Mayeux R, Marder K, et al. Mitochondrial DNA mutations in complex I and tRNA genes in Parkinson's disease. Neurology 2000;54:703–709.

108. Richter G, Sonnenschein A, Grunewald T, et al. Novel mitochondrial DNA mutations in Parkinson's disease. J Neural Transm 2002;109:721–729.

109. Swerdlow RH, Parks JK, Miller SW, et al. Origin and functional consequences of the complex I defect in Parkinson's disease. Ann Neurol 1996;40:663–671.

110. Gu M, Cooper JM, Taanman JW, Schapira AH. Mitochondrial DNA transmission of the mitochondrial defect in Parkinson's disease. Ann Neurol 1998;44:177–186.

111. Lin MT, Simon DK, Ahn CH, et al. High aggregate burden of somatic mtDNA point mutations in aging and Alzheimer's disease brain. Hum Mol Genet 2002;11:133–145.

112. Fedorow H, Tribl F, Halliday G, et al. Neuromelanin in human dopamine neurons: comparison with peripheral melanins and relevance to Parkinson's disease. Prog Neurobiol 2005;75:109–124.

113. Tribl F, Gerlach M, Marcus K, et al. "Subcellular proteomics" of neuromelanin granules isolated from the human brain. Mol Cell Proteomics 2005;4:945–957.

114. Marsden CD. Neuromelanin and Parkinson's disease. J Neural Transm Suppl 1983;19:121–141.

115. Youdim MB, Ben-Shachar D, Riederer P. The enigma of neuromelanin in Parkinson's disease substantia nigra. J Neural Transm Suppl 1994;43:113–122.
116. Double, Ben-Shachar D, Youdim M, et al. Influence of neuromelanin on oxidative pathways within the human substantia nigra. Neurotoxicol Teratol 2002;24:621.
117. Wilms H, Rosenstiel P, Sievers J, et al. Activation of microglia by human neuromelanin is NF-kappaB dependent and involves p38 mitogen-activated protein kinase: implications for Parkinson's disease. FASEB J 2003;17:500–502.
118. Shamoto-Nagai M, Maruyama W, Akao Y, et al. Neuromelanin inhibits enzymatic activity of 26S proteasome in human dopaminergic SH-SY5Y cells. J Neural Transm 2004;111:1253–1265
119. Double K, Rowe DB, Halliday GM, et al. A biochemical test to diagnose Parkinson's disease. J Neural Transm Suppl 2002;109(III).
120. Gerlach M, Reichmann H, Riederer P. Die Parkinson-Krankheit. Springer-Verlag, Wien, 2003.
121. Grünblatt E, Mandel S, Jacob-Hirsch J, et al. Gene expression profiling of parkinsonian substantia nigra pars compacta; alterations in ubiquitin-proteasome, heat shock protein, iron and oxidative stress regulated proteins, cell adhesion/cellular matrix and vesicle trafficking genes. J Neural Transm 2004;111:1543–1573.

Chapter 7
Neurotransmitter and Neurotrophic Factor-Secreting Cell Line Grafting for the Treatment of Parkinson's Disease

Isao Date, Tetsuro Shingo, and Takao Yasuhara

Introduction

Parkinson's disease is a neurological disorder that manifests as involuntary movement. It is characterized by chronic progressive degeneration of the nigrostriatal dopaminergic system. Although stereotactic surgery such as deep brain stimulation has been performed as surgical therapy, intracerebral cell grafting has recently attracted increasing attention as a new surgical therapy. Intracerebral grafting of a dopamine-secreting cell line to compensate for dopamine deficiency and of a neurotrophic factor-secreting cell line to protect or regenerate the host intrinsic dopaminergic system have been investigated. This chapter summarizes neurotransmitter and neurotrophic factor-secreting cell line grafting for the treatment of Parkinson's disease.

Cell Grafting for Parkinson's Disease

The purpose of intracerebral cell grafting for Parkinson's disease can be divided into three goals: (1) to provide dopamine into the host by grafting; (2) to reconstruct the neural circuit by extending neurites from the donor cells; and (3) to provide neurotrophic factors from the donor cells to protect or to regenerate intrinsic neurons in the host. The first donor cells used for these purposes were fetal nigral cells, which are dopamine-producing cells [1]. Although these donor cells not only are the source of dopamine but also extend neurites into the host to reconstruct neuronal circuits, there are ethical issues related to the use of fetal tissue and immunological problems related to allogeneic grafts.

To avoid these issues related to fetal cell grafting, basic and clinical studies using autologous tissues, such as sympathetic ganglion cells and adrenal chromaffin cells, have been performed [2]. Chromaffin cells are not only the source of

I. Date
Department of Neurological Surgery, Okayama University, Okayama 700-8558, Japan

A. Fisher et al. (eds.), *Advances in Alzheimer's and Parkinson's Disease*,
© Springer 2008

catecholamines such as dopamine, they also produce a large number of neuro-trophic factors. Thus, various methods to increase the survival of grafted chro-maffin cells have been investigated [3,4]. Despite these merits, there is a possibility that these autologous cells may be injured by aging or the disease itself.

Cell Line Grafting and Encapsulation

The merits of using cell lines as donor cells for intracerebral grafting are that their supply is theoretically unlimited and genetic manipulation is easy to per-form. At the present time, because most of the cell lines used as stable source of neurotransmitters and neurotrophic factors are derived from species other than humans, immunological rejection and tumor formation are the main issues to be solved. To overcome these issues, cell lines have been encapsulated in hollow fibers consisting of a semipermeable polymer membrane, and grafted into the host brain; basic studies as well as clinical application have been published [5,6]. Because the capsule is semipermeable, oxygen and nutrients can be provided to the capsule, neurotransmitters and neurotrophic factors can be secreted from the capsule, but antibodies or immunocytes cannot get into the capsule; thus, the inside of the capsule is immunologically isolated. In addition, the rigid nature of the capsule can prevent tumor formation of the cells. By attaching a silicone tether, the capsule can be retrieved from the brain whenever necessary, demonstrating the safety of this procedure.

Neurotransmitter-Secreting Cell Line Grafting

As a dopamine-secreting cell line, PC12 cells, which are derived from rat pheochromocytoma, have been most commonly used as a donor cell line. It has been confirmed that when PC12 cells are grafted into the striatum of parkinsonian model rats the host animals demonstrated histological, neuro-chemical, and behavioral recovery. When PC12 cells were grafted into the xenogeneic monkey brain without encapsulation, these cells were completely rejected by 8 weeks after grafting, which was confirmed by magnetic resonance imaging (MRI) and histological analysis [7].

We have performed a long-term primate study using encapsulated PC12 cells for the treatment of Parkinson's disease [8]. Left intracarotid injection of 1-methyl-4-phenyl-1,2,3,6-tetrahydropyridine (MPTP), which is a neurotoxin for dopaminergic neurons, was performed to create a hemiparkinsonian model monkey. Encapsulated PC12 cells were grafted into the left striatum. Neuro-chemical analysis revealed that dopamine secretion from the grafted capsules was maintained for 12 months, and histological analysis demonstrated very good survival of grafted PC12 cells at 12 months after grafting. Although immunosuppressant was not administered, there were no signs of immunological rejection in the host brain, and reactive gliosis surrounding the capsule was minimal. Hematological analysis did not show abnormality in the white blood cell count or CD4/CD8 value. The monkeys receiving encapsulated PC12 cells showed behavioral improvement for 12 months. Results of a similar experiment have been published from another laboratory, in which positron emission tomography (PET) showed dopamine secretion from the grafted PC12 capsule [9].

Neurotrophic Factor-Secreting Cell Line Grafting

It has been reported that protection and repair of neurons can be expected by intracerebral administration of neurotrophic factors for various neurological disorders. Meanwhile, owing to the development of molecular biology techniques it is now possible to establish cell lines that secrete various types of neurotrophic factors. When these cell lines are encapsulated and grafted into the brain, stable delivery of neurotrophic factors can be achieved.

Intracerebral grafting of cell lines secreting neurotrophic factors, such as nerve growth factor (NGF), ciliary neurotrophic factor (CNTF), glial cell line-derived neurotrophic factor (GDNF), basic fibroblast growth factor (bFGF), and vascular endothelial growth factor (VEGF), have been reported so far [3,5,10–13].

GDNF is a potent neurotrophic factor for dopaminergic neurons. Using genetic manipulation we established a cell line that secretes GDNF and then encapsulated this cell line and grafted it in the right striatum of host rats. The host rats received 6-hydroxydopamine (6-OHDA), a neurotoxin for dopaminergic neurons, in the right striatum either before or after transplantation of the GDNF-secreting cell line; long-term evaluation was then performed. GDNF was secreted continuously from the grafted capsule for a period of 6 months, and good survival of GDNF-secreting cells was obtained in the capsule. Good survival and regeneration of host dopaminergic fibers in the striatum and dopaminergic cell bodies in the substantia nigra was observed, and host animals showed behavioral recovery [11,14].

Simultaneous Delivery of Neurotransmitter and Neurotrophic Factor by Cell Line Grafting

Regarding cell grafting for Parkinson's disease, activation of the host's intrinsic dopaminergic system by delivering neurotrophic factors from the grafted cells, in addition to dopamine replacement from the grafted cells, has been investigated. Although these approaches have been performed separately thus far, we are trying to deliver both dopamine and GDNF simultaneously using the encapsulated cell line grafting technique.

We have established a cell line named PC12-GDNF by inserting the GDNF gene into PC12 cells, with the cationic liposome method. Although conventional PC12 cells secrete dopamine in a stable fashion but do not secrete GDNF, the newly established PC12-GDNF cell line secretes both dopamine and GDNF. When the PC12-GDNF cell line was grafted into the parkinsonian model rats, the animals demonstrated histological, neurochemical, and behavioral recovery.

As shown above, more effective therapy can be provided if simultaneous delivery of neurotransmitters and neurotrophic factors into the host brain is achieved.

Intracerebral Grafting of Human-Derived Cell Line

When we consider clinical application of intracerebral grafting of cell lines secreting neurotransmitters and neurotrophic factors, it is essential to investigate human-derived cell lines as well. Human amniotic epithelial cells comprise one group of cells that have been studied for the purpose of cell grafting and regenerative medicine [15]. These cells are known to secrete neurotransmitters such as dopamine and acetylcholine and various types of neurotrophic factors.

The human amniotic epithelial cell line was immortalized, encapsulated, and grafted into the striatum of unilateral parkinsonian model rats. Dopamine was continuously secreted from the grafted capsule, and bFGF and transforming growth factor-beta (TGFβ), which are neurotrophic factors for dopaminergic neurons, were also secreted from the capsule. The regenerative effect of the graft on the host dopaminergic system and functional recovery of the host were observed.

These studies demonstrated that human-derived cell line can be grafted into the brain using the encapsulation technique without immunological rejection or tumor formation.

Control of Neurotransmitter and Neurotrophic Factor Secretion After Grafting into the Brain

When we perform intracerebral grafting of cell lines, it would be ideal if the amount of neurotransmitters and neurotrophic factors can be controlled after grafting. The authors established a new cell line named PC12th Tet-Off cells; the amount of dopamine secreted from PC12th Tet-Off cells can be controlled by stimulation from the outside. By inserting genes of tyrosine hydroxylase (a limiting enzyme for dopamine production) under Tet-Off regulatory system, the amount of dopamine production can be controlled when tetracycline or its analog doxycycline is added from the outside. In vitro study showed that dopamine secretion from PC12th Tet-Off cells could be reduced by adding doxycycline to the culture medium in a dose-dependent fashion. Moreover, when encapsulated PC12th Tet-Off cells were grafted into the striatum of parkinsonian model rats, the apomorphine-induced rotational behavior could be controlled by administering doxycycline to the hosts.

Because control of neurotransmitter and neurotrophic factor secretion from the cell line after grafting can be achieved, the issues related to the clinical application of cell line grafting are being gradually overcome.

Clinical Reports of Encapsulated Cell or Cell Line Grafting for Neurological Disorders

Regarding clinical application of encapsulated cell grafting for the treatment of neurological disorders, three trials have been reported. Buchser et al. [16] performed intrathecal grafting of encapsulated bovine chromaffin cells in patients with cancer pain, with the result that oral and epidural morphine intake by the patients could be reduced after surgery. In another study, an encapsulated CNTF-producing cell line was grafted intrathecally in patients with amyotrophic lateral sclerosis. Human CNTF protein, which was never detected preoperatively, was then detected in the cerebrospinal fluid of these patients for long periods of time. In addition, there was no apparent systemic immunological reaction in these patients, thus demonstrating the safety of this encapsulated cell grafting procedure [5]. The protocol of encapsualted CNTF-producing cell grafting into the brain of patients with Huntington's disease has also been published [17].

Conclusion

The idea of using cell lines as donors has been popular among researchers in the field of neural transplantation, but how to control immunological rejection and tumor formation has been an important issue to resolve. Because encapsulated cell line grafting has been available and various types of cell lines secreting neurotransmitters and neurotrophic factors can be created using genetic manipulation, the research using this technique is expected to develop as a potential therapy for many neurological disorders including Parkinson's disease.

The encapsulated cell grafting technique is safe because the grafts can be retrieved easily from the host brain when necessary. When cell transplantation using embryonic stem cells or other types of stem cell is ultimately applied clinically, encapsulated cell grafting technique can be used as a first step to demonstrate the safety of using those cells for grafting purposes.

References

1. Bjorklund A, Stenevi U. Reconstruction of the nigrostriatal dopamine pathway by intracerebral transplants. Brain Res 1979;177:555–560.
2. Date I. Parkinson's disease, trophic factors, and adrenal medullary chromaffin cell grafting: basic and clinical studies. Brain Res Bull 1996;40:1–19.
3. Date I, Ohmoto T, Imaoka T, et al. Cografting with polymer-encapsulated human nerve growth factor-secreting cells and chromaffin cell survival and behavioral recovery in hemiparkinsonian rats. J Neurosurg 1996;84:1006–1012.
4. Date I, Imaoka T, Miyoshi Y, et al. Chromaffin cell survival and host dopaminergic fiber recovery in a patient with Parkinson's disease treated by cografts of adrenal medulla and pretransected peripheral nerve: case report. J Neurosurg 1996;84:685–689.
5. Aebischer P, Schluep M, Deglon N, et al. Intrathecal delivery of CNTF using encapsulated genetically modified xenogeneic cells in amyotrophic lateral sclerosis patients. Nat Med 1996;2:696–699.
6. Emerich DF, Winn SR, Christenson L, et al. A novel approach to neural transplantation in Parkinson's disease: use of polymer-encapsulated cell therapy. Neurosci Biobehav Rev 1992;16:437–447.
7. Yoshida H, Date I, Shingo T, et al. Stereotactic transplantation of a dopamine-producing capsule into the striatum for treatment of Parkinson disease: a preclinical primate study. J Neurosurg 2003;98:874–881.
8. Date I, Shingo T, Yoshida H, et al. Grafting of encapsulated dopamine-secreting cells in Parkinson's disease: long-term primate study. Cell Transplant 2000;9:705–709.
9. Subramanian T, Emerich DF, Bakay RA, et al. Polymer-encapsulated PC-12 cells demonstrate high-affinity uptake of dopamine in vitro and ^{18}F-dopa uptake and metabolism after intracerebral implantation in nonhuman primates. Cell Transplant 1997;6:469–477.
10. Date I, Shingo T, Ohmoto T, et al. Long-term enhanced chromaffin cell survival and behavioral recovery in hemiparkinsonian rats with co-grafted polymer-encapsulated human NGF-secreting cells. Exp Neurol 1997;147:10–17.
11. Date I, Shingo T, Yoshida H, et al. Grafting of encapsulated genetically modified cells secreting GDNF into the striatum of parkinsonian model rats. Cell Transplant 2001;10:397–401.
12. Fujiwara K, Date I, Shingo T, et al. Reduction of infarct volume and apoptosis by grafting of encapsulated basic fibroblast growth factor-secreting cells in a model of middle cerebral artery occlusion in rats. J Neurosurg 2003;99:1053–1062.
13. Yasuhara T, Shingo T, Kobayashi K, et al. Neuroprotective effects of vascular endothelial growth factor (VEGF) upon dopaminergic neurons in a rat model of Parkinson's disease. Eur J Neurosci 2004;19:1494–1504.
14. Shingo T, Date I, Yoshida H, et al. Neuroprotective and restorative effects of intrastriatal grafting of encapsulated GDNF-producing cells in a rat model of Parkinson's disease. J Neurosci Res 2002;69:946–954.
15. Uchida S, Suzuki Y, Araie M, et al. Factors secreted by human amniotic epithelial cells promote the survival of rat retinal ganglion cells. Neurosci Lett 2003;341:1–4.
16. Buchser E, Goddard M, Heyd B, et al. Immunoisolated xenogenic chromaffin cell therapy for chronic pain: initial clinical experience. Anesthesiology 1996;85:1005–1012.
17. Bachoud-Levi AC, Deglon N, Nguyen JP, et al. Neuroprotective gene therapy for Huntington's disease using a polymer encapsulated BHK cell line engineered to secrete human CNTF. Hum Gene Ther 2000;11:1723–1729.

Chapter 8
Novel Therapeutic Target in PD: Experimental Models

Francesco Fornai

Introduction

Parkinson's disease (PD) is a progressive, late-onset neurodegenerative disorder that is characterized by the presence of selective motor symptoms and neuro-pathological features, including nigrostriatal dopaminergic denervation and development of cytoplasmic inclusions. This disorder has a mean age of onset of 60 years, with the incidence increasing dramatically with age.

Despite the rarity of familial forms of PD, recent genetic discoveries on single genes linked to the disease led to greater insight into the molecular pathways involved in PD pathogenesis. In fact, the similarity between the genetic and sporadic forms of PD indicates that genetic information allows us to dissect the key biochemical pathways of this degenerative disorder and to study alterations occurring also in sporadic parkinsonism. Recently, environmental factors have also received attention as risk factors for PD. Epidemiological studies have suggested that environmental chemicals, such as pesticides, might be crucial factors in PD pathogenesis [1,2]. These agents induce dysfunction of mitochondria that leads to the production of reactive oxygen species (ROS) and oxidative damage, which is implicated in the pathogenesis of this disorder [3]. Another possibility is that dopamine metabolism may be considered an endogenous toxin responsible for production of ROS and induction of PD [4]. In fact, a growing body of evidence indicates that the presence of high levels of cytoplasmic dopamine (DA) and DA metabolites are an important source of oxidative stress, suggesting that DA might play a role in promoting neuronal damage and inclusion formation [5–9].

One major point in understanding the pathogenesis of PD consists in being able to reproduce the behavioral, biochemical, and pathological features that occur during the natural course of this neurological disorder by producing an experimental model that benefits from the same therapeutic agents that give symptomatic relief to PD patients. One hallmark of PD is the presence of

F. Fornai
Department of Human Morphology and Applied Biology, University of Pisa, 56126 Pisa, Italy

A. Fisher et al. (eds.), *Advances in Alzheimer's and Parkinson's Disease*,
© Springer 2008

inclusion bodies called Lewy bodies, which consist in neuronal cytoplasmic aggregates of proteinaceous material featuring various sizes and shapes, depending on the brain area. These inclusions appear as pale eosinophilic structures containing ubiquitin and ubiquitin proteasome (UP) system-related proteins such as parkin, an E3 ubiquitin-protein ligase [10], ubiquitin-C-terminal hydrolase isozyme L1 (Uch-L1) [11], and α-synuclein [12], which does not belong to the UP system but interacts with ubiquitin as a substrate for UP [13,14]. α-Synuclein is considered the best immunological marker of Lewy bodies, where it is found more frequently than ubiquitin [12,15], and mutations of the α-synuclein gene are associated with rare familial cases of PD [16]. The presence of these proteins in cytoplasmic inclusion bodies and the demonstration that genetic mutations of ubiquitin, α-synuclein, parkin, and Uch-L1 are associated with inherited forms of PD suggested involvement of the UP system in the pathogenesis of this neurodegenerative disease [17].

On the basis of these findings, novel models of PD were produced by expressing α-synuclein in *Drosophila melanogaster* [18] or overexpressing mutant or wild-type α-synuclein in transgenic mice [19,20]. Recent in vivo and in vitro studies demonstrated a linkage between genetic mutations, oxidative stress, and an impaired UP system, suggesting a new experimental model based on the inhibition of this pathway [7,21,22].

Several experimental models of PD make use of dopaminergic toxins. Among all of them, the toxins most used are 1-methyl-4-phenyl-1,2,3,6-tetrahydropyridine (MPTP) and methamphetamine (MA). More recently also rotenone and paraquat, which have mechanisms of action similar to that of MPTP or 3,4-methylenedioxymethamphetamine (MDMA, "ecstasy'), a derivative of MA, are used. Other models are obtained by microinfusions of 6-hydroxydopamine (6-OHDA) in the striatum, substantia nigra, or median forebrain bundle (MFB).

Ideal Model of Parkinson's Disease

Considering the various features that characterize PD, an ideal model of the disorder should have the following features: (1) decreased striatal dopamine (DA) levels; (2) a gradual loss of DA neurons in the substantia nigra pars compacta (SNpc) particularly in its ventrolateral part; (3) a significant motor deficit; (4) the loss of neurons in the pontine nucleus of the locus coeruleus; and (5) occurrence of neuronal inclusions, which contain ubiquitin and α-synuclein, known as Lewy bodies (LB), in selective areas of the brain.

6-Hydroxydopamine

Available models of PD have contributed greatly to our understanding of the pathophysiology and therapeutics of the disease. 6-Hydroxydopamine (6-OHDA) is a classic animal model of PD that is associated with SNpc DA neuronal death. 6-OHDA toxicity is rather selective for monoaminergic neurons, and it is due to the preferential uptake by DA and norepinephrine (NE) transporters [23]. 6-OHDA must be administered by local stereotactic infusion because it cannot cross the blood–brain barrier. It is injected into the SN or the MFB, which carry ascending dopaminergic, noradrenergic, and serotoninergic projections to the forebrain, or the striatum. This area is the target of the nigrostriatal dopaminergic pathway, where the injection of 6-OHDA produces retrograde nigrostriatal degeneration, which occurs within 1 to 3 weeks [24]. Accumulation of 6-OHDA in DA neurons provokes the formation of ROS and generates quinones [25]. The degree of unilateral lesion produced by 6-OHDA infusions in the striatum can be assayed by assessing the unilateral turning behavior in the rats, which correlates with the extent of the nigrostriatal lesion [24,26]. The presence of quantifiable motor deficit (turning) is a significant advantage of this model because it allows pharmacological screening and assaying the outcome of striatal transplants or gene therapy to repair the damaged nigrostriatal pathway [27]. This model offers a variety of limits, however: It is unilateral, does not produce the entire neuropathology of PD, and provokes mechanical damage due to the invasive procedure.

MPTP

A classic model that reproduces Parkinson's disease consists of administering the dopaminergic neurotoxin MPTP [28], which inhibits the respiratory chain at the level of mitochondrial complex I and induces DA neurodegeneration. The exposure of humans and nonhuman primates to this neurotoxin causes an irreversible, severe syndrome that mimics the symptoms of PD, including tremor, rigidity, slowness of movements, and postural instability. This model was discovered accidentally at the beginning of the 1980s as a cause of toxic parkinsonism in humans. In fact, at the beginning of the 1980s a group of intravenous drug users developed irreversible parkinsonism after injection of MPTP, a contaminant of synthetic meperidine [28]. Despite the tragic event, this presented us with a revolutionary tool for understanding Parkinson's disease, as a chemical compound when administered systemically was able to produce the massive loss of nigrostriatal DA fibers concomitant with striatal DA depletion. Moreover, this caused severe behavioral impairment, which was improved by administering either levo-dopamine (L-dopa) or DA agonists. The use of this neurotoxin as an experimental model of PD in various animal species allowed the selection of important dysfunctions in cellular metabolism that may be involved in the pathogenesis of PD, such as oxidative stress, mitochondrial dysfunction, and abnormal protein aggregation.

Inhibition of complex I of the mitochondrial electron-transfer chain by MPTP leads to decreases in adenosine triphosphate (ATP) content, loss of mitochondrial membrane potential, and induction of ROS production. At present, oxidative stress is considered a leading hypothesis of the pathogenesis of PD [29]. MPTP is a highly lipophilic compound; it crosses the blood-brain barrier and, once in the brain, is oxidized by monoamine oxidase B (MAO-B) in glia, to 1-methyl-4-phenyl-2,3-dihydropiridinium (MPDP+) and 1-methyl-4-phenylpyridinium (MPP+) [30]. Subsequently, it is taken up by dopamine, serotonin, and norepinephrine transporters (DAT, SERT, NET) that carry it into the cells [5,31]. The uptake by DAT might explain the selectivity of the nigrostriatal damage induced by MPTP. Also neuromelanin, contained in dopaminergic neurons, may contribute to the neurodegeneration in PD and in MPTP-treated monkeys by catalyzing ROS formation through interaction with iron [32]. The neurotoxin MPTP has been used as a model of PD in both monkeys and mice. In particular, the monkey MPTP model is used to estimate new strategies and drugs for the treatment of PD symptoms, whereas the MPTP mouse model generally has been used to investigate the molecular mechanism of dopaminergic degeneration. In both of these species, MPTP damages the dopaminergic pathways, induces a massive loss of nigral dopaminergic neurons and striatal dopamine, and causes an irreversible parkinsonian syndrome. Nonetheless, the finest pathological feature that characterizes PD (i.e., the occurrence of neuronal cytoplasmic inclusions) does not appear following acute MPTP administration. In fact, although neuronal inclusions similar to Lewy bodies were described in an old java monkey [33], classic Lewy bodies have not been demonstrated in MPTP-intoxicated patients or in monkeys exposed to acute administration of MPTP [34]. This may be consistent with the fact that the dopaminergic neurons are rapidly degenerated. Moreover, the acute MPTP model does not consistently induce loss from other monaminergic nuclei, such as the locus coeruleus, which is typically involved in PD [33,34].

Prolonged administration of low doses of MPTP to mice induces cell death of dopaminergic neurons, morphologically defined as apoptosis [35]. Recent in vitro findings on different neuronal cell lines link the effects of MPTP/MPP+ with molecules typically involved in apoptotic pathways [36]. In PC12 cells, which produce DA, the MPP+ induces cell death. This may be accomplished by the formation of toxic DA quinone and melanin due to MPP+-stimulated DA oxidation [37]. A model using continuous MPTP administration closely mimics what occurs in PD (see later).

Rotenone

A model based on the continuous injection of rotenone, a member of the family of rotenoids (used widely as insecticides), has been developed. This model is characterized by progressive degeneration of nigrostriatal neurons, the appearance of motor deficits, and the presence of inclusion bodies. Rotenone binds, as MPP+, to mitochondrial complex I, which is thus inhibited. This toxin induces neuronal degeneration by oxidative stress. Rotenone is extremely hydrophobic and crosses biological membranes easily independent of the transporter (unlike MPP+), and it reaches the brain rapidly. Further studies, carried out by administering rotenone, reproduced the occurrence of neuronal inclusions and reported a mixed degeneration of nigral DA and striatal γ-aminobutyric acid (GABA) neurons, thus featuring neuropathological findings of multisystemic degeneration [38]. This differs from classic PD, where striatal GABA neurons are spared. Further studies, carried out by Greenamyre and collaborators [39], demonstrated that, depending on the rotenone dose, various degrees of selective neurotoxicity can be achieved [40]. Rotenone has often been used as a mitochondrial inhibitor in cell cultures but has been used less frequently in animals.

When acute rotenone is administered orally [41], intravenously [42], stereotactically [43], or subcutaneously [44], it produces damage to basal ganglia. Hirsch and collaborators [38] found that intrafemoral venous infusion of rotenone to rats produces nigrostriatal dopaminergic neurodegeneration [38]. However, continuous administration is required to build neuronal inclusions.

Greenamyre's group [1] developed a new in vivo model of PD by exposing rats continuously to rotenone by implanting osmotic minipumps filled with a solution of rotenone or saline. Their results indicated that systemic or partial inhibition of mitochondrial complex I reproduces behavioral, neurochemical, and neuropathological features of PD. Inhibition of complex I by rotenone, similar to MPTP, stimulates the production of ROS [45]. The concentration of rotenone and the degree of inhibition were uniform in different brain areas. Nevertheless, this chronic and uniform inhibition of complex I caused selective degeneration of dopaminergic neurons, suggesting that these neurons are preferentially sensitive to the inhibition of complex I [39]. The authors observed that the neuronal degeneration began in the nerve terminal and progressed to the cell bodies. Moreover, selective degeneration of dopaminergic nigrostriatal neurons was accompanied by the formation of cytoplasmic inclusion bodies, which were both α-synuclein-positive and ubiquitin-positive [1]. In this model, osmotic minipumps were implanted under the skin on the back of each animal, and a cannula was placed in the jugular vein and attached to a subcutaneous minipump [1].

In vitro study of organotypic cultures treated with rotenone was used to investigate how mitochondrial dysfunction can lead to neuronal damage. The results of this study showed that low concentrations of rotenone, over weeks, resulted in slow loss of dopaminergic cell processes, altered cell morphology, decreased tyrosine hydroxylase (TH) protein, and increased oxidative stress [46]. In the rotenone model, there is increased expression of α-synuclein which may become insoluble, and may form cytoplasmic inclusions [47]. Rotenone treatment does not increase α-synuclein levels by an increase in gene transcription because the mRNA levels do not change [48]. Rather, the aggregation appears to be related to oxidative damage; the α-synuclein may be directly modified by oxidative damage or may form adducts with DA quinone [13,39].

Using three models of increasing complexity, Sherer and colleagues [40] demonstrated the involvement of oxidative damage in rotenone toxicity. In these experiments they used SK-N-MC, human neuroblastoma cells, a chronic midbrain slice culture model, and finally brains from rotenone-treated animals. Rotenone caused dose-dependent ATP depletion, oxidative damage, and cell death in all these models. In brains from rats it was observed that the effects induced by rotenone were most notably in the midbrain and olfactory bulb-dopaminergic regions affected by Parkinson's disease [40].

Preliminary studies indicated that rotenone treatment also affects the UP system. The rotenone model thus represents an advantageous experimental model of PD and the first model in which an environmental agent was able to induce the formation of inclusion bodies in association with nigrostriatal dopaminergic damage, which so far was been induced by either 6-OHDA or MPTP. Moreover, it might be a good model to investigate the molecular basis of the formation of inclusion bodies and their link with neuronal cell death.

New In Vivo Environmental Models of Parkinson's Disease

Subcutaneous Rotenone Exposure

In a subsequent article [49], Sherer demonstrated that the same features of PD can be reproduced by chronic, systemic exposure to rotenone following implantation of subcutaneous osmotic pumps. Chronic subcutaneous exposure to low doses of rotenone (2.0–3.0 mg/kg/day) caused highly selective nigrostriatal dopaminergic lesions. Moreover, rotenone-treated animals with demonstrable dopaminergic lesions showed reduced motor activity, flexed posture, and in some cases rigidity. Subcutaneous rotenone exposure caused α-synuclein-positive cytoplasmic aggregates in nigral neurons. Also in vitro, the complex I inhibition due to rotenone exposure elevated α-synuclein levels and caused α-synuclein aggregates [48]. It is not clear how rotenone exposure caused α-synuclein aggregates in dopaminergic neurons. The systemic rotenone infusion results in uniform complex I inhibition throughout the brain,1 suggesting that the nigrostriatal dopaminergic system may have intrinsic sensitivity to complex I impairment. However, the exact mechanisms that underlie rotenone-induced degeneration remain unclear. It appears that a bioenergetic defect cannot solely explain rotenone-induced degeneration, but oxidative damage, which has been implicated in PD pathogenesis, may be involved in rotenone toxicity. In vitro, chronic rotenone treatment causes progressive oxidative damage [48]. Rotenone treatment caused selective nigrostriatal dopaminergic degeneration without affecting neurons from other brain regions.

Continuous Administration of MPTP

When rotenone is administered discontinuously, it is known to produce basal ganglia damage [42,43]. However, continuous administration is required to build neuronal inclusions, as shown by Greenamyre's group [1].This represents a crucial point that goes beyond pharmacokinetic implications, suggesting that in order to build up neuronal inclusions mild, prolonged inhibition of the mitochondrial respiratory chain is needed up to a level that does not lead to sudden cell death. If this is the case, one should expect that inhibitors of the mitochondrial respiratory chain would induce similar effects when delivered continuously at a mild dosage.

In mice, PD is classically modeled by application of the neurotoxin MPTP, which is transported to brain where it is metabolized to MPP +. MPP + is selectively taken up into dopaminergic neurons [50,51]. As it has been previously described, when injected into mice MPTP causes loss of nigral dopaminergic neurons and striatal DA but does not induce the neuronal inclusions characteristic of PD, also after repeated injections [52,53]. However, only continuous administration of rotenone triggers formation of neuronal inclusions [1], whereas discontinuous administration of rotenone damages the basal ganglia but produces no inclusions [42,43].

Recently our group [22] examined whether continuous application of MPTP in mice produces formation of neuronal inclusions and leads to impairment of the UP system; the aim was to test whether mild, prolonged inhibition of the mitochondrial respiratory chain causes a chronic decrease in UP activity. We implanted mice chronically with osmotic minipumps filled with MPTP and measured the pathogenic effects of continuous delivery of MPTP in comparison with sporadic injections. Mice exposed to continuously administered MPTP exhibit a dose-dependent PD-like syndrome with the formation of neuronal inclusions containing α-synuclein and decreased UP activity [22].

Moreover, we found that chronic, continuous MPTP infusions induce effects also on the locus coeruleus. In particular, we observed alterations in pontine NE neurons, where there was a marked NE cell loss visualized by dopamine β-hydroxylase immunostaining, and a decrease in NE levels in various target areas of the locus coeruleus. Moreover, surviving neurons contained inclusions immunopositive for ubiquitin and α-synuclein [22].

These results demonstrate that inhibition of mitochondrial complex I increases oxidative damage and damages the UP system. The impairments of this system represent a final common pathway in the pathogenesis of PD-like syndrome [2,7,17,38]. As witnessed by the composition of Lewy bodies, which contain both ubiquitin and α-synuclein, it is consistent with a general failure of the UP system. This is further substantiated by the fact that mutations in enzymes of the UP system are found in familial forms of PD [2,54,55].

Genes Associated with Parkinson's Disease and Genetic Models

The effects of both genetic and environmental factors appear to contribute to PD. The identification of families in which typical parkinsonism is inherited sheds light on genes that cause features that resemble sporadic PD. These genes are involved in molecular pathways that lead to degeneration of the nigrostriatal system. The genes associated with either PD or Parkinson-related syndromes include α-synuclein, parkin, ubiquitin C-terminal hydrolase isozyme L1 (Uch-L1), DJ-1, and nuclear receptor-related factor 1. There is increasing evidence that genes involved in inherited forms of the disease may act as susceptibility factors also in the sporadic form of PD [56].

Apart from these results, new models of PD were produced in vivo on transgenic animals and in vitro on transfected cell cultures. Both of these models are important because they replicate important features of PD, including progressive motor symptoms and protein aggregation, and they are helpful in elucidating the pathophysiology of PD.

Parkin

Mutations in the gene that codes for parkin cause a recessively inherited parkinsonism. The changes include point mutations and exon rearrangements, including both deletions and duplications [57,58]. This form of the disease has been termed autosomal recessive juvenile parkinsonism because of its early onset. In fact, parkin mutations were found in patients with onset before age 30 and with a family history that presented as recessive inheritance. Parkin-related PD is characterized by the loss of SNpc dopaminergic neurons, but it is not associated with Lewy bodies [59]. Parkin belongs to a family of proteins with a conserved ubiquitin-like and RIG finger motif, and it has been reported that it functions as a ubiquitin E3 protein ligase, a component of the UP system, which identifies and targets misfolded proteins to the proteasome for degradation. This suggests that impairment of this activity is probably the cause of PD [10,17]. Some reports suggest a link between parkin and α-synuclein function, and this interaction may involve the proteasome. In fact, the E3 ligase activity of parkin modulates the sensitivity of cells to cell death induced by both the proteasome inhibitor and the mutant α-synuclein [60].

A parkin-knockout mouse model has recently been described. In this model, similar to parkin-induced PD, there was no inclusion body formation. Moreover, surprisingly, in contrast to previous studies, striatal catecholamine levels were normal, and no evidence was found for nigrostriatal, cognitive, or noradrenergic dysfunction [61]. However, it has been found that when parkin is up-regulated it may suppress toxicity induced by unfolded protein [62]. This potential neuroprotective role of parkin in the pathogenesis of PD deserves particular attention. In a recent study, Lo Bianco and colleagues [63], using a rat lentiviral model of PD that expressed mutated human α-synuclein, found that animals overexpressing parkin showed a significant reduction of α-synuclein-induced neuropathology. In these animals, therefore, nigrostriatal degeneration and loss of dopaminergic cells in the SN were inhibited. These experimental models can advance our knowledge of PD, suggesting that parkin gene therapy may represent a prominent candidate for treatment of this neurodegenerative disorder.

Ubiquitin C-Terminal Hydroxylase-L1

Genomic studies in familial PD identified a dominant mutation (Ile93Met) in ubiquitin C-terminal hydroxylas-L1 (Uch-L1) in one family affected by inherited PD [64]. This enzyme belongs to a family of enzymes responsible for the degradation of polyubiquitin chains to ubiquitin monomers, and the Ile93Met mutation decreases the activity of this de-ubiquitinating enzyme. Although mutations in Uch-L1 are rare in PD, this enzyme is present in LB, suggesting that it may be crucial for the process of inclusions maturation. In fact, a dysfunction of Uch-L1 could impair the overall efficacy of the ubiquitin system and increase accumulation of damaged proteins [17].

DJ-1

DJ-1 mutations are associated with PARK7, a monogenic form of autosomal recessive human parkinsonism [65]. DJ-1 was previously associated with several biological processes. More recently it was identified as a hydroperoxide-responsive protein that shifts to a more acidic isoform under oxidative conditions, suggesting that it works as a sensor for oxidative stress [66]. Two types of mutation in DJ-1 have been described in patients with PD. One is a deletion of several of its exons, which prevents synthesis of the protein, and the other is a point mutation at a highly conserved residue (Leu166Pro) that makes the protein less stable and promotes its degradation through the UP system, thereby reducing the amount of DJ-1 levels [65].

DJ-1-deficient mice were used as an experimental model of PD. These mice had a normal number of dopaminergic neurons in the SN, and the DA levels in the striatum were normal. However, DJ-1 mice showed hypolocomotion when subjected to an amphetamine challenge and increased striatal denervation and dopaminergic neuron loss induced by MPTP [67]. Embryonic cortical neurons from DJ-1-deficient mice showed increased sensitivity to oxidative damage. In these experiments, the restoration of DJ-1 expression to DJ-1 mice or cells via adenoviral vector delivery reversed all effects induced by DJ-1 loss. These findings indicate that loss of DJ-1 function leads to neurodegeneration. In fact, wild-type mice that received adenoviral delivery of DJ-1 resisted MPTP-induced striatal damage, and neurons overexpressing DJ-1 were protected from oxidative stress in vitro. The results of Kim and colleagues [67] demonstrate that DJ-1 protects against neuronal oxidative stress and that loss of DJ-1 may lead to PD by conferring hypersensitivity to dopaminergic insults.

α-*Synuclein*

α-Synuclein is a neuronal protein involved in several neurodegenerative disorders [68] that share some pathological features consisting of aggregates of both normal and altered α-synuclein in specific neurons. In PD, α-synuclein is a constant component of neuronal inclusions called Lewy bodies, and it is considered the best immunological marker of these inclusions. The presence of α-synuclein in neuronal inclusions suggests that this protein might play a relevant role in PD, and it is suggestive of specific biochemical steps that might be crucial during early stages of the formation of inclusion bodies [12,69].

During the late 1990s it was found that missense punctiform mutations A53T, A30P, and E46K in the gene coding for α-synuclein were responsible for altered structure of the protein and were linked to rare early-onset autosomal dominant inherited forms of PD [16,70,71]. A role for α-synuclein was suggested in membrane-associated processes at the presynaptic level through interaction with the DA transporter [72,73], and it was postulated that α-synuclein is important for regulating the trafficking of synaptic vesicles [74]. Recent epidemiological and neurological studies have demonstrated that humans bearing a triplicate of the normal α-synuclein gene and producing overexpression of the normal protein were affected by severe inherited parkinsonism [75,76]. Overexpression of normal α-synuclein leads to dopaminergic neuronal death.

These genetic studies on various types of α-synuclein-related PD suggest that α-synuclein may be relevant per se as a cause of this movement disorder, either by being present in the normal amount with an altered structure or via overexpression with a normal configuration. In keeping with this concept, one can hypothesize that in sporadic forms of PD normal α-synuclein might become neurotoxic when (1) the conditions of dopaminergic cells lead to nongenetic structural alterations or (2) when specific environmental agents affecting DA neurons cause overexpression of α-synuclein, which induces morphological alterations and cell death. For instance, one environmental factor that increases α-synuclein in human DA neurons is cocaine [77], and previous exposure to cocaine is known to enhance the deleterious effects of subsequent exposure to DA neurotoxins [78].

Several genetic models of PD have been based on the overexpression of mutant or wild-type forms of α-synuclein in transgenic mice or flies [18-20]. The central role of α-synuclein in Lewy body formation was confirmed by findings obtained using transgenic Drosophila, which normally do not express α-synuclein. When this protein was expressed, transgenic flies developed neuronal loss and intracellular aggregates resembling Lewy bodies in PD [18]. Masliah and colleagues [19], for the first time, generated mice transgenic for human α-synuclein to elucidate the role of this protein in neurodegenerative disorders. Since then, various transgenic α-synuclein mice have been generated that overexpress wild-type α-synuclein, as well as mutated forms of human α-synuclein [20]. Again, injection of either human wild-type or mutant α-synuclein-expressing viral vectors into nigrostriatal pathways of transgenic mice caused the formation of inclusions that contain α-synuclein [77]. Also in cell models, overexpression of either wild-type or mutant α-synuclein produced the formation of cytoplasmic aggregates [79].

These transgenic studies are advantageous because they demonstrate that a component of cellular toxicity is derived from overexpression or mutations of the α-synuclein. Moreover, they allow one to identify the role of α-synuclein in Lewy body formation. Nevertheless, they do not faithfully replicate all the features of PD.

Inhibition of the UP System: One Cause of Parkinson's Disease

The UP system is a ubiquitous, multienzymatic proteolytic pathway that removes misfolded ubiquitinated proteins [80]. Recently, genetic mutations of proteins of the UP system were found to be responsible for inherited forms of PD. The pathogenesis of a defective protein degradation pathway has been implicated in PD, suggesting that failure of this multienzymatic complex might be involved in sporadic cases of PD. Lending substance to this hypothesis, McNaught and colleagues [81] found an impairment of proteasomal activity and reduced expression of proteasomal subunits in nigral postmortem tissue from sporadic-PD patients. Recently, it was demonstrated in vivo and in vitro that exogenous inhibition of the UP system reproduces features of PD including cytoplasmic inclusions and selective degeneration of the nigrostriatal pathway [7,82,83].

Parkinson's disease has been associated with oxidative stress and mitochondrial dysfunction; in fact, dopaminergic neurons are particularly subject to increased oxidative stress due to production of free radicals during DA auto-oxidation and DA metabolis [74]. Oxidative stress can directly or indirectly affect the UP system. Proteasomal components can be directly oxidized, or increased production of oxidized proteins can overwhelm the UP system, resulting in toxic accumulation of damaged proteins in the cell. Dopamine can also induce inhibition of the UP system in PC12 cells, which is dependent on free radical production and DA uptake [84]. Moreover, in recent years, genetic mutations responsible for inherited PD were found to decrease the enzymatic activity of the UP system, suggesting a tight association between impairment of the UP system and the occurrence of selective damage to the nigrostriatal pathway featured by subcellular alterations similar to those seen with PD. In a recent study, Fornai and colleagues [7] demonstrated that intrastriatal administration of lactacystin or epoxymycin causes neuronal damage that selectively affects striatal DA neurons, inducing their degeneration and a decrease in striatal DA levels. Striatal UP system inhibition also resulted in retrograde degeneration of nigral neurons and formation of cytoplasmic α-synuclein-, parkin-, E1-, and ubiquitin-immunostained inclusions in spared nigral neurons. Similarly, the same effects were found in vitro in the PC12 cell line. This simple model allows one to investigate the parallelism between inclusions obtained in vitro and those observed in vivo; the parallelism between proteasome inhibition and inclusion formation; the fine ultrastructural constitution of inclusions; why the proteasome inhibition leads to damage that is particularly severe for DA-producing cells; and the biochemical mechanisms underlying selective neurotoxicity. The last point is particularly crucial because failure of the UP activity in genetic PD is ubiquitous, whereas the neuropathology of PD mostly affects nigrostriatal DA neurons [17,85].

Morover, in vivo and in vitro results indicate that proteasome inhibition-induced toxicity was suppressed by inhibiting DA synthesis. On the other hand, the toxicity of proteasome inhibition was enhanced by drugs augmenting DA availability, demonstrating that impairment of the UP system produces cell death and neuronal inclusions that depend on the occurrence of endogenous DA [7]. Recently, it has been reported that low levels of chronic proteasomal inhibition in neuroblastoma cells can alter mitochondrial homeostasis by increasing free radicals and oxidative stress [86]. Thus, UP system inhibition is able to reproduce important features of PD, confirming the involvement of an impaired activity of this system in the pathogenesis of PD.

Novel Therapeutic Target in Parkinson's Disease

Although Lewy bodies were originally described solely in PD, in recent years these inclusions have been described in a spectrum of neurodegenerative disorders. These data suggest that common biochemical alterations leading to the formation of intracellular inclusions might underlie various pathological conditions. Consequently, the knowledge of the biochemical steps involved in the formation of neuronal inclusions could represent a key to developing new therapeutic strategies. The data accumulated during the last few years demonstrate that PD research is now converging on the dysfunction of a specific biochemical complex as a final common pathway for both the inherited and the sporadic forms of this disease. It is evident that several degenerative diseases are connected with impairment of the UP system—either primary impairment (genetic mutations of its components, presence of proteasome inhibitors) or an excess of UP substrates (increased amount of ubiquitinated misfolded proteins) [87].

In this manner, development of pharmacological modulators of the endogenous proteasome activators may provide a therapeutic tool able to clear the cell from an excess of misfolded proteins [87]. Improved enzymatic efficacy of the UP might be achieved by enhancing the activity of the proteasome activators or might be obtained by supporting the activity of high-molecular-weight heat shock proteins (HSPs). Important support of this hypothesis is evident from the observation that HSPs promote refolding of abnormal proteins and enhanced proteolytic activity of the UP.[88] In recent years, pharmacological agents that increase the clearance of neuronal aggregates have been described. In summary, an effective strategy in PD and in other neurodegenerative disorders might be represented by the use of specific drugs that, by enhancing the activity of enzymes of the UP system, increase cell viability and remove inclusion bodies.

References

1. Betarbet R, Sherer TB, MacKanzie G, et al. Chronic systemic pesticide exposure reproduces features of Parkinson's disease. Nat Neurosci 2000;3:1301–1306.
2. Greenamyre JT, Hastings TG. Parkinson's-divergent causes, convergent mechanisms. Science 2004:304:1120–1122.
3. Jenner P. The MPTP-treated primate as a model of motor complications in PD: primate model of motor complications. Neurology 2003:61:S4–S11.
4. Cohen G. Oxy-radical toxicity in catecholamine neurons. Neurotoxicology 1984;5:77–82.
5. Miller GW, Gainetdinov RR, Lavey AI, Caron MG. Dopamine transporter and neuronal injury. Trends Pharmacol Sci 1999;20:424–429.
6. Sulzer D, Bogulavsky J, Larsen KE, et al. Neuromelanin biosynthesis is driven by excess cytosolic catecholamines not accumulated by synaptic vesicles. Proc Natl Acad Sci U S A 2000;97:11869–11874.
7. Fornai F, Lenzi P, Gesi M, et al. Fine structure and biochemical mechanisms underlying nigrostriatal inclusions and cell death after proteasome inhibition. J Neurosci 2003;23:8955–8966.
8. Fornai F, Lenzi P, Gesi M, et al. Methamphetamine produces neuronal inclusions in the nigrostriatal system and in PC12 cells. J Neurochem 2004;88:114–123.
9. Fornai F, Lenzi P, Frenzilli G, et al. DNA damage and ubiquitinated neuronal inclusions in the substantia nigra and striatum of mice following MDMA (ecstasy). Psychopharmacology 2004;173:353–363.
10. Shimura H, Hattori N, Kubo S, et al. Familial Parkinson disease gene product, parkin, is a ubiquitin-protein ligase. Nat Genet 2000;25:302–305.
11. Liu Y, Fallon L, Lashuel HA, et al. The UCH-L1 gene encodes two opposing enzymatic activities that affect alpha-synuclein degradation and Parkinson's disease. Cell 2002;18:209–218.
12. Spillantini MG, Schmidt ML, Lee VM, et al. Alpha-synuclein in Lewy bodies. Nature 1997;388:839–840.
13. Conway KA, Rochet JC, Bieganski RM, Lansbury PT Jr. Kinetic stabilization of the alpha-synuclein protofibril by a dopamine alpha-synuclein adduct. Science 2001;294:1346–1349.
14. Sulzer D. Alpha-synuclein and cytosolic dopamine: stabilizing a bad situation. Nat Med 2001;7:1280–1282.
15. Fornai F, Lenzi P, Ferrucci M, et al. Occurrence of neuronal inclusions combined with increased nigral expression of α-synuclein within dopaminergic neurons following treatment with amphetamine derivatives in mice. Brain Res Bull 2005;65:405–413.
16. Polymeropoulos MH, Lavedan C, Leroy E, et al. Mutation in the alpha-synuclein gene identified in families with Parkinson's disease. Science 1997;276:2045–2047.
17. Chung KK, Dawson VL, Dawson TM. The role of the ubiquitin-proteosomal pathway in Parkinson's disease and other neurodegenerative disorders. Trends Neurosci 2001;24:S7–S14.
18. Feany MB, Bender WW. A Drosophila model of Parkinson's disease. Nature 2000;404:394–398.

19. Masliah E, Rockenstein E, Veinbergs I, et al. Dopaminergic loss and inclusion body formation in alpha-synuclein mice: implication for neurodegenerative disorders. Science 2000;287:1265–1269.
20. Van der Putten H, Wiederhold KH, Probst A, et al. Neuropathology in mice expressing human alpha-synuclein. J Neurosci 2000;20:6021–6029.
21. Schapira AH, Olanow CW. Neuroprotection in Parkinson disease: mysteries, myths, and misconceptions. JAMA 2004;291:358–364.
22. Fornai F, Schluter OM, Lenzi P, et al. Parkinson-like syndrome induced by continuous MPTP infusion: convergent roles of the ubiquitin-proteasome system and alpha-synuclein. Proc Natl Acad Sci U S A 2005;102:3413–3418.
23. Luthman J, Fredriksson A, Sundstrom E, et al. Selective lesion of central dopamine or noradrenaline neuron systems in the neonatal rat: motor behavior and monoamine alterations at adult stage. Behav Brain Res 1989;33:267–277.
24. Przedborski S, Levivier M, Jiang H, et al. Dose-dependent lesions of the dopaminergic nigrostriatal pathway induced by intrastriatal injection of 6-hydroxydopamine. Neuroscience 1995;67:631–647.
25. Cohen G, Werner P. Free radicals, oxidative stress, and neurodegeneration. In: Calne DB (ed) Neurodegenerative Diseases. Saunders, Philadelphia, 1994, pp 139–161.
26. Hefti F, Melamed E, Wurtman RJ. Partial lesions of the dopaminergic nigrostriatal system in rat brain: biochemical characterization. Brain Res 1980;195:123–126.
27. Bjorklund LM, Sanchez-Pernaute R, Chung S, et al. Embryonic stem cells develop into functional dopaminergic neurons after transplantation in a Parkinson rat model. Proc Natl Acad Sci U S A 2002;99:2344–2349.
28. Langston JW, Ballard P, Tetrud JW, Irwin I. Chronic Parkinsonism in humans due to a product of meperidine-analog synthesis. Science 1983;219:979–980.
29. Przedborski S, Jackson-Lewis V, Vila M, et al. Free Radicals and nitric oxide toxicity in Parkinson's disease. In: Gordin A, Kaakkola S, TerñÊinen H (eds). Parkinson's Disease. Lippincott, Philadelphia, 2003, pp 83–94.
30. Markey SP, Johannessen JN, Chiueh CC, et al. Intraneuronal generation of a pyridinium metabolite may cause drug-induced parkinsonism. Nature 1984;311:464–467.
31. Mayer RA, Kindt MV, Heikkila RE. Prevention of the nigrostriatal toxicity of 1-methyl-4-phenyl-1,2,3,6-tetrahydropyridine by inhibitors of 3,4-dihydroxyphenylethylamine transport. J Neurochem 1986;47:1073–1079.
32. D'Amato RJ, Lipman ZP, Snyder SH. Selectivity of the parkinsonian neurotoxin MPTP: toxic metabolite MPP+ binds to neuromelanin. Science 1986;231:987–989.
33. Forno LS, Langston JW, DeLanney LE, et al. Locus ceruleus lesions and eosinophilic inclusions in MPTP-treated monkeys. Ann Neurol 1986;20:449–455.
34. Forno LS, DeLanney LE, Irwin I, Langston JW. Similarities and differences between MPTP-induced parkinsonism and Parkinson's disease: neuropathologic considerations. Adv Neurol 1993;60:600–608.
35. Tatton NA, Kish SJ. In situ detection of apoptotic nuclei in the substantia nigra compacta of 1-methyl-4-phenyl-1,2,3,6-tetrahydropyridine-treated mice using terminal deoxynucleotidyl transferase labelling and acridine orange staining. Neuroscience 1997;77:1037–1048.
36. Nicotra A, Parvez S. Apoptotic molecules and MPTP-induced cell death. Neurotoxicol Teratol 2002;24:599–605.
37. Lee CS, Song EH, Park SY, Han ES. Combined effect of dopamine and MPP+ on membrane permeability in mitochondria and cell viability in PC12 cells. Neurochem Int 2003;43:147–154.
38. Hoglinger GU, Feger F, Prigent A, et al. Chronic systemic complex I inhibition induces a hypokinetic multisystem degeneration in rats. J Neurochem 2003;84:491–503.
39. Greenamyre JT, Betarbet R, Sherer TB. The rotenone model of Parkinson's disease: genes, environment and mitochondria. Parkinsonism Relat Disord 2003;9:S59-S64.
40. Sherer TB, Betarbet R, Testa CM, et al. Mechanism of toxicity in rotenone models of Parkinson's disease. J Neurosci 2003;23:10756–10764.

41. Marking L. Oral toxicity of rotenone to mammals. US Fish Wildlife Serv Invest Fish Control 1988;4:1.
42. Ferrante RJ, Schulz JB, Kowall NW, Beal MF. Systemic administration of rotenone produces selective damage in the striatum and globus pallidus, but not in the substantia nigra. Brain Res 1997l;753:157–162.
43. Heikkila RE, Nicklas WJ, Vyas I, Duvoisin RC. Dopaminergic toxicity of rotenone and the 1-methyl-4-phenylpyridinium ion after their stereotaxic administration to rats: implication for the mechanism of 1-methyl-4-phenyl-1,2,3,6-tetrahydropyridine toxicity. Neurosci Lett 1985;62:389–394.
44. Thiffault C, Langston JW, Di Monte DA. Increased striatal dopamine turnover following acute administration of rotenone to mice. Brain Res 2000;885:283–288.
45. Hensley K, Pye QN, Maidt ML, et al. Interaction of alpha-phenyl-N-tert-butyl nitrone and alternative electron acceptors with complex I indicates a substrate reduction site upstream from the rotenone binding site. J Neurochem 1998;71:2549–2557.
46. Testa CM, Sherer TB, Greenamyre JT. Rotenone induces oxidative stress and dopaminergic neuron damage in organotypic substantia nigra cultures. Mol Brain Res 2005;134:109–118.
47. Lee HJ, Shin SY, Choi C, et al. Formation and removal of alpha-synuclein aggregates in cells exposed to mitochondrial inhibitors. J Biol Chem 2002;277:5411–5417.
48. Sherer TB, Betarbet R, Stout AK, et al. An in vitro model of Parkinson's disease: linking mitochondrial impairment to altered alpha-synuclein metabolism and oxidative damage. J Neurosci 2002;22:7006–7015.
49. Sherer TB, Kim JH, Betarbet R, Greenamyre TJ. Subcutaneous rotenone exposure causes highly selective dopaminergic degeneration and α-synuclein aggregation. Exp Neurol 2003;179:9–16.
50. Nicklas WJ, Vyas I, Heikkila RE. Inhibition of NADH-linked oxidation in brain mitochondria by 1-methyl-4-phenyl-pyridine, a metabolite of the neurotoxin, 1-methyl-4-phenyl-1,2,5,6-tetrahydropyridine. Life Sci 1985;36:2503–2508.
51. Ramsay RR, Salach JI, Singer TP. Uptake of the neurotoxin 1-methyl-4-phenylpyridine (MPP+) by mitochondria and its relation to the inhibition of the mitochondrial oxidation of NAD+-linked substrates by MPP+. Biochem Biophys Res Commun 1986;134:743–748.
52. Dauer W, Kholodilov N, Vila M, et al. Resistance of alpha-synuclen null mice to the parkinsonian neurotoxin MPTP. Proc Natl Acad Sci U S A 2002;99:14524-14529.
53. Drolet RE, Behrouz B, Lookingland KJ, Goudreau JL. Mice lacking alpha-synuclein have an attenuated loss of striatal dopamine following prolonged chronic MPTP administration. Neurotoxicology 2004;25:761-769.
54. Lansbury PT Jr, Brice A. Genetics of Parkinson's disease and biochemical studies of implicated gene products. Curr Opin Cell Biol 2002;14:653–660.
55. Dauer W, Przedborski S. Parkinson's disease: mechanisms and models. Neuron 2003;39:889–909.
56. Kruger R. Genes in familial parkinsonism and their role in sporadic Parkinson's disease. J Neurol 2004;S6:2–6.
57. Kitada T, Asakawa S, Hattori N, et al. Mutations in the parkin gene cause autosomal recessive juvenile parkinsonism. Nature 1998;392:605–608.
58. Oliveira SA, Scott WK, Martin ER, et al. Parkin mutations and susceptibility alleles in late-onset Parkinson's disease. Ann Neurol 2003;53:624–629.
59. Mizuno Y, Hattori N, Mori H, et al. Parkin and Parkinson's disease. Curr Opin Neurol 2001;14:477–482.
60. Petrucelli L, O'Farrell C, Lockhart PJ, et al. Parkin protects against the toxicity associated with mutant alpha-synuclein: proteasome dysfunction selectively affects catecholaminergic neurons. Neuron 2002;36:1007–1019.
61. Perez FA, Palmiter RD. Parkin-deficent mice are not a robust model of parkinsonism. Proc Natl Acad Sci U S A 2005;102:2174–2179.

62. Imai Y, Soda M, Takahashi R. Parkin suppresses unfolded protein stress-induced cell death through its E3 ubiquitin-protein ligase activity. J Biol Chem 2000;275:35661–35664.

63. Lo Bianco C, Schneider BL, Bauer M, et al. Lentiviral vector delivery of parkin prevents dopaminergic degeneration in an alpha-synuclein rat model of Parkinson's disease. Proc Natl Acad Sci U S A 2004;101:17510–17515.

64. Leroy E, Boyer R, Auburger G, et al. The ubiquitin pathway in Parkinson's disease. Nature 1998;395:451–452.

65. Bonifati V, Rizzu P, van Baren MJ, et al. Mutations in the DJ-1 gene associated with autosomal recessive early-onset parkinsonism. Science 2003;299:256–259.

66. Mitsumoto A, Nakagawa Y. DJ-1 is an indicator for endogenous reactive oxygen species elicited by endotoxin. Free Radic Res 2001;35:885–893.

67. Kim RH, Smith PD, Aleyasin H, et al. Hypersensitivity of DJ-1-deficient mice to 1-methyl-4-phenyl-1,2,3,6-tetrahydropyrindine (MPTP) and oxidative stress. Proc Natl Acad Sci U S A 2005;102:5215–5120.

68. Dev KK, Hofele K, Barbieri S, et al. Part II. α-Synuclein and its molecular pathophysiological role in neurodegenerative disease. Neuropharmacology 2003;45:14–44.

69. Larsen EK, Sulzer D. Autophagy in neurons: a review. Histol Histopathol 2002;17:897–908.

70. Kruger R, Kuhn W, Mueller T, et al. Ala30Pro mutation in the gene encoding α-synuclein in Parkinson's disease. Nat Genet 1998;18:106–108.

71. Zarranz JJ, Alegre J, Gomez-Esteban JC, et al. The new mutation, E46K, of α-synuclein causes Parkinson and Lewy body dementia. Ann Neurol 2004;55:164–173.

72. Recchia A, Debetto P, Negro A, et al. α-Synuclein and Parkinson's disease. FASEB J 2004;18:617–626.

73. Sidhu A, Wersinger C, Vernier P. Does α-synuclein modulate dopaminergic synaptic content and tone at the synapse? FASEB J 2004;18:637–647.

74. Lotharius J, Brundin P. Impaired dopamine storage resulting from alpha-synuclein mutations may contribute to the pathogenesis of Parkinson's disease. Hum Mol Gen 2002;11;2395–2407.

75. Singleton AB, Farrer M, Jhonson J, et al. Alpha-synuclein locus triplication causes Parkinson's disease. Science 2003;302:841.

76. Farrer M, Kachergus J, Forno L, et al. Comparison of kindreds with parkinsonism and alpha-synuclein genomic multiplications. Ann Neurol 2004;55:174–179.

77. Kirik D, Rosenblad C, Burger C, et al. Parkinson-like neurodegeneration induced by targeted overexpression of alpha-synuclein in the nigrostriatal system. J Neurosci 2002;22:2780–2791.

78. Kleven MS, Seiden LS. Repeated injection of cocaine potentiates methamphetamine-induced toxicity to dopamine-containing neurons in rat striatum. Brain Res 1991;557:340–343.

79. Stefanis L, Larsen KE, Rideout J, et al. Expression of A53T mutant but not wild-type α- synuclein in PC12 cells induces alterations of the ubiquitin-dependent degradation system, loss of dopamine release, and autophagic cell death. J Neurosci 2001;21: 9549–9560.

80. Ciechanover A, Orian A, Schwartz AL. Ubiquitin-mediated proteolysis: biological regulation via destruction. Bioessay 2000;22:442–451.

81. McNaught KS, Belizaire R, Isacson O, et al. Altered proteasomal function in sporadic Parkinson's disease. Exp Neurol 2003;179:38–46.

82. McNaught KS, Belizaire R, Jenner P, et al. Selective loss of 20S proteasome alpha-subunits in the substantia nigra pars compacta in Parkinson's disease. Neurosci Lett 2002;326:155–158.

83. McNaught KS, Bjorklund LM, Belizaire R, et al. Proteasome inhibition causes nigral degeneration with inclusion bodies in rats. Neuroreport 2002;13:1437–1441.

84. Betarbet R, Sherer TB, Greenamyre JT. Ubiquitin-proteasome system and Parkinson's diseases. Exp Neurol 2005;191:S17–S27.
85. Forno LS. Neuropathology of Parkinson's disease. J Neuropathol Exp Neurol 1996;55:259–272.
86. Sullivan PG, Dragicevic NB, Deng JH, et al. Proteasome inhibition alters neuronal mitochondrial homeostasis and mitochondria turnover. J Biol Chem 2004;279:20699–20707.
87. Fornai F, Lenzi P, Gesi M, et al. Recent knowledge on molecular components of Lewy bodies discloses future therapeutic strategies in Parkinson's disease. Curr Drug Targets CNS Neurol Disord 2003;2:149–152.
88. Hartl FU, Hayer-Hartl M. Molecular chaperones in the cytosol: from nascent chain to folded protein. Science 2002;295:1852–1858.

Chapter 9
LRRK2 (Leucine-Rich Repeat Kinase 2) Gene on PARK8 Locus in Families with Parkinsonism

Zbigniew K. Wszolek, Alexander Zimprich, Saskia Biskup, Ryan J. Uitti, Donald B. Calne, A. Jon Stoessl, Akiko Imamura, Matthew J. Farrer, Judith Miklossy, Thomas Meitinger, Thomas Gasser, Dennis W. Dickson, and Patrick L. McGeer

Introduction

It is estimated that about 10% to 30% of Parkinson's disease (PD) cases are familial [1]. Eleven PD loci/mutations have already been identified [2] (Table 1). The PARK8 locus on chromosome 12p11.2-q13.1 was first found in 2002 in a large Japanese kindred known as the Sagamihara family [3]. The linkage analysis studies performed by our group on 21 caucasian families showed probable linkage to this locus in 10 kindreds [4]. In late April of 2004, we found the first mutation in the leucine-rich repeat kinase 2 *(LRRK2)* gene in family D (western Nebraska). Two weeks later the second mutation in this gene for family A (German-Canadian) [5] was discovered. Other groups have confirmed our discovery in a number of other families [6–9].

Clinical and Pathological Characterization of Families with the *LRRK2* Gene

In this chapter we provide a short clinical and pathological description of families with *LRRK2* gene mutations, including family D, family A, and family 469. These longitudinal studies spanning 14 years of genealogical and clinical research eventually led to the genetic discoveries.

Family D (Western Nebraska)

Most family D members reside in western Nebraska, where their ancestors settled in the early 19th century [10,11]. The family is most likely of English

Z.K. Wszolek
Department of Neurology, Mayo Clinic Jacksonville, Jacksonville, FL 32224, USA

A. Fisher et al. (eds.), *Advances in Alzheimer's and Parkinson's Disease*,
© Springer 2008

Table 1 Parkinson's disease mutations and genetic loci

Locus	Chromosomal location	Mutation	Inheritance
PARK1	4q21	α-Synuclein	AD
PARK2	6q25.2-27	Parkin	AR
PARK3	2p13	Unknown	AD
PARK4	4p15	Reassigned (PARK1)	AD
PARK5	4p14	UCH-L1	AD
PARK6	1p35-36	PINK1	AR
PARK7	1p36	DJ-1	AR
PARK8	12p11.2-q13.1	LRRK2/lrrk2	AD
PARK9	1p36	Unknown (atypical parkinsonism)	AR
PARK10	1p32	Unknown	?
PARK11	2p36-37	Unknown	AD

AD, autosomal dominant inheritance; AR, autosomal recessive inheritance PARK9 is a form of atypical PD. The *tau* gene and other gene loci associated with atypical parkinsonism are not listed here.

extraction but no definite origin has been established. The pedigree contains 190 individuals spanning six generations with 22 affected members (Fig. 1).

Both sexes are affected, with a male/female ratio of 8:14. The average age at onset of symptoms in the affected individuals was 65 years (range 48–78 years). The average survival after onset of symptoms was 13 years (range 4–26 years). The shortest survival time (4 years) occurred in individual III-14, who died in 1955 of a massive cerebral hemorrhage complicating surgical treatment of her parkinsonism. Excluding this patient, the average survival time was 14 years (range 5–26 years). The most common initial symptoms and signs included bradykinesia (60%, 9 of 15 patients for whom information is available) and unilateral resting hand tremor (40%, 6 of 15 patients). One family member presented with both bradykinesia and resting tremor. Ten patients received carbidopa/levodopa therapy, and the response was excellent in all. Long-term complications related to dopaminomimetic therapy, such as dyskinesias and motor fluctuations, were seen in five individuals. Motor complications such as cerebellar dysfunction, pyramidal signs, dysautonomia, dystonia, or chorea were not observed. Vertical gaze palsy was documented in one affected individual (III-21) at age 83 years, about 5 years from the onset of the symptomatic disease. This person had no other atypical clinical features and remained responsive to carbidopa/levodopa therapy until demise (see below, case description). Dementia was not documented in any affected individuals, including III-21.

The results of routine laboratory tests such as complete blood cell count (CBC), urinalysis, and serum electrolytes; liver, renal, and thyroid function tests; and vitamin B_{12} and folate levels were all normal or negative. Cerebrospinal fluid examination was performed on two affected individuals (III-20, III-21). Normal findings were seen in individual III-21, but III-20 had abnormal findings related to

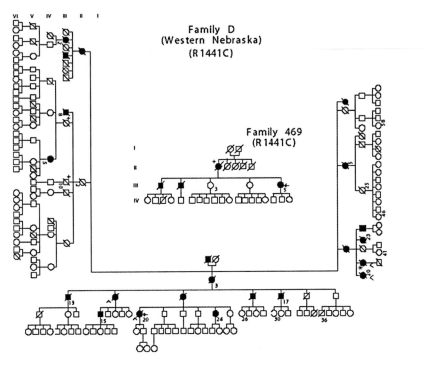

Fig. 1 Pedigrees of family D (western Nebraska) and family 469. Squares, males; circles, females. Filled squares or circles represent subjects affected by parkinsonism. A diagonal line through a square or circle indicates a deceased person. An arrow indicates a proband. A star indicates a supranuclear gaze palsy. A caret indicates an autopsy

traumatic subarachnoid hemorrhage. Neuroimaging studies including head computed tomography (CT) and/or magnetic resonance imaging (MRI) were done on three affected individuals (III-20, III-21, IV-20), and no structural pathology was identified. Electromyography was performed in an affected individual (III-21) and demonstrated the presence of peripheral neuropathy thought to be related to longstanding diabetes mellitus. Electroencephalography (EEG) recordings were not performed.

Four patients died during the study period, and three autopsies were performed (in one case autopsy could not be arranged). Autopsy material from individual III-14, who died in 1955, was retrieved and reexamined. Therefore, four brains were available for this study.

The four autopsied cases are briefly described.

- *Case III-14.* This right-handed woman presented with bradykinesia at age 48 years as the youngest affected individual in this family. Later she developed a resting tremor in both hands, which was worse on the left side. Therapy with anticholinergic medications provided limited benefit, so she was referred for surgical treatment of her parkinsonism at age 50 (in 1953). She underwent bilateral anterior choroidal artery ligation without significant improvement. At age 52, she was bedridden and severely immobilized. She was operated on again. She received an ethanol injection to the globus pallidus, which was followed by cerebral hemorrhage and death. On autopsy, neuronal loss (NL) and gliosis of the substantia nigra (SN) were present. Lewy bodies (LBs) and Lewy neurites were found in the SN, pons, medullary tegmentum, and basal nucleus of Meynert. Neurofibrillary tangles (NFTs), senile plaques (SPs), and amyloid deposits were not seen. The final diagnosis was parkinsonism due to brain stem LB pathology, LB disease (LBD).

- *Case III-20.* This right-handed woman developed a resting tremor in the left hand at age 68 years. Neurological examination performed shortly thereafter showed mild parkinsonism characterized by bradykinesia, rigidity, and resting tremor in both hands, which was more marked in the left hand. No other abnormalities were seen. Cognitive functions were normal. Treatment with carbidopa/levodopa alleviated her symptoms. At age 80, she developed postural instability but could ambulate with a wheeled walker. At that time, dyskinesias affecting her neck and shoulders were observed. A year later, she fell and suffered a small, traumatic subarachnoidal hemorrhage. Because of her increasing motor disability, she required nursing home placement. Her cognitive functions remained intact. She had no known history of rapid eye movement sleep disorder or hallucinations. She died of pneumonia at age 88.

  Autopsy revealed mild cortical atrophy and ventricular dilation. There was nearly total loss of pigmentation of the SN and locus ceruleus. Microscopically, the SN showed marked NL and gliosis with extraneuronal melanin. There was moderate NL in the locus ceruleus. SPs were seen in the frontal, temporal, parietal, and enthorinal cortices and in the amygdala and hippocampus. NFTs were seen only in the subthalamic nucleus, the midbrain tectum, and the pontine tegmentum. Immunohistochemical evaluation showed LBs and Lewy neurites in the cingulate gyrus, frontal and insular cortices, SN, pons, medulla, and basal nucleus of Meynert. The final diagnosis was parkinsonism due to diffuse LBD.

- *Case III-21.* This patient was a right-handed woman in whom parkinsonism appeared at age 78. Her first symptoms were slowness of movement and difficulty walking. She subsequently developed a right hand resting tremor. She was treated with carbidopa/levodopa a year later, with significant alleviation of her symptoms. She denied having falls. An examination performed when she was 83 years old demonstrated obvious parkinsonism. Her face lacked expression, with reduced blinking. Her voice was soft. Supranuclear gaze palsy developed, with the vertical gaze somewhat impaired in both directions. Horizontal gaze and oculocephalic maneuvers were normal. There was minimal appendicular rigidity, which was more pronounced in the extremities on the right side. No nuchal rigidity was seen. Her posture was stooped, but her balance was normal. She had mild weakness in distal leg muscles bilaterally, with reduced sensation in both feet and legs, which was thought to be related to her long-standing type II diabetes mellitus. At age 85, she became wheelchair-bound because of progressive parkinsonism and peripheral neuropathy. Her Mini Mental State Examination (MMSE) was 22 of 30. She died of pneumonia at age 89.

The SN and locus ceruleus had decreased pigmentation. Microscopically, the SN had moderate to marked NL and gliosis with extraneuronal melanin. The SN NL was patchy but most marked in ventral and lateral cell groups. Tauopathy was present, with a moderate number of NFTs and pretangles in the SN, basal nucleus of Meynert, subthalamic nucleus, and oculomotor complex. No SP was seen. Immunohistochemistry studies showed no LBs or Lewy neurites in any part of the brain. The final diagnosis was parkinsonism due to tauopathy.

- *Case IV-20.* This patient was a right-handed woman who developed left arm dysfunction at age 57, characterized by a tendency to hold the arm in a flexed position against the body while walking. Her head CT was normal, and PD was diagnosed. Carbidopa/levodopa therapy produced definite improvement in motor function. By age 59, micrographia had developed, and her speech was becoming soft. Examination demonstrated moderately severe bilateral rigidity and bradykinesia but no tremor. The gait showed mild parkinsonian features. There was postural instability. With adjustment of the carbidopa/levodopa dosage, function improved once again, but fluctuations in motor performance developed over time, with peak-dose dyskinesia and wearing-off phenomenon. In 1989, brain MRI was essentially normal, showing only moderately decreased signal intensity in the globus pallidus and, to a lesser extent, in the putamen bilaterally. By 67 years of age, she had had frequent episodes of freezing and more difficulties with balance but not falling. Her cognition was fully preserved, with an MMSE score of 29 of 30 possible points at an evaluation performed 1 week before her death. She died of pneumonia at age 68.

  Grossly, mild cortical atrophy and ventricular dilatation were seen. There was total loss of pigmentation in the SN and locus ceruleus. Microscopically, the SN showed marked NL and gliosis, with extraneuronal melanin. Immunohistochemistry revealed no LBs or Lewy neurites in any part of the brain. The final diagnosis was parkinsonism due to nigral degeneration without any distinctive histopathology.

Genetic analysis of these four affected individuals (III-14, III-20, III-21, IV-20) showed the presence of the *LRRK2* 4321C > T (R1441C) missense mutation.

Family A (German-Canadian)

We have been investigating family A since 1991 [12,13]. The initial family assessment was performed in Canada. The proband's mother immigrated to Canada from northern Germany at the beginning of the 20th century. Genealogical studies traced the family's origin to northern Germany, where additional family members were found. At the present time, the family members reside in Canada, the United States, Germany, and France. The current pedigree contains 16 affected members spanning six generations (Fig. 2). Both sexes are affected, with a male/female ratio of 2:1. The average age at onset of symptoms was 53 years (range 35–65 years). The average survival after onset of symptoms was 12 years (range 5–18 years). The initial sign was unilateral resting tremor. Beneficial response to carbidopa/levodopa therapy was seen. Long-term complications of carbidopa/levodopa therapy were present in two patients, including end-of-dose wearing-off, peak-dose dyskinesia, and unpredictable on-off

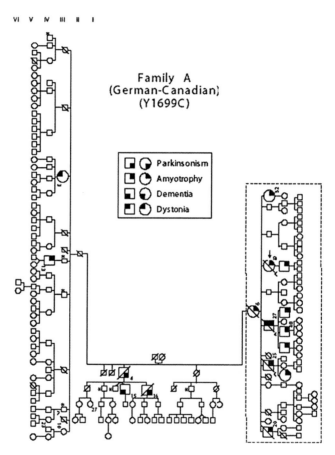

Fig. 2 Pedigree of family A. Squares represent males. Affected individual symbols are enlarged for better visualization of the types of clinical involvement presented in a solid box. A diagonal line through a square or a circle indicates a deceased person. An arrow indicates a proband. A caret indicates an autopsy. A star indicates a person lost during World War II. Family members enclosed in a dotted box were born and resided/reside in Canada and the United States, except II-6, who was born in Germany and emigrated to Canada. All other family members were born and resided/reside in Germany, Denmark, and France

swings. Postural instability was noted in two. Dementia alone was present in two family members from the German branch (III-3, III-15). Dystonia (Meige syndrome) was documented in one family member (IV-13), also from the German branch. Amyotrophy was present in one affected individual (III-27) from the Canadian branch and was characterized by distal limb muscle weakness, muscle atrophy, and the presence of fasciculations.

Laboratory tests, including CBC, urinalysis, and serum electrolytes; liver, renal, and thyroid function tests; serum protein electrophoresis; and vitamin B_{12} and folate levels were negative or normal. Head CT of two individuals (III-29, IV-48) were normal. Head MRI of one patient (IV-48) revealed minimally enlarged ventricles and the presence of several small foci of signal abnormality in periventricular white matter. EEG performed on the same individual (IV-48) was normal.

Autopsies were performed on three affected individuals (III-25, III-27, III-29) from the Canadian branch. Patient III-25 died in 1975 before the initiation of our research on this family. However, we were able to retrieve the pathology report and one glass slide showing the basis pontis. Two other patients (III-27, III-29) underwent detailed neuropathological examination (Fig. 3). Severe loss of pigmented neurons with much extracellular melanin was seen in the SN. NL and gliosis in the SN with ubiquitin-immunoreactive neuronal lesions were present. LBs were not seen in the SN, but α-synuclein immunohistochemistry revealed Lewy bodies and Lewy neurites in limbic gray matter in III-27. This individual also had concurrent Alzheimer disease (AD)-type pathology. The other individual (III-29) had mild motor neuron disease. The presence of clinical amyotrophy characterized by muscle weakness, atrophy, and fasciculations and pathological evidence of anterior horn cell degeneration, reactive microglia, and reactive astrocytes are suggestive of the possible coexistence of PD and amyotrophic lateral sclerosis (ALS).

The genetic studies demonstrated the presence of a *LRRK2* 5096A > G (Y1699C) missense mutation.

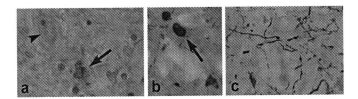

Fig. 3 Neuropathology of two members of family A. One individual has nonspecific neuronal loss in the substantia nigra (a) with extraneuronal neuromelanin in macrophages (arrow). Residual neurons have Marinesco bodies (arrowhead). Other individuals (b, c) have severe nigral neuronal loss without obvious Lewy bodies, yet cortical Lewy bodies were found in the limbic cortex (b) and Lewy neurites in the hippocampus (c) with α-synuclein immunohistochemistry

Family 469

Family 469 resides in the United States. The pedigree contains 31 individuals spanning four generations with four affected family members (Fig. 1) and a male/female ratio of 1.6:1.0.5 The average age at onset was 56 years (range 52–60 years). The average survival after onset of symptoms was 23 years (range 21–26 years). The initial sign was resting tremor in three affected family members (in the upper extremity in two and in the lower extremity in one) and rigidity in one (III-2). All affected individuals exhibited cardinal parkinsonian signs including bradykinesia, rigidity, resting tremor, and postural instability with a good response to carbidopa/levodopa. One family member (II-1) developed cognitive impairments during the course of her illness.

Laboratory tests, including CBC, urinalysis, and serum electrolytes; liver, renal, and thyroid function tests; serum protein electrophoresis; and vitamin B12, ceruloplasmin, and folate levels were negative or normal. One case (III-2) was studied with head CT and MRI and with EEG without any abnormalities being identified.

No autopsies have been performed so far.

Molecular genetic studies demonstrated the presence of a *LRRK2* 4321C > T (R1441C) missense mutation. This is the same mutation found in family D. However, the genealogical investigations performed in these two families demonstrated no common ancestors. No sharing of haplotypes for the closest flanking microsatelite repeat markers was identified on genetic analysis. This suggests that the mutations in these two families either arose independently or are very ancient.

PET Studies

Seven affected family members were studied with positron emission tomography (PET): four from family D (IV-15, IV-21, IV-24, IV-25) and three from family A (IV-47, IV-48, IV-50) [12,14]. One family D member and one family A member underwent PET with 18F-6-fluoro-L-dopa (FD) only. One individual had PET with FD and ^{11}C-raclopride (RAC), and four had PET with FD, [11]C-($\pm$)α-dihydrotetrabenazine (^{11}C-DTBZ), and [11]C-*d-threo*-methylphenidate (^{11}C-MP). These studies revealed the presence of PET abnormalities similar to those of sporadic PD with impairment of presynaptic dopamine function affecting the putamen more than the caudate and preserved striatal dopamine receptors.

LRRK2/ROCO Gene

The predicted product of the *LRRK2* gene is a large protein, lrrk2 (2527 amino acids encoded by 51 exons). This protein is a member of a recently defined ROCO protein family [5], as predicted by the presence of several conserved domains: a Roc (Ras of complex proteins) domain that belongs to the Ras/ GTPase superfamily; and a COR domain (C-terminal of Roc) that is also characteristic for this protein family [15,16]. The Ras-like small GTPases act as molecular switches regulating diverse cellular processes, including mitogenic signaling, cytoskeletal reorganization, vesicle trafficking, and nucleocytoplasmic transport. In addition, three further conserved domains are encoded, 12 leucine-rich repeats (LRR), a nonreceptor tyrosine kinase-like domain, and a WD40 domain. The presence of LRR and WD40 motifs suggests that protein–protein interactions play an important role in the function of lrrk2, which may function as a component of a multiprotein complex. The tyrosine kinase-like MAPKKKs are primary effectors of small GTPases; the presence of similar domains in lrrk2 raises the possibility that its kinase activity may undergo GTP-dependent autoactivation. *LRRK2*/lrrk2 is official nomenclature approved by the Human Genome Nomenclature Organization.

Additional *LRRK2* Families

G2019S Families

The *LRRK2* 6055G > A (G2019S) mutation was found in different populations. Our study identified 22 patients from 13 families and 3 nonfamilial caucasian patients from Europe (Norway, Ireland, and Poland) and North America [17]. Their average age at onset was 57 years (range 39–78 years). The penetrance of the mutation was found to be highly age-dependent, increasing from 17% at age 50 to 85% at age 70. Cardinal parkinsonian features were seen in the affected patients including a good response to dopaminomimetic therapy. Additional clinical features suggestive of a parkinsonism-plus disorder were absent. All patients, despite their different geographical origins, were found to have the same haplotype around the mutation highlighting a common but evidently ancient ancestry. In a study by Gilks et al. (2005), eight PD patients from southeastern England carrying the G2019S mutation were identified [9]. Their average age at symptomatic disease onset was 57 years (range 41–70 years). Their symptoms were described as similar to those seen in patients with the typical sporadic form of PD. Pathology examinations were done in three of these patients showing nigral cell loss, pigment incontinence, and LBs. In two of them, LBs were demonstrated in the limbic cortices. In one, neocortical SPs and occasional NFTs were seen.

Three additional studies identified 51 G2019S mutation carriers with parkinsonism from 26 genealogically unrelated families [7,8,18]. These patients were from Brazil, Italy, Portugal, and the United States. The average ages at onset were 61, 50, and 57 years, respectively. All cardinal signs of PD were present with an excellent response to carbidopa/levodopa. No autopsies have been performed so far.

The most recent report, published only in abstract form, analyzes the frequency of the G2019S mutation in 200 index autosomal dominant PD cases from Europe and North Africa [19]. The frequency of this mutation in North African families was found to be greater than in the European families. The clinical features of 21 identified affected mutation carriers resembled those seen in typical late-onset PD cases. However, some cognitive dysfunction (low MMSE scores) was present. The authors also found 15 unaffected mutation carriers with ages ranging from 32 to 74 years, further substantiating the evidence that the penetrance in this particular *LRRK2* gene mutation is indeed age-dependent.

Basque Families and British Kindred

Four separately ascertained families (UGM003, UGM004, UGM005, UGM006) from the Basque region in Spain with 19 affected family members spanning five generations were described as carrying a dardarin R1396G mutation [6,20,21]. Dardarin R1396G is *LRRK2* 4321C<G (R1441G); the gene structure presented by Paisàn-Ruiz et al. (2004) is incomplete [6]. In addition, this is the same nucleotide mutated as in family D and family 469, albeit with a C<T transition encoding a R1441C substitution. The average age at onset was 62 years (range 51-80 years). The average survival after onset of symptoms was 14 years. Affected individuals usually presented with a unilateral resting tremor and later developed all the cardinal signs of PD. They had a good response to carbidopa/levodopa treatment with the subsequent appearance of levodopa-related complications. No cognitive decline has been observed. An additional 21 Spanish PD patients, including 10 Basques, carrying the R1396G mutation were identified. Six of them had a positive family history. No other clinical or pathological details of these cases are available at this juncture.

Another mutation (dardarin, Y1654C) identical to *LRRK2* 5096A > G (Y1699C) described in family A was identified in a British family known as the PL kindred. This family contains 23 family members with 12 affected individuals spanning four generations [6]. The phenotype of the PL kindred was not specifically described but was reported to be similar to that of the Basque families, with a benign course and an excellent response to low doses of carbidopa/levodopa. No other clinical or pathological details of this kindred are known at the present time. Because all seven mutation carriers from whom blood samples were available for analysis were clinically affected and all seven unaffected individuals did not carry the mutation, the penetrance was estimated to be 100% in this family.

Sagamihara Family

The large Sagamihara family was initially reported in 1978 [22]. All family members reside in Sagamihara (Kanagawa Prefecture), Japan. The pedigree contains 47 affected family members spanning seven generations. Both sexes are affected. The mean age at onset was 54 years [23]. In 10 individuals definitely affected with PD, the initial symptoms and signs were either gait disturbance (five patients) or resting tremor. With disease progression, all of the cardinal signs of PD were present, with good response to levodopa. While on levodopa, two patients developed psychosis [22]. There were neither pyramidal or cerebellar signs nor dementia. Autopsies were performed on four affected family members. Depigmentation of the SN with NL was seen, but LBs were absent.

Genetic studies revealed the presence of a *LRRK2* 6059T > C (I202T) missense mutation in affected members of this family.

Discussion

The discovery of the *LRRK2* gene is a significant step forward toward better understanding the basic mechanisms of neurodegeneration in familial parkinsonism. Mutations of the *LRRK2* gene are common, probably even more common than those in the parkin gene. The *LRRK2* gene affects different ethnic populations and so far has been identified in 151 affected individuals from 47 families residing in North and South America, Europe, and Africa. The youngest average age at onset (46 years) was present in family SAO [8]. The oldest average age at onset (66 years) was seen in families UGM003/03 and P-394. The average disease duration in all families was 16 years (range 13–23 years). The most common initial sign was resting tremor. The second most common initial sign was bradykinesia. With disease progression, the cardinal signs of PD were almost universally present. A good response to carbidopa/levodopa therapy was observed. However, cognitive impairment, including dementia, supranuclear gaze palsy, hyperreflexia, dystonia, and amyotrophy, were seen in some affected individuals from certain families. Complications of carbidopa/levodopa therapy were common (Table 2).

The pathological data are limited. NL and gliosis of the SN are present in all autopsied cases to date. However, LBs can be absent or widespread; and additional pathology, including tau deposition, has been documented. More brains from *LRRK2* mutation carriers need to be examined to obtain a better characterization of pathological findings. Hopefully, specific antibodies to lrrk2 will be developed in the near future. We already know that there was some difficulty characterizing intranuclear and intracytoplasmic inclusions in the brains of affected individuals from family A. Having specific antibodies would undoubtedly help to describe features of these inclusions and their localization.

The molecular genetic analysis so far identified seven distinct missense *LRRK2* gene mutations (3342A > G, I1122V, R1441C, R1441G, Y1699C, G2019S, I2020T). The G2019S substitution appears to be the most common among them. These coding changes are located in the LRR domain (3342A > G,

Table 2 Clinical, pathological, and molecular genetic features of larger LRRK2 gene families

Family/Reference	Affected members	Average age at onset(years)	Average disease duration (years)	Initial signs	RT	B	R	Response to levodopa	Other features (number of cases)	Pathological features	Mutation
A[12]	16	53	12	RT	+	+	+	+	Amyotrophy (2) Dementia (2)	NL, Gl in SN, NFT, SP, LB in cortex, axonal spheroids and NL in AH	Y1699C
D[10]	22	65	13	RT, B	+			+	Supranuclear gaze palsy (1)	NL, Gl in SN, LB in brainstem, tau positive glia	R1441C
469[5]	4	56	23	RT	+	+	+	+	–	NA	R1441C
32[5]	7	54	20	B	+	+	+	+	–	NA	I2020T
Sagamihara	47	54	NA	RT,GD	+	+	+	+	–	NL in SN, no LB	I2020T
UGM003/03[20]	9	66	18	RT	+	+	+	+	–	NA	R1441G
UGM004/04[20]	4	62	NA	RT	+	+	+	+	Hyperreflexia Babinski sign Delirium (1)	NA	R1441G
UGM005/05[20]	8	58	10	RT	+	+	+	+	Writer's cramp(1) Foot dystonia (1)	NA	R1441G

Table 2 (continued)

Family/ Reference	Affected members	Average age at onset(years)	Average disease duration (years)	Initial signs	RT	B	R	Response to levodopa	Other features (number of cases)	Pathological features	Mutation
UGM006/ 06[20]	15	61	NA	RT	+	+	+	+	–	NA	R1441G
PL[6]	12	NA	NA	RT	*	*	*	*	–	NA	Y1699C
P-089[18]	4	59	NA	NA	†	†	†	+	–	NA	G2019S
P-104[18]	4	58	NA	NA	†	†	†	+	–	NA	G2019S
F-05[18]	5	64	NA	NA	†	†	†	+	–	NA	G2019S
P-394[18]	4	66	NA	NA	†	†	†	+	–	NA	G2019S
292[16]	7	58	NA	NA	+	+	+	+	–	NA	G2019S
415[16]	6	54	NA	NA	+	+	+	+	–	NA	G2019S
IT-025[8]	5	49	11	NA	+	+	+	+	dystonia (1)	NA	G2019S
SAO[8]	5	46	15	NA	+	+	+	+	–	NA	G2019S
LISB[8]	5	54	14	NA	+	+	+	+	Sleep disturbance(3) dystonia (2)	NA	G2019S

This table includes only large families (those with at least four affected individuals) and with published pedigrees. RT; resting tremor, B; bradykinesia, R; rigidity, P; postural instability, GD; gait disturbance, NA; not available, nigral cell loss, Gl; gliosis, SN; substantia nigra, NFT; neurofibrillary tangles, SP; senile plaques, AH; anterior horn, LB; Lewy bodies, +; present, –; absent, * for PL family based on reported

I1122V), Roc domain (R1441C, R1441G), COR domain (Y1699C), and MAPKKK domain (G2019S, I2020T). Based on our experience with genes encoding *tau* and *parkin,* it is quite likely that more *LRRK2* gene mutations will be identified in the near future.

The *LRRK2* PD phenotype resembles more closely the phenotype seen in sporadic PD than any other form of familial parkinsonism described so far. It is hoped that the creation of transgenic animals carrying *LRRK2* mutations will help foster curative drug therapies for patients suffering from familial and sporadic forms of PD.

References

1. Payami H, Zareparsi S. Genetic epidemiology of Parkinson's disease, J Geriatr Psychiatry Neurol 1998;11(2):98–106.
2. Healy DG, Abou-Sleiman PM, and Wood NW, PINK, PANK, or PARK? A clinicians guide to familial parkinsonism. Lancet Neurol 2004;3(11):652–662.
3. Funayama M, Hasegawa K, Kowa H, et al. A new locus for Parkinson's disease (PARK8) maps to chromosome 12p11.2-q13.1. Ann Neurol 2002;51(3):296–301.
4. Zimprich A, Muller-Myhsok B, Farrer M, et al. The PARK8 locus in autosomal dominant parkinsonism: confirmation of linkage and further delineation of the disease-containing interval. Am J Hum Genet 2003;74(1):11–19.
5. Zimprich A, Biskup S, Leitner P, et al. Mutations in LRRK2 cause autosomal-dominant parkinsonism with pleomorphic pathology. Neuron 2004;44(4):601–607.
6. Paisan-Ruiz C, Jain S, Evans EW, et al. Cloning of the gene containing mutations that cause PARK8-linked Parkinson's disease. Neuron 2004;44(4):595–600.
7. Nichols WC, Pankratz N, Hernandez D, et al. Genetic screening for a single common LRRK2 mutation in familial Parkinson's disease. Lancet 2005;365:410–412.
8. Di Fonzo A, Rohe CF, Ferreira J, et al. A frequent LRRK2 gene mutation associated with autosomal dominant Parkinson's disease. Lancet 2005;365:412–415.
9. Gilks WP, Abou-Sleiman PM, Gandhi S, et al. A common LRRK2 mutation in idiopathic Parkinson's disease. Lancet 2005;365:415–416.
10. Wszolek ZK, Pfeiffer B, Fulgham JR, et al. Western Nebraska family (family D) with autosomal dominant parkinsonism. Neurology 1995;45:502–505.
11. Wszolek ZK, Pfeiffer RF, Tsuboi Y, et al. Autosomal dominant parkinsonism associated with variable synuclein and tau pathology. Neurology 2004;62:1619–1622.
12. Wszolek ZK, Vieregge P, Uitti RJ, et al. German-Canadian family (family A) with parkinsonism, amyotrophy, and dementia-longitudinal observations. Parkinsonism Relat Disord 1997;3:125–139.
13. Wszolek ZK, Tsuboi Y, Uitti RJ, et al. PARK8 locus is associated with late-onset autosomal dominant parkinsonism: clinical, pathological and linkage analysis study of family A & D. Neurology 2003;60(suppl 1);282–283.
14. Pal PK, Wszolek ZK, Uitti R, et al. Positron emission tomography of dopamine pathways in familial parkinsonian syndromes. Parkinsonism Relat Disord 2001;8(1):51–56.
15. Bosgraaf L, Van Haastert PJ. Roc, a Ras/GTPase domain in complex proteins. Biochim Biophys Acta 2003;1643:5–10.
16. Shen J. Protein kinases linked to the pathogenesis of Parkinson's disease. Neuron 2004;44(4):575–577.
17. Kachergus J, Mata IF, Hulihan M, et al. Identification of a novel LRRK2 mutation linked to autosomal dominant parkinsonism: evidence of a common founder across European populations. Am J Hum Genet 2005;76(4):672–680.

18. Hernandez DG, Paisan-Ruiz C, McInerney-Leo A, et al. Clinical and positron emission tomography of Parkinson's disease caused by LRRK2. Ann Neurol 2005;57(3):453–456.
19. Lesage S, Ibanez P, Lohmann E, et al. The G2019S LRRK2 mutation in autosomal dominant European and North African Parkinson's disease is frequent and its penetrance is age-dependent. Neurology 2005;64(suppl 1):1826.
20. Paisan-Ruiz C, Saenz A, de Munain AL, et al. Familial Parkinson's disease: clinical and genetic analysis of four Basque families. Ann Neurol 2005;57(3):365–372.
21. Perez Tur J. Clinical and molecular findings in LRRK2 Spanish and UK families. Parkinsonism Relat Disord 2005;11(suppl 2):70.
22. Nukada H, Kowa H, Saitoh T, et al. A big family of paralysis agitans. Clin Neurol 1978;18:627–634.
23. Funayama M, Hasegawa K, Ohta E, et al. An LRRK2 mutation as a cause for the parkinsonism in the original PARK8 family. Ann Neurol 2005;57(6):918–921.

Chapter 10
Microglia Activation and Gene Expression of Cytokines in Parkinson's Disease

Makoto Sawada, Kazuhiro Imamura, and Toshiharu Nagatsu

Introduction

We previously reported increasing levels of proinflammatory cytokines and decreasing levels of neurotrophins in the nigrostriatal region of postmortem brain and/or ventricular or spinal cerebospinal fluid in Parkinson's disease (PD) or in animal models of PD induced by 1-methyl-4-phenyl-1,2,3,6-tetrahydropyridine (MPTP) or 6-hydroxydopamine [1–5]. Increasing levels of proinflammatory cytokines and decreasing levels of neurotrophins may produce apoptotic death of the nigrostriatal dopamine neurons, a characteristic feature in PD [2,4,6–9].

Increasing levels of proinflammatory cytokines suggest inflammatory processes in the brain, termed neuroinflammation, in PD patients, in which activated microglia participate. During inflammation in the brain, activated microglia produce proinflammatory cytokines. As the first features of inflammation in PD, McGeer et al [10,11]. reported an increasing number of major histocompatibility complex (MHC) class II antigen [human leukocyte antigen-DR (HLA-DR)]-positive microglial cells in the substantia nigra.

The presence of α-synuclein-positive intracellular inclusions, called Lewy bodies, in dopamine neurons in the substantia nigra is another feature of PD. In Lewy body disease (LBD)/dementia with Lewy bodies (DLB), in which both parkinsonian movement disorder and dementia are observed, Lewy bodies are widely distributed not only in the brain stem but also in the cerebral cortex and amygdala [12]. As an approach to elucidate the relation among cytokines, activated microglia, and damage to neurons and neurites, we further explored, in postmortem brains, the question: Are changes in cytokine levels neurotoxic factors that mediate cell death by induction of apoptosis, or are they compensatory and neuroprotective responses? For these studies, we examined the distribution pattern of activated microglia and the gene expression of cytokines

M. Sawada
Department of Brain Life Science, Research Institute of Environmental Medicine, Nagoya University, Nagoya, Aichi 464-8601, Japan

A. Fisher et al. (eds.), *Advances in Alzheimer's and Parkinson's Disease*,
© Springer 2008

and neurotrophins in brains from normal control subjects, PD patients, and LBD patients.

Increased Levels of Cytokines from Activated Microglia in the Putamen in PD

We identified proteins of tumor necrosis factor-α (TNFα) and IL-6 in the putamen and peripheral blood mononuclear cells from PD patients by Western blot analysis, confirming our previous results by enzyme-linked immunosorbent assay (ELISA) [1,3]. We then proved, using immunohistochemistry, the coexistence of TNFα, interleukin-6 (IL-6), and MHC class II (CR3/43) in induced endothelial cell adhesion molecule-I (ICAM-I)- and leukocyte function associated antigen-I (LFA-I)-positive activated microglia in the putamen from PD patients. These results confirm that TNFα and IL-6 are produced by activated microglia in the putamen in PD.

Increased Activated Microglia Associated With Damaged Dopamine Neurons During the Progression of PD

We examined, using immunohistochemistry, the changes in activated microglia in the brains of PD patients at both early and advanced stages. The number of MHC class II-positive activated microglia that were positive for TNFα and IL-6 increased in the substantia nigra and putamen as the neuronal degeneration proceeded. MHC class II-, TNFα-, and IL-6-positive activated microglia were associated with damaged tyrosine hydroxylase (TH)-positive DA neurons and neurites. These results suggest that activated microglia producing cytokines may be associated with disease progression. The activated microglia may act either for neurotoxicity or neuroprotection.

Activated Microglia in the Substantia Nigra, Putamen, and Various Regions of the Brain in PD

In normal brains, many Ki-M1p-positive resting microglia, but only a few MHC class II (CR3/43)-positive activated microglia, were observed in the substantia nigra and putamen. Imamura et al [13]. of our group found that the number of MHC II-positive activated microglia increased in PD not only in the substantia nigra and putamen but also in the hippocampus, transentorhinal cortex, cingulate cortex, and temporal cortex.

Activated microglia were associated with degenerated DA neurons in the substantia nigra, suggesting that they may be neurotoxic. However, in the putamen and pallidum, activated microglia were associated with neurites without degeneration, suggesting that they may not be neurotoxic but possibly neuroprotective. These results suggest that activated microglia could be either neurotoxic or neuroprotective in different regions of the PD brain.

Different Expression of Cytokines in the Hippocampus and Putamen in PD and LBD

Activated microglia have multiple roles. First, MHC II-positive activated microglia induce antigen formation. Second, activated microglia phagocytose damaged cells. Third, activated microglia produce neurotoxic substances such as superoxide anions, nitric oxide, and glutamate. Fourth, activated microglia also produce neurotrophic substances such as neurotrophins and proinflammatory cytokines, which are pleiotropic and either neurotoxic or neuroprotective.

To examine whether activated microglia in the substantia nigra and putamen and in the hippocampus are neurotoxic or neuroprotective, we compared the expression of cytokines and neurotrophins in the hippocampus and putamen in postmortem PD and LBD brains and those of normal controls.

In normal controls, neuronal loss and activated microglia were not observed in the hippocampus CA 2/3 region with intense staining of brain-derived neurotrophic factor (BDNF). Immunohistochemical examination of the hippocampus CA 2/3 region from PD showed that the number of such MHC II (R3/43)-positive microglia increased, which were also ICAM-I (CD54)-, LFA-1 (CD11a)-, TNFα-, and IL-6-positive. α-Synuclein-positive cells were also observed. However, the number of BDNF-positive cells were slightly decreased compared with normal controls. In the hippocampus from LBD patients, the number of MHC-II (CR3/43)-positive microglia and α-synuclein-positive cells increased more than those in PD patients.

Expression of mRNA of cytokines and neurotrophins was examined by reverse transcription polymerase chain reaction (RT-PCR) in the hippocampus and putamen in normal controls and PD and LBD cases. In the hippocampus, mRNA levels of IL-6 and TNFα increased in PD and LBD cases, but mRNA levels of BDNF greatly decreased in the LBD cases compared with that in normal controls and PD cases. In the putamen, mRNA levels of IL-6 were markedly increased in both PD and LBD brains. In contrast, mRNA expression of BDNF increased in PD brains but greatly decreased in LBD brains.

As described above, activated microglia were observed in the putamen and hippocampus in both PD and LBD cases. However, neuronal loss was observed mainly in the putamen, but not in the hippocampus in PD cases. In contrast, neuronal degeneration was observed in both the hippocampus and the putamen in LBD brains.

These results suggest that the function of activated microglia may be different in the hippocampus and putamen and also in PD and LBD. Activated microglia in the hippocampus in PD are not neurotoxic but are speculated to be neuroprotective as significant neurodegeneration was not observed. In the LBD hippocampus, increasing IL-6 levels and decreasing BDNF levels suggest a neurotoxic role of activated microglia to induce apoptotic neuronal death. Neurodegeneration was observed in the putamen of both PD and LBD brains. In the putamen of both PD and LBD cases, mRNA levels of IL-6 were increased. In contrast, mRNA levels of BDNF increased in PD but decreased in LBD. These changes in IL-6 and BDNF levels in the putamen in PD and LBD again suggest that microglia activation may be either neuroprotective or neurotoxic.

Neuroprotective and Neurotoxic Subtypes of Activated Microglia in the PD Brain

We found two types of activated microglia by immunohistochemistry [13]: one associated with neurodegeneration and the other not associated with neuronal degeneration in the autopsied brains from PD patients; the former was found in the substantia nigra and the latter in the putamen. These results suggest the presence of neurotoxic or nontoxic/neuroprotective subtypes of activated microglia in PD. The results on different expression of cytokines and neurotrophins in the hippocampus and putamen from PD and LBD patients also support the hypothesis on the presence of neuroprotective or neurotoxic subtypes of activated microglia. As other supporting evidence of this hypothesis, Sawada and coworkers [14] found that transduction of the HIV-1 Nef gene into activated microglia caused a change in the microglia from the nontoxic type to the toxic type, accompanied by robust expression of NADPH oxidase, which produces toxic oxygen free radicals. Hirsch et al [7]. also proposed separate populations of microglia; one subpopulation of glial cells may play a neuroprotective role by metabolizing dopamine and scavenging oxygen free radicals and another that may deteriorate dopamine neurons by producing nitric oxide and proinflammatory cytokines.

Conclusion

We examined microglial activation and gene expression of cytokines and neurotrophins in the hippocampus and putamen from patients with PD or LBD and compared them with those of normal controls. Activated microglia were observed in the putamen as well as in the hippocampus from both PD and LBD patients. Increased expression of IL-6 and TNFα mRNAs was observed in the putamen and hippocampus in both PD and LBD brains. In contrast, the expression of BDNF increased in the putamen in PD cases but decreased in the hippocampus and putamen in LBD cases. These results suggest that there may exist either neuroprotective or neurotoxic subset(s) of activated microglia in PD, depending on the circumstances present at a particular time during the progression of the disease. Activated microglia are speculated to be neuroprotective at least at an early stage of PD but may assume a deleterious role after chronic activation over the course of the disease. This concept is supported by the in vitro finding by Sawada and coworkers [14] that different microglia populations may act in either a protective or a destructive capacity, depending on the cytokines and other factors released in the microenvironment.

There remains a need for in vivo proof of neuroprotective and neurotoxic subsets of microglia and toxic or protective changes in microglia.

References

1. Mogi M, Nagatsu T. Neurotrophins and cytokines in Parkinson's disease. Adv Neurol 1999;80:135–139.
2. Nagatsu T, Mogi M, Ichinose H, et al. Cytokines in Parkinson's disease. Neurosci News 1999;2:88–90.
3. Nagatsu T, Mogi M, Ichinose H, Togari A. Cytokines in Parkinson's disease. J Neural Transm Suppl 2000;58:143–151.
4. Nagatsu T. Parkinson's disease: changes in apoptosis-related factors suggesting possible gene therapy. J Neural Transm 2002;109:731–745.
5. Nagatsu T, Sawada M. Inflammatory process in Parkinson's disease: role for cytokines. Curr Pharmacol Design 2005;11:999–1016.
6. Hartmann S, Hunot PP, Michel PP, et al. Caspase-3: a vulnerability factor and a final effector in the apoptotic cell death of dopaminergic neurons in Parkinson's disease. Proc Natl Acad Sci U S A 2000;97:2875–2880.
7. Hirsch EC, Hunot S, Damier P, Faucheux B. Glial cells and inflammation in Parkinson's disease: a role in neurodegeneration? Ann Neurol 1998;44(suppl 1):S115–S120.
8. Hirsch EC, Hunot S, Faucheux B, et al. Dopaminergic neurons degenerate by apoptosis in Parkinson's disease. Mov Disord 1999;14:383–385.
9. Hirsch EC. Inflammatory changes and apoptosis in Parkinson's disease. Adv Behav Biol 2002;51:259–263.
10. McGeer PL, Itagaki S, Boyes BE, McGeer EG. Reactive microglia are positive for HLA-DR in the substantia nigra of Parkinson's disease and Alzheimer's disease brain. Neurology 1988;38:1285–1291.
11. McGeer PL, McGeer EG. The inflammatory response system of brain: implications for therapy of Alzheimer and other neurodegenerative diseases. Brain Res Rev 1995;21:195–218.
12. Kosaka K. Diffuse Lewy body disease. Neuropathology 2002;20(suppl):73–78.
13. Imamura K, Hashikawa N, Sawada M, et al. Distribution of major histocompatibility class II-positive microglia and cytokine profile of Parkinson's disease brains. Acta Neuropathol (Berl) 2003;106:518–526.
14. Vilhardt F, Plastre O, Sawada M, et al. The HIV-1 Nef protein and phagocyte NADPH oxidase activation. J Biol Chem 2002;277:42136–42143.

Chapter 11
Dissecting the Biochemical Pathways Mediated by Genes Implicated in Parkinson's Disease: Induction of DJ-1 Expression in A30P α-Synuclein Mice

Mark Frasier, Shanti Frausto, Daniel Lewicki, Lawrence Golbe, and Benjamin Wolozin

Introduction

Parkinson's disease (PD), a common age-dependent movement disorder, is characterized by loss of dopaminergic neurons in the substantia nigra and the presence of Lewy bodies in the remaining dopaminergic neurons. Lewy bodies are proteinaceous cytoplasmic inclusions that are present in many of the remaining neurons. Both environmental and genetic components appear to contribute to the pathophysiology of the disease. The ability of complex I inhibitors to selectively induce nigral degeneration in animal models and rare human cases implicates environmental toxins in the pathophysiology of PD. Laboratory studies show that pesticides, such as 1-methyl-4-phenyl-1,2,3,6-tetrahydropyridine (MPTP) and rotenone, can induce syndromes that partially resemble PD [1–3]. The role of the environment in PD is supported by multiple epidemiological studies that have shown a higher prevalence of sporadic PD in rural areas. This elevated rate of PD is hypothesized to be due to exposure to agricultural pesticides and herbicides [4–7]. A common feature among these pesticides is that they inhibit mitochondrial complex I and increase reactive oxygen species. The oxidative stress imposed by the agents is thought to play a particularly important role in the pathophysiology of PD.

Genetic mutations are associated with rare familial cases of PD. Mutations in α-synuclein, parkin, UCH-L1, PINK1, and DJ-1 have been identified in patients with familial PD [8–13]. The α-synuclein protein has attracted particular attention because it contributes to both sporadic and familial PD. α-Synuclein is one of the main components of the Lewy bodies found in sporadic PD. Mutations in α-synuclein have also been identified in several kindreds with familial PD. Mutations at A53T, A30P, and E46K and triplication of the α-synuclein gene

M. Frasier
Michael J. Fox Foundation, NewYork, NY 10163, USA

A. Fisher et al. (eds.), *Advances in Alzheimer's and Parkinson's Disease*,
© Springer 2008

have been reported in families that have early-onset PD [8–10,14]. These muta-tions enhance the tendency for α-synuclein to form fibrils, which are hypothe-sized to be toxic and stimulate the neurodegeneration associated with PD [15–18]. Overexpressing A53T or A30P α-synuclein in mice leads to a syndrome characterized by progressive motor impairment and the formation of intracel-lular inclusions containing aggregated α-synuclein [19–23]. The mechanism of toxicity of the α-synuclein fibrils is unknown, but free radical production and proteasomal inhibition have both been implicated [15–18].

α-Synuclein appears normally to function as a chaperone protein that has a particular affinity for lipids. It binds lipids, is abundant in synapses, and is hypothesized to be involved in vesicular trafficking associated with synaptic and lysosomal function [24–28]. It shares significant homology with 14-3-3 proteins, binds proteins that associate with 14-3-3, and can promote protein refolding [29–31]. The proteins shown to associate with α-synuclein include extracellular regulated kinase, tyrosine hydroxylase, phospholipase C, protein kinase C, the dopamine transporter, sept4, and S6' proteasomal protein [26,29,32–39].

DJ-1 is another PD-related gene that is attracting increasing attention. Two mutations in the DJ-1 gene have been reported in autosomal recessive early onset PD. One mutation is a point mutation at L166P, and the other is a deletion of exons 1–5 [40]. This mutation leads to rapid degradation of the protein and a corresponding reduction in the amount of homo-dimeric DJ-1, which is thought to be the active complex [41–43]. The function of DJ-1 is not well understood. Previous studies indicate that DJ-1 is an oncogene that shows sequence homology to protein chaperones, proteases, and catalase [44]. A recent study suggests that DJ-1 can inhibit the aggregation of α-synuclein [45]. Other studies suggest that DJ-1 protects against multiple stresses, including oxidative stress [46,47]. Block-ing DJ-1 expression via siRNA confers susceptibility to multiple stresses includ-ing oxidative insults [46]. The action of DJ-1 appears to be linked to redox activity because cysteine 106 becomes acidified upon exposure to oxidative stress, and this acidification targets DJ-1 to the mitochondria [48]. These data suggest that DJ-1 may play a critical role in the oxidative stress response in PD patients.

Increasing evidence suggests that the pathologies associated with the various proteins linked to neurodegenerative diseases overlap. Tau is a cytoskeletal protein associated with neurofibrillary tangles in Alzheimer's disease that also accumulates in Pick bodies in Pick's disease and in dystrophic neurons in brains of patients harboring the A53T α-synuclein mutation [49]. Fibrillization of α-synuclein enhances tau fibrillization in vitro and both tau fibrillization and hyperphosphorylation in vivo [50,51]. Pathological overlap also exists between DJ-1 and tau. DJ-1 inclusions and tau fibrils are co-localized in numerous tauopathies and possibly in PD [52,53]. Because of the overlap between the pathologies associated with α-synuclein and DJ-1, we investigated whether mice overexpressing A30P α-synuclein also show DJ-1 pathology. We found that DJ-1 is elevated in the diseased mice. Immunocytochemical studies in human tissues from patients harboring the A53T α-synuclein mutation show that the DJ-1 is present in reactive astrocytes. Knockdown of DJ-1 in cell culture increases the vulnerability to rotenone, a mitochondrial complex I inhibitor, a finding that concurs with other reports suggesting that DJ-1 might be linked to a stress response.

Materials and Methods

Animals

The human A30P α-synuclein transgene had been injected into hybrid B6/DBA oocytes [22]. Founders were extensively (7–10 generations) back-crossed into a C57Bl/6 background. Intercrossing of the highest expressing line 31 yielded a stable colony of homozygous 31H mice [22]. These were the animals used in the present study. The mice develop symptoms at 6 to 14 months. The symptoms begin with a tremor and progress to an end-stage phenotype characterized by muscular rigidity, postural instability, and ultimately paralysis. Mice that progressed to end-stage symptoms were sacrificed by cervical dislocation, and their brains were hemi-sectioned. Asymptomatic transgenic mice were of the same A30P α-synuclein transgenic line as the symptomatic mice and were sacrificed with aged-matched symptomatic animals. The right hemisphere was fixed in formalin and subsequently embedded in paraffin for immunohisto-chemistry study. The left hemisphere was flash-frozen in methyl-2-butane and kept at $-80\,^\circ$C for subsequent immunoblot analysis.

Antibodies and Immunohistochemistry

The brains were sectioned sagitally at a thickness of 4 μm and mounted on Superfrost-plus slides (Fisher Scientific, Allentown, PA, USA). Monoclonal DJ-1 antibody was used to detect DJ-1 by immunoblot (MBL Laboratories, 1:1000). Polyclonal antibodies were made to the DJ-1 epitope AQVKAPLVLKD and used to detect DJ-1 by immunohistochemistry (1:500). For immunofluorescence study, sections were first treated with 70% formic acid at room temperature for 15 minutes, rinsed in phosphate-buffered saline (PBS) for 10 minutes, and then blocked for 20 minutes in 2% fetal bovine serum and 1% normal goat serum in PBS at room temperature. Primary antibody was incubated on the sections overnight at 4 °C and followed by two 10-minute washes in PBS. Fluorescent cy2 anti-mouse (1:2500) and rhodamine anti-rabbit (1:200) secondary antibodies (Jackson Immunology) were incubated on the sections for 1 hour at room temperature in the dark. Following two more 10-minute washes in PBS, coverslips were applied using the Fluormount G mounting media (Electron Microscopy Sciences, Washington, PA, USA).

Brain Tissue Extraction

Brain tissue extraction was performed as previously described by Sahara et al. [54], with minor modifications. Left hemi-brains were weighed and homogenized in 3 volumes of TBS pH 7.4 (1 mM EDTA, 5 mM sodium pyrophosphate, 30 mM glycerol 2-phosphate, 30 mM sodium fluoride, 1 mM EDTA) containing a protease inhibitor cocktail (Sigma-Aldrich). Protein content was determined via the BCA method, and total protein and concentration were adjusted with the homogenization buffer to be equal among samples. Samples were then centrifuged at 150,000 g for 15 minutes at 4 °C in a Beckman TLA 1004 rotor (Beckman, Palo Alto, CA, USA). The resulting pellet was rehomogenized in three volumes of a salt sucrose buffer (10 mM Tris HCl pH 7.4, 0.8 M NaCl, 10% sucrose, 1 mM EDTA) with a protease inhibitor cocktail added immediately before use; the supernatant was labeled S1 and frozen at –80 °C. The resuspended pellet was again centrifuged at 150,000 g for 15 minutes at 4 °C; this time the pellet was discarded, and the supernatant was brought to a 1% sarkosyl solution and incubated at 37 °C for 1 hour with gentle shaking. After centrifugation (150,000 g for 30 minutes at 4 °C) the supernatant was kept, labeled S2, and frozen at –80 °C. The pellet was resuspended in Tris/EDTA (10 mM Tris HCl pH 7.4, 1 mM EDTA), washed once more with 1% sarkosyl solution, and frozen at –80 °C.

Immunoblot Analysis

Samples were separated on a gradient 8% to 16% sodium dodecyl sulfate polyacrylamide gel electrophoresis (SDS-PAGE), transferred to polyvinyl diisopropyl fluoride (PVDF) membranes and stained with a monoclonal DJ-1 antibody (MBL Laboratories, 1:1000) followed by anti-mouse horseradish peroxidase (HRP)-conjugated secondary antibody (Santa Cruz Biotechnology, Santa Cruz, CA, USA), and detected with SuperSignal West Pico Enhanced Chemiluminescence Kit (Pierce, Minneapolis, MN, USA), followed by exposure to Kodak film.

Quantification of Immunoblots

Densitometry analysis was performed using 'Unscan It' software (Silk Scientific, Orem, UT, USA). Bands were analyzed and normalized to actin levels. Unpaired Student's t-tests were used to determine statistical significance between expression levels in the animals ($n = 5$ for both the transgenic symptomatic and nontransgenic animals).

Carbonyl Detection

Protein carbonyl formation was determined by the oxyblot protocol (Chemicon, Temecula, CA, USA). Brain stem homogenate (20 mg) was added to the 2,4-dinitrophenylhydrazine (DNPH) solution and incubated at room temperature for 15 minutes in 20 ml of a final solution containing 6% SDS. Neutralization solution (7.5 ml) was added to stop the reaction, and samples were loaded directly on a 8% to 12% SDS gradient gel. After being transferred to a PVDF membrane and left overnight at 4 °C, the membranes were blocked in 5% milk in TBS-T for 1 hour and probed with anti-dinitrophenylhydrazone rabbit polyclonal antibody overnight at 4 °C. The membranes were washed with TBS-T four times for 10 minutes each time and added to anti-rabbit HRP-conjugated antibody and visualized with chemiluminescence.

Immunoprecipitation

A 500-µg aliquot of protein from each sample was diluted to a final volume of 1 ml in lysis solution. Then 25 µl of protein G (Sigma) was added to preclear nonspecific binding proteins and left for 2 hours at 4 °C. Samples were centrifuged and supernatants transferred to new sterile microcentrifuge tubes. Next, 5 µl of polyclonal anti-DJ-1 (3864) antibody was added to each sample, followed by overnight incubation at 4 °C. Another 25 µl of protein G was added to each sample, and binding to immune complexes was performed during 2 hours of rocking at 4 °C with gentle agitation. The negative control consisted of preimmune serum from the same animal from which the antibody was made, brain homogenates, and protein G in lysis solution. Samples were washed/centrifuged four times in a wash solution of 0.3% Triton-X in PBS pH 7.4 plus protease inhibitors. After the final washing and centrifugation, samples were resuspended in protein loading buffer, heat denatured at 90 °C for 90 seconds, centrifuged again, and resolved by 12% PAGE at 85 V for 90 minutes. Western blots were performed as detailed above.

DJ-1 siRNA Transfection

BE-M17 cells, approximately 90% confluent in 24 well dishes, were transfected with the DJ-1 siRNA target sequence AAGGAAAUGGAGACGGUCAUG at a working concentration of 20 pmol/ml. Serum-free Opti-MEM (Invitrogen, Carlsbad, CA, USA) lacking all antimicrobial drugs was used as the transfection medium. Transfection medium (50 µl) was combined with 2 µl Lipofectamine 2000 per transfection in one tube. In a separate tube, 50mL transfection medium was combined with 7 mL of siRNA (20 pmol/mL). After 5 minutes at room temperature, the content of the tube 1 were added to tube 2. The samples were then incubated 20 minutes at room temperature to facilitate association of RNA with Lipofectamine 2000. After rinsing the cells twice with serum-free transfection medium, 400 μl of transfection medium was added to tube 2, and its contents were added to a single well of the 24-well plate. Cells were incubated at 37°C for 3 hours before normal serum-containing medium replaced the transfection medium. Cells were then allowed to grow for 96 hours before plating on 96-well dishes.

LDH assay

Lactate dehydrogenase (LDH) release was measured according to the instructions in the nonradioactive cytotoxicity assay (Promega, Madison, WI, USA). BE-M17 cells were plated in 96-well dishes at 5000 cells/well in 100 μl phenol-free growth medium. Following 48 hours of pharmacological treatment, 50 ml of medium was removed and added to 50 ml of reaction mixture. The plates were incubated for 30 minutes at room temperature in the dark and read on an absorbance plate reader at 490 nM. The absorbance of wells containing only medium and the reaction mixture were subtracted from the measurements.

Results

Symptomatic Mice Exhibit Elevated DJ-1 Levels

Transgenic mice overexpressing mutant A30P α-synuclein develop motor impairments with age. The motor dysfunction begins with abnormal moving tremor and progresses to bradykinesia, hindlimb and forelimb impairment, and finally full paralysis over a period of 3 to 4 weeks. Motor impairment does not occur until high-molecular-weight aggregates of α-synuclein develop in the brains of these animals, with the severity of motor impairment usually correlating with the amount of α-synuclein aggregation present [51]. To understand whether aggregation of α-synuclein affects the biology of DJ-1, we examined the expression of DJ-1 in brains from mice overexpressing α-synuclein. Symptomatic animals were sacrificed at various stages of motor impairment, and their brains were immediately dissected and flash-frozen for biochemical analyses. Brain stems from symptomatic mice exhibited significantly higher levels of DJ-1 than aged-matched nontransgenic animals (Fig. 1A). Both DJ-1 monomer and dimers were resolved by SDS-PAGE and detected by the antibody used. The DJ-1 dimer was also elevated in the symptomatic mice (Fig. 1A). Figure 1B shows the densitometric analysis of the DJ-1 monomer and reveals significant differences in DJ-1 levels between nontransgenic and symptomatic transgenic animals ($n = 3$–8, $p < 0.05$). These results demonstrate that DJ-1 expression is increased in brain tissue exhibiting α-synuclein aggregation.

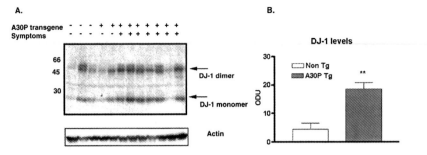

Fig. 1 Increased expression of DJ-1 in brains of transgenic mice over-expressing A30P α-synuclein. Brain stem from symptomatic A30P α-synuclein mice showed increased expression of DJ-1 monomer and dimer (arrows, left panel). Quantification of the reactivity showed a more than threefold increase in DJ-1 reactivity (right panel). **$p < 0.05$, $n = 3-8$

Symptomatic Mice Have Increased Oxidation

J-1 is reported to be involved in the oxidative stress response in cells [46]. We wanted to determine if oxidative stress is elevated in the brain stems of the symptomatic animals exhibiting increased DJ-1 levels. As a measurement of oxidative stress, we employed the Oxyblot detection kit (Chemicon) that detects Fig. 2 carbonyl groups of proteins. The protein carbonyl groups are derivitized to 2,4-dinitrophenylhydrazone (DNP-hydrazone), separated by SDS-PAGE, and visualized with an antibody directed against the DNP-hydrazone moiety. To determine if oxidized proteins were elevated in symptomatic transgenic mice, we performed the Oxyblot detection on brain stem samples from symptomatic transgenic mice and non-transgenic mice. Oxidized proteins were increased in brains from symptomatic transgenic animals compared to aged-matched nontransgenic controls; oxidation was particularly apparent in the high-molecular-weight proteins (Fig. 2, left panel). Densitometric analysis of the immunoblots revealed over a twofold increase in oxidized protein in the symptomatic transgenic animals compared to age-matched nontransgenic controls (Fig. 2, right panel). Actin was used as a protein-loading control.

DJ-1 Is Present in Reactive Astrocytes in Brain Sections from the Contursi Kindred

To determine where DJ-1 was elevated in the brain of these animals, we examined the brains of the symptomatic A30P α-synuclein mice and control nontransgenic mice by immunocytochemistry. No immunohistochemical reactivity was observed in mouse tissue (data not shown). To gain information despite the inability to detect DJ-1 by immunocytochemistry in mouse tissue, Fig. 3 we analyzed sections of human cingulate cortex that were

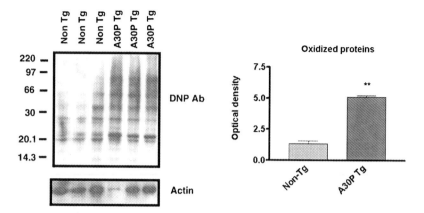

Fig. 2 Oxyblot of tissue from nontransgenic (Non Tg) and symptomatic transgenic A30P α-synuclein mice (A30P Tg) shows increased free-radical damage. Quantification of the data in the left panel showed a more than twofold increase in oxidized protein (right panel). **$p < 0.01$, $n = 3$

obtained from symptomatic subjects from the Contursi kindred, who harbor the A53T α-synuclein mutation associated with familial PD. Immunofluorescence performed with anti-DJ-1 antibody showed different patterns of reactivity in sections from patients or neurologically normal controls. Anti-DJ-1 staining of Contursi patient tissue revealed strong labeling of reactive astrocytes but no DJ-1 labeling of neurons (Fig. 3A). The glial nature of the

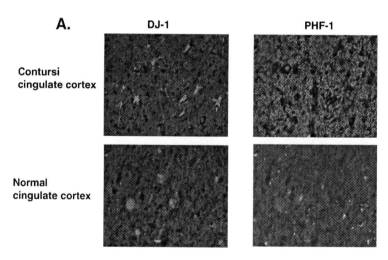

Fig. 3 Immunocytochemistry of DJ-1 in human brain tissue. A. Sections of cingulated cortex obtained from subjects in the Contursi kindred or neurologically normal controls were double-stained with either DJ-1 antibody or PHF-1 antibody. DJ-1 co-localized with PHF-1 in the normal brain but not in diseased brain

staining was confirmed by staining of adjacent sections with antibody against glial fibrillary acid protein (GFAP), which showed a pattern and location of staining similar to that observed with anti-DJ-1 antibody (Fig. 3C). In contrast, DJ-1 staining in sections from neurologically normal cingulate cortical tissue showed a different pattern of reactivity that co-localized with anti-phospho-tau staining in normal tissue using the PHF-1 antibody (Fig. 3B). Serial sections of normal tissue stained with PHF-1 antibody emphasize a staining pattern similar to that of DJ-1 (Fig. 3D). These results suggest that the increased DJ-1 reactivity observed by immunoblot in diseased tissue from symptomatic mice occurs in glia that are present in the brain stem of symptomatic mice. Such glia might be part of the reactive astrocytosis that occurs in response to oxidative stress in the symptomatic animals [51].

DJ-1 Co-immunoprecipitates with Tau and Phospho-tau in Brain Homogenates

We have previously demonstrated elevated phosphorylated tau on the PHF-1 epitope in symptomatic transgenic mice [51]. In other studies, DJ-1 has been shown to co-localize with tau fibrils in various tauopathies [52,53]. These same studies showed that the immunocytochemical patterns observed with various DJ-1 antibodies vary depending on the antibody and raise the possibility that our immunocytochemical studies were not detecting DJ-1 present in neurons, perhaps because of conformational changes in DJ-1 or hidden epitopes [52,53]. To investigate whether our immunocytochemical studies might be missing tau-associated DJ-1 present in neurons, we investigated whether DJ-1 and tau could be immunoprecipitated together from brain homogenates in the symptomatic mice.

Immunoprecipitation was performed to pull down either DJ-1 or tau protein; immunoprecipitation using preimmune serum was performed on the negative controls. We observed significant co-immunoprecipitation between DJ-1 and tau (Fig. 4). Using anti-DJ-1 antibody for the immunoprecipitation, we were able to detect an association of DJ-1 and tau using either anti-tau antibody (tau 5) (Fig. 4A, top left panel) or anti-phospho-tau antibody (PHF-1) (Fig. 4A, bottom left panel) for immunoblotting. There was a higher association of DJ-1 with tau in the lysates from symptomatic mice than from asymptomatic or nontransgenic mice (Fig. 4A), although the differences could result from the elevated levels of DJ-1 in tissues from symptomatic mice and appear to be proportional to the total amount of DJ-1 immunoprecipitated (Fig. 4A, top right panel). The complex could also be observed by immunoprecipitating with tau-5 and probing for DJ-1 (Fig. 4B).

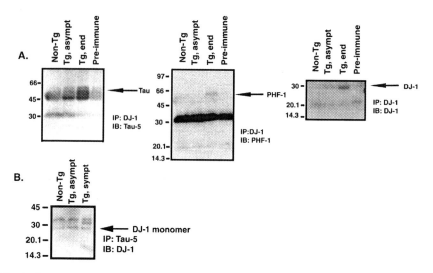

Fig. 4 Co-association of DJ-1 with tau. **A.** Brain tissue from nontransgenic mice or transgenic mice overexpressing A30P α-synuclein were immunoprecipitated with anti-DJ-1 antibody. Symptomatic mice showed increased tau-DJ-1 complex, as shown by immunoblotting of the pull-downs with either tau-5 antibody (top left panel) or PHF-1 antibody (bottom left panel). A small amount of nonspecific association was observed with the preimmune serum. Expression of DJ-1 is highest in the symptomatic mice, which resulted in increased pull-down of DJ-1 (top right panel). A strong, nonspecific band is present in the PHF-1 immunoblot at approximately 30 kDa. **B.** The DJ-1 tau complex could also be observed by immunoprecipitating tau with tau-5 antibody and immunoblotting with anti-DJ-1 antibody

DJ-1 siRNA Knockdown Confers Susceptibility to Oxidative Insult in Cell Culture

To explore whether DJ-1 was playing a protective function, we developed an siRNA protocol to block DJ-1 protein synthesis in neuroblastoma cells. Because the half-life of DJ-1 is reported to be 4 days [55], we transfected BE-M17 neuroblastoma cells with either DJ-1 or scrambled the DJ-1 sequence siRNA and plated them in 96-well dishes 4 days later. Lysates were run on a Western blot and probed with an antibody directed against DJ-1. The siRNA was successful in decreasing ~80% of DJ-1 protein levels (Fig. 5A). In parallel cultures, we treated the cells with 500 nM of the mitochondrial complex I inhibitor rotenone for 48 hours. We then determined the LDH levels released from the cells as a measurement of toxicity to the cells. Cells transfected with the DJ-1 siRNA had significantly increased (> twofold higher) LDH release when treated with rotenone compared to the scrambled siRNA-treated cells (Fig. 5B). These experiments suggest that DJ-1 is involved in the protective oxidative stress response in cells.

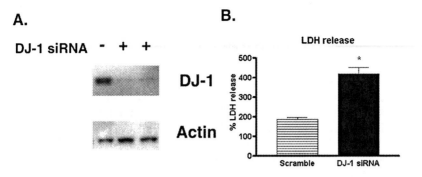

Fig. 5 Knockdown of DJ-1. BE-M17 cells were treated with RNAi for DJ-1. **A.** Immunoblots of the cells treated with RNAi for 4 days showed significant knockdown of DJ-1. **B.** Cells treated with 500 nM rotenone for 48 hours showed increased toxicity when DJ-1 was knocked down, as indicated by the release of LDH into the medium. $*p < 0.05, n = 3$

Discussion

Studies using molecular genetics have identified mutations in multiple genes that are associated with familial PD. One area of intense interest is the identification of biochemical pathways through which these genes contribute to PD. A related question is whether the genes implicated in familial PD share biochemical pathways. α-Synuclein is one of the proteins most strongly implicated in PD. Mutations in α-synuclein are associated with familial PD, and α-synuclein is one of the major proteins that accumulates in Lewy bodies in PD. Transgenic mice overexpressing A30P α-synuclein provide an in vivo model for investigating the pathological changes associated with α-synuclein aggregation. As these mice develop age-dependent α-synuclein aggregation, they show progressive motor impairment. Recent results indicate that the α-synuclein pathology stimulates pathological changes in tau protein [50,51]. Our results now extend this picture by demonstrating that DJ-1 levels also increase as the α-synuclein pathology progresses. Immunocytochemical studies using human tissue show that the DJ-1 increases in astrocytes, which indicates that the changes in DJ-1 are present in both the mouse and humans. The function of DJ-1 is poorly understood, but our studies using RNAi suggest that DJ-1 might contribute to cytoprotection because knockdown renders cells more vulnerable to the complex I inhibitor rotenone. Together these results raise the possibility that DJ-1 is induced in response to the oxidative stress present in brains exhibiting α-synuclein pathology.

Identifying where DJ-1 is expressed in brains exhibiting α-synuclein pathology remains an important question. The results obtained from immunocytochemical experiments with anti-DJ-1 antibodies appear to differ depending on the type of pathology. Results from our study and others indicate that DJ-1 expression seen with synucleinopathies occurs mainly in reactive astrocytes [44,53]. On the other hand, under some conditions, DJ-1 appears to associate with tau protein because DJ-1 staining co-localizes with tau inclusions in tauopathies [52,53].

The ability of some studies to detect DJ-1 in neurons prompted us to explore biochemical evidence of neuronal DJ-1 expression. We investigated whether tau and DJ-1 also associate in the brains of A30P α-synuclein mice; we also investigated whether DJ-1 associates with α-synuclein but did not observe any co-association (data not shown). We observed that DJ-1 co-precipitated with both tau and phosphorylated tau, and that this association only occurs in symptomatic animals. These data must be interpreted with caution because they could result from a specific physiological interaction of DJ-1 with tau due to binding in intact neurons. On the other hand, the process of immunoprecipitation utilizes a homogenization step, which could allow astrocytic DJ-1 to come into contact with neuronal tau protein and result in a nonspecific interaction between DJ-1 and tau. Further studies using methods capable of detecting interactions within cells (e.g., fluorescence resonance technology) could explore this putative association in more detail.

The properties of both DJ-1 and tau are known to change under pathological conditions. DJ-1 undergoes a pH shift, becoming more acidic [48]. Tau becomes hyperphosphorylated and can undergo a conformational change [56,57]. Assuming that DJ-1 does interact with tau in neurons, the complex might be identifiable in symptomatic mice because of the increased DJ-1 expression stimulated by the pathological processes present in the symptomatic A30P α-synuclein mice. The reason detection is difficult may result from the steric hindrance induced by the association of DJ-1 with tau. Future studies with different antibodies directed against different epitopes in DJ-1 might clarify this issue.

Increasing data suggest that DJ-1 plays a role in the stress response. Cysteine 105 of DJ-1 becomes acidified under conditions of oxidative stress [48]. In addition, increases or decreases in DJ-1 levels directly correlate with the degree of protection against oxidative stressors [46]. The particular type of stress that appears most closely linked to the pathophysiology of PD is inhibition of mitochondrial complex I [58]. Rats exposed to chronic treatment with the complex I inhibitor rotenone exhibit motor impairment and loss of midbrain dopaminergic neurons [1]. Cell culture studies also indicate that rotenone can induce α-synuclein aggregation [59]. In addition, the complex I inhibitor MPTP causes a parkinsonian syndrome in humans and animals. Based on this connection, we investigated whether DJ-1 affects the vulnerability to rotenone using cell culture. We observed that loss of DJ-1 due to siRNA rendered the cells more susceptible to rotenone toxicity, as measured by LDH release. These results are consistent with the prior study showing that DJ-1 protects against exposure to oxidative conditions [46,47]. Protection against mitochondrial toxicity is also consistent with results showing that DJ-1 localizes at the mitochondria under conditions of oxidative stress [48]. Because rotenone is a toxin selective for mitochondria, the sensitivity of the DJ-1-depleted cells to rotenone supports a mitochondrial localization of DJ-1 and also supports prior data suggesting that DJ-1 protects against oxidative stress.

Discovery of genes associated with familial neurodegenerative diseases provides new avenues for understanding the pathophysiology of the more common neurodegenerative diseases occurring in the general population. A particularly intriguing observation is that many of the proteins causing familial neurodegenerative diseases normally interact with each other. These proteins appear to function as part of the pathophysiological cascade of events associated with neurodegeneration. α-Synuclein interacts with tau in vitro and stimulates its phosphorylation by protein kinase A [60]. α-Synuclein also induces fibrillization and stimulates phosphorylation of tau [50,51]. Tau and α-synuclein co-localize in inclusions in human neurodegenerative disorders and mouse models [49,50,61,62]. Prior results showed co-localization of DJ-1 with pathological tau [52,53,63]. Our current report strengthens the linkage between DJ-1, α-synuclein, and tau by showing that DJ-1 expression is stimulated by α-synuclein pathology, and that DJ-1 might interact with tau.

Tau also appears to associate with other proteins implicated in PD. Parkin, an E3 ligase, binds CHIP, which is known to ubiquitinate tau protein [64–66]. Taken together, these results suggest that an interrelated cascade of events contributes to the pathophysiology of neurodegenerative diseases such as PD and that mutations in any of the multiple proteins associated with this cascade can precipitate premature neurodegeneration. Conversely, any protein that can modulate the cascade of events that occur during neurodegeneration represents a potentially valuable therapeutic target.

Acknowledgments We thank Philippe Kahle and Christian Haass (University of Munich, Germany) for providing the A30P α-synuclein mice; Peter Heutnik (Erasmus Medical College, The Netherlands) and Peter Davies for providing antibodies; and John Trojanowski for help in aquiring tissue from the Contursi kindred as well as for helpful advice. We also thank Nancy Muma (Loyola University, Chicago) for her useful comments. This research was supported by grants to B.W. from NINDS (NS41786-01), NIA (AG17485), and USAMRC (DAMD17-01-1-0781).

References

1. Betarbet R, Sherer TB, MacKenzie G, et al. Chronic systemic pesticide exposure reproduces features of Parkinson's disease. Nat Neurosci 2000;3(12):1301–1306.
2. Thiruchelvam M, Richfield EK, Bagg, RB, et al. The nigrostriatal dopaminergic system as a preferential target of repeated exposures to combined paraquat and maneb: implications for Parkinson's disease. J Neurosci 2000;20(24):9207–9214.
3. Manning-Bog AB, McCormack AL, Li J, et al. The herbicide paraquat causes up-regulation and aggregation of alpha-synuclein in mice: paraquat and alpha-synuclein. J Biol Chem 2002;277(3):1641–1644.
4. Engel LS, Checkoway H, Keifer MC, et al. Parkinsonism and occupational exposure to pesticides. Occup Environ Med 2001;58(9):582–589.

5. Gorell J, Rybicki B, Cole-Johnson C, Peterson E. Occupational metal exposures and the risk of Parkinson's disease. Neuroepidemiology 1999;18:303–308.

6. Liou HH, Tsai MC, Chen CJ, et al. Environmental risk factors and Parkinson's disease: a case-control study in Taiwan. Neurology 1997;48(6):1583–1588.

7. Golbe LI, Farrell TM, Davis PH. Case-control study of early life dietary factors in Parkinson's disease. Arch Neurol 1988;45(12):1350–1353.

8. Polymeropoulos MH, Lavedan C, Leroy E, et al. Mutation in the alpha-synuclein gene identified in families with Parkinson's disease. Science 1997;276(5321):2045–2047.

9. Kruger R, Kuhn W, Muller T, et al. Ala30Pro mutation in the gene encoding α-synuclein in Parkinson's disease. Nat Gen 1998;18:106–108.

10. Zarranz JJ, Alegre J, Gomez-Esteban JC, et al. The new mutation, E46K, of alpha-synuclein causes parkinson and Lewy body dementia. Ann Neurol 2004;55(2):164–173.

11. Kitada T, Asakawa S, Hattori, N, et al. Mutations in the parkin gene cause autosomal recessive juvenile parkinsonism. Nature 1998;392:605–608.

12. Bonifati, V, Rizzu, P, Squitieri, F, et al. DJ-1(PARK7), a novel gene for autosomal recessive, early onset parkinsonism. Neurol Sci 2003;24(3):159–160.

13. Valente EM, Abou-Sleiman PM, Caputo V, et al. Hereditary early-onset Parkinson's disease caused by mutations in PINK1. Science 2004;304(5674):1158–1160.

14. Singleton AB, Farrer M, Johnson J, et al. α-Synuclein locus triplication causes Parkinson's disease. Science 2003;302(5646):841.

15. Hashimoto M, Takeda A, Hsu L, et al. Role of cytochrome c as a stimulator of α-synuclein aggregation in Lewy body disease. J Biol Chem 1999;274:28849–8852.

16. Paik S, Shin H, Lee J, et al. Copper(II)-induced self-oligomerization of α-synuclein. Biochem J 1999;340:821–828.

17. Paik SR, Shin HJ, Lee JH. Metal-catalyzed oxidation of alpha-synuclein in the presence of copper(II) and hydrogen peroxide. Arch Biochem Biophys 2000;378(2):269–277.

18. Ostrerova-Golts N, Petrucelli L, Hardy J, et al. The A53T α-synuclein mutation increases iron-dependent aggregation and toxicity. J Neurosci 2000;20:6048–6054.

19. Masliah E, Rockenstein E, Veinbergs I, et al. Dopaminergic loss and inclusion body formation in alpha-synuclein mice: implications for neurodegenerative disorders. Science 2000;287(5456):1265–1269.

20. Giasson BI, Duda J, Quinn SM., et al. Neuronal alpha-synucleinopathy with severe movement disorder in mice expressing A53T human alpha-synuclein. Neuron 2002;34(4):521–533.

21. Lee M, Stirling W, Xu Y, et al. Human α-synuclein-harboring familial Parkinson's disease-linked Ala-53 to Thr mutation causes neurodegenerative disease with α-synuclein aggregation in transgenic mice. Proc Natl Acad Sci U S A 2002;99:8968–8973.

22. Kahle PJ, Neumann M, Ozmen L, et al. Subcellular localization of wild-type and Parkinson's disease-associated mutant alpha-synuclein in human and transgenic mouse brain. J Neurosci 2000;20(17):6365–6373.

23. Kahle PJ, Neumann M, Ozmen L, et al. Selective insolubility of alpha-synuclein in human Lewy body diseases is recapitulated in a transgenic mouse model. Am J Pathol 2001;159(6):2215–2225.

24. Cuervo AM, Stefanis L, Fredenburg R, et al. Impaired degradation of mutant alpha-synuclein by chaperone-mediated autophagy. Science 2004;305(5688):1292–1295.

25. Murphy DD, Rueter SM, Trojanowski, JQ, Lee VM. Synucleins are developmentally expressed, and alpha-synuclein regulates the size of the presynaptic vesicular pool in primary hippocampal neurons. J Neurosci 2000;20(9):3214–3220.

26. Sharon R, Goldberg MS, Bar-Josef I, et al. α-Synuclein occurs in lipid-rich high molecular weight complexes, binds fatty acids, and shows homology to the fatty acid-binding proteins. Proc Natl Acad Sci U S A 2001;98(16):9110–9115.

27. Perrin RJ, Woods WS, Clayton DF, George JM. Interaction of human α-synuclein and Parkinson's disease variants with phospholipids: structural analysis using site-directed mutagenesis. J Biol Chem 2000;275:34393–34398.

28. Perrin RJ, Woods WS, Clayton DF, George JM. Exposure to long chain polyunsaturated fatty acids triggers rapid multimerization of synucleins. J Biol Chem 2001;276(45): 41958–41962.
29. Ostrerova N, Petrucelli L, Farrer M, et al. α-Synuclein shares physical and functional homology with 14-3-3 proteins. J Neurosci 1999;19:5782–5791.
30. Xu J, Kao S, Lee F, et al. Dopamine-dependent neurotoxicity in α-synuclein: a mechanism for selective neurodegeneration in Parkinson disease. Nat Med 2002;5:600–606.
31. Souza, JM, Giasson BI, Lee VM, Ischiropoulos H. Chaperone-like activity of synucleins. FEBS Lett 2000;474(1):116–119.
32. Jenco J, Rawlingson A, Daniels B, Morris A. Regulation of phospholipase D2: Selective inhibition of mammalian phospholipase D isoenzymes by α and β synucleins. Biochemistry 1998;37:4901–4909.
33. Engelender S, Kaminsky Z, Guo X, et al. Synphilin-1 associates with alpha-synuclein and promotes the formation of cytosolic inclusions. Nat Genet 1999;22:110–114.
34. Pronin AN, Morris AJ, Surguchov A, Benovic JL. Synucleins are a novel class of substrates for G protein-coupled receptor kinases. J Biol Chem 2000;275:26515–26522.
35. Choi P, Snyder H, Petrucelli L, et al. SEPT5_v2 is a parkin-binding protein. Brain Res Mol Brain Res 2003;117(2):179–189.
36. Lee FJ, Liu F, Pristupa ZB, Niznik HB. Direct binding and functional coupling of alpha-synuclein to the dopamine transporters accelerate dopamine-induced apoptosis. FASEB J 2001;15(6):916–926.
37. Ihara M, Tomimoto H, Kitayama H, et al. Association of the cytoskeletal GTP-binding protein Sept4/H5 with cytoplasmic inclusions found in Parkinson's disease and other synucleinopathies. J Biol Chem 2003;278(26):24095–24102.
38. Snyder H, Mensah K, Theisler C, et al. Aggregated and monomeric alpha-synuclein bind to the S6' proteasomal protein and inhibit proteasomal function. J Biol Chem 2003;278(14):11753–11759.
39. Lindersson E, Beedholm R, Hojrup P, et al. Proteasomal inhibition by alpha-synuclein filaments and oligomers. J Biol Chem 2004;279(13):12924–12934.
40. Bonifati V, Rizzu P, van Baren MJ, et al. Mutations in the DJ-1 gene associated with autosomal recessive early-onset parkinsonism. Science 2003;299(5604):256–259.
41. Honbou K, Suzuki NN, Horiuchi M, et al. The crystal structure of DJ-1, a protein related to male fertility and Parkinson's disease. J Biol Chem 2003;278(33):31380–31384.
42. Tao X, Tong L. Crystal structure of human DJ-1, a protein associated with early onset Parkinson's disease. J Biol Chem 2003;278:(33)31372–31379.
43. Moore DJ, Dawson VL, Dawson TM. Genetics of Parkinson's disease: What do mutations in DJ-1 tell us? Ann Neurol 2003;54(3):281–282.
44. Bandopadhyay R, Kingsbury AE, et al. The expression of DJ-1 (PARK7) in normal human CNS and idiopathic Parkinson's disease. Brain 2004;127(Pt 2):420–430.
45. Shendelman S, Jonason A, Martinat C, et al. DJ-1 is a redox-dependent molecular chaperone that inhibits α-synuclein aggregate formation. PLoS Biol 2004;2:e362.
46. Yokota T, Sugawara K, Ito K, et al. Down regulation of DJ-1 enhances cell death by oxidative stress, ER stress, and proteasome inhibition. Biochem Biophys Res Commun 2003;312(4):1342–1348.
47. Martinat C, Shendelman S, Jonason A, et al. Sensitivity to oxidative stress in DJ-1-deficient dopamine neurons: and ES-derived cell model of primary parkinsonism. PLoS Biol 2004;2:e327.
48. Canet-Aviles RM, Wilson MA, Miller DW, et al. The Parkinson's disease protein DJ-1 is neuroprotective due to cysteine-sulfinic acid-driven mitochondrial localization. Proc Natl Acad Sci U S A 2004;101(24):9103–9108.
49. Duda JE, Giasson BI, Mabon ME, et al. Concurrence of alpha-synuclein and tau brain pathology in the Contursi kindred. Acta Neuropathol (Berl) 2002;104(1):7–11.

50. Giasson BI, Forman MS, Higuchi M, et al. Initiation and synergistic fibrillization of tau and alpha-synuclein. Science 2003;300(5619):636–640.
51. Frasier M, Walzer M, McCarthy L, et al. Tau phosphorylation increases in symptomatic mice overexpressing A30P alpha-synuclein. Exp Neurol 2005;192(2):274–287.
52. Rizzu P, Hinkle DA, Zhukareva V, et al. DJ-1 colocalizes with tau inclusions: a link between parkinsonism and dementia. Ann Neurol 2004;55(1):113–118.
53. Neumann M, Muller V, Gorner K, et al. Pathological properties of the Parkinson's disease-associated protein DJ-1 in alpha-synucleinopathies and tauopathies: relevance for multiple system atrophy and Pick's disease. Acta Neuropathol (Berl) 2004;107(6): 489–496.
54. Sahara N, Lewis J, DeTure M, et al. Assembly of tau in transgenic animals expressing P301L tau: alteration of phosphorylation and solubility. J Neurochem 2002;83(6): 1498–1508.
55. Olzmann JA, Brown K, Wilkinson KD, et al. Familial Parkinson's disease-associated L166P mutation disrupts DJ-1 protein folding and function. J Biol Chem 2004;279(9):8506–8515.
56. Wang JZ, Wu Q, Smith A, et al. Tau is phosphorylated by GSK-3 at several sites found in Alzheimer disease and its biological activity markedly inhibited only after it is prephosphorylated by A-kinase. FEBS Lett 1998;436(1):28–34.
57. Biernat J, Mandelkow EM, Schroter C, et al. The switch of tau protein to an Alzheimer-like state includes the phosphorylation of two serine-proline motifs upstream of the microtubule binding region. EMBO J 1992;11(4):1593–1597.
58. Dauer W, Przedborski S. Parkinson's disease: mechanisms and models. Neuron 2003;39(6):889–909.
59. Sherer TB, Betarbet R, Testa CM, et al. Mechanism of toxicity in rotenone models of Parkinson's disease. J Neurosci 2003;23(34):10756–10764.
60. Jensen PH, Hager H, Nielsen MS, et al. Alpha-synuclein binds to tau and stimulates the protein kinase A-catalyzed tau phosphorylation of serine residues 262 and 356. J Biol Chem 1999;274(36):25481–25489.
61. Forman MS, Schmidt ML, Kasturi S, et al. Tau and alpha-synuclein pathology in amygdala of parkinsonism-dementia complex patients of Guam. Am J Pathol 2002;160(5):1725–1731.
62. Ishizawa T, Mattila P, Davies P, et al. Colocalization of tau and alpha-synuclein epitopes in Lewy bodies. J Neuropathol Exp Neurol 2004;62(4):389–397.
63. Bandyopadhyay S, Cookson MR. Evolutionary and functional relationships within the DJ1 superfamily. BMC Evol Biol 2004;4(1):6.
64. Petrucelli L, Dickson D, Kehoe K, et al. CHIP and Hsp70 regulate tau ubiquitination, degradation and aggregation. Hum Mol Genet 2004;13(7):703–714.
65. Shimura H, Schwartz D, Gygi SP, Kosik KS. CHIP-Hsc70 complex ubiquitinates phosphorylated tau and enhances cell survival. J Biol Chem 2004;279(6):4869–4876.
66. Imai Y, Soda M, Hatakeyama S, et al. CHIP is associated with parkin, a gene responsible for familial Parkinson's disease, and enhances its ubiquitin ligase activity. Mol Cell 2002;10(1):55–67.

Chapter 12
Neuroinflammation in Early Stages of Alzheimer's Disease and Parkinson's Disease

Piet Eikelenboom, A. Crosswell, C. van Engen, M. Limper, J.J.M. Hoozemans, R. Veerhuis, W.A. van Gool, and J.M. Rozemuller

Introduction

Neurodegenerative disorders are characterized by depositions of abnormal folded proteins and a focal loss of neurons with reactive gliosis at a neuropathological level. One of the disputed mechanisms for the reactive gliosis and neuronal loss is an inflammatory response.

The neuroinflammatory response in neurodegenerative diseases has been studied extensively in Alzheimer's disease (AD) and prion disease [1]. Both disorders are characterized by extracellular amyloid deposits of abnormal folded proteins, Aβ in AD and PrP in prion disease. The amyloid deposits in AD are associated with inflammation-related proteins—including complement factors, acute-phase proteins, and proinflammatory cytokines—and clusters of activated microglia [1,2]. In AD, the absence of immunoglobulins and T-cell subsets within or in the vicinity of the amyloid deposits indicates that humoral or classic cellular immune-mediated responses are not involved in cerebral amyloid formation, which supports the view that the amyloid deposits are closely associated with a locally induced chronic inflammatory response. The idea that the conformationally changed peptides can induce an innate immunity response is supported by the finding that fibrillar Aβ, but not the monomeric Aβ, can activate the classic complement pathway in an antibody-independent fashion [3].

More recently the role of microglia has been become the focus of attention in the pathogenesis of neurodegenerative disorders characterized by intracellular accumulation of abnormal folded proteins, such as Parkinson's disease (PD). This disorder is characterized by intraneuronal proteinaceous accumulations, with conformationally changed α-synuclein as the main component.

The precise role of the neuroinflammatory response in the pathogenesis of neurodegenerative disorders is poorly understood; and, in particular, the

P. Eikelenboom
Department of Neurology, Academic Medical Center, University of Amsterdam, Amsterdam, The Netherlands and Department of Psychiatry, Vrije Universiteit Medical Center

A. Fisher et al. (eds.), *Advances in Alzheimer's and Parkinson's Disease*,
© Springer 2008

question of beneficial versus detrimental effects of microglial activation is highly topical. Animal studies have shown that microglial activation, in principle, aims at central nervous system (CNS) protection and neuronal repair [4]. Otherwise, failed microglial engagement due to excessive or sustained activation could significantly contribute to acute and chronic brain diseases. Dysregulation of microglial cytokine production could thereby promote harmful actions of the defense mechanisms, resulting in direct neurotoxicity as well as disturbed neural functions as they are sensitive in cytokine signaling [5]. Here we review the evidence for a relation between neuroinflammation and regenerative pathways in relatively early stages of AD and provide some new data on microglia activation in the early stages of PD pathology.

Aβ Deposition and the Neuroinflammatory Response in AD

The view that altered metabolism of β-amyloid precursor (βAPP) with progressive deposition of its Aβ fragment should be considered the key step in the molecular pathogenesis of AD is strongly supported by the fact that all three causal genes (*APP, PS-1, PS-2*) increase the propensity of Aβ to aggregate and to form insoluble fibrils. These findings have stimulated the controversial concept that AD may be a purely "amyloid-driven" process. However, AD most likely results from a complex sequence of steps involving multiple factors beyond the production of Aβ alone. The diffuse plaque, the overwhelming plaque type in the nondemented elderly, is characterized by low-grade aggregated Aβ deposits without glial or neuritic alteration. In contrast to the classic neuritic plaque, which consists of high fibrillar Aβ, the diffuse plaques show very weak or no immunolabeling for complement factors, and there is no clustering of activated microglia [6].

Next to the plaques in the neuroparenchyma, fibrillar Aβ deposits can be found in the walls of cerebral vessels in AD brains. The most common type of cerebral amyloid angiopathy is amyloid deposition in the walls of meningeal and medium-sized arteries or, more rarely, microcapillary amyloid angiopathy with fine radiating deposits of amyloid into the neuropil (so-called dysphoric angiopathy). Immunostaining shows strong staining for early and late complement in both large vessel and microcapillary amyloid angiopathy. Clustering of macophages/microglia is associated with microcapillary amyloid but not with amyloid deposits in the walls of larger vessels [7].

In transgenic mice that express human βAPP harboring the pathological familial autosomal dominant mutations, especially the fibrillar compact amyloid plaques, are closely associated with microglial activation and complement proteins [8,9]. The transgenic βAPP with the vasculotropic Dutch/Iowa mutant shows a robust accumulation of cerebral micovascular Aβ deposition associated with a prominent neuroinflammatory response [10]. These findings in several transgenic mouse models reflect the findings in human AD brains that a neuroinflammatory response is associated with the congophilic plaques in the neuroparenchyma and with microcapillary amyloid angiopathy.

Neuroinflammation and Regenerative Pathways in AD

Recent neuropathological and neuroradiological studies have shown that microglial activation is a relatively early event in the pathogenesis of AD and that it precedes the process of severe neuropil destruction. Clinicopathological studies performed in clinically well evaluated patients have indicated that in the early stages of AD amyloid/microglia complex formation is significantly increased in cerebral neocortex brain and precedes extensive tau-related neurofibrillary pathology [11,12]. These findings are in agreement with a positron emission tomography (PET) study using the peripheral benzodiazepine ligand PK11195 as a marker for activated microglia [13]. In this study it was found in AD patients that activation of microglia precedes brain atrophy. Similarly, in prion disease, which is also characterized by extracellular deposition of amyloid, the onset of microglial activation was found to coincide with the earliest changes in cerebral morphology. In scrapie-affected mice, microglial activation occurs many weeks before neuronal loss and subsequent clinical signs become apparent [14].

Early investigators noted that AD brains are not only undergoing degeneration but also show signs of a regenerative process [15,16]. Regulation of tissue degradation and remodeling involves a complex network that includes proteases and protease inhibitors, cytokines, integrins, and adhesion molecules. Dystrophic neurites associated with fibrillar Aβ deposits are decorated for the growth-associated protein GAP-43 and βAPP, and these neurites are also outlined by the adhesion molecules laminin, collagen IV, and various β1-integrins including the laminin receptor VLA-6 [17]. Laminin and collagen IV were shown to interact with APP in vitro [18]. In addition, low amounts of extracellular matrix components and Aβ were demonstrated to interact with neurite outgrowth in a dose-dependent manner [19]. In contrast to the plaque-associated dystrophic neurites, the tau-positive neuropil threads outside plaques in AD brains are not immunolabeled for the growth-associated and adhesion factors mentioned above. In the cerebral cortex of nondemented individuals, neurites associated with congophilic amyloid plaques are positive for growth-associated proteins but negative for tau staining [17]. These plaques are also characterized by activated complement factors and microglial clustering [20]. Taken together, these findings indicate that in the early stage of AD fibrillar Aβ deposition is associated with both an aberrant regenerative response of sprouting neurons and a microglia-mediated neuroinflammatory response. Recently, we have shown that during the initial stage of AD pathology a neuronal up-regulation of cyclooxygenase-2 (COX-2) and cell cycle proteins is found, and in the late stage of AD pathology, characterized by neurofibrillary changes, there is down-regulation in the expression of COX-2 and cell cycle proteins [21,22]. With gene expression microarrays, Blalock and colleagues found that in brains of incipient AD cases the main categories of up-regulated genes included genes that encode the regulation of cell proliferation and differentiation, genes for cell adhesion, genes for complement factors, and genes for prostaglandin synthesis pathways [23]. Although RNA microarray analysis is a relatively new and emerging technique, it confirms the immuno-histochemical findings of the early involvement of inflammatory and regenera-tive pathways in AD pathogenesis.

Inflammation and the Pathological Cascade in AD

The involvement of a broad variety of inflammatory proteins in the pathologi-cal cascade is not related to a single pathogenic event but to a number of subsequent steps [24]. These steps include the role of interleukin-1 (possibly together with other cytokines) in the regulation of APP synthesis; the role of Aβ-associated proteins (most of which are acute-phase proteins) in fibril for-mation and deposition and removal of the Aβ peptide; and the role of activated microglia in the production of potentially toxic products (e.g., reactive oxygen species, proinflammatory cytokines, excitotoxins, proteases) that could damage the neighboring neurites. The present view is that the neuroinflammatory response is a double-edge sword that has both beneficial and deleterious effects on the progression of the disease process [25]. This is not surprising because both elimination f pathogenic stimuli (e.g., the removal of fibrillar Aβ deposits) and tissue repair are essential characteristics of the inflammatory process.

In this context, it is interesting that there are two phases in AD neurodegen-eration. The first phase involves increased neuronal COX-2 and cell cycle proteins expression as a response to the induction of neuronal plasticity. The second phase is one in which neurons fail to cope with the increasing presence of misfolded proteins and undergo neurofibrillary degeneration. Aβ deposition, inflammation, and neuroregenerative neurites are also found in the transgenic mouse model for AD, whereas "later" neurodegenerative characteristics (e.g., neurofibrillary degeneration) are not seen in these models. The precise relation at a mechanistic level between the neuroregenerative and neurodegen-erative events in AD pathology remains elusive.

The notion of a close relation between the inflammatory response and the aberrant neuroregenerative response with dystrophic neurites around the fibril-lar Aβ is supported by immunization studies with the Aβ peptide or by passive anti-Aβ immunotherapy. One study in APP transgenic mice showed that after treatment with anti-Aβ the plaques disappear, and after 3 days a significant reduction is seen in the number and size of amyloid-associated dystrophic neuritis [26]. Human pathology after immunization with Aβ in an AD patient showed the presence of extensive areas of the neocortex with very few Aβ plaques or plaque-associated dystrophic neurites, which contain densities of tangles, neuropil threads, and amyloid angiopathy similar to those seen in AD patients who were not treated [27]. However, passive immunization studies with Aβ antibodies in APP transgenic mice also showed that the clearance of parenchymal Aβ deposits is accompanied by an increase in vascular amyloid and microhemorrhages associated with the amyloid-laden vessels. At the moment, it is unknown if these microhemorrhages are caused by an inflammatory response induced by an increase of the vascular fibrillar Aβ load.

Inflammation and the Pathological Cascade in PD

McGeer and coworkers reported initially that the loss of dopaminergic neurons in the substantia nigra of PD patients is associated with large numbers of HLA-DR-reactive microglia [28]. The extraneuronal deposition of neuromelanin released from degenerating dopaminergic neurons in the substantia nigra is thought to be a potential trigger for microglial activation, with subsequent release of neurotoxic mediators that contribute to further progression of neurodegeneration with loss of dopaminergic neurons [29]. Microglial activation has also been reported to be associated with loss of dopaminergic neurons in animal models of PD induced by 1-methyl-4-phenyl-1,2,3,6-tetrahydropyridine (MTPT), rotenone, annonacine, and lipopolysaccharide [30]. These models seem to suggest that PD could be the consequence of interactions between environmental factors, microglial activation, and the unique properties of the dopaminergic neurons in the substantia nigra [31]. Because the substantia nigra belongs to brain regions with the highest density of microglia [32], dopaminergic neurons in the substantia nigra may encounter an excessively high level of oxidative stress during brain inflammation [31].

Recent studies show that the substantia nigra is not the induction site in the brain of the neurodegenerative process underlying PD. Instead, autopsy studies with incident Lewy body pathology indicate that the pathology associated with PD starts first with the formation of immunoreactive Lewy body neurites and Lewy bodies in the noncatecholaminergic neurons of the dorsal glossopharyngeus-vagus complex [33,34]. To study the involvement of activated microglia in early stages of PD pathology, we investigated the presence of activated microglia at several stages of PD pathology. We quantified the amount of CD-68 immuno-positive activated microglia in the medulla oblongata (dorsal glossopharyngeus-vagus complex) and mesencephalon (substantia nigra) in postmortem brain specimens from 24 PD patients and 20 controls without α-synuclein brain pathology. Microglia activation in the substantia nigra could be found in patients with less than 2 years of clinical PD symptoms and in brain regions with nondopaminergic neurons (Fig. 1) (Rozemuller et al., manuscript in preparation). The activated microglia in the affected brain

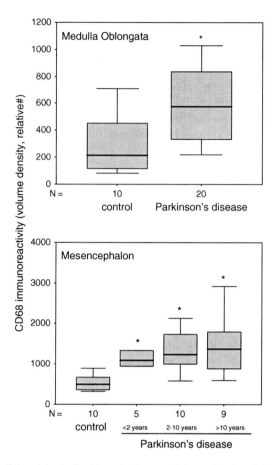

Fig. 1 Increased microglial activity in Parkinson's disease. Volume density of CD68 immu-noreactivity was quantified in the medulla oblongata and mesencephalon of Parkinson's disease and control cases. Results are expressed as box plots: The interquartile range contains 50% of the values, the whiskers extend from the box to the highest and lowest values, and the line across the box indicates the median. *p > 0.05 versus control cases (Student's *t*-test)

regions were widely dispersed and did not show clustering, as is observed around amyloid plaques in AD brains.

Conclusions

The present findings show that micoglial activation is an early event in the pathogenesis of both AD and PD. In AD, clusters of activated microglial cells are associated with extracelluar deposits of fibrillar Aβ. In PD, activated microglial cells are found in brain areas with intraneuronally located α-synuclein deposits. In contrast to the microglia in AD, the activated microglia are not found as clusters, are distributed throughout the affected brain regions of PD patients, and no clustering of activated microglia cells is seen. The presence of clustering of activated microglial cells around extracellular fibrillar deposits in AD and the absence of such clustering in PD brains, characterized by intraneuronal fibrillar deposits, indicate an essential difference between the neuroinflammatory responses in the two diseases. The amyloid plaques as miliary foci with complement activation, focal accumulation of activated microglia,, and regenerative tissue response meet the classic histopathological criteria for an inflammatory response [35]. The challenge for this response is the extracellular deposition of fibrillar Aβ, which could be considered a "foreign" substance. The recent findings that removal of this substance by anti-Aβ antibodies leads to clearance of the Aβ deposits with subsequently reduction of the dystrophic neurites support this notion.

The distribution of activated microglia without clustering in brain areas with α-synuclein-positive neurons suggests that activated microglia are involved in the pathological processes of brains other than those involved with AD (e.g., PD). Experimental animal studies, such as the facial nerve axotomy paradigm, show that neuronal damage triggers activation of the surrounding microglia as part of a neuronal repair program [4]. The presence of activated microglia at the sites of initial PD pathological changes could reflect a mechanism for α-synuclein-positive neurons, that is similar to that which is observed in injured neurons that elicit a microglia-mediated neuronal regeneration program. Whether prolonged microglia activation in these areas or in dopaminergic neuron-rich brain regions (e.g., the substantia nigra) could contribute to further neurodegeneration remains elusive.

Acknowledgments Recent work discussed in this review was supported financially by the Internationale Stichting Alzheimer Onderzoek (ISAO): grant 04503 to J.J.M.H. and grant 03509 to R.V.

References

1. Eikelenboom P, Bate C, van Gool WA, et al. Neuroinflammation in Alzheimer's disease and prion disease. Glia 2002;40:232–239.
2. Veerhuis R, Boshuizen RS, Familian A. Amyloid associated proteins in Alzheimer's and prion disease. Curr Drug Targets CNS Neurol Disord 2005;4:235-248.
3. Rogers J, Cooper NR, Webster S, et al. Complement activation by β-amyloid in Alzheimer's disease. Proc Natl Acad Sci U S A 1992;89:10061–10020.
4. Streit WJ. Microglia as neuroprotective, immunocompetent cells of the CNS. Glia 2002;40:133–139.
5. Hanisch U-K. Microglia as a source and target of cytokines. Glia 2002;40:140–155.
6. Eikelenboom P, Veerhuis R. The role of complement activation and microglia in the pathogenesis of Alzheimer's disease. Neurobiol Aging 1996;17:673–680.
7. Rozemuller JM, van Gool WA, Eikelenboom P. The neuroinflammatory response in plaque and amyloid angiopathy. Curr Drug Targets CNS Neurol Disord 2005;4:223–233.

 8. Benzing WC, Wujek JR, Ward EK, et al. Evidence for glial-mediated inflammation in aged APP(SW) transgenic mice. Neurobiol Aging 1999;20:581–589.
 9. Matsuaka Y, Picciano M, Malester B, et al. Inflammatory responses to amyloidosis in a transgenic mouse model for Alzheimer's disease. Am J Pathol 2001;158:1345–1354.
10. Miao J, Xu F, Davis J, et al. Cerebral microvascular amyloid-β protein deposition induces vascular degeneration and neuroinflammation in transgenic mice expressing human vasculotropic mutant amyloid-β precursor protein. Am J Pathol 2005;167:505–515.
11. Arends YM, Duyckaerts C, Rozemuller JM, et al. Microglia, amyloid and dementia in Alzheimer's disease: a correlative study. Neurobiol Aging 2000;21:39–47.
12. Vehmas AK, Kawas CH, Stewart WF, Troncoso JC. Immunoreactive cells and cognitive decline in Alzheimer's disease. Neurobiol Aging 2003;24:321–331.
13. Cagnin DJ, Brooks AM, Kennedy RN, et al. In-vivo measurement of microglia in dementia. Lancet 2001;358:461–467.
14. Williams AE, van Dam AM, Ritchie D, et al. Immunohistochemical appearance of cytokines, prostaglandin E2, and lipocortin-1 in the CNS during the incubation period of murine scrapie correlates with progressive PrP accumulations. Brain Res 1997;754:171–180.
15. Fischer O. Miliare Nekrosen mit drusigen Wucherungen der Neurofibrillen, eine regelmässige Veränderung der Hirnrinde bei seniler Demenz. Monatsch Psychiatr Neurol 1907;22:361–372.
16. Bouman L. Senile plaques. Brain 1934;57:128–142.
17. Zhan SS, Kamphorst W, VanNostrand WE, Eikelenboom P. Distribution of neuronal growth-promoting factors and cytoskeletal proteins in altered neurites and non-demented elderly. Acta Neuropathol (Berl) 1995;89:365–372.
18. Breen KC. APP-collagen interaction is mediated by a heparin bridge mechanism. Mol Chem Neuropathol 1992;16:109–121.
19. Koo EH, Park L, Selkoe DJ. Amyloid-β protein as substrate interacts with extracellular matrix to promote neurite outgrowth. Proc Natl Acad Sci U S A 1994;90:1564–1568.
20. Veerhuis R, van Breemen MJ, Hoozemans JJM, et al. Amyloid β-associated proteins C1q and SAP enhance the Aβ1-42 peptide-induced cytokine secretion by adult human microglia in vitro. Acta Neuropathol (Berl) 2003;105:135–144.
21. Hoozemans JJM, Bruckner MK, Rozemuller AJM, et al. Cyclin D1 and cyclin E are colocalized with cyclo-oxygenase 2 (COX-2) in pyramidal neurons in Alzheimer disease temporal cortex. J Neuropathol Exp Neurol 2002;61:678–688.
22. Hoozemans JJM, Veerhuis R, Arendt T, Eikelenboom P. Neuronal COX-2 expression and phosphorylation of pRb precede p38 MAPK activation and neurofibrillar changes in AD temporal cortex. Neurobiol Dis 2004;15:492–499.
23. Blalock EM, Geddes JW, Chen KC, et al. Inciepient Alzheimer's disease: microarray correlation analyses reveal major transcriptional and tumor suppressor responses. Proc Natl Acad Sci U S A 2004;101:2173–2178.
24. Eikelenboom P, Zhan SS, van Gool WA, Allsop D. Inflammatory mechanisms in Alzheimer's disease. Trends Pharmacol Sci 1994;15:447–450.
25. Wyss-Coray T, Mucke L. Inflammation in neurodegenerative disease: a double-edged sword. Neuron 2002;35:419–432.
26. Brendza RP, Bacskai BJ, Cirrito JR, et al. Anti-Aβ antibody treatment promotes the rapid recovery of amyloid-associated neuritic dystrophy in PDAPP transgenic mice. J Clin Invest 2005;115:428–433.
27. Nicoll JAR, Wilkinson D, Holmes C, et al. Neuropathology of human Alzheimer's disease after immunization with amyloid-β peptide; a case report. Nat Med 2003;9:448–452.
28. McGeer PL, Itagaki S, Boyes BE, McGeer EG. Reactive microglia are positive for HLA-DR in the substantia nigra of Parkinson's and Alzheimer's disease brains. Neurology 1988;38:1285–1291.

29. Wilms H, Rosenstiel P, Sievers J, et al. Activation of microglia by human neuromelanin is NF-κB dependent and involves p38 mitogen-activated protein kinase: implications for Parkinson's disease. FASEB J 2003;17:500–502.

30. Hirsch EC, Hunot S, Hartmann A. Neuroinflammatory processes in Parkinson's disease. Parkinsonism Relat Dis 2005;11:S1–S9.

31. Gao HM, Jiang J, Wilson B, et al. Microglial activation-mediated delayed and progressive degeneration of rat nigral dopaminergic neurons: relevance to Parkinson's disease. J Neurochem 2002;81:1285–1297.

32. Lawson L, Perry VH, Dri P, Gordon G. Heterogeneity in the distribution and morphology of microglia in the mouse brain. Neuroscience 1990;39:151–170.

33. Del Tredici K, Rüb U, de Vos RAI, et al. Where does Parkinson's disease begin in the brain? J Neuropathol Exp Neurol 2002;61:413–426.

34. Braak H, Del Tredici K, Rüb U, et al. Staging of brain pathology related to sporadic Parkinson's disease. Neurobiol Aging 2003;24:197–211.

35. Metchinikoff E. Leçons sur la Pathologie Comparée de l'Inflammation. Masson, Paris, 1892.

Chapter 13
Alzheimer's Disease, Parkinson's Disease, and Frontotemporal Dementias: Different Manifestations of Protein Misfolding

John Q. Trojanowski, Mark S. Forman, and Virginia M-Y. Lee

Introduction[1]

Alzheimer's disease (AD) and many other aging-related neurodegenerative disorders appear to share common disease mechanisms as they are associated with the pathological aggregation of proteins that misfold and accumulate as fibrillar amyloid deposits in selectively vulnerable regions of the central nervous system (CNS) (for recent reviews, see refs. 1–4). For example, neurofibrillary tangles (NFTs) and senile plaques (SPs), the two diagnostic hallmark lesions of AD, are formed by intraneuronal accumulations of abnormal tau filaments and extracellular deposits of Aβ fibrils, respectively [1–4]. A growing body of evidence supports the view that these lesions compromise the function and viability of neurons, including the discovery of mutations in the genes encoding tau, A βprecursor proteins (APPs), and presenilins, which are pathogenic for familial neurodegenerative disorders. Nonetheless, the exact mechanisms whereby brain degeneration results from NFTs and SPs remain incompletely understood, although abundant NFT-like inclusions formed by filamentous aggregates of tau also are the dominant neuropathological findings in a subset of sporadic and familial frontotemporal dementias (FTDs) collectively referred to as tauopathies [1–3].

Furthermore, although Lewy bodies (LBs) are the signature lesions of Parkinson's disease (PD), it was uncertain if LBs contributed to neurodegeneration in PD until pathogenic mutations in the α-synuclein gene (*SNCA*) were discovered

---

J.Q. Trojanowski
The Center for Neurodegenerative Disease Research, Department of Pathology
and Laboratory Medicine, and Institute on Aging, The University of Pennsylvania School
of Medicine, Philadelphia, PA 19104, USA

[1] Due to the explosive growth in new information on the genetic and molecular pathology of the neurodegenerative diseases discussed here, which have been the subject of numerous reviews from many diverse perspectives, only a limited number of recent primary publications are listed and additional earlier literature citations are available in the recent reviews cited here.

A. Fisher et al. (eds.), *Advances in Alzheimer's and Parkinson's Disease*,
© Springer 2008

and it was demonstrated that abnormal α-synuclein filaments are the principal components of LBs (for recent reviews, see refs. 5–9). These and other recent findings clearly implicate LBs in the pathogenesis of PD, although a growing number of other neurodegenerative disorders have been shown to be characterized by filamentous α-synuclein lesions, collectively called synucleinopathies [5–9].

Because the misfolding and fibrillization of normal brain proteins result in their accumulation as disease-specific fibrillar inclusions in many neurodegenerative disorders, pathological amyloidogenic processes appear to be a common mechanism underlying the onset and progression of AD, PD, FTDs, and related tauopathies and synucleinopthies. Thus, the progressive conversion of normal soluble tau, Aβ, and α-synuclein into insoluble oligomers, protofibrils, and fully formed amyloid fibrils is increasingly the focus of drug discovery efforts to identify disease-modifying therapies for these and other aging-related neurodegenerative disorders characterized by brain amyloidosis [4].

Alzheimer's Disease and Related Tauopathies

As reviewed in more detail elsewhere [4], the Aβ-centric focus of most current AD drug discovery efforts reflects remarkable progress in understanding the fibrillization and aggregation of Aβ to form SPs as well as insights into how these events contribute to the pathogenesis of AD. The most direct evidence implicating APPs and Aβ in the pathogenesis of AD has come from studies of genetic mutations that cause autosomal dominantly inherited forms of familial AD (FAD) as most but not all of these mutations lead to increased production and accumulation of specific Aβ species (Aβ42). This occurs either through effects on the APP itself or through effects on presenilin 1 or 2, which form part of γ-secretase, one of the proteolytic complexes that cleaves APP to generate Aβ [2–4]. These and many other observations support the Aβ amyloid cascade hypothesis of AD that was proposed more than a decade ago and predicts that increased production, aggregation, and accumulation of Aβ initiates a cascade of events leading to neurotoxicity and eventually to the clinical manifestation of AD [3,4]. Accordingly, most drugs currently in development or in clinical trials for AD have been designed to target one or more mechanisms leading to Aβ amyloidosis, such as by inhibiting or reducing the production of amyloidogenic Aβ peptides or by promoting the clearance of SPs and Aβ fibrils [3,4]. Thus, inhibitors of γ- and β-secretases, passive and active Aβ vaccines, metal-binding drugs (e.g., clioquinol), statins, nonsteroidal antiinflammatory drugs (NSAIDs) (e.g., flurbiprofen), and glycosaminoglycan mimetics are among a few of the compounds targeting Aβ that are in various stages of preclinical or clinical testing [4].

Alzheimer's disease also has been proposed to result from the accumulation of filamentous tau inclusions. The tau hypothesis of AD neurodegeneration emerged from studies of the pathobiology of NFTs as well as from research into the normal biology and functions of tau, which are far better understood than those of other neurodegenerative disease proteins [2–4,10]. For example, it was known that normal tau binds to and stabilizes microtubules (MTs), which are essential for axonal transport in neurons [2–4]. The phosphorylation of tau was known to regulate negatively the binding of all six normal brain tau isoforms to MTs. Furthermore, because the subunit proteins of the paired helical filaments (PHFs) that form AD NFTs are abnormally phosphorylated forms of tau (known as PHFtau), it was not surprising that PHFtau proteins were found to be incapable of binding to and stabilizing MTs; this loss of function defect could be reversed by enzymatic dephosphorylation of PHFtau [4,10]. Based on the wealth of information on the normal biology of tau and MTs as well as on the pathobiology of NFTs, the tau hypothesis of AD neurodegeneration predicts that by converting normal tau to functionally impaired PHFtau MTs become unstable and depolymerize. This in turn disrupts MT-based axonal transport and compromises the viability of affected neurons [4,10]. Because these events could culminate in neuronal degeneration, leading to the onset/ progression of AD, abrogating this cascade of events has become a target for discovering novel disease-modifying therapies to treat AD and related tauopathies [4].

Although controversial initially, the tau hypothesis of AD neurodegeneration garnered support from a series of discoveries showing that mutations in the gene encoding tau were pathogenic for hereditary FTD with parkinsonism linked to chromosome 17 (FTDP-17). This is a heterogeneous group of disorders characterized by prominent tau pathologies in the absence of SPs or other disease-specific amyloid lesions [11–14]. Similar to the discoveries of genetic mutations pathogenic for FAD, elucidation of the genetic basis of FTDP-17 provided compelling evidence that tau abnormalities were sufficient to cause neurodegenerative disease (for a recent review, see ref. 15). Indeed, this notion was supported further by follow-up studies shortly thereafter [16], which showed that a number of tau gene mutations pathogenic for FTDP-17 resulted in losses of tau function (i.e., loss of the ability to bind to and promote the assembly of MTs) and/or gains of potentially toxic properties by mutant tau isoforms (i.e., increased amyloidogenic propensity). An additional catalyst for dispelling doubt about the importance of tau pathologies as mediators of neurodegeneration in AD and other neurodegenerative tauopathies has been the successful development of worm, fly, and mouse models of tauopathies that show persuasive verisimilitude to their authentic human counterparts (for a recent review, see ref. 17). For these and other reasons, sporadic and familial tauopathies and the role of tau abnormalities in mechanisms of brain degeneration have become an increasingly intense focus of basic as well as clinical research [15,18]. Indeed, rapid progress in this research arena has prompted renewed efforts to harmonize or rationalize the current nosology of FTDs and to link the nosology of FTDs to their genetic and neuropathological underpinnings; this has led to renewed efforts to develop a consensus on the clinical, genetic, and neuropathological definitions of various FTDs [19]. However, it should be emphasized that the concepts underlying these definitions are still evolving as new insights from research on AD, FTDs, and neurodegenerative tauopathies continue to emerge.

Because of rapid progress in research on the role of tau in mechanisms of neurodegeneration, tau abnormalities in AD and related tauopathies appear increasingly to be tractable or "drug-able" targets for drug discovery. For example, recent studies suggest that MT-stabilizing drugs (e.g., paclitaxel) may have potential therapeutic benefit in these disorders by offsetting the loss of tau function due to its sequestration in NFTs and/or its excessive phosphorylation [20], and other studies suggest that MT-stabilizing drugs also ameliorate the neurotoxic effects of Aβ in AD and that they can be modified to enhance penetration of the blood–brain barrier [21,22]. Other strategies to develop novel therapies that target tau abnormalities are being investigated using high throughput screening (HTS) to identify drugs in large compound libraries that block the fibrillization and aggregation of tau or reduce tau protein levels [23–25]. Preliminary data from HTS efforts that target inhibition of tau phosphorylation also appear promising, and proof of concept studies using LiCl to ameliorate tau pathology by inhibiting glycogen synthase kinase-3 (GSK-3) in a mouse model of a neurodegenerative tauopathy suggest that this is a fruitful avenue for drug discovery [26]. Notably, like MT-stabilizing drugs, GSK-3 inhibitors may be able to block pathways that lead to the formation of both Aβ and tau amyloid lesions in AD [27].

Parkinson's Disease, Related Synucleinopathies, and Other Parkinsonian Movement Disorders

The dramatic advances summarized above from research on AD and other tauopathies have been paralleled by similar progress in understanding PD (for recent reviews see refs. 2, 3, 5–9, 28). For example, the identification of α-synuclein gene (SNCA) mutations in two kindreds with autosomal dominant PD [29] and the rapid identification of fibrillar α-synuclein as the principal component of LBs and Lewy neurites (LNs) shortly thereafter [30] provided strong support for the concept that LBs defined classic PD. Moreover, this notion was supported by the subsequent identification of additional SNCA mutations and LB pathology in kindreds with other clinical manifestations of disease ranging from typical PD to parkinsonism with concurrent dementia known as PD dementia (PDD) and dementia with LBs (DLB) or LB dementia [31–33]. Thus, although LBs were recognized as distinct intraneuronal inclusions formed by densely packed filaments 10 to 20 nm in diameter, whether their significance to PD extended beyond their role as diagnostic markers was controversial until the discoveries summarized above. Accordingly, these and other advances in PD research clearly implicate the deposition of pathological α-synuclein fibrils in LBs and LNs not only as diagnostic signatures of PD and related synucleinpathaies but also as integral to the onset and progression of these disorders. Moreover, these advances emphasized the oft-neglected point that although the nigrostriatal system degenerates in PD multiple additional brain regions are always affected in PD and other parkinsonian disorders, so the clinical manifestations of these conditions commonly extend beyond those attributable to the pathology affecting the nigrostriatal system [8]. In this regard, there are remarkable parallels between sporadic and familial neurodegenerative parkinsonian disorders caused by accumulations of fibrillar tau amyloid (e.g., progressive supranuclear palsy, FTDP-17) and those caused by similar accumulations of fibrillary α-synuclein inclusions (e.g., sporadic and hereditary forms of PD, PDD, and DLB).

Nonetheless, the mechanism whereby α-synuclein leads to neurodegeneration is unclear, and understanding these mechanisms is hampered by gaps in our understanding of the normal functions of this abundant synaptic protein. However, because LB pathology is composed of fibrillar α-synuclein and several autosomal dominant mutations in *SNCA* lead to enhanced rates of protein fibrillization, it is plausible that these conformational changes in the structure of α-synuclein may render it neurotoxic or that small, prefibrillar oligomers of α-synuclein are the toxic species leading to neuron dysfunction and degeneration [5,8]. Thus, similar to Aβ-centric and tau-focused drug discovery efforts, there is increasing interest in the design of novel therapeutic interventions for PD and related synucleinopathies that target the misfolding, fibrillization and aggregation of α-synuclein [3,4,18,34–36].

Numerous additional genetic loci have been implicated in familial PD; and because LBs are not the hallmarks of all forms of neurodegenerative parkinsonism, other pathological mechanisms must result in a PD-like neurodegenerative disorder [8]. For example, loss of function mutations in the *PARK2 (parkin)* gene were reported in 1998 in autosomal recessive juvenile parkinsonism (ARJP), which shares features of LB PD [37], including neuron loss and gliosis in the substantia nigra but no LBs or LNs [5,8]. *Parkin* encodes an E3 protein ubiquitin ligase involved in the proteolysis of misfolded proteins by the proteasome, and the loss of parkin function in ARJP may lead to toxic accumulations of one or more of its substrates [5,8]. Furthermore, pathogenic autosomal recessive mutations were identified in *PARK7 (DJ1)* and *PINK1* genes within the PARK7 and PARK6 loci, respectively [38,39]. Notably, patients with mutations in *DJ1* or *PINK1* appear clinically similar to those with ARJP, but the neuropathology in these patients remains to be defined, as is the case for the normal functions performed by DJ1 and Pink1 [5,8]. Emerging data suggest that loss of DJ-1 sensitizes neurons to oxidative stress and/or proteasomal inhibition, which may affect mitochondrial function; and PINK1, a serine/threonine protein kinase containing an amino-terminal mitochondrial localization signal, may protect cells against cell stress. However, further studies are needed to determine if mutations in both *DJ-1* and *PINK1* cause a PD-like disorder by impairing mitochondrial function.

Adding further to the genotypic and phenotypic complexity of PD and PD-like movement disorders, autosomal dominant mutations in the *LRRK2* gene at the PARK8 locus have been described in patients with a familial disorder characterized by L-dopa-responsive parkinsonism in association with dementia, amyotrophy, or both together with nigrostriatal degeneration accompanied by LBs, NFTs, or ubiquitin-positive but tau- and α-synuclein-negative filamentous lesions [40,41]. Although the function of LRRK2 is unknown, it contains a number of well-defined domains including Ras/GTPase, tyrosine kinase, WD40, and leucine-rich repeat domains; the most common mutation identified to date, [G]2019[S], is located within a critical region of the kinase domain [5,8]. Accordingly, it is critical to elucidate the basis for the clinical and neuropathological phenotypic variability associated with *LRRK2* mutations, as it could help clarify why sporadic PD and AD co-occur more commonly than predicted by chance alone and why familial AD mutations in APP and both presenilin genes frequently result in a triple brain amyloidosis characterized by SP, NFTs, and LBs [42].

Finally, although it was well known that heterozygous mutations in the glucocerebrosidase gene (*GBA*) cause autosomal recessive Gaucher's disease, recent studies identified heterozygous mutations in the *GBA* gene that increase the risk for LB PD. They also showed that patients who present with Gaucher's disease often develop parkinsonism, which occasionally is accompanied by the features of typical LB PD [43,44]. Because α-synuclein binds lipid membranes and lipids may modulate α-synuclein toxicity, it is plausible that mutations in *GBA* alter the membrane lipid composition in ways that predispose α-synuclein to fibrillize and acquire neurotoxic properties [45].

Degeneration of the substantia nigra is common to all forms of parkinsonism, but LBs are present only in a subset of parkinsonian disorders; hence, parkinsonism is a clinical entity reflective of nigrostriatal degeneration that appears to result from diverse underlying mechanisms and pathologies [8]. Thus, there is need for a unifying nosology of these parkinsonian disorders that integrates current concepts of the genetic and neuropathological basis of classic PD and other forms of neurodegenerative parkinsonism.

It is likely that a new nosology will continue to evolve in the future as new insights about mechanisms of parkinsonian disorders continue to emerge. However, a more contemporary nomenclature must be developed now, acknowledging that a variety of disease mechanisms and attendant pathologies associated with nigrostriatal degeneration can underlie this group of heterogeneous movement disorders. Indeed, although LB pathology is by far the most common hallmark brain lesion linked to nigrostriatal degeneration (as exemplified by LB PD), other pathologies (including α-synuclein-positive glial cytoplasmic inclusions and NFTs) also underlie a clinical phenotype similar to that of classic PD. Accordingly, a rational nosology for parkinsonian disorders tightly linked to their molecular and genetic underpinnings is critical to facilitate mechanistically based drug discovery efforts, the informed design of animal models of each form of parkinsonism, and efforts to determine how genetic and environmental factors differentially contribute to these disorders. For example, efforts are underway to exploit insights into the role of LBs and LNS in neurodegeneration so as to pursue drug targets related to the fibrillization of α-synuclein that are relevant to LB, PD, and related synucleinopathies; however, other strategies are needed for PD-like disorders in which there are no LBs, LNs, or similar filamentous α-synuclein inclusions [34–36]. To accomplish this and enhance prospects for developing more effective therapies to treat PD and related neurodegenerative movement disorders in the near future, we recently proposed a revised nosology of "Parkinson diseases" that is grounded in the dramatic advances in understanding these disorders summarized above [8].

Conclusions

The recent studies summarized here provide novel insights into common mechanisms underlying the formation of pathological inclusions in AD, PD, FTDs, and related synucleinopathies as well as tauopathies and other forms of dementia and movement disorders. Importantly, converging data highlight the shared mechanisms underlying many of these disorders that are linked to the misfolding, fibrillization, and aggregation of disease proteins. These new insights also provide a strong rationale for pursuing efforts to identify therapeutic agents that could directly or indirectly inhibit the formation of amyloid and progressive brain amyloidosis, regardless of the amyloidogenic disease protein, thereby offering hope of developing novel drugs that show efficacy for intervening in multiple neurodegenerative brain amyloidoses, be they single, double, or triple brain amyloidoses.

Acknowledgments We thank our colleagues for their contributions to the work summarized here, which has been supported by grants from the NIH (K08 AG20073, P01 AG09215, P30 AG10124, P01 AG11542, P01 AG14382, P01 AG14449, P01 AG17586, PO1 AG-19724, and PO1 NS-044233), the Marian S. Ware Alzheimer Program, the Benaroya family, and the Picower Foundation. V.M.Y.L. is the John H. Ware 3rd Professor for Alzheimer's Disease Research and J.Q.T. is the William Maul Measy-Truman G. Schnabel Jr., MD Professor of Geriatric Medicine and Gerontology.

References

1. Dobson CM. Protein folding and misfolding. Nature 2003;426:884–890.
2. Forman MS, Trojanowski JQ, Lee VM-Y. Neurodegenerative diseases: a decade of discoveries paves the way for therapeutic breakthroughs. Nat Med 2004;10:1055–1063.
3. Selkoe DJ. Cell biology of protein misfolding: the examples of Alzheimer's and Parkinson's diseases. Nat Cell Biol 2004;6:1054–1061.
4. Skovronsky DM., Lee VM-Y, Trojanowski, JQ. Neurodegenerative diseases: new concepts of pathogenesis and their therapeutic implications. Annu Rev Pathol Mech Dis 2005;1:151–170.
5. Cookson MR. The biochemistry of Parkinson's disease. Annu Rev Biochem 2005;74:29–52.

6. Dawson TM, Dawson VL. Molecular pathways of neurodegeneration in Parkinson's disease. Science 2003;302:819–822.

7. Feany MB. New genetic insights into Parkinson's disease. N Engl J Med 2004;351:1937–1940.

8. Forman MS, Lee VM-Y, Trojanowski JQ. Nosology of Parkinson's disease: looking for the way out of a quackmire. Neuron 2005;47:479–482.

9. Greenamyre JT, Hastings TG. Parkinson's: divergent causes, convergent mechanisms. Science 2004;304:1120–1122.

10. Lee VM-Y, Daughenbaugh R, Trojanowski JQ. Microtubule stabilizing drugs for the treatment of Alzheimer's disease. Neurobiol Aging 1994;15:S87–S89.

11. Clark LN, Poorkaj P, Wszolek Z, et al. Pathogenic implications of mutations in the tau gene in pallido-ponto-nigral degeneration and related neurogenerative disorders linked to chromosome 17. Proc Natl Acad Sci U S A 1998;95:13103–13107.

12. Hutton M, Lendon CL, Rizzu P, et al. Association of missense and 5'-splice-site-mutations in tau with the inherited dementia FTDP-17. Nature 1998;393:702–705.

13. Poorkaj P, Bird TE, Wijsman E, et al. Tau is a candidate gene for chromosome 17 frontotemporal dementia. Ann Neurol 1998;43:815–825.

14. Spillantini MG, Murrell TR, Goedert M, et al. Mutation in the tau gene in familial multiple system tauopathy with presenile dementia. Proc Natl Acad Sci USA 1998;95:7737–7741.

15. Forman MS, Lee VM-Y, Trojanowski JQ. Hereditary tauopathies and idiopathic frontotemporal dementias. In: Esiri M, Lee VM-Y, Trojanowski JQ (eds) The Neuropathology of Dementia, 2nd ed. Cambridge University Press, Cambridge, UK, 2004, pp 257–288.

16. Hong M, Zhukareva V, Vogelsberg-Ragaglia V, et al. Mutation-specific functional impairments in distinct tau isoforms of hereditary FTDP-17. Science 1998;282:1914–1917.

17. Lee VM-Y, Kenyon TK, Trojanowski JQ. Transgenic animal models of tauopathies. Biochim Biophys Acta 2005;1739:251–259.

18. Fillit HM, Refolo LM. Advancing drug discovery for Alzheimer's disease. Curr Alzheimer Res 2005;2:105–109.

19. McKhann GM, Albert MS, Grossman M, et al. Clinical and pathological diagnosis of frontotemporal dementia: report of the Work Group on Frontotemporal Dementia and Pick's Disease. Arch Neurol 2001;58:1803–1809.

20. Zhang B, Maiti A, Shively S, et al. Microtubule binding drugs offset tau sequestration by stabilizing microtubules and reversing fast axonal transport deficits in a murine neurodegenerative tauopathy model. Proc Natl Acad Sci U S A 2005;102:227–231.

21. Michaelis ML, Ansar S, Chen Y, et al. Beta-amyloid-induced neurodegeneration and protection by structurally diverse microtubule-stabilizing agents. J Pharmacol Exp Ther 2005;312:659–668.

22. Rice A, Liu Y, Michaelis ML, et al. Chemical modification of paclitaxel (Taxol) reduces P-glycoprotein interactions and increases permeation across the blood-brain barrier in vitro and in situ. J Med Chem 2005;48:832–838.

23. Dickey CA, Eriksen J, Kamal A, et al. Development of a high throughput drug screening assay for the detection of changes in tau levels: proof of concept with HSP90 inhibitors. Curr Alzheimer Res 2005;2:231–239.

24. Pickhardt M, Gazova Z, von Bergen M, et al. Anthraquinones inhibit tau aggregation and dissolve Alzheimer's paired helical filaments in vitro and in cells. J Biol Chem 2005;280:3628–3635.

25. Pickhardt M, von Bergen M, Gazova Z, et al. Screening for inhibitors of tau polymerization. Curr Alzheimer Res 2005;2:219–226.

26. Noble W, Planel E, Zehr C, et al. Inhibition of glycogen synthase kinase-3 by lithium correlates with reduced tauopathy and degeneration in vivo. Proc Natl Acad Sci USA 2005;102:6990–6995.

27. Phiel CJ, Wilson CA, Lee VM-Y, Klein PS. GSK-3 alpha regulates production of Alzheimer's disease amyloid-beta peptides. Nature 2003;423:435–439.
28. Giasson BI, Lee VM-Y, Trojanowski JQ. Parkinson's disease, dementia with Lewy bodies, multiple system atrophy and the spectrum of diseases with alpha-synuclein inclusions. In: Esiri M, Lee VM-Y, Trojanowski JQ (eds) The Neuropathology of Dementia, 2nd ed. Cambridge University Press, Cambridge, UK, 2004, pp 353–375.
29. Polymeropoulos MH, Lavedan C, Leroy E, et al. Mutation in the alpha-synuclein gene identified in families with Parkinson's disease. Science 1997;276:2045–2047.
30. Spillantini MG, Schmidt ML, Lee VM-Y, et al. Alpha-synuclein in Lewy bodies. Nature 1997;388:839–840.
31. Kruger R, Kuhn W, Muller T, et al. Ala30Pro mutation in the gene encoding alpha-synuclein in Parkinson's disease. Nat Genet 1998;18:106–108.
32. Singleton AB, Farrer M, Johnson J, et al. Alpha-synuclein locus triplication causes Parkinson's disease. Science 2003;302:841.
33. Zarranz JJ, Alegre J, Gomez-Esteban JC, et al. The new mutation, E46K, of alpha-synuclein causes Parkinson and Lewy body dementia. Ann Neurol 2004;55:164–173.
34. Lansbury PT Jr. Back to the future: the 'old-fashioned' way to new medications for neurodegeneration. Nat Med 2004;10(suppl):S51–S57.
35. Li J, Zhu M, Rajamani S, et al. Rifampicin inhibits alpha-synuclein fibrillation and disaggregates fibrils. Chem Biol 2004;11:1513–1521.
36. Zhu M, Rajamani S, Kaylor J, et al. The flavonoid baicalein inhibits fibrillation of alpha-synuclein and disaggregates existing fibrils. J Biol Chem 2004;279:26846–26857.
37. Kitada T, Asakawa S, Hattori N, et al. Mutations in the parkin gene cause autosomal recessive juvenile parkinsonism. Nature 1998;392:605–608.
38. Bonifati V, Rizzu P, van Baren MJ, et al. Mutations in the DJ-1 gene associated with autosomal recessive early-onset parkinsonism. Science 2003;299:256–259.
39. Valente EM, Abou-Sleiman PM, Caputo V, et al. Hereditary early-onset Parkinson's disease caused by mutations in PINK1. Science 2004;304:1158–1160.
40. Paisan-Ruiz C, Jain S, Evans EW, et al. Cloning of the gene containing mutations that cause PARK8-linked Parkinson's disease. Neuron 2004;44:595–600.
41. Zimprich A, Biskup S, Leitner P, et al. Mutations in LRRK2 cause autosomal-dominant parkinsonism with pleomorphic pathology. Neuron 2004;44:601–607.
42. Giasson BI, Forman MS, Higuchi M, et al. Initiation and synergistic fibrillization of tau and alpha-synuclein. Science 2003;300:636–640.
43. Aharon-Peretz J, Rosenbaum H, Gershoni-Baruch R. Mutations in the glucocerebrosidase gene and Parkinson's disease in Ashkenazi Jews. N Engl J Med 2004;351:1972–1977.
44. Lwin A, Orvisky E, Goker-Alpan O, et al. Glucocerebrosidase mutations in subjects with parkinsonism. Mol Genet Metab 2004;81:70–73, 2004.
45. Welch K, Yuan J. Alpha-synuclein oligomerization: a role for lipids? Trends Neurosci 2003;26:517–519, 2003.

Chapter 14
Role of Oxidative Insult and Neuronal Survival in Alzheimer's and Parkinson's Diseases

Akihiko Nunomura, Paula I. Moreira, Xiongwei Zhu, Adam D. Cash, Mark A. Smith, and George Perry

Introduction

Neuronal death and protein aggregation are common features of Alzheimer's disease (AD), Parkinson's disease (PD), and other age-associated neurodegenerative disorders. In chronic neurodegenerative disorders, neuronal death occurs over years, not the minutes of classically defined apoptosis and necrosis. Vulnerable neurons show both responses of apoptosis and regeneration, evidenced by accumulated oxidative insult and attempts at cell cycle reentry. Oxidation of nucleic acids, proteins, and lipids are all seen in postmortem brains from patients with AD and PD. Some of the oxidatively modified macromolecules are identified also in biological fluids from the patients with AD and PD as well as cellular or animal models of these diseases (reviewed in refs. 1,2). Indeed, the oxidative modifications are seen at early stages of the diseases and have been demonstrated to precede the formation of the pathological hallmark lesions, such as senile plaques, neurofibrillary tangles (NFTs), and Lewy bodies. These structures consist of aggregation of disease-specific proteins such as amyloid-β (Aβ), tau, and α-synuclein, respectively, and have generally been assumed to be responsible for neurodegeneration and subsequent neuronal death. However, in the context of chronic degeneration in AD and PD, accumulation of disease-specific structures may, in fact, be responses to increased oxidative stress to prolong neuronal survival.

Neuronal Death in AD and PD

Neurodegenerative disorders are characterized clinically by an insidious onset and slowly progressive course. Selective loss of a particular subset of neurons is a common feature of these disorders. In the superior temporal sulcus of AD, loss of 40% of the neuronal population occurs within 10 years [3]. Similarly, in

A. Nunomura

Department of Psychiatry and Neurology, Asahikawa Medical College, Asahikawa 078-8510, Japan

A. Fisher et al. (eds.), *Advances in Alzheimer's and Parkinson's Disease,*
© Springer 2008

the caudal substantia nigra of PD, loss of 45% of the neuronal population occurs within 10 years [4]. In striking contrast with this chronic process of cell death in neurodegenerative disorders, an apoptotic pathway of physiologically programmed cell death takes only several hours or, at most, a few days for completion (Table 1). In the development of the lateral motor column of the chick embryo, loss of 40% of the population occurs within 3 days [5]. The fact that only 5% to 6% of the population in the lateral motor column is undergoing apoptosis at any particular time during this period [5] indicates that the apoptotic pathway requires less than 24 hours for completion. Therefore, if we assume that neuronal death actually occurs in the manner of classic apoptosis in chronic diseases such as AD and PD, we should encounter only one in several thousands of neurons actively undergoing apoptosis in postmortem brain samples with these diseases [6]. Indeed, extremely rare apoptotic neuronal death compatible with chronic progression was demonstrated in the hippocampus in AD. The hippocampal neurons showing condensed, fragmented nuclei and shrunken cytoplasm as well as labeling for activated caspase-3, the effector enzyme of the terminal apoptotic cascade, were observed at a frequency of 1 in 2600 to 1 in 5650 counted neurons (0.02–0.04%) [7]. On the other hand, a relatively high percentage of neurons showing activated caspase-3 positivity without morphological features of apoptosis was reported in the parahippocampal gyrus in AD [8] as well as the nigral dopaminergic neurons in PD [9]. An electron microscopic study did not reveal any morphological features of apoptosis, such as chromatin condensation in the activated caspase-3-positive neurons, suggesting that caspase-3 activation in these neurons marks the beginning of the effector phase of apoptosis [9]. Therefore, neurons living in a "sick state" for many years in the chronic neurodegenerative disorders may represent abortosis (abortive apoptosis), where apoptotic mechanisms play a role in the process of degeneration although actual completion of apoptosis is absent [10]. The term *aposklesis* (withering) [11] and the term *paraptosis* (next to, or related to apoptosis) [12] are also proposed to refer to an alternative nonapoptotic form of cell death in neurodegeneration.

Table 1 Cell death in development (apoptosis) and in chronic neurodegeneration

Parameter	Apoptosis[a]	Neuronal death in AD and PD[b]
Time for 40% cell loss	3 Days	~ 10 Years
% Neuronal population undergoing apoptosis or degeneration	5–6%[c]	Unknown

AD, Alzheimer's disease; PD, Parkinson's disease.
[a]In the lateral motor column of the chick embryo [3].
[b]In the superior temporal sulcus of AD4 and the pars compacta of the caudal substantia nigra of PD [5].
[c]By counting pyknotic cells.

Oxidative Insults and Cell Cycle Reentry at an Early Stage of Neurodegeneration

Association of Oxidative Stress with Genetic Mutations and Risk Factors for AD and PD

The two most common neurodegenerative disorders are AD and PD, and advanced age is the major risk factor for both [13,14]. The prevalence of AD, especially, increases exponentially throughout aging, with about half of the population afflicted by the age of 95 [13]. As in other organ systems, cells in the brain encounter a cumulative burden of oxidative and metabolic stress that may be a universal feature of the aging process as well as a major causal factor of senescence. Each of the macromolecules, including nucleic acids, proteins, and lipids, is oxidatively modified during aging. Indeed, the brain is especially vulnerable to free radical damage because of its high oxygen consumption rate, abundant lipid content, and relative paucity of antioxidant enzymes compared with other organs [15,16].

Not only advanced age but also genetic mutations causing AD and PD are associated with oxidative stress. Recently, an increasing number of in vitro and in vivo studies have suggested that oxidative stress has an involvement in autosomal dominant familial AD with mutations in the amyloid-β protein precursor (AβPP), presenilin-1 (PS-1), or presenilin-2 (PS-2) gene. Indeed, increased oxidative stress, elevated vulnerability to oxidative stress-induced cell death, and/or reduced antioxidant defenses have been demonstrated in: (1) cell lines expressing mutant human AβPP, PS-1, or PS-2 [17–20]; (2) transgenic mice expressing mutant human AβPP and/or PS-1 as well as knock-in mice expressing mutant human PS-1 [21–27]; (3) fibroblasts and lymphoblasts from familial AD patients with AβPP or PS-1 gene mutations [28]; and (4) cerebral cortex of autopsied brain samples from patients with AβPP or PS-1 gene mutations [29,30]. Mutations in the parkin gene causing autosomal recessive familial PD have also been shown to increase the formation of reactive oxygen and nitrogen species [31]. Moreover, the possession of one or both apolipoprotein E4 (ApoE4) alleles, the major genetic risk factor for early-to late-onset sporadic and familial AD, is associated with oxidative stress. In vitro, ApoE shows allele-specific antioxidant activity, with ApoE2 the most effective and ApoE4 the least effective [32]. Indeed, oxidative damage in an ApoE genotype-dependent manner has been demonstrated in autopsied brain samples of AD patients [33–35].

Medical risk factors for AD include traumatic brain injury, cerebral infarcts, diabetes mellitus, hypercholesterolemia, and hyperhomocysteinemia. Environmental and lifestyle-related risk factors for AD include aluminum exposure, smoking, high calorie intake, lack of exercise, and lack of intellectual activities [16,36–38]. Traumatic brain injury, exposure to metals (aluminum, copper, iron, lead, manganese, mercury, and their combination) and pesticides, and high calorie intake are associated with an increased risk of PD [39–41]. Indeed, all these risk factors are associated with an increase in oxidative stress [16,42–49]. With this notion, it is not surprising that agents or nutrients inhibiting free radical formation reduce the incidence of AD and PD. Indeed, not only agents or nutrients such as vitamins E and C, but also estrogen, nonsteroidal antiinflammatory drugs (NSAIDs), statins, n-3 poly-unsaturated fatty acids, and wine have been proven to have an antioxidant activity and to reduce the incidence of AD [37,38,50–54].

Vitamin E and NSAIDs may also be associated with a reduced risk of PD [55–57]. Furthermore, calorie restriction, exercise, and intellectual activity have been proven to promote neuronal survival through decreased oxidative stress in experimental animals [16,36].

Known genetic mutations and risk factors for AD and PD that cause or promote oxidative damage as well as agents, nutrients, and behaviors that prevent or attenuate oxidative damage are summarized in Fig. 1. This evidence strongly suggests that oxidative stress is universally involved in the pathogenesis of AD and PD, especially upstream of the pathological cascade.

Primary Role of Oxidative Stress in AD and PD

An early involvement of oxidative stress in the pathogenesis of AD is demon-strated more directly by recent studies on cell culture models, transgenic animal models, postmortem brains from patients with AD and Down's syndrome, and biological fluids from patients with AD and subjects with mild cognitive impairment (MCI). We selected an in situ approach to identify markers of nucleic acid oxidation and protein oxidation in postmortem brain samples. Surprisingly, not only is the oxidative damage more prominent in AD cases with lesser amounts of Aβ deposition or shorter disease duration [58], but the oxidative damage precedes Aβ deposition in a series of Down's syndrome brains, a model of AD neuropathology with known predictable chronology [59]. Our observation corresponds with the results of increased nucleic acid oxidation in cerebrospinal fluid (CSF) from AD cases, in which the shorter the disease duration, the greater the oxidative damage [60]. Moreover, individuals with MCI who, at least in part, represent the prodromal stage of AD, show significantly increased levels of lipid peroxidation and nucleic acid oxidation in peripheral samples [61,62] as well as decreased levels of plasma antioxidants [63]. These data, obtained from human subjects, clearly indicate early involve-ment of oxidative stress in AD pathogenesis, which is supported by experimen-tal studies using cell culture models and transgenic animal models of AD. Increased lipid peroxidation or protein oxidation precedes Aβ plaque deposi-tion or Aβ fibril formation in transgenic mouse or the C. elegans model of AD amyloidosis [26,64]. Indeed, oxidative stress induces intracellular Aβ accumula-tion and tau phosphorylation in cell cultures [65,66], and vitamin E reduces Aβ

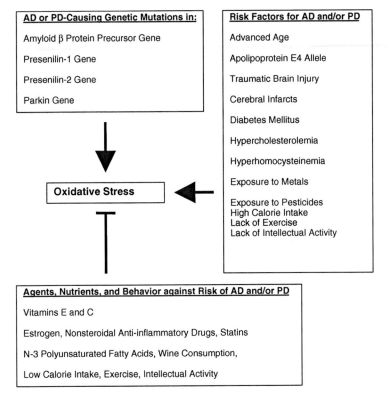

Fig. 1 Genetic, medical, environmental, and lifestyle-related factors for Alzheimer's disease (AD) and Parkinson's disease (PD): relation to oxidative stress. Most of the known genetic mutations and risk factors for AD and PD are associated with an increase of oxidative stress. In contrast, several agents and nutrients that are known to reduce the incidence of AD and/or PD have antioxidant properties per se or help to prevent/reduce free radical generation/propagation. Low-calorie diet as well as physical and intellectual activities are suggested to enhance the production of antioxidant enzymes in the brain

and tau lesions in transgenic animals [67,68]. Furthermore, dietary copper stabilizes brain copper/zinc superoxide dismutase (SOD) activity and reduces Aβ production in AβPP transgenic mice [69]. Moreover, AβPP mutant mice crossed with manganese SOD heterozygous knockout mice show increased Aβ plaque deposition in brain [70].

Several lines of evidence support early involvement of oxidative stress also in PD. Studies demonstrating oxidative dimer formation as the critical rate-limiting step for fibrillogenesis of α-synuclein have revealed that overproduction of reactive oxygen species and/or impairments of cellular antioxidative mechanisms are primary events in the initiation and propagation of PD [71]. A reduced level of glutathione is found in brains of patients with incidental Lewy body disease, which is supposed to represent presymptomatic PD [72]. In the substantia nigra of patients with PD, there is a 176% increase in total iron content, with a shift of the Fe(III)/Fe(II) ratio in favor of Fe(III) [73]. Data from transcranial ultrasonography studies imply that iron accumulation occurs very early in the disease [74]. Of note, an increased level of iron leads to amplification of reactive oxygen species via the Fenton reaction. Furthermore, a primary role of mitochondrial dysfunction and oxidative damage in PD is supported by an animal model of PD with chronic infusion of the complex I inhibitor rotenone. In this model, infusion of rotenone produces selective loss of substantia nigra dopaminergic neurons as well as cytoplasmic α-synuclein-immunoreactive inclusions closely resembling Lewy bodies [75]. A subsequent in vitro experiment with rotenone clearly indicates that the mechanisms of the neuronal loss and the inclusion formation involve oxidative damage [76]. Indeed, evidence for mitochondrial dysfunction in the substantia nigra of PD patients comes from a decrease in complex I activity and an increase in mitochondrial deletion/ rearrangements [77,78]. Therefore, mitochondrial dysfunction and metal dysregulation are key features associated with oxidative insult in PD that are common to the pathophysiology of AD [79–81].

Cell Cycle Reentry in AD and PD

Recently, the harlequin mutant mouse showing progressive degeneration of cerebellar and retinal neurons provided a genetic model of oxidative stress-mediated neurodegeneration with a direct connection between cell cycle reentry and oxidative stress in the aging central nervous system (CNS) [82]. In this model, oxidative DNA damage is likely to precede cell cycle reentry, although the exact mechanism(s) by which oxidative stress induces cell cycle abnormalities in postmitotic neurons is unknown.

In vulnerable neurons in AD, various components of the cell cycle machinery are activated: cyclin-dependent kinases, cyclins, cyclin-dependent kinase inhibitors, and proliferation-associated nuclear proteins (reviewed in ref 83). Interestingly, some of the cell cycle-related proteins are expressed in 5% to 10% of neurons in the vulnerable regions not only in patients with AD but also in MCI cases [84]. These findings suggest that cell cycle events are involved at an early stage of neurodegeneration and that cell cycle-induced death should be abortive in the absence of actual completion at least for months if not years. A recent study on transgenic mice expressing nonmutant human tau shows that the transgenic mice exhibit neurofibrillary tangle (NFT) formation and extensive neuronal death and that the neurodegeneration is independent of NFT formation but actually involves reexpression of cell cycle regulatory proteins [85].

Compared with AD, there are fewer reports on cell cycle dysregulation in PD. However, α-synuclein overexpression in PC12 cells has been demonstrated to increase the levels of phosphorylated extracellular signal-regulated kinases, with subsequent enhancement of the proliferation rate and enrichment of the cells in the S phase of the cell cycle [86]. Furthermore, cyclin B immunoreactivity is found in Lewy bodies in PD [86]. Of note, genetic factors of AD and PD are likely to be associated with cell cycle events. AβPP, PS-1, and PS-2 have important roles in cell cycle control; and the disruption caused by familial AD mutations may impair cell cycle control (reviewed in ref. 87). Cyclin E is a substrate of the parkin-ubiquitin-ligase complex, and parkin deficiency potentiates the accumulation of cyclin E [88].

Pathological Hallmarks and Neuronal Survival Response

Pathological Hallmarks: Pathogenic, Incidental, or a Beneficial Coping Response?

Alzheimer's disease is defined pathologically by extraneuronal deposition of Aβ senile plaques and cytoplasmic inclusion of NFTs composed of tau. PD is defined pathologically by cytoplasmic inclusion of Lewy bodies composed of α-synuclein. From the time of their original description, a major focus has been to understand the role that these lesions play in the pathogenesis of AD and PD. However, it is still questionable whether senile plaques and/or NFTs in AD or Lewy bodies in PD cause associated neurodegeneration or neurological, behavioral, and cognitive deficits that accompany the disease. Recently, in a cellular model of Huntington's disease (a neurodegenerative disorder caused by abnormal polyglutamine expansion) the formation of neuronal inclusion bodies has been demonstrated to be associated with improved neuronal survival and decreased levels of mutant huntingtin throughout the neuron. Inclusion bodies may sequester the diffuse form of mutant huntingtin. Indeed, many neurons die without forming an inclusion body. Thus, the formation of neuronal inclusion bodies can function as a coping response to toxic mutant huntingtin [89]. This interesting finding encourages us to reconsider roles of intraneuronal and even extraneuronal aggregates of specific protein in neurodegenerative disorders.

Senile plaques and NFTs are present in a considerable percentage of brains of cognitively normal elderly subjects. Surprisingly, a study investigating autopsied subjects between 69 and 100 years of age who were cognitively normal revealed that 49% of those normal subjects met the Khachaturian criteria for AD based on senile plaque density, 25% met the CERAD criteria based on senile plaque density, and 24% were in stages IV to VI of the Braak and Braak staging of AD based on NFT density [90]. Furthermore, it is well known that there is no correlation, or at best a poor correlation, between neuronal loss and senile plaque density as well as between disease severity and senile plaque density in AD [91]. Accordingly, AβPP transgenic mice showing massive deposition of Aβ plaques in the brain lack consistent widespread neuronal loss [92]. By contrast, neuronal loss and clinical severity correlate with NFT density. However, the amount of neuronal loss largely exceeds the amount of NFTs [3]. Neuronal death that occurs independently of NFT formation is observed in transgenic mice expressing nonmutant human tau [85]. Moreover, an ultrastructural analysis demonstrated that a reduction in number and total length of microtubules seen in pyramidal neurons in AD was unrelated to the presence of NFTs [93]. Indeed, neurons with NFTs are estimated to be able to survive for decades [94], which suggests that NFTs themselves are not obligatory for neuronal death in AD.

The appearance of Lewy bodies is not uncommon in brains of the normal elderly, which is not necessarily linked with neuronal loss [95]. Incidental Lewy bodies in brains from individuals without clinical PD show an age-specific prevalence rising from 3.8% to 12.8% between the sixth and ninth decades [96]. Because the prevalence of PD has been estimated at 0.6% for individuals between the ages of 65 and 69 years and 2.6% for those between 85 and 89 years [97], the prevalence of incidental Lewy bodies in the population is much higher than that of PD. In an investigation of the prevalence of Lewy body pathology in a community-based population, 15.1% of nondemented, non-PD subjects show Lewy body formation in the cortical and subcortical regions. Surprisingly, there is no significant difference in Lewy body density scored by the consensus guideline between these subjects with incidental Lewy bodies and patients with Lewy body dementia [98]. Indeed, analysis of surviving nigral neurons that lack or contain Lewy bodies show no quantifiable difference in viability estimated by cell size, nucleolar size, and messenger RNA level of the low-molecular-weight neurofilament subunit [99,100]. Furthermore, mutations in the parkin gene cause juvenile PD exhibiting nigral degeneration largely in the absence of Lewy bodies [101]. In α-synuclein transgenic animals, overexpression of α-synuclein causes the formation of inclusions but does not always cause neuronal loss [102,103].

Therefore, senile plaques and NFTs in AD and Lewy bodies in PD may not be indispensable for death of vulnerable neurons. Interestingly, recent findings in a cellular system are consistent with a toxic oligomer hypothesis or toxic protofibril hypothesis, which provides ideas that oligomer or protofibril of specific proteins such as Aβ or α-synuclein may be responsible for cell death and that the fibrillar form typically observed at autopsy may actually be neuroprotective [104,105]. Considering the chronological primacy of oxidative stress in the pathological cascade, we cannot exclude the possibility that the processes of the pathology formation are involved in compensatory changes against oxidative insult in AD and PD. Yet it is also important to consider that oligomers have not been convincingly demonstrated as an in vivo intermediate.

Possible Protective Function of Aβ, Tau, and α-Synuclein Against Oxidative Stress

Although high concentrations of Aβ, in a micromolar range, can lead to oxidative stress in various biological systems (reviewed in ref. 106), it is apparent from cell [65], animal [26,64,68–70], and human [59] studies that oxidative stress chronologically precedes Aβ deposition. Moreover, increases in the density of Aβ plaque deposition are associated with decreased levels of nucleic acid oxidation in neurons in postmortem brains from patients with sporadic and familial AD and Down's syndrome [30,58,59]. These findings indicate that the process of Aβ plaque formation is a neuronal protective response against oxidative stress. Recently, in vitro and in vivo studies have demonstrated an antioxidant activity of Aβ. Monomeric Aβ(1-40) and Aβ(1-42) have been shown to protect cultured neurons from iron- and copper-induced toxicity [107]. In contrast, co-injection of iron and Aβ(1-42) into rat cerebral cortex is significantly less toxic than injection of iron alone [108]. Furthermore, addition of physiological concentrations (in a low nanomolar range) of Aβ(1-40) and Aβ(1-42) has been shown to protect lipoproteins from oxidation in CSF and plasma [109]. These Aβ peptides fail to prevent metal-independent oxidation; and Aβ(25-35), lacking a metal-binding site located in the N-terminal part (histidine at positions 6, 13, and 14 and tyrosine at position 10), is less effective at inhibiting oxidation. Therefore, it is likely that the mechanism by which Aβ inhibits oxidation is via chelating metal ions [109]. Indeed, copper, iron, and zinc are elevated in the rims and cores of Aβ plaques in postmortem AD brains [110,111], We suppose that chelation of redox-active copper and iron is an important mechanism of the protective function of Aβ, and that elevation of zinc, a redox-inert antioxidant, may be a homeostatic response to oxidative stress, which subsequently accelerates the formation of Aβ plaques [112].

This aspect of Aβ that actually plays an important role in antioxidant defenses may apply equally to tau. Cellular [66], animal [67], and human [59] studies suggest that oxidative stress chronologically precedes NFT formation. Oxidative stress activates several kinases, including glycogen synthase kinase-3 and mitogen-activated protein kinases, which are activated in AD and are capable of phosphorylating tau. Once phosphorylated, tau becomes particularly vulnerable to oxidative modification and consequently aggregates into fibrils (reviewed in ref. [113]). Therefore, NFT formation is likely to be a result of neuronal oxidation, which is accompanied by induction of the antioxidant enzyme heme oxygenase-1 [114]. In fact, in postmortem AD brains, neuronal oxidative damage to nucleic acid is actually decreased by the presence of NFTs when neurons with NFTs are compared to neurons free of NFTs [58]. Although the exact mechanism of how NFT formation opposes oxidative stress is unknown, redox-active iron accumulation is strikingly associated with NFTs [115] and tau is found capable of binding to iron and copper and thereby possibly exerts antioxidant activity [116].

α-Synuclein also may play a protective role against oxidative stress. Both in vitro [71,76] and in vivo [75] models have suggested a primary role in oxidative stress in α-synuclein aggregation. The herbicide paraquat causes oxidative stress and α-synuclein aggregation as well as nigral neurodegeneration in mouse brains; and surprisingly, mice overexpressing α-synuclein display α-synuclein aggregation but are completely protected against neurodegeneration [117]. Similarly, in an α-synuclein-transfected neuronal cell line under oxidative stress conditions, increased a-synuclein expression protects cells from oxidative stress via inactivation of c-Jun N-terminal kinase, a member of the mitogen-activated protein kinase family [19]. More specifically, it is proposed that α-synuclein can limit oxidative damage to cells involving its methionines (at positions 1, 5, 116, and 127), serving as a natural scavenger of reactive oxygen species [118]. Furthermore, Lewy bodies can sequester redox-active iron [119], being similar to senile plaques and NFTs.

As we have reviewed here, every disease-specific protein potentially plays a protective role against oxidative stress. However, the efficiency of the protective function may be dependent on the aggregation state of the protein [107]. Recently, an increasing body of evidence has been collected to support the hypothesis that oligomers, not monomers or fibrils, are the toxic form of the protein (reviewed in refs. [104,105]). Therefore, further study is required to assess the relation between oxidative stress and oligomer formation, which may provide an important clue to early therapeutic intervention in neurodegenerative disorders.

Conclusions

Most of the known genetic, medical, environmental, and lifestyle-related factors of AD and PD are associated with increased oxidative stress. In contrast, human cases at the preclinical stages of AD and PD (e.g., subjects with MCI, young adults with Down's syndrome, and subjects with incidental Lewy bodies) as well as cellular and animal models of the diseases provide consistent evidence that oxidative insult is a significant early event in the pathological cascade of AD and PD. In contrast to the general aspects of the pathological hallmarks, aggregation of the disease-specific protein may be involved in a compensatory response against the oxidative insult. Although a common underlying mechanism for the protective response is presently obscure, sequestration of redox-active metals by the disease-specific structures may be associated with the neuronal survival response against oxidative insult.

Acknowledgments Work in the authors' laboratories has been supported by funding from the Alzheimer's Association, Philip Morris USA Inc., and Philip Morris International.

References

1. Smith MA, Rottkamp CA, Nunomura A, et al. Oxidative stress in Alzheimer's disease. Biochim Biophys Acta 2000;1502(1):139–144.
2. Jenner P. Oxidative stress in Parkinson's disease. Ann Neurol 2003;53(suppl 3):S26–S36; discussion S36–S28.
3. Gomez-Isla T, Hollister R, West H, et al. Neuronal loss correlates with but exceeds neurofibrillary tangles in Alzheimer's disease. Ann Neurol 1997;41(1):17–24.

4. Fearnley JM, Lees AJ. Ageing and Parkinson's disease: substantia nigra regional selectivity. Brain 1991;114 (Pt 5):2283–2301.

5. Hamburger V. Cell death in the development of the lateral motor column of the chick embryo. J Comp Neurol 1975;160(4):535–546.

6. Perry G, Nunomura A, Smith MA. A suicide note from Alzheimer disease neurons? Nat Med 1998;4(8):897–898.

7. Jellinger KA, Stadelmann C. Problems of cell death in neurodegeneration and Alzheimer's disease. J Alzheimers Dis 2001;3(1):31–40.

8. Gastard MC, Troncoso JC, Koliatsos VE. Caspase activation in the limbic cortex of subjects with early Alzheimer's disease. Ann Neurol 2003;54(3):393–398.

9. Hartmann A, Hunot S, Michel PP, et al. Caspase-3: a vulnerability factor and final effector in apoptotic death of dopaminergic neurons in Parkinson's disease. Proc Natl Acad Sci U S A 2000;97(6):2875–2880.

10. Raina AK, Hochman A, Zhu X, et al. Abortive apoptosis in Alzheimer's disease. Acta Neuropathol (Berl) 2001;101(4):305–310.

11. Graeber MB, Grasbon-Frodl E, Abell-Aleff P, Kosel S. Nigral neurons are likely to die of a mechanism other than classical apoptosis in Parkinson's disease. Parkinsonism Relat Disord 1999;5(4):187–192.

12. Sperandio S, de Belle I, Bredesen DE. An alternative, nonapoptotic form of programmed cell death. Proc Natl Acad Sci U S A 2000;97(26):14376–14381.

13. Hy LX, Keller DM. Prevalence of AD among whites: a summary by levels of severity. Neurology 2000;55(2):198–204.

14. Fahn S, Sulzer D. Neurodegeneration and neuroprotection in Parkinson disease. NeuroRx 2004;1(1):139–154.

15. Coyle JT, Puttfarcken P. Oxidative stress, glutamate, and neurodegenerative disorders. Science 1993;262(5134):689–695.

16. Mattson MP, Chan SL, Duan W. Modification of brain aging and neurodegenerative disorders by genes, diet, and behavior. Physiol Rev 2002;82(3):637–672.

17. Guo Q, Sopher BL, Furukawa K, et al. Alzheimer's presenilin mutation sensitizes neural cells to apoptosis induced by trophic factor withdrawal and amyloid beta-peptide: involvement of calcium and oxyradicals. J Neurosci 1997;17(11):4212–4222.

18. Eckert A, Steiner B, Marques C, et al. Elevated vulnerability to oxidative stress-induced cell death and activation of caspase-3 by the Swedish amyloid precursor protein mutation. J Neurosci Res 2001;64(2):183–192.

19. Hashimoto M, Hsu LJ, Rockenstein E, et al. Alpha-synuclein protects against oxidative stress via inactivation of the c-Jun N-terminal kinase stress-signaling pathway in neuronal cells. J Biol Chem 2002;277(13):11465–11472.

20. Marques CA, Keil U, Bonert A, et al. Neurotoxic mechanisms caused by the Alzheimer's disease-linked Swedish amyloid precursor protein mutation: oxidative stress, caspases, and the JNK pathway. J Biol Chem 2003;278(30):28294–28302.

21. Smith MA, Hirai K, Hsiao K, et al. Amyloid-beta deposition in Alzheimer transgenic mice is associated with oxidative stress. J Neurochem 1998;70(5):2212–2215.

22. Guo Q, Sebastian L, Sopher BL, et al. Increased vulnerability of hippocampal neurons from presenilin-1 mutant knock-in mice to amyloid beta-peptide toxicity: central roles of superoxide production and caspase activation. J Neurochem 1999;72(3):1019–1029.

23. Leutner S, Czech C, Schindowski K, et al. Reduced antioxidant enzyme activity in brains of mice transgenic for human presenilin-1 with single or multiple mutations. Neurosci Lett 2000;292(2):87–90.

24. Takahashi M, Dore S, Ferris CD, et al. Amyloid precursor proteins inhibit heme oxygenase activity and augment neurotoxicity in Alzheimer's disease. Neuron 2000;28(2):461–473.

25. Matsuoka Y, Picciano M, La Francois J, Duff K. Fibrillar beta-amyloid evokes oxidative damage in a transgenic mouse model of Alzheimer's disease, Neuroscience 2001;104(3):609–613.

26. Pratico D, Uryu K, Leight S, et al. Increased lipid peroxidation precedes amyloid plaque formation in an animal model of Alzheimer amyloidosis. J Neurosci 2001;21(12):4183–4187.

27. LaFontaine MA, Mattson MP, Butterfield DA. Oxidative stress in synaptosomal proteins from mutant presenilin-1 knock-in mice: implications for familial Alzheimer's disease. Neurochem Res 2002;27(5):417–421.

28. Cecchi C, Fiorillo C, Sorbi S, et al. Oxidative stress and reduced antioxidant defenses in peripheral cells from familial Alzheimer's patients. Free Radic Biol Med 2002;33(10):1372–1379.

29. Bogdanovic N, Zilmer M, Zilmer K, et al. The Swedish APP670/671 Alzheimer's disease mutation: the first evidence for strikingly increased oxidative injury in the temporal inferior cortex. Dement Geriatr Cogn Disord 2001;12(6):364–370.

30. Nunomura A, Chiba S, Lippa CF, et al. Neuronal RNA oxidation is a prominent feature of familial Alzheimer's disease. Neurobiol Dis 2004;17(1):108–113.

31. Hyun DH, Lee M, Hattori N, et al. Effect of wild-type or mutant parkin on oxidative damage, nitric oxide, antioxidant defenses, and the proteasome. J Biol Chem 2002;277(32):28572–28577.

32. Miyata M, Smith JD. Apolipoprotein E allele-specific antioxidant activity and effects on cytotoxicity by oxidative insults and beta-amyloid peptides. Nat Genet 1996;14(1):55–61.

33. Montine KS, Reich E, Neely MD, et al. Distribution of reducible 4-hydroxynonenal adduct immunoreactivity in Alzheimer disease is associated with APOE genotype. J Neuropathol Exp Neurol 1998;57(5):415–425.

34. Ramassamy C, Averill D, Beffert, U et al. Oxidative damage and protection by antioxidants in the frontal cortex of Alzheimer's disease is related to the apolipoprotein E genotype. Free Radic Biol Med 1999;27(5-6):544–553.

35. Tamaoka A, Miyatake F, Matsuno S, et al. Apolipoprotein E allele-dependent antioxidant activity in brains with Alzheimer's disease. Neurology 2000;54(12):2319–2321.

36. Mattson MP. Gene-diet interactions in brain aging and neurodegenerative disorders. Ann Intern Med 2003;139(5 Pt 2):441–444.

37. Mayeux R. Epidemiology of neurodegeneration. Annu Rev Neurosci 2003;26:81–104.

38. Haan MN, Wallace R. Can dementia be prevented? Brain aging in a population-based context. Annu Rev Public Health 25:1–24.

39. Logroscino G, Marder K, Cote L, et al. Dietary lipids and antioxidants in Parkinson's disease: a population-based case-control study. Ann Neurol 39(1):89–94.

40. Lai BC, Marion SA, Teschke K, Tsui JK. Occupational and environmental risk factors for Parkinson's disease. Parkinsonism Relat Disord 8(3):297–309.

41. Gorell JM, Peterson EL, Rybicki BA, Johnson CC. Multiple risk factors for Parkinson's disease. J Neurol Sci 217(2):169–174.

42. Preston AM. Cigarette smoking-nutritional implications. Prog Food Nutr Sci 15(4):183–217.

43. Moriel P, Plavnik FL, Zanella MT, et al. Lipid peroxidation and antioxidants in hyperlipidemia and hypertension. Biol Res 33(2):105–112.

44. Maritim AC, Sanders RA, Watkins JB 3rd. Diabetes, oxidative stress, and antioxidants: a review. J Biochem Mol Toxicol 17(1):24–38.

45. Perna AF, Ingrosso D, De Santo NG. Homocysteine and oxidative stress. Amino Acids 25(3-4):409–417.

46. Bramlett HM, Dietrich WD. Pathophysiology of cerebral ischemia and brain trauma: similarities and differences. J Cereb Blood Flow Metab 24(2):133–150.

47. Gupta VB, Anitha S, Hegde ML, et al. Aluminium in Alzheimer's disease: are we still at a crossroad? Cell Mol Life Sci 62(2):143–158.

48. Valko M, Morris H, Cronin MT. Metals, toxicity and oxidative stress. Curr Med Chem 12(10):1161–1208.

49. Abdollahi M, Ranjbar A, Shadnia S, et al. Pesticides and oxidative stress: a review. Med Sci Monit 2004;10(6):RA141–RA147.

50. Hamburger SA, McCay PB. Spin trapping of ibuprofen radicals: evidence that ibuprofen is a hydroxyl radical scavenger. Free Radic Res Commun 1990;9(3-6):337–342.

51. Behl C, Skutella T, Lezoualc'h F, et al. Neuroprotection against oxidative stress by estrogens: structure-activity relationship. Mol Pharmacol 1997;51(4):535–541.

52. Commenges D, Scotet V, Renaud S, et al. Intake of flavonoids and risk of dementia. Eur J Epidemiol 2000;16(4):357–363.

53. Green P, Glozman S, Weiner L, Yavin E. Enhanced free radical scavenging and decreased lipid peroxidation in the rat fetal brain after treatment with ethyl docosahexaenoate. Biochim Biophys Acta 2001;1532(3):203–212.

54. Stoll LL, McCormick ML, Denning GM, Weintraub NL. Antioxidant effects of statins. Drugs Today (Barc) 2004;40(12):975–990.

55. De Rijk MC, M. Breteler MM, den Breeijen JH, et al. Dietary antioxidants and Parkinson disease: the Rotterdam Study. Arch Neurol 1997;54(6):762–765.

56. Chen H, Zhang SM, Hernan MA, et al. Nonsteroidal anti-inflammatory drugs and the risk of Parkinson disease. Arch Neurol 2003;60(8):1059–1064.

57. Etminan M, Gill SS, Samii A. Intake of vitamin E, vitamin C, and carotenoids and the risk of Parkinson's disease: a meta-analysis. Lancet Neurol 2005;4(6):362–365.

58. Nunomura A, Perry G, Aliev G, et al. Oxidative damage is the earliest event in Alzheimer disease. J Neuropathol Exp Neurol 2001;60(8):759–767.

59. Nunomura A, Perry G, Pappolla MA, et al. Neuronal oxidative stress precedes amyloid-beta deposition in Down syndrome. J Neuropathol Exp Neurol 2000;59(11): 1011–1017.

60. Abe T, Tohgi H, Isobe C, et al. Remarkable increase in the concentration of 8-hydroxyguanosine in cerebrospinal fluid from patients with Alzheimer's disease. J Neurosci Res 2002;70(3):447–450.

61. Pratico D, Clark CM, Liun F, et al. Increase of brain oxidative stress in mild cognitive impairment: a possible predictor of Alzheimer disease. Arch Neurol 2002;59(6):972–976.

62. Migliore L, Fontana I, Trippi F, et al. Oxidative DNA damage in peripheral leukocytes of mild cognitive impairment and AD patients. Neurobiol Aging 2005;26(5):567–573.

63. Rinaldi P, Polidori MC, Metastasio A, et al. Plasma antioxidants are similarly depleted in mild cognitive impairment and in Alzheimer's disease. Neurobiol Aging 2003;24(7):915–919.

64. Drake J, Link CD, Butterfield DA, Oxidative stress precedes fibrillar deposition of Alzheimer's disease amyloid beta-peptide (1-42) in a transgenic Caenorhabditis elegans model. Neurobiol Aging 2003;24(3):415–420.

65. Misonou H, Morishima-Kawashima M, Ihara Y. Oxidative stress induces intracellular accumulation of amyloid beta-protein (Abeta) in human neuroblastoma cells. Biochemistry 2000;39(23):6951–6959.

66. Gomez-Ramos A, Diaz-Nido J, Smith MA, et al. Effect of the lipid peroxidation product acrolein on tau phosphorylation in neural cells. J Neurosci Res 2003;71(6):863–870.

67. Nakashima H, Ishihara T, Yokota O, et al. Effects of alpha-tocopherol on an animal model of tauopathies. Free Radic Biol Med 2004;37(2):176–186.

68. Sung S, Yao Y, Uryu K, et al. Early vitamin E supplementation in young but not aged mice reduces Abeta levels and amyloid deposition in a transgenic model of Alzheimer's disease. FASEB J 2004;18(2):323–325.

69. Bayer TA, Schafer S, Simons A, et al. Dietary Cu stabilizes brain superoxide dismutase 1 activity and reduces amyloid Abeta production in APP23 transgenic mice. Proc Natl Acad Sci U S A 2003;100(24):14187–14192.

70. Li F, Calingasan NY, Yu F, et al. Increased plaque burden in brains of APP mutant MnSOD heterozygous knockout mice. J Neurochem 2004;89(5):1308–1312.

71. Krishnan S, Chi EY, Wood SJ, et al. Oxidative dimer formation is the critical rate-limiting step for Parkinson's disease alpha-synuclein fibrillogenesis. Biochemistry 2003;42(3):829–837.

72. Jenner P, Dexter DT, Sian J, et al. Oxidative stress as a cause of nigral cell death in Parkinson's disease and incidental Lewy body disease; The Royal Kings and Queens Parkinson's Disease Research Group. Ann Neurol 1992;32(suppl):S82–S87.

73. Sofic E, Riederer P, Heinsen H, et al. Increased iron (III) and total iron content in post mortem substantia nigra of parkinsonian brain. J Neural Transm 1988;74(3):199–205.

74. Berg D, Roggendorf W, Schroder U, et al. Echogenicity of the substantia nigra: association with increased iron content and marker for susceptibility to nigrostriatal injury. Arch Neurol 2004;59(6):999–1005.

75. Betarbet R, Sherer TB, MacKenzie G, et al. Chronic systemic pesticide exposure reproduces features of Parkinson's disease. Nat Neurosci 2000;3(12):1301–1306.

76. Sherer TB, Betarbet R, Stout AK, et al. An in vitro model of Parkinson's disease: linking mitochondrial impairment to altered alpha-synuclein metabolism and oxidative damage. J Neurosci 2002;22(16):7006–7015.

77. Schapira AH, Cooper JM, Dexter D, et al. Mitochondrial complex I deficiency in Parkinson's disease. J Neurochem 1990;54(3):823–827.

78. Gu G, Reyes PE, Golden GT, et al. Mitochondrial DNA deletions/rearrangements in Parkinson disease and related neurodegenerative disorders. J Neuropathol Exp Neurol 2002;61(7):634–639.

79. 7 Loeffler DA, Connor JR, Juneau PL, et al. Transferrin and iron in normal, Alzheimer's disease, and Parkinson's disease brain regions. J Neurochem 1995;65(2):710–724.

80. Hirai K, Aliev G, Nunomura A, et al. Mitochondrial abnormalities in Alzheimer's disease. J Neurosci 2001;21(9):3017–3023.

81. Mezzetti A, Pierdomenico SD, Costantini F, et al. Copper/zinc ratio and systemic oxidant load: effect of aging and aging-related degenerative diseases. Free Radic Biol Med 1998;25(6):676–681.

82. Klein JA, Longo-Guess CM, Rossmann MP, et al. The harlequin mouse mutation downregulates apoptosis-inducing factor. Nature 2002;419(6905):367–374.

83. Arendt T. Alzheimer's disease as a disorder of dynamic brain self-organization. Prog Brain Res 2005;147:355–378.

84. Yang Y, Mufson EJ, Herrup K. Neuronal cell death is preceded by cell cycle events at all stages of Alzheimer's disease. J Neurosci 2003;23(7):2557–2563.

85. Andorfer C, Acker CM, Kress Y, et al. Cell-cycle reentry and cell death in transgenic mice expressing nonmutant human tau isoforms. J Neurosci 2005;25(22):5446–5454.

86. Lee SS, Kim YM, Junn E, et al. Cell cycle aberrations by alpha-synuclein over-expression and cyclin B immunoreactivity in Lewy bodies. Neurobiol Aging 2003;24(5):687–696.

87. Zhu X, Raina AK, Perry G,. Smith MA. Alzheimer's disease: the two-hit hypothesis. Lancet Neurol 2004;3(4):219–226.

88. Staropoli JF, McDermott C, Martinat C, et al. Parkin is a component of an SCF-like ubiquitin ligase complex and protects postmitotic neurons from kainate excitotoxicity. Neuron 2003;37(5):735–749.

89. Arrasate M, Mitra S, Schweitzer ES, et al. Inclusion body formation reduces levels of mutant huntingtin and the risk of neuronal death. Nature 2004;431(7010):805–810.

90. Davis DG, Schmitt FA, Wekstein DR, Markesbery WR. Alzheimer neuropathologic alterations in aged cognitively normal subjects. J Neuropathol Exp Neurol 1999;58(4):376–388.

91. Neve RL, Robakis NK. Alzheimer's disease: a re-examination of the amyloid hypothesis. Trends Neurosci 1998;21(1):15–19.

92. Irizarry MC, McNamara M, Fedorchak K, et al. APPSw transgenic mice develop age-related A beta deposits and neuropil abnormalities, but no neuronal loss in CA1. J Neuropathol Exp Neurol 1997;56(9):965–973.

93. Cash AD, Aliev G, Siedlak SL, et al. Microtubule reduction in Alzheimer's disease and aging is independent of tau filament formation. Am J Pathol 2003;162(5):1623–1627.

94. Morsch R, Simon W, Coleman PD. Neurons may live for decades with neurofibrillary tangles. J Neuropathol Exp Neurol 1999;58(2):188–197.

95. Perry RH, Irving D, Tomlinson BE. Lewy body prevalence in the aging brain: relationship to neuropsychiatric disorders, Alzheimer-type pathology and catecholaminergic nuclei. J Neurol Sci 1990;100(1-2):223–233.

96. Lowe JS, Leigh N. Disorders of movement and system degenerations. In: Graham DI, Lantos PL (eds) Greenfield's Neuropathology. Arnold, London, 2002, pp 325–430.

97. De Rijk MC, Launer LJ, Berger K, et al. Prevalence of Parkinson's disease in Europe: a collaborative study of population-based cohorts; Neurologic Diseases in the Elderly Research Group. Neurology 2000;54(11 suppl 5):S21–S23.

98. Wakisaka Y, Furuta A, Tanizaki Y, et al. Age-associated prevalence and risk factors of Lewy body pathology in a general population: the Hisayama study. Acta Neuropathol (Berl) 2003;106(4):374–382.

99. Gertz HJ, Siegers A, Kuchinke J. Stability of cell size and nucleolar size in Lewy body containing neurons of substantia nigra in Parkinson's disease. Brain Res 1994;637(1-2): 339–341.

100. Bergeron C, Petrunka C, Weyer L, Pollanen MS. Altered neurofilament expression does not contribute to Lewy body formation. Am J Pathol 1996;148(1):267–272.

101. Kitada T, Asakawa S, Hattori N, et al. Mutations in the parkin gene cause autosomal recessive juvenile parkinsonism. Nature 1998;392(6676):605–608.

102. Matsuoka Y, Vila M, Lincoln S, et al. Lack of nigral pathology in transgenic mice expressing human alpha-synuclein driven by the tyrosine hydroxylase promoter. Neurobiol Dis 2001;8(3):535–539.

103. Lo Bianco C, Ridet JL, Schneider BL, et al. Alpha-synucleinopathy and selective dopaminergic neuron loss in a rat lentiviral-based model of Parkinson's disease. Proc Natl Acad Sci U S A 2002;99(16):10813–10818.

104. Caughey B, Lansbury PT. Protofibrils, pores, fibrils, and neurodegeneration: separating the responsible protein aggregates from the innocent bystanders. Annu Rev Neurosci 2003;26:267–298.

105. Walsh DM, Selkoe DJ. Oligomers on the brain: the emerging role of soluble protein aggregates in neurodegeneration. Protein Pept Lett 2004;11(3):213–228.

106. Kontush A. Amyloid-beta: an antioxidant that becomes a pro-oxidant and critically contributes to Alzheimer's disease. Free Radic Biol Med 2001;31(9):1120–1131.

107. Zou K, Gong JS, Yanagisawa K, M. Michikawa M. A novel function of monomeric amyloid beta-protein serving as an antioxidant molecule against metal-induced oxidative damage. J Neurosci 2002;22(12):4833–4841.

108. Bishop GM, Robinson SR. Human Abeta1-42 reduces iron-induced toxicity in rat cerebral cortex. J Neurosci Res 2003;73(3):316–323.

109. Kontush A, Berndt C, Weber W, et al. Amyloid-beta is an antioxidant for lipoproteins in cerebrospinal fluid and plasma. Free Radic Biol Med 2001;30(1):119–128.

110. Lovell MA, Robertson JD, Teesdale WJ, et al. Copper, iron and zinc in Alzheimer's disease senile plaques. J Neurol Sci 1998;158(1):47–52.

111. Dong J, Atwood CS, Anderson VE, et al. Metal binding and oxidation of amyloid-beta within isolated senile plaque cores: Raman microscopic evidence. Biochemistry 2003;42(10):2768–2773.

112. Cuajungco MP, Goldstein LE, Nunomura A, et al. Evidence that the beta-amyloid plaques of Alzheimer's disease represent the redox-silencing and entombment of abeta by zinc. J Biol Chem 2000;275(26):19439–19442.

113. Lee HG, Perry G, Moreira PI, et al. Tau phosphorylation in Alzheimer's disease: pathogen or protector? Trends Mol Med 2005;11(4):164–169.

114. Takeda A, Smith MA, Avila J, et al. In Alzheimer's disease, heme oxygenase is coincident with Alz50, an epitope of tau induced by 4-hydroxy-2-nonenal modification. J Neurochem 2000;75(3):1234–1241.

115. Smith MA, Harris PL, Sayre LM, Perry G. Iron accumulation in Alzheimer disease is a source of redox-generated free radicals. Proc Natl Acad Sci U S A 1997;94(18):9866–9868.

116. Sayre LM, Perry G, Harris PL, et al. In situ oxidative catalysis by neurofibrillary tangles and senile plaques in Alzheimer's disease: a central role for bound transition metals. J Neurochem 2000;74(1):270–279.
117. Manning-Bog AB, McCormack AL, Purisai MG, et al. Alpha-synuclein overexpression protects against paraquat-induced neurodegeneration. J Neurosci 2003;23(8):3095–3099.
118. Uversky VN, Yamin G, Souillac PO, et al. Methionine oxidation inhibits fibrillation of human alpha-synuclein in vitro. FEBS Lett 2002;517(1-3):239–244.
119. Castellani RJ, Siedlak SL, Perry G, Smith MA. Sequestration of iron by Lewy bodies in Parkinson's disease. Acta Neuropathol (Berl) 2000;100(2):111–114.

Chapter 15

Redox Proteomics Identification of Oxidatively Modified Proteins in Alzheimer's Disease Brain and in Brain from a Rodent Model of Familial Parkinson's Disease: Insights into Potential Mechanisms of Neurodegeneration

Rukhsana Sultana, H. Fai Poon, and D. Allan Butterfield

Introduction

Oxidative stress, an imbalance between the oxidant and antioxidant systems, has been implicated in the pathogenesis of numerous neurodegenerative diseases [1]. Among all the body organs, the brain is particularly vulnerable to oxidative damage because of its high utilization of oxygen, increased levels of polyunsaturated fatty acids, and relatively high levels of redox transition metal ions in certain brain regions; in addition, the brain has relatively low levels of antioxidants [2–6]. The presence of iron ion in an oxygen-rich environment can further lead to enhanced production of superoxide radicals and ultimately to a cascade of oxidative events. Either the oxidant directly or the products of oxidative stress could trigger the oxidative modification of a number of cellular macromolecular targets, including proteins, lipids, DNA, RNA, and carbohydrates, which may lead to impairment of cellular functions [2,3,5,7–9].

Among the earliest of these changes following an oxidative insult are increased levels of toxic carbonyls, 3-nitrotyrosine (3-NT), and 4-hydroxy-2-trans-nonenal (HNE) [2,4,7,10–13]. HNE is derived from free radical attack on unsaturated acyl chains of phospholipids, particularly arachidonic acid. Oxidation leads to introduction of carbonyl groups to proteins [14]. Carbonyl groups are incorporated into proteins by direct oxidation of certain amino acid side chains, peptide backbone scission, or Michael addition reactions with products of lipid peroxidation or glycol oxidation [4,15,16]. Protein carbonyls can be detected by the derivatization of the carbonyl group with 2,4-dinitrophenylhydrazine (DNPH), followed by immunochemical detection of the hydrazone product [14]. Oxidative stress can stimulate additional damage via overexpression of inducible nitric oxide synthase (iNOS) and the action of constitutive

R. Sultana
Department of Chemistry; Sanders-Brown Center on Aging, University of Kentucky, Lexington, KY 40506-0055, USA

A. Fisher et al. (eds.), *Advances in Alzheimer's and Parkinson's Disease*,
© Springer 2008

neuronal NOS (nNOS), which leads to increased levels of 3-NT. The levels of thiobarbituric acid reactive substance (TBARS), free fatty acid release, HNE and acrolein formation, and iso- and neuroprostane formation are the most commonly used parameters to index lipid peroxidation. DNA and RNA oxidation are measured by formation of 8-OH-2α-deoxyguanosine and other oxidized bases as well as altered DNA repair mechanisms.

Proteomics

Oxidatively modified brain proteins were initially identified using immunopre-cipitation methods [17,18]. However, there are serious limitations to the use of this technique. For example, prior knowledge about the identity of the protein of interest is required, the availability of the particular antibody for the protein of interest is necessary, and the time-consuming and laborious nature of the process is a hindrance. In addition, posttranslational modification of protein may change the structure of proteins, thereby preventing the formation of the appropriate antigen–antibody complex. Redox proteomics has enabled us to identify a large number of oxidatively modified proteins in cells, tissues, and other biological samples that were previously undetected by other methods such as immunoprecipitation [19–21]. Unlike gene analysis and mRNA analysis, proteomics provides a broad spectrum of information that allows insights into the mechanisms of disease and identification of disease-associated markers and may also help to identify selected targets for specific therapy (Fig. 1).

Redox proteomics couples two-dimensional (2D) gel electrophoresis separa-tion of proteins and 2D Western blots with mass spectrometric techniques that

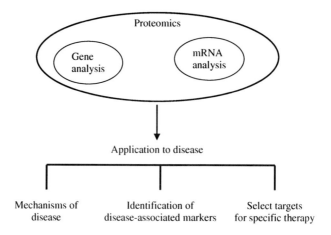

Fig. 1 Proteomics: global analysis of cellular proteins

allow facile identification of oxidatively modified proteins without consuming as much time and effort as immunochemical methods [19–21]. Two-dimensional polyacrylamide gel electrophoresis (2D-PAGE) allows the investigator to analyze complex protein mixtures based on two important physicochemical properties: isoelectric focusing (IEF), which separates proteins based on their isoelectric points (pI); and separation of proteins based on their relative mobility (Mr) on sodium dodecyl sulfate (SDS)-PAGE in the second dimension [22]. Normally a single spot on the 2D gel represents a single protein [23]. This property allows separation of thousands of different protein spots on one gel. In addition, 2D-PAGE is used to catalog proteins and create databases [24].

2D-PAGE is a sensitive, reliable method with high reproducibility, although many challenges still exist. The first serious limitation of 2D is the solubilization process for membrane proteins [25] as ionic detergents would introduce a charge to the protein, thereby interfering with IEF. The inability to detect low-abundance proteins is the second limitation of 2D-PAGE; and the third limitation is the insensitivity to proteins of high lysine and arginine content (which leads to small tryptic peptides that could be lost on a gel). The use of chaotropic agents such a urea and thiourea coupled with nonionic or zwitterionic detergents can solubilize proteins and also avoid protein precipitation during the IEF and the SDS gel processes [26]. The use of immobilized pH IEF strips (immobilized pH gradient, or IPG, strips) improves the reproducibility of proteins maps and also eliminates the typical cathodic drift associated with previously used tube gels [27]. The use of narrow-range IPG strips enables protein separation over a wide pH range but within 1 pH unit. However, the normally employed IEF strip pH range (i.e., 3–10) limits the identification of highly basic proteins. If a protein from a low-abundance protein group were involved in the pathogenesis of a disease, it would be difficult to use this technique for detection.

In our laboratory we coupled redox proteomics techniques with immunochemical detection of protein carbonyls derivatized by 2,4-dinitrophenylhydrazine (DNPH), nitrated proteins indexed by 3-NT, and protein adducts of HNE, followed by mass spectrometric (MS) analysis (as shown in Fig. 2) to identify oxidatively modified proteins from Alzheimer's disease (AD) brain and related models. With this method we employ a parallel analysis: The 2D Western blots and 2D gel images are matched by computer-assisted image analysis, and the anti-DNP/nitrotyrosine/HNE immunoreactivity of individual proteins are normalized to their content, obtained by measuring the intensity of colloidal Coomassie Blue staining or SYPRO ruby-stained spots (Figs. 3, 4). Such analysis allows comparison of levels of oxidatively modified brain proteins in experimental versus control subjects. Once the protein is identified as oxidatively modified, it is digested in gel with a protease (e.g., trypsin) that not only cleaves the protein into small peptides but produces sequence-specific proteolysis. These mass fingerprints are modified proteins from AD brain and related models; they are characteristic of a particular protein, which facilitates the identification of a particular protein using a

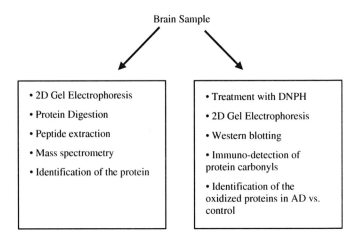

Fig. 2 Redox proteomics to identify oxidatively modified brain proteins in Alzheimer's disease

suitable database (Table 1) that compares the experimental masses with theoretical masses of trypsin-generated protein sequences.

Mass spectrometry determines the peptide masses and can determine the amino acid sequence for the proteins of interest. Modern MS instruments use softer ionization techniques than previously, and they can provide a precise peptide mass. The two most commonly employed MS techniques are MALDI

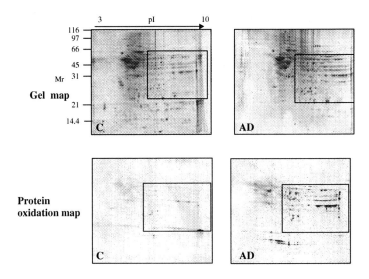

Fig. 3 Two-dimensional maps of brain proteins from controls (C) and Alzheimer's disease patients (AD)

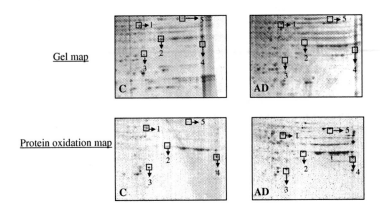

Fig. 4 Oxidatively modified proteins in Alzheimer's disease brain (AD) identified by redox proteomics using expanded two-dimensional oxyblots from Figure 3. 1, enolase; 2, glyceraldehyde 3-phosphate dehydrogenase; 3, carbonic anhydrase II; 4, voltage-dependent anion channel protein-1; 5, ATP synthase α chain

(matrix-assisted laser desorption/ionization) and ESI (electrospray ionization). With MALDI analysis the peptide sample is mixed with a matrix, usually α-cyano-4-hydroxycinnamic acid or 2,5-dihydroxybenzoic acid, and deposited on a plate that is subjected to laser radiation. The matrix absorbs the energy, which is then transferred to the peptides. The peptides then vaporize as detectable MH^+ ions by an unknown mechanism related to the acidic nature of the matrix.

In contrast to MALDI MS, ESI permits direct MS analysis of the samples from high-performance liquid chromatography (HPLC) columns for characterization. towing to the high potential difference between the capillary and the MS instrument, the inlet sample is dispersed as small droplets. These droplets undergo solvent evaporation until droplet fission occurs, because of the high charge-to-surface tension ratio, finally leading to the formation of a single detectable ion per droplet. The best online preseparation of peptides with HPLC and MS requires low salt concentration. In addition, reducing the flow time to nanoliters per minute can increase the time for analysis. Tandem MS/MS provides better isolation and fragmentation of a specific ion. This tandem MS/MS technique provides further information about the sequence

Table 1 Mass spectrometry search engines for peptide mass fingerprinting

Search engine and URL
Mascot—http://www.matrixscience.com
MOWSE—http://www.hgmp.mrc.ac.uk/Bioinformatics/Webapp/mowse
Profound—http://www.prowl.rocketfeller.edu/sgi-bin/profound
MS-fit—http://www.prospector.ucsf.edu/ucsfhtlm3.4/msfit.htm
Peptident—http://www.expasy.ch.ch/tools/peptident.html

of the protein [28]. With MS/MS analysis, the isolation of a single ion is achieved by scanning all of the ions that were generated from a sample, followed by application of a wide range of frequencies, except for the resonating frequency of the ion of interest. Fragmentation of the isolated ion, which provides additional information for protein identification or for evaluation of possible protein modification, is the final step in MS/MS.

The identity of proteins is determined by employing online databases following MS analysis. SwissProt, the most commonly used database for protein identification, is based on computer algorithms [29]. SwissProt and other databases are available through the Internet; they are listed in Table 1. These search engines provide theoretical protease digestion of the proteins contained in the database, to which are compared the experimental masses obtained by MS. The successful protein identification using these databases also accounts for several factors, such as protein size and the probability that a single peptide occurs in the whole database. The search engine produces a probability score for each entry, which is calculated by a mathematical algorithm specific for each search engine. Any hit with a score higher that the one specific for significance of the particular search engine is considered statistically significant and has a valid chance to be the protein cut from a given spot. In addition, the molecular weight and pI of the protein is calculated based on the position in the 2D map to avoid any false identification. In many cases, validation of protein identification is achieved by immunochemical means [13,18,30–34].

In this chapter, we review the redox proteomics identification of oxidatively modified proteins in AD and Parkinson's disease (PD), two age-related neurodegenerative disorders that involve deposition of aggregated proteins (Aβ, synuclein, and parkin) as pathological hallmarks of the respective disorders.

Alzheimer's Disease

Alzheimer's disease is an age-related neurodegenerative disorder characterized by progressive loss of memory and cognition, accumulation of extracellular amyloid plaques (Aβ) and intracellular neurofibrillary tangles (NFTs), and loss of synaptic connections in selective brain regions. NFTs consist of aggregates of hyper-phosphorylated microtubule-associated protein tau that form paired helical filaments and related straight filaments [35]. Amyloid α-peptide (Aβ), a 40- to 42-amino acid peptide derived from proteolytic cleavage of an integral membrane protein known as amyloid precursor protein (APP) by the action of β- and γ-secretases, is the main amyloid component of senile plaque (SP). Ab is thought to play a casual role in the development and progression of AD [36]. Furthermore, a number of studies suggest that the small oligomers of Aβ are the actual toxic species of this peptide rather than Aβ fibrils [37–40].

Several mechanisms have been proposed to explain AD pathogenesis. These mechanisms include amyloid cascade, excitoxicity, oxidative stress, and inflammation. We previously showed that regions of AD brain rich in β have increased protein oxidation, whereas Aβ-poor cerebellum does not [41]. Protein carbonyls, HNE, and 3-nitrotyrosine levels were found to be elevated in AD brain and cerebrospinal fluid (CSF) [9,37,41], results that support the oxidative stress hypothesis of AD. Moreover, the observation that vitamin E in cell culture diminishes Aβ(1-42)-induced oxidative stress and neurotoxicity further supports a role of oxidative stress in AD pathology [2,42,43]. Aβ-induced lipid peroxidation leads to increased formation of HNE in vitro and also was observed in AD brain and CSF [8,16,17,44]. Using immunoprecipitation techniques, Lauderback et al. showed the HNE-mediated oxidative modification of glutamate transporter (GLT-1) in AD brain. GLT-1 is involved in regulating the levels of glutamate outside the neuron. These researchers also observed that synaptomes treated with β(1-42) demonstrated HNE-modified GLT-1 [17]. This oxidative modification leads to altered structure [45] and loss of function of the transport protein, which could eventually lead to excitotoxic neuronal death (Fig. 5) [46].

One of the mechanisms for removal of HNE from neurons is by conjugation to GSH, followed by the action of glutathione S-transferase (GST) and the multidrug resistant protein-1 (MRP-1) to efflux this conjugate from the cell [47]. However, in AD brain, GST and MRP-1 were demonstrated to have excessively bound HNE and showed reduced activity, supporting the idea that oxidative modification leads to loss of functionality [18].

As noted above, there are several serious limitations to the use of immuno-precipitation to identify proteins, including the requirement of prior identification of the protein of interest, the availability of a specific antibody for this protein, and the extensive time needed for this process. Moreover, sometimes a posttranslational modification can change the structure of proteins, thereby preventing the formation of the appropriate antigen–antibody complex. Redox

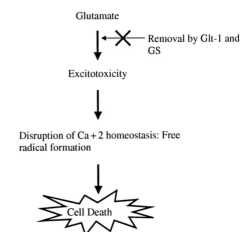

Fig. 5 The glutamate transporter Glt-1 and glutamine synthase (GS) modulate glutamate-induced excitotoxicity. Conversely, if Glt-1 and GS are oxidatively modified and lose functionality, neuronal death can occur

Table 2 Proteomic identification of specifically oxidatively modified proteins in AD brain

- Energy dysfunction—creatine kinase; α-enolase; γ-enolase; triose phosphate isomerase; phosphoglycerate mutase 1
- Excitotoxicity—glutamine synthase; glutamate transported by EAAT2
- Proteasomal dysfunction—ubiquitin carboxy-terminal hydrolase L-1
- Lipid abnormalities and cholinergic dysfunction—neuropolypeptide h3
- Neuritic abnormalities—dihydropyrimidinase-related protein 2; β-actin
- Tau hyperphosphorylation—peptidyl prolyl *cis*-trans isomerase
- Synaptic abnormalities—γ-soluble *N*-ethylmaleimide-sensitive factor attachment protein
- pH buffering and CO_2 transport—carbonic anhydrase II

proteomics, which couples 2D gel electrophoresis separation of proteins and 2D Western blots with MS techniques, is highly successful in identifying oxidatively modified brain proteins [19–21]. Proteomics has enabled us to identify a large number of oxidatively modified proteins in AD brain and models thereof.

The identification of oxidatively modified proteins in the AD inferior parietal lobule (IPL) and hippocampus was accomplished by proteomics [10,13,33,34,48,49]. This research has provided insights into the role of oxidative stress in AD and has helped to unravel the mechanisms associated with AD pathology [10,13,33,34,48,49]. Oxidatively modified brain proteins identified by our redox proteomics approach include creatine kinase BB (CK), glutamine synthase (GS), ubiquitin carboxy-terminal hydrolase L-1 (UCH L-1), triose phosphate isomerase (TPI), neuropolypeptide h3, dihydropyrimidinase-related protein 2 (DRP2), α-enolase, phosphoglycerate mutase 1 (PGM1), γ-soluble NSF attachment protein (SNAP), carbonic anhydrase II (CA-II), and peptidyl prolyl cis-trans isomerase (Pin 1). No oxidatively modified proteins were identified in cerebellum [34], confirming earlier studies [41].

The oxidatively modified proteins in AD brain are involved in known dysfunctional processes in AD. The identified oxidatively modified proteins were grouped based on their functions (Table 2) and were linked to AD pathology, symptomatology, and loss of enzyme activity, consistent with a plausible mechanism of neurodegeneration [19,20].

Energy Dysfunction

Creatine kinase, TPI, ATP synthase-α, GAPDH, VDAC-1, PGM1, and α-enolase are enzymes involved in energy metabolism and were identified as oxidized proteins with reduced activity in AD brain, that could be linked to the observed decreased ATP production in AD [50] and could be detrimental to neurons [10,13,30,34,48,49,51]. Decreased ATP production would lead to impaired ion-motive ATPases, altered protein synthesis, and maintenance of synaptic transmission, all of which are hallmarks of AD [10,13,30,34,48,49,52]. Decreased ATP production could induce hypothermia, leading to abnormal tau hyperphosphorylation through differential inhibition of kinase and phosphatase activities, ion pumps, electrochemical gradients, cell potential, and voltage-gated ion channels [53].

Excitotoxicity

Glutamine synthase and EAAT2 (GLT-1) are involved in regulating extraneuronal glutamate levels and neurotransmission. GS and EAAT2 oxidation could lead to accumulation of glutamate in the synaptic cleft, leading to influx of calcium into the cell via activation of N-methyl-D-aspartate (NMDA) and α-amino-3-hydroxy-5-methyl-4-isoxazole propionic acid (AMPA) receptors and causing neuronal excitotoxic death [54]. As noted above, HNE, a lipid peroxidation product, has been shown to modify oxidatively the glutamate transporter EAAT2 in AD brain and synaptosomes treated with Aβ(1-42) [17].

Proteasomal Dysfunction

When proteins are damaged or aggregated, they become ubiquitinylated, as a polyubiquitin chain. Such poly (ubiquitin) chains can be as large as 70 units [55]. The poly (ubiquitin) chain is a marker that targets the damaged protein to the 26S proteasome for subsequent degradation. UCH L-1 removes ubiquitin from the poly (ubiquitin) chain, one ubiquitin unit at a time from the carboxyl terminal end before insertion of the damaged protein into the core of the proteasome [56]. This has the effect of maintaining the pool of ubiquitin in the brain. Oxidative modification of UCH L-1 was found in AD brain [34,48,57]. Presumed resultant decreased UCH L-1 activity in AD brain could lead to depletion of the free pool of ubiquitin or cause saturation of the proteasome with polyubiquitin chains and accumulation of damaged proteins, leading to synaptic deterioration and degeneration. Decreased activity of UCHL-1 would lead to increased protein ubiquitinylation, decreased proteasomal activity, and accumulation of damaged and aggregated proteins, all of which are observed in AD brain [20,58]. A recent in vitro study showed that HNE decreases hydrolase activity of recombinant UCH-L1 [54,59], and that the HNE-bound protein and crosslinked proteins could impair proteasomal function [59]. Others recently confirmed the oxidative modification of UCH L-1 in AD brain using proteomics [57]. Interestingly, if UCH L-1 is dysfunctional, as it is in the gracile axonal dystrophic mouse, oxidative modification of important brain proteins occurs [60].

Lipid Abnormalities and Cholinergic Dysfunction

Neuropolypeptide h3 (also known as phosphatidylethanolamine-binding protein, or PEBP) may play an important role in maintaining phospholipid asymmetry of the membrane [61]. Oxidative modification of neuropolypeptide h3 has been observed in AD brain [10]. Because this protein is indirectly involved in the production of choline acetyltransferase, oxidative modification of neuro-polypeptide h3 could lead to altered choline acetyltransferase levels. Moreover, in its role as PEBP, its oxidative modification could lead to apoptosis by the exposure of phosphatidylserine to the outer bilayer leaflet of the membrane, leading to cell death and observed cognitive decline in AD [62]. Aβ(1-42) and HNE added to synaptosomes lead to a loss of phospholipid asymmetry [63,64].

Neuritic Abnormalities

DRP-2 is involved in axonal outgrowth and pathfinding through transmission and modulation of extracellular signals, [65–67] and β-actin is involved in cell integrity. Decreased expression of DRP-2 protein was observed in AD, adult Down's syndrome (DS) [68], fetal DS, [69] schizophrenia, and affective disorders.70 Oxidation of DRP-2 and β-actin, as observed in AD brain, [10,37,49] could be related to the observed shortening of dendrites and synapse loss in AD brain.71 Shortened dendrites would be predicted to lead to less efficient inter-neuronal communication, which could be important in a cognitive and memory disorder.

Tau Hyperphosphorylation

Peptidyl-prolyl isomerases (PPIases or Pin 1) catalyze the conversion of the cis to trans conformation and vice versa of proteins between given amino acids and a proline [72]. Also, PPIases have been shown to be necessary for entry into mitosis, and they interact with cell cycle regulating proteins (e.g., p53, Myt1, Wee1, Cdc25C). We determined by proteomics that PPIase (Pin 1) is oxidized in AD brain [33,34]. This modification conceivably could cause dramatic structural modifications, which could affect the properties of targeted proteins. One target for Pin1 is tau, a neuronal cytoskeletal protein, which is hyperphosphorylated in AD [73]. Recent studies show an inverse relation of Pin 1 activity and co-localization with phosphorylated tau in AD brain [74–76]. In addition, the cell cycle machinery of AD neurons was reported to be altered in AD brain [77,78]. Pin 1 oxidation and decreased activity could therefore be involved in the initial events that trigger tangle formation, cell cycle-related abnormalities, and oxidative damage [33,34,79]. All these effects can lead to memory loss.

Synaptic Abnormalities and LTP

Oxidation of γ-SNAP, a member of the N-ethylmaleimide-sensitive factor (NSF) attachment proteins (SNAPs), could impair vesicular transport in the constitutive secretory pathway as well as in neurotransmitter release, hormone secretion, and mitochondrial organization [34,80,81]. This, in turn, could lead to impaired learning and memory processes and altered neurotransmitter systems in AD brain.

pH Buffering and CO_2 Transport

Carbonic anhydrase II plays an important role in regulating cellular pH, CO_2 and HCO_3^- transport, and maintaining H_2O and electrolyte balance [82]. CA-II deficiency leads to cognitive defects, varying from disabilities to severe mental retardation, in addition to osteoporosis, renal tubular acidosis, and cerebral calcification. Oxidation and decreased activity of this protein was observed in AD brain compared to age-matched controls [13,34,83]. Oxidization of CA-II may lead to an imbalance of both the extracellular and intracellular pH in the cell, mitochondrial alterations in oxidative phosphorylation, and impaired synthesis of glucose and lipids. Moreover, altered neuronal pH could contribute to the known protein aggregation in AD brain.

AD Models for $A\beta(1\text{-}42)$

Identification of oxidatively modified AD brain proteins was substantially recapitulated in vitro and in vivo by action of human $A\beta(1\text{-}42)$ in neuronal cell cultures, synaptosomes, intracerebral injection into rat basal forebrain, and expression in Caenorhabditis elegans [31,84–87]. These findings are consistent with the notion that $A\beta(1\text{-}42)$ (Fig. 6) contributes to the observed oxidative stress and oxidative modification of proteins in AD brain [4,19,20].

Parkinson's Disease

Parkinson's disease (PD) is the second most common neurodegenerative disorder affecting the population of age 65 and older [88]. Clinical symptoms of PD, such as bradykinesia, resting tremor, cogwheel rigidity, and postural instability, result from loss of dopaminergic neurons in the substantia nigra compacta. Mutations in α-synuclein, parkin, DJ-1, and PINK1 contribute to early-onset familial PD [89]. Four mutations of α-synuclein have been identified in familial PD: A53T, A30P, E46A, and genomic duplication [90,91].

Oxidative damage is a well known pathological change in PD brains [92–95]. Overexpression of wild-type or mutant α-synuclein induces toxicity that is associated with oxidative stress [96]. Moreover, oxidative stress in PD is linked to cell death in PD brains by mitochondrial dysfunction, excitotoxicity, and the toxic effects of nitric oxide [94].

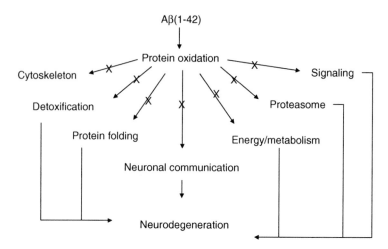

Fig. 6. Potential dysfunction of proteomics-identified β(1-42)-induced oxidized proteins. Protein oxidative modification is similar to that in Alzheimer's disease brain

Redox Proteomics in PD

α-Synuclein has a strong tendency to aggregate, leading to neurotoxicity. Expression of mutant α-synuclein in cells produces increased oxidative parameters and accelerated cell death in response to oxidative insult [97]. Symptoms in A30P α-synuclein transgenic mice occur in parallel with the aggregation of α-synuclein, [98] and these mice develop an age-dependent accumulation of α-synuclein in neurons of the brain stem [99,100], suggesting that α-synuclein aggregation-associated oxidative stress is involved in the pathology in A30P α-synuclein transgenic mice.

Using redox proteomics, several significantly oxidatively modified brain stem proteins were identified in symptomatic mice with overexpression of a A30P mutation in α-synuclein compared to the brain proteins from the nontransgenic mice. These proteins were identified as carbonic anhydrase 2 (CA-II), α-enolase (ENO1), and lactate dehydrogenase 2 (LDH2) [101]. The activities of these enzymes were also significantly decreased in the A30P α-synuclein transgenic mice brains when compared to the brain proteins from nontransgenic control [101]. This observation is consistent with the notion that oxidative modification of proteins leads to loss of their activities [16–18,41].

Carbonic anhydrase II, which, as noted above, is an oxidatively modified protein in AD brain, is a Zn^{2+} metalloenzyme that catalyzes reversible hydration of CO2 to bicarbonate (HCO3–). CA-II shares high (68%) similarity to the mitochondrial counterpart carbonic anhydrase 5a (CA-5a) and 5b (CA-5b), implicating the potential coupling or interaction with each other to function in metabolic processes, cellular transport, gluconeogenesis, and mitochondrial metabolism [102,103]. Oxidative modification of CA-II may lead to loss of the buffering system in brains with resultant aggregation of synuclein and subsequent neurodegeneration.

LDH2 is a subunit of lactate dehydrogenase (LDH), a glycolytic protein that catalyzes the reversible interconversion of pyruvate to lactate. Lactate is the predominant monocarboxylate oxidized by mitochondria for intracellular lactate transport [104]. Therefore, oxidative inactivation of LDH may contribute to mitochondrial dysfunction in PD patients.

Eno1 is a subunit of enolase that interconverts 2-phosphoglycerate and phosphoenolpyruvate during glycolysis. Enolase was identified in an intermembrane space/outer mitochondrial membrane fraction [105]. These studies suggest that enolase is present in mitochondria and contributes to mitochondrial function. Therefore, oxidative inactivation of enolase may alter normal glycolysis and mitochondrial function in brains and may contribute to the alteration of energy metabolism in PD. Interestingly, LDH2 and ENO1 (possibly CA II) are associated with mitochondrial function. Increasing data implicate mitochondrial dysfunction and oxidation in PD [94,106–108]. Furthermore, 1-methyl-4-phenyl-1,2,3,6-tetrahydropyridine (MPTP) and rotenone lead to mitochondrial dysfunction with increased oxidative modification of proteins and α-synuclein aggregation [109–112]. Moreover, DJ-1, PINK1, and parkin all appear to modulate mitochondrial function [113–115]. The observation that each of the oxidatively modified brain proteins in A30P mutant synuclein mice is associated with mitochondria provides strong evidence for the notion that mitochondrial dysfunction contributes to the toxicity of PD and implicates mitochondrial pathology in toxicity associated with aggregated synuclein. This suggests that the oxidative stress-mediated mitochondrial dysfunction may be responsible, at least partially, for the neurodegeneration in the brains of A30P α-synuclein transgenic mice. Furthermore, this oxidative stress-mediated impaired energy metabolism and mitochondrial dysfunction is contributed by the oxidative inactivation of ENO1, LDH2, and CA-II. Therefore, the mitochondria dysfunction in familial PD may be associated with the oxidative inactivation of ENO1, LDH2, and CA-II.

Conclusions

The application of redox proteins to AD brain revealed important targets of protein oxidation. The use of in vivo and in vitro models of AD with human Aβ(1-42), which led to the identification of oxidatively modified proteins similar to those found in AD brain, provided strong evidence of the oxidative stress and neurotoxicity associated with Aβ(1-42) in AD brain. The use of relevant models for AD could be a powerful tool to investigate the role and mechanisms of Aβ(1-42) in the pathogenesis of AD. Furthermore, the use of animal models together with redox proteomics approaches have provided potential insights into the mechanisms of neurodegeneration in AD and PD and may also be of value in the development of therapeutic approaches to prevent or delay these neurodegenerative disorders.

Acknowledgments This work was supported in part by NIH grants to D.A.B. (AG-05119, AG-10836). We thank Dr. Benjamin Wolozin for providing brains from A30P mutant mice.

References

1 Butterfield DA. Amyloid beta-peptide (1-42)-induced oxidative stress and neurotoxicity: implications for neurodegeneration in Alzheimer's disease brain—a review. Free Radic Res 2002;36(12):1307–1313.

2 Butterfield DA, CastegnaA, Lauderback CM, Drake J. Evidence that amyloid beta-peptide-induced lipid peroxidation and its sequelae in Alzheimer's disease brain contribute to neuronal death. Neurobiol Aging 2002;23(5):655–664.

3 Butterfield DA, Drake J, Pocernich C, Castegna A. Evidence of oxidative damage in Alzheimer's disease brain: central role for amyloid beta-peptide. Trends Mol Med 2001;7(12):548–554.

4 Butterfield DA, Lauderback CM, Lipid peroxidation and protein oxidation in Alzheimer's disease brain: potential causes and consequences involving amyloid beta-peptide-associated free radical oxidative stress. Free Radic Biol Med 2002;32(11):1050–1060.

5 Markesbery WR. Oxidative stress hypothesis in Alzheimer's disease. Free Radic Biol Med 1997;23(1):134–147.

6 Pocernich CB, Cardin AL, Racine CL, et al. Glutathione elevation and its protective role in acrolein-induced protein damage in synaptosomal membranes: relevance to brain lipid peroxidation in neurodegenerative disease. Neurochem Int 2001;39(2):141–149.

7 Lovell MA, Xie C, Markesbery WR, Acrolein is increased in Alzheimer's disease brain and is toxic to primary hippocampal cultures. Neurobiol Aging 2001;22(2):187–194.

8 Mark RJ, Lovell MA, Markesbery WR, et al. A role for 4-hydroxynonenal, an aldehydic product of lipid peroxidation, in disruption of ion homeostasis and neuronal death induced by amyloid beta-peptide. J Neurochem 1997;68(1):255–264.

9 Smith MA, Richey PL, Taneda S, et al. Advanced Maillard reaction end products, free radicals, and protein oxidation in Alzheimer's disease. Ann N Y Acad Sci 1994;738:447–454.

10 Castegna A, Thongboonkerd V, Klein JB, et al. Proteomic identification of nitrated proteins in Alzheimer's disease brain. J Neurochem 2003;85(6):1394–1401.

11 Smith MA, Richey Harris PL, Sayre LM, et al. Widespread peroxynitrite-mediated damage in Alzheimer's disease. J Neurosci 1997;17(8):2653–2657.

12 Smith MA, Sayre LM, Monnier VM, Perry G. Oxidative posttranslational modifications in Alzheimer disease: a possible pathogenic role in the formation of senile plaques and neurofibrillary tangles. Mol Chem Neuropathol 1996;28(1-3):41–48.

13 Sultana R, Poon HF, Cai J, et al. Identification of nitrated proteins in Alzheimer's disease brain using redox proteomics approach. Neurobiol Dis 2006;22(1):76–87.

14 Butterfield DA, Stadtman ER. Protein oxidation processes in aging brain. Adv Cell Aging Gerontol 1997;2:161–191.

15 Berlett BS, Stadtman ER, Protein oxidation in aging, disease, and oxidative stress. J Biol Chem 1997;272(33):20313–20316.

16 Butterfield DA, Hensley K, Cole P, et al. Oxidatively induced structural alteration of glutamine synthetase assessed by analysis of spin label incorporation kinetics: relevance to Alzheimer's disease. J Neurochem 1997;68(6):2451–2457.

17 Lauderback CM, Hackett JM, Huang FF, et al. The glial glutamate transporter, GLT-1, is oxidatively modified by 4-hydroxy-2-nonenal in the Alzheimer's disease brain: the role of Abeta1-42. J Neurochem 2001;78(2):413–416.

18 Sultana R, Butterfield DA, Oxidatively modified GST and MRP1 in Alzheimer's disease brain: implications for accumulation of reactive lipid peroxidation products. Neurochem Res 2004;29:2215–2220.

19 Butterfield DA. Proteomics: a new approach to investigate oxidative stress in Alzheimer's disease brain. Brain Res 2004;1000(1-2):1–7.

20 Butterfield DA, Perluigi M, Sultana R. Oxidative stress in Alzheimer's disease brain: new insights from redox proteomics. Eur J Pharmacol 2006;545(1):39–50.

21 Dalle-Donne I, Scaloni A, Butterfield DA, Redox Proteomics: From protein Modifications to Cellular Dysfunction and Diseases. Wiley, Hoboken, NJ, 2006.

22 Rabilloud T. Two-dimensional gel electrophoresis in proteomics: old, old fashioned, but it still climbs up the mountains. Proteomics 2002;2:3–10.

23 Tilleman K, Stevens T, Spittaels I, et al. Differential expression of brain proteins in glycogen synthase kinase-3 transgenic mice: a proteomics point of view. Proteomics 2002;2:94–104.

24 Kaji H, Tsuji T, Mawuenyega KG, et al. Profiling of Caenorhabditis elegans proteins using two-dimensional gel electrophoresis and matrix assisted laser desorption/ionization-time of flight-mass spectrometry. Electrophoresis 2000;21(9):1755–1765.

25 Santoni V, Molloy M, Rabilloud T. Membrane proteins and proteomics: un amour impossible? Electrophoresis 2000;21(6):1054–1070.

26 Herbert B. Advances in protein solubilization for two-dimensional gel electrophoresis. Electrophoresis 1999;20:660–663.

27 Molloy MP. Two-dimensional electrophoresis of membrane proteins using immobilized pH gradients, Anal Biochem 2000;280(1):1–10.

28 Aebersold R, Goodlett DR, Mass spectrometry in proteomics. Chem Rev 2001;101(2):269–295.

29 Hoogland C, Sanchez C, Tonella L, et al. The 1999 SWISS-2DPAGE database update. Nucleic Acids Res 2000;28(1):286–288.

30 Aksenova M, Butterfield DA, Zhang SX, et al. Increased protein oxidation and decreased creatine kinase BB expression and activity after spinal cord contusion injury. J Neurotrauma 2002;19(4):491–502.

31 Boyd-Kimball D, Castegna A, Sultana R, et al. Proteomic identification of proteins oxidized by Abeta(1-42) in synaptosomes: implications for Alzheimer's disease. Brain Res 2005;1044(2):206–215.

32 Butterfield DA, Poon HF, St Clair D, et al. Redox proteomics identification of oxidatively modified hippocampal proteins in mild cognitive impairment: insights into the development of Alzheimer's disease. Neurobiol Dis 2006;22(2):223–232.

33 Sultana R, Boyd-Kimball D, Poon HF, et al. Oxidative modification and down-regulation of Pin 1 Alzheimer's disease hippocampus: a redox proteomics analysis. Neurobiol Aging 2006;27(7):918–925.

34 Sultana R, Boyd-Kimbal Dl, Poon HF, et al. Redox proteomics identification of oxidized proteins in Alzheimer's disease hippocampus and cerebellum: an approach to understand pathological and biochemical alterations in AD. Neurobiol Aging 2006;27:1564–1576.

35 Grundke-Iqbal I, Iqbal K, Quinlan M, et al. Microtubule-associated protein tau: a component of Alzheimer paired helical filaments. J Biol Chem 1986;261(13):6084–6089.

36 Selkoe DJ. Presenilin, notch, and the genesis and treatment of Alzheimer's disease. Proc Natl Acad Sci U S A 2001;98(20):11039–11041.

37 Aksenov MY, Aksenova MV, Butterfield DA, et al. Protein oxidation in the brain in Alzheimer's disease. Neuroscience 2001;103(2):373–383.

38 Drake J, Link CD, Butterfield DA, Oxidative stress precedes fibrillar deposition of Alzheimer's disease amyloid beta-peptide (1-42) in a transgenic Caenorhabditis elegans model. Neurobiol Aging 2003;24(3):415–420.

39 Lambert MP, Viola KL, Chromy BA, et al. Vaccination with soluble Abeta oligomers generates toxicity-neutralizing antibodies. J Neurochem 2001;79(3):595–605.

40 Oda T, Wals P, Osterburg HH, et al. Clusterin (apoJ) alters the aggregation of amyloid beta-peptide (Abeta1-42) and forms slowly sedimenting Abeta complexes that cause oxidative stress. Exp Neurol 1995;136(1):22–31.

41 Hensley K, Hall K, Subramaniam R, et al. Brain regional correspondence between Alzheimer's disease histopathology and biomarkers of protein oxidation. J Neurochem 1995;65(5):2146–2156.

42 Boyd-Kimball D, Sultana R, Mohmmad-Abdul H, Butterfield DA, Rodent Abeta(1-42) exhibits oxidative stress properties similar to those of human Abeta(1-42): implications for proposed mechanisms of toxicity. J Alzheimers Dis 2004;6(5):515–525.

43 Yatin SM, Varadarajan S, Butterfield DA. Vitamin E prevents Alzheimer's amyloid beta-peptide (1-42)-induced neuronal protein oxidation and reactive oxygen species production. J Alzheimers Dis 2000;2(2):123–131.

44 Markesbery WR, Lovell MA. Four-hydroxynonenal, a product of lipid peroxidation, is increased in the brain in Alzheimer's disease. Neurobiol Aging 1998;19(1):33–36.

45 Subramaniam R, Roediger F, Jordan B, et al. The lipid peroxidation product, 4-hydroxy-2-trans-nonenal, alters the conformation of cortical synaptosomal membrane proteins. J Neurochem 1997;69(3):1161–1169.

46 Butterfield DA, Pocernich CB. The glutamatergic system and Alzheimer's disease: therapeutic implications. CNS Drugs 2003;17(9):641–652.

47 Paumi CM, Wright M, Townsend AJ, Morrow CS, Multidrug resistance protein (MRP) 1 and MRP3 attenuate cytotoxic and transactivating effects of the cyclopentenone prostaglandin, 15-deoxy-delta(12,14)prostaglandin J2 in MCF7 breast cancer cells. Biochemistry 2003;42(18):5429–5437.

48 Castegna A, Aksenov M, Aksenova M, et al. Proteomic identification of oxidatively modified proteins in Alzheimer's disease brain. Part I. Creatine kinase BB, glutamine synthase, and ubiquitin carboxy-terminal hydrolase L-1. Free Radic Biol Med 2002;33(4):562–571.

49 Castegna A, Aksenov M, Thongboonkerd V, et al. Proteomic identification of oxidatively modified proteins in Alzheimer's disease brain. Part II. Dihydropyrimidinase-related protein 2, alpha-enolase and heat shock cognate 71. J Neurochem 2002;82(6):1524–1532.

50 Small GW, Okonek A, Mandelkern MA, et al. Age-associated memory loss: initial neuropsychological and cerebral metabolic findings of a longitudinal study. Int Psychogeriatr 1994;6(1):23–44; discussion 60–22.

51 Mazzola JL, Sirover MA, Reduction of glyceraldehyde-3-phosphate dehydrogenase activity in Alzheimer's disease and in Huntington's disease fibroblasts. J Neurochem 2001;76(2):442–449.

52 Hoyer S. Causes and consequences of disturbances of cerebral glucose metabolism in sporadic Alzheimer disease: therapeutic implications. Adv Exp Med Biol 2004;541:135–152.

53 Planel E, Miyasaka T, Launey T, et al. Alterations in glucose metabolism induce hypothermia leading to tau hyperphosphorylation through differential inhibition of kinase and phosphatase activities: implications for Alzheimer's disease. J Neurosci 2004;24(10):2401–2411.

54 Masliah E, Alford M, DeTeresa R, et al. Deficient glutamate transport is associated with neurodegeneration in Alzheimer's disease. Ann Neurol 1995;40:759-766.

55 Pickart CM, Fushman D, Polyubiquitin chains: polymeric protein signals. Curr Opin Chem Biol 2004;8(6):610–616.

56 Wilkinson KD, Tashayev VL, O'Connor LB, et al. Metabolism of the polyubiquitin degradation signal: structure, mechanism, and role of isopeptidase T. Biochemistry 1995;34(44):14535–14546.

57 Choi J, Levey AI, Weintraub ST, et al. Oxidative modifications and down-regulation of ubiquitin carboxyl-terminal hydrolase L1 associated with idiopathic Parkinson's and Alzheimer's diseases. J Biol Chem 2004;279(13):13256–13264.

58 Butterfield DA, Boyd-Kimball D, Castegna A. Proteomics in Alzheimer's disease: insights into potential mechanisms of neurodegeneration. J Neurochem 2003;86(6):1313–1327.

59 Hyun DH, Lee MH, Halliwell B, Jenner P. Proteasomal dysfunction induced by 4-hydroxy-2,3-trans-nonenal, an end-product of lipid peroxidation: a mechanism contributing to neurodegeneration? J Neurochem 2002;83(2):360–370.

60 Castegna A, Thongboonkerd V, Klein J, et al. Proteomic analysis of brain proteins in the gracile axonal dystrophy (GAD) mouse, a syndrome that emanates from dysfunctional ubiquitin carboxyl-terminal hydrolase L-1, reveals oxidation of key proteins. J Neurochem 2004;88(6):1540–1546.

61 Daleke SL, Lyles JV. Identification and purification of aminophospholipid flippases. Biochim Biophys Acta 2000;1486(1):108–127.

62 Davies P. Challenging the cholinergic hypothesis in Alzheimer disease. JAMA 1999;281(15):1433–1434.

63 Castegna A, Lauderback CM, Mohmmad-Abdul H, Butterfield DA. Modulation of phospholipid asymmetry in synaptosomal membranes by the lipid peroxidation products, 4-hydroxynonenal and acrolein: implications for Alzheimer's disease. Brain Res 2004;1004(1-2):193–197.

64 Mohmmad-Abdul H, Butterfield D. Protection against amyloid beta-peptide (1-42)-induced loss of phospholipid asymmetry in synaptosomal membranes by tricyclodecan-9-xanthogenate (D609) and ferulic acid ethyl ester: implications for Alzheimer's disease. Biochim Biophys Acta 2005;1741(1-2):140–48.

65 Hamajima N, Matsuda K, Sakata S, et al. A novel gene family defined by human dihydropyrimidinase and three related proteins with differential tissue distribution. Gene 1996;180:157–163.

66 Kato K, Hamajima N, Inagaki H, et al. Post-meiotic expression of the mouse dihydropyrimidinase-related protein 3 (DRP-3) gene during spermiogenesis. Mol Reprod Dev 1998;51(1):105–111.

67 Wang LH, Strittmatter SM. A family of rat CRMP genes is differentially expressed in the nervous system. J Neurosci 1996;16(19):6197–6207.

68 Lubec G, Nonaka M, Krapfenbauer K, et al. Expression of the dihydropyrimidinase related protein 2 (DRP-2) in Down syndrome and Alzheimer's disease brain is down-regulated at the mRNA and dysregulated at the protein level. J Neural Transm Suppl 1999;57:161–177.

69 Weitzdoerfer R, Fountoulakis M, Lubec G. Aberrant expression of dihydropyrimidinase related proteins-2, -3 and -4 in fetal Down syndrome brain. J Neural Transm Suppl 2001;61:95–107.

70 Johnston-Wilson NL, Sims CD, Hofmann JP, et al. Disease-specific alterations in frontal cortex brain proteins in schizophrenia, bipolar disorder, and major depressive disorder; the Stanley Neuropathology Consortium. Mol Psychiatry 2000;5(2):142–149.

71 Coleman PD, Flood DG, Neuron numbers and dendritic extent in normal aging and Alzheimer's disease. Neurobiol Aging 1987;8(6):521–545.

72 Schutkowski M, Bernhardt A, Zhou XZ, et al. Role of phosphorylation in determining the backbone dynamics of the serine/threonine-proline motif and Pin1 substrate recognition. Biochemistry 1998;37(16):5566–5575.

73 Zhou XZ, Kops O, Werner A, et al. Pin1-dependent prolyl isomerization regulates dephosphorylation of Cdc25C and tau proteins. Mol Cell 2000;6(4):873–883.

74 Holzer M, Gartner U, Stobe A, et al. Inverse association of Pin1 and tau accumulation in Alzheimer's disease hippocampus. Acta Neuropathol (Berl) 2002;104(5):471–481.

75 Kurt MA, Davies DC, Kidd M, et al. Hyperphosphorylated tau and paired helical filament-like structures in the brains of mice carrying mutant amyloid precursor protein and mutant presenilin-1 transgenes. Neurobiol Dis 2003;14(1):89–97.

76 Ramakrishnan P, Dickson DW, Davies P. Pin1 colocalization with phosphorylated tau in Alzheimer's disease and other tauopathies. Neurobiol Dis 2003;14(2):251–264.

77 Arendt T. Synaptic plasticity and cell cycle activation in neurons are alternative effector pathways: the 'Dr. Jekyll and Mr. Hyde concept' of Alzheimer's disease or the yin and yang of neuroplasticity. Prog Neurobiol 2003;71(2-3):83–248.

78 Butterfield DA,. Abdul HM, Opii W, et al. Pin1 in Alzheimer's disease. J Neurochem 2006;98(6):1697–1706.

79 Sultana R, Butterfield DA. Regional expression of key cell cycle proteins in brain from subjects with amnestic mild cognitive impairment. Neurochem Res 2007;32:655–662.

80 Beckers CJ, Block MR, Glick BS, et al. Vesicular transport between the endoplasmic reticulum and the Golgi stack requires the NEM-sensitive fusion protein. Nature 1989;339(6223):397–398.

81 Stenbeck G. Soluble NSF-attachment proteins. Int J Biochem Cell Biol 1998;30(5):573-577.

82 Sly WS, Hu PY. Human carbonic anhydrases and carbonic anhydrase deficiencies. Annu Rev Biochem 1995;64:375–401.

83 Meier-Ruge W, Iwangoff P, Reichlmeier K. Neurochemical enzyme changes in Alzheimer's and Pick's disease. Arch Gerontol Geriatr 1984;3(2):161–165.

84 Boyd-Kimball D, Poon HF, Lynn BC, et al. Proteomic identification of proteins specifically oxidized in Caenorhabditis elegans expressing human Abeta(1-42): implications for Alzheimer's disease. Neurobiol Aging 2006;27(9):1239–1249.

85 Boyd-Kimball D, Sultana R, Poon HF, et al. Proteomic identification of proteins specifically oxidized by intracerebral injection of Abeta(1-42) into rat brain: implications for Alzheimer's disease. Neuroscience 2005;132(2):313–324.

86 Boyd-Kimball D, Sultana R, Poon HF, et al. Gamma-glutamylcysteine ethyl ester protection of proteins from Abeta(1-42)-mediated oxidative stress in neuronal cell culture: a proteomics approach. J Neurosci Res 2005;79(5):707–713.

87 Sultana R, Newman SF, Abdul HM, et al. Protective effect of D609 against amyloid-beta1-42-induced oxidative modification of neuronal proteins: redox proteomics study. J Neurosci Res 2006;84(2):409–417.

88 Eriksen JL, Dawson TM, Dickson DW, Petrucelli L. Caught in the act: alpha-synuclein is the culprit in Parkinson's disease. Neuron 2003;40(3):453–456.

89 Dawson TM, Dawson VL. Molecular pathways of neurodegeneration in Parkinson's disease. Science 2003;302(5646):819–822.

90 Kruger R, Kuhn W, Muller T, et al. Ala30Pro mutation in the gene encoding alpha-synuclein in Parkinson's disease. Nat Genet 1998;18(2):106–108.

91 Polymeropoulos MH, Lavedan C, Leroy E, et al. Mutation in the alpha-synuclein gene identified in families with Parkinson's disease. Science 1997;276(5321):2045–2047.

92 Alam ZI, Daniel SE, Lees AJ, et al. A generalised increase in protein carbonyls in the brain in Parkinson's but not incidental Lewy body disease. J Neurochem 1997;69(3):1326–1329.

93 Floor E, Wetzel MG. Increased protein oxidation in human substantia nigra pars compacta in comparison with basal ganglia and prefrontal cortex measured with an improved dinitrophenylhydrazine assay. J Neurochem 1998;70(1):268–275.

94 Jenner P. Oxidative stress in Parkinson's disease. Ann Neurol 2003;53(suppl 3):S26–S36.

95 Yoritaka A, Hattori N, Uchida K, et al. Immunohistochemical detection of 4-hydroxynonenal protein adducts in Parkinson disease. Proc Natl Acad Sci USA 1996;93(7):2696–2701.

96 Ostrerova-Golts N, Petrucelli L, Hardy J, et al. The A53T alpha-synuclein mutation increases iron-dependent aggregation and toxicity. J Neurosci 2000;20(16):6048–6054.

97 Lee M, Hyun D, Halliwell B, Jenner P. Effect of the overexpression of wild-type or mutant alpha-synuclein on cell susceptibility to insult. J Neurochem 2001;76(4):998–1009.

98 Neumann M, Kahle PJ, Giasson BI, et al. Misfolded proteinase K-resistant hyperphosphorylated alpha-synuclein in aged transgenic mice with locomotor deterioration and in human alpha-synucleinopathies. J Clin Invest 2002;110(10):1429–1439.

99 Giasson BI, Duda JE, Quinn SM, et al. Neuronal alpha-synucleinopathy with severe movement disorder in mice expressing A53T human alpha-synuclein. Neuron 2002;34(4):521–533.

100 Kahle PJ, Neumann M, Ozmen L, et al. Selective insolubility of alpha-synuclein in human Lewy body diseases is recapitulated in a transgenic mouse model. Am J Pathol 2001;159(6):2215–2225.

101 Poon HF, Frasier M, Shreve N, et al. Mitochondrial associated metabolic proteins are selectively oxidized in A30P alpha-synuclein transgenic mice: a model of familial Parkinson's disease. Neurobiol Dis 2005;18(3):492–498.

102 Heck RW, Tanhauser SM, Manda R, et al. Catalytic properties of mouse carbonic anhydrase V. J Biol Chem 1994;269(40):24742–24746.

103 Shah GN, Hewett-Emmett D, Grubb JH, et al. Mitochondrial carbonic anhydrase CA VB: differences in tissue distribution and pattern of evolution from those of CA VA suggest distinct physiological roles. Proc Natl Acad Sci U S A 2000;97(4):1677–1682.

104 Kasischke KA, Vishwasrao HD, Fisher PJ, et al. Neural activity triggers neuronal oxidative metabolism followed by astrocytic glycolysis. Science 2004;305(5680):99–103.

105 Giege P, Heazlewood JL, Roessner-Tunali U, et al. Enzymes of glycolysis are functionally associated with the mitochondrion in Arabidopsis cells. Plant Cell 2003;15(9):2140–2151.

106 Schapira AH. Mitochondrial dysfunction in neurodegenerative disorders. Biochim Biophys Acta 1998;1366(1-2):225–233.

107 Schapira AH. Causes of neuronal death in Parkinson's disease. Adv Neurol 2001;86:155–162.

108 Sherer TB, Betarbet R, Greenamyre JT. Environment, mitochondria, and Parkinson's disease. Neuroscientist 2002;8(3):192–197.

109 Ferrante RJ, Hantraye P, Brouillet E, Beal MF. Increased nitrotyrosine immunoreactivity in substantia nigra neurons in MPTP treated baboons is blocked by inhibition of neuronal nitric oxide synthase. Brain Res 1999;823(1-2):177–182.

110 Pennathur S, Jackson-Lewis V, Przedborski S, Heinecke JW. Mass spectrometric quantification of 3-nitrotyrosine, ortho-tyrosine, and o,o'-dityrosine in brain tissue of 1-methyl-4-phenyl-1,2,3, 6-tetrahydropyridine-treated mice, a model of oxidative stress in Parkinson's disease. J Biol Chem 1999;274(49):34621–34628.

111 Sherer TB, Betarbet R, Kim JH, Greenamyre JT. Selective microglial activation in the rat rotenone model of Parkinson's disease. Neurosci Lett 2003;341(2):87–90.

112 Sherer TB, Kim JH, Betarbet R, Greenamyre JT. Subcutaneous rotenone exposure causes highly selective dopaminergic degeneration and alpha-synuclein aggregation. Exp Neurol 2003;179(1):9–16.

113 Canet-Aviles RM, Wilson MA, Miller DW, et al. The Parkinson's disease protein DJ-1 is neuroprotective due to cysteine-sulfinic acid-driven mitochondrial localization. Proc Natl Acad Sci U S A 2004;101(24):9103–9108.

114 Palacino JJ, Sagi D, Goldberg MS, et al. Mitochondrial dysfunction and oxidative damage in parkin-deficient mice. J Biol Chem 2004;279(18):18614–18622.

115 Valente EM, Abou-Sleiman PM, Caputo V, et al. Hereditary early-onset Parkinson's disease caused by mutations in PINK1. Science 2004;304(5674):1158–1160.

Chapter 16
Biomarkers for Alzheimer's Disease and Parkinson's Disease

John H. Growdon, Michael C. Irizarry, and Clemens Scherzer

Introduction

Alzheimer's disease (AD) and Parkinson's disease (PD) are neurodegenerative diseases that depend on skilled examination and expert application of clinical criteria to establish an accurate diagnosis. Even in the best hands, diagnostic accuracy rarely exceeds 90%, and misdiagnoses are common. A high priority research goal therefore is to identify and then validate one or more biological measures that would enable physicians to diagnose AD and PD easily and accurately. In this chapter, we broadly define biomarkers as noninvasive biochemical or molecular measures that reflect the pathophysiological processes of AD or PD. In addition to having the potential of aiding in diagnosis, biomarkers hold the promise of predicting those individuals at high risk for disease and serve as surrogate markers of treatment effects. Biomarkers are abundant in general medicine: Prostate-specific antigen (PSA) screens for prostate cancer, blood sugar for diabetes mellitus, cholesterol panels to assess risk for vascular disease, and CD4 counts to monitor human immunodeficiency virus (HIV) therapy. In each case, a biological measure identifies the disease or disease process and generally leads to further evaluation and treatment initiation. To date, there are no accepted biological markers for any neurodegenerative disease. To stimulate progress toward developing biomarkers for neurodegenerative diseases, a workshop on biological markers for AD was held in 1998 [1] and specified some features of an ideal biomarker (Table 1).

In addition to AD, the general principles of biomarker qualities hold for PD and other neurodegenerative disease as well. Moreover, biomarkers could serve several purposes, ranging from a way to screen populations for disease to aiding in diagnosis, to serving as a surrogate outcome measure in clinical drug trials (Table 2).

J.H. Growdon
Department of Neurology, Partners HealthCare System, Inc. Boston, MA. USA

A. Fisher et al. (eds.), *Advances in Alzheimer's and Parkinson's Disease*, 169
© Springer 2008

Table 1 Some Qualities of an Ideal Biomarker

An ideal biomarker should:
- Reflect an important and possibly unique feature of the illness
- Be validated in neuropathologically confirmed cases
- Have a sensitivity and specificity > 80%
- Be easy to perform and highly reliable
- Be noninvasive and acceptable to patients
- Be inexpensive

Table 2 Uses of Biomarkers

- Epidemiological screening: higher sensitivity, but lower specificity is acceptable
- Predictive testing: higher specificity more important than sensitivity
- Aid to disease diagnosis: sensitivity and specificity > 85%
- Surrogate outcomes: sensitivity and specificity ~100%.

Although it is theoretically possible that a single biomarker could serve comprehensively as a predictive marker, a diagnostic markerm and a marker of progression, it is much more likely that it will be necessary to use a number of markers, either separately or in combination, to fulfill different biomarker roles. Now, 7 years after the workshop publication, there are still no completely satisfactory markers for any of the four potential uses. There has been enough progress, however, at least in the AD field, to give optimism and justify a review, taking stock of the current status of biomarkers as diagnostic aids.

Definitions

This chapter focuses on biomarkers measured in body fluids and used as diagnostic aids. Neuroimaging biomarkers for AD and PD have been comprehensively reviewed elsewhere [2–4]. The clinical usefulness of a biomarker depends on its accuracy for detecting disease, as indicated by sensitivity and specificity. Sensitivity refers to the capacity of a biomarker to identify a substantial percentage of patients with the disease. A sensitivity of 100%, for example, would indicate that a marker can accurately identify in a heterogeneous group every person in that group who has either AD or PD. Specificity refers to the capacity of the test to distinguish AD (or PD) from normal aging and other causes of cognitive impairment and dementia. A test with a specificity of 100% would be capable of differentiating AD from other cases of dementia in every case. A test that is sensitive enough to identify every person with disease in a group would likely include some individuals who do not have the disease.

A test that is highly specific for the disease would likely exclude some individuals who in fact have the disease.

Reports on the genetic basis of AD and PD illustrate these points. For AD, causative mutations in the *APP, PS1*, and *PS2* genes are 100% specific for identifying individuals with familial early-onset AD; carriers with these mutations have virtually 100% certainty of developing AD. These mutations are, however, rare and account for < 1% of individuals with AD. Consequently, genetic screens for these mutations are not generally part of a routine diagnostic workup for AD because the sensitivity is so low: Among any 100 or 1000 individuals with bona fide AD, there may be 0 to 1 who carries a causative mutation. In this case the sensitivity would be < 1% although the specificity may be 100%. The same situation obtains for PD; mutations in *SNCA* and *LRRK2* and homozygous mutations in *PARK2, DJ-1*, or *PINK1*, may be 100% specific for familial PD, but because these abnormalities are so rare genetic screens are not time- or cost-effective for routine diagnostic assessments because the sensitivity is well below 1%.

These comments are not meant to imply that the explosion of information regarding the genetic components of AD and PD is not important. To the contrary, discovering the causal gene mutations in AD helped generate and provided strong support for the amyloid hypothesis for AD [5]. Similarly, the rare families with mutations in the gene for α-synuclein spurred investigations leading to the discovery that α-synuclein is a principal component of Lewy bodies [6,7], whether from a familial or nonfamilial case of PD. In the same vein, the recent discovery of *LRRK2* mutations [8] draws attention to biochemical abnormalities as potential factors leading to PD.

Amyloid Fragments and Tau in AD

The characteristic histopathological abnormalities leading to a definitive diagnosis of AD are numerous neurofibrillary tangles and amyloid plaques in brain tissue. Neurofibrillary tangles are composed primarily of hyperphosphorylated tau protein. Tau circulates in the cerebrospinal fluid (CSF), and assays have been developed to measure both total tau and hyperphosphorylated tau. The mature senile plaque consists of degenerating neurites surrounding an amyloid core. The amyloid components of senile plaques are 40 to 43 amino acid cleavage products, termed Aβ, of a large amyloid precursor protein. Aβ fragments derive from sequential β- and γ-secretase cleavages; Aβ42 appears to be the most neurotoxic fragment and is most concentrated in the mature neuritic plaque. Aβ fragments also circulate normally in the CSF and can be measured in plasma as well. It was proposed more than a decade ago that detecting abnormalities in either Aβ or tau in CSF might provide a window on the neuropathological processes associated with AD and known to occur in brain. There is now widespread consensus from multiple reports that a characteristic CSF profile exists that correlates with the clinical diagnosis of probable AD: increased levels of tau and phosphotau and decreased levels of Aβ42 [9]. This formula has at least 90% sensitivity and specificity for distinguishing individuals with a clinical diagnosis of probable AD from nondemented control subjects. Compared to other neurological diseases characterized by dementia, the sensitivity and specificity of this pattern for identifying AD generally fall into the 80% to 85% range (Table 3).

Because lumbar puncture with examination of CSF is not a routine part of the diagnostic workup for dementia in many countries, we sought to determine whether we could measure Aβ in plasma of patients with AD and determine whether these measures differ among AD, mild cognitive impairment (MCI), nondemented controls, and PD patients without dementia. A secondary goal was to determine whether plasma Aβ measures correlated with the extent of cognitive impairment, degree of dementia severity, medication use, or apolipoprotein E (ApoE) genotype. We were aware of prior reports indicating that levels of Aβ42 were increased in familial-onset autosomal dominant AD [10] and also reported to be elevated in blood of individuals with MCI 3 years prior to their conversion to AD [11].

To this end, we collected plasma samples from 146 AD patients, 37 MCI patients, 96 patients with PD with normal cognitive test scores, and 92 normal control subjects. We measured Aβ40 and Aβ42 levels by enzyme-linked immunosorbent assay (ELISA) using the capture antibody DNT77 and the detector antibodies BA27 and BC05. The main finding was that the mean Aβ40 and Aβ42 levels increased significantly with age in each diagnostic group. When co-varied for age, mean plasma levels for Aβ40 and Aβ42 did not differ significantly among the four diagnostic groups [12]. Our data are consistent with other reports in the literature indicating that the plasma levels of Aβ40 and 42 are neither sensitive nor specific diagnostic markers for AD. Furthermore, we found that these levels did not correlate with duration of dementia, cognitive

Table 3 Summary of CSF Biomarkers of AD

Disease	Total tau	Phospho-tau	Aβ42
Normal aging	Normal	Normal	Normal
Alzheimer's disease	Increased	Increased	Decreased
Depression	Normal	Normal	Normal
Creutzfieldt-Jakob disease	Greatly increased	Normal to slightly increased	Normal to slightly decreased
Frontotemporal dementia	Normal to slightly increased	Normal	Decreased
Lewy body disease	Normal to slightly increased	Normal to slightly increased	Decreased

Adapted from Blennow [9].

test scores, family history of AD ApoE genotype, or drug use, including acetylcholinesterase inhibitors, statins, and estrogen.

Biomarkers in Body Fluids of PD Patients

To date, there are no biochemical or molecular markers of PD that approach the potential usefulness of Aβ and tau measures in CSF of AD patients. Dopamine and its metabolites are reduced in brains of PD patients, but the biochemical determinations of dopamine and its metabolites in CSF are neither sensitive nor specific for the diagnosis of PD. As part of an effort to identify potential biomarkers of PD in the bloodstream, we measured total homocysteine levels in the same cohort of AD, MCI, PD, and normal control individuals who donated blood samples for the Aβ measures. Although mean homocysteine levels did not differ across the AD, PD, and control cases, the level was significantly ($p < 0.002$) higher in the 73 PD patients treated with levodopa than in the 20 who did not take this medication. This result was not unexpected, as metabolism of levodopa and dopamine generates homocysteine. Furthermore, total homocysteine levels in PD patients were significantly correlated ($p < 0.05$) with test scores on the information, memory, and concentration subscales of the Blessed Dementia Scale score: the higher the total homocysteine level, the more errors on the mental status test. These data confirm the prior report of a modest increase in plasma homocysteine following levodopa initiation [13] and extend the findings by linking high homocysteine levels in nondemented PD patients to impaired test performance on a mental status examination. These results indicate that measuring total homocysteine levels in plasma is not a useful biomarker for PD (or AD for that matter) but may serve to identify a subset of PD patients at risk for developing dementia.

The fields of genomics, proteomics, and metabolomics open new opportunities for biomarker development. We have begun a PD gene discovery program based on microarray technology. In these studies, we extract RNA from blood samples and prepare labeled cRNA, which is then hybridized to Affymetrix chips that contain more than 20,000 gene transcripts. Using the preliminary data based on 100 blood samples (50 PD patients and 50 control subjects matched for differential blood counts), we have identified > 20 differentially expressed genes that were significantly linked to PD, including cellular quality control genes, neuronal genes, and mitochondrial genes. In the next step, we will conduct a prospective study focused on the 5 to 10 most promising genes that give the largest signal. The overall goal of this project is to develop a genomic fingerprint that can accurately identify patients with idiopathic PD and distinguish them from individuals with atypical parkinsonian syndromes as well as individuals free from neurological disease.

References

1. Ronald and Nancy Reagan Research Institute of the Alzheimer's Association and the National Institute on Aging Working Group. Consensus report of the Working Group on Molecular and Biochemical Markers of Alzheimer's Disease. Neurobiol Aging 1998;19:109–116.
2. Kantarci K, Clifford RJ. Quantitative magnetic resonance techniques as surrogate markers of Alzheimer's disease. NeuroRx 2004;1:196–205.
3. Jagust W. Molecular neuroimaging in Alzheimer's disease. NeuroRx 2004;1:206–212.
4. Brooks DJ. Neuroimaging in Parkinson's disease NeuroRx 2004;1:243–254.
5. Selkoe DJ. Alzheimer's disease: genotypes, phenotypes and treatments. Science 1997;271:630–631.
6. Spillantini MG, Schmidt MI, Lee MY, et al. α-Synuclein in Lewy bodies. Nature 1997;388:839–840.
7. Irizarry M, Growdon W, Gomez-Isla T, et al. Nigral and cortical Lewy bodies and dystrophic nigral neurites in Parkinson's disease and cortical Lewy body disease contain α-synuclein immunoreactivity. J Neuropathol Exp Neurol 1998;57:334–337.
8. Zimprich A, Biskup S, Leitner P, et al. Mutations in LRRK2 cause autosomal-dominant parkinsonism with pleomorphic pathology. Neuron 2004;4:601–607.
9. Blennow K. Cerebrospinal fluid protein biomarkers for Alzheimer's disease. NeuroRx 2004;1:213–225.
10. Scheuner D, Eckman C, Jensen M, et al. Secreted amyloid beta-protein similar to that in the senile plaques of Alzheimer's disease is increased in vivo by the presenilin 1 and 2 and APP mutations linked to familial Alzheimer's disease. Nat Med 1996;2:864–870.
11. Mayeux R, Tang MX, Jacobs DM, et al. Plasma amyloid beta-peptide 1–42 and incipient Alzheimer's disease. Ann Neurol 1999;46:412–416.
12. Fukumoto H, Tennis M, Locascio J, et al. Age but not diagnosis is the main predictor of plasma amyloid β-protein levels. Arch Neurol 2003;60:958–964.
13. O'Suilleabhain PE, Bottiglieri T, Dewey RB, et al. Modest increase in plasma homocysteine follows levodopa initiation in Parkinson's disease. Mov Disord 2004;19:1403–1408.

Chapter 17
Bioluminescent Imaging of Excitotoxic and Endotoxic Brain Injury in Living Mice

Jian Luo, Amy H. Lin, and Tony Wyss-Coray

Introduction

Bioluminescence has been used recently to monitor and quantify gene activity repeatedly in live animals and has been successfully used to study disease progression in peripheral organs [1,2]. The bioluminescence imaging technique is quantitative and can faithfully report gene activation if appropriate fusion gene constructs are used. Therefore, monitoring activation of an injury responsive gene by bioluminescence imaging could provide valuable information about injury severity and disease progression.

Members of the transforming growth factor-β (TGFβ) family of proteins are ubiquitous and essential regulators of many important cellular and physiological processes—such as proliferation, differentiation, migration, cell survival, and inflammatory responses—and have been implicated in tumorigenesis, fibrosis, inflammation, and neurodegeneration [3,4]. The three isoforms of TGFβ (TGFβ1, 2, and 3) have key roles in injury responses and wound repair in various tissues [5]. TGFβ1 is up-regulated in brain injury and has been implicated in brain trauma, stroke, multiple sclerosis, and neurodegenerative disorders including Alzheimer's disease [6].

The Smad proteins mediate signaling for the TGFβ superfamily of cytokines [7]. Smad2 and Smad3 are activated by TGFβ and activin, and bone morphogenetic proteins (BMPs) recruit Smad1, Smad5, and Smad8 in combination with Smad4 after binding to BMP receptors [3,8]. TGFβ binding leads to phosphorylation of Smad2 or Smad3. Once phosphorylated, these Smads form heteromeric complexes with Smad4 [3,8]. These complexes then migrate to the nucleus, where they can bind to the TGFβ-responsive promoter either directly through the Smad-binding elements (SBEs) or in association with other sequence-specific transcription factors [9]. Global gene expression analysis of

J. Luo
Department of Neurology and Neurological Sciences, Stanford University School of Medicine, Stanford, CA, USA

A. Fisher et al. (eds.), *Advances in Alzheimer's and Parkinson's Disease*,
© Springer 2008

cells or tissues revealed that Smad2/3 may regulate expression of several hundred immediate-early and immediate genes [10].

We have engineered transgenic reporter mice that harbor a luciferase reporter gene fused to a promoter consisting of SBE [11]. After injecting a luciferase substrate, we can use bioluminescence imaging to obtain and follow optical signatures in a spatial and temporal manner in living mice over time. In the present experiment we demonstrated that SBE-luc mice can be used to monitor brain injury noninvasively in living mice. Administration of the excitotoxin kainic acid (KA) and the endotoxin lipopolysaccharide (LPS) resulted in reporter gene activation that could be quantified over the skull using bioluminescence imaging.

Methods

Mice and Injury Models

Transgenic mouse lines harboring a SBE-luc transgene have been described previously [11]. This transgene consists of 12 SBE repeats fused to a herpes simplex virus thymidine kinase minimal promoter upstream of firefly luciferase and an SV40 late polyadenylation signal [11]. Heterozygous, 2- to 3-month-old mice with detectable luciferase activity in tail biopsies ("responder" mice [11]) were used in this study. For excitotoxin-induced injury, KA 30 mg/kg (Tocris, Ellisville, MO, USA) dissolved in phosphate-buffered saline (PBS) was injected intraperitoneally. Seizure activity was scored from 0 to 5, with 0 showing no behavioral changes and 5 showing constant rearing and falling [12]. All KA-injected mice reached at least stage 3. Control mice were given injections of PBS. Acute endotoxic injury was induced by intraperitoneal injection of LPS (*Escherichia coli* serotype 055:B5, 2 mg/kg; Sigma-Aldrich, St, Louis, MO, USA) [11]. All animal handling was performed in accordance with institutional guidelines and approved by the local Institutional Animal Care and Use Committee (IACUC).

Stereotactic Injections

Mice were anesthetized using isoflurane and immobilized in a stereotactic frame for intrahippocampal injections. A single, 1-µl injection of 1D11 or TGFβ1 (both from R&D Systems, Minneapolis, MN, USA) was delivered at 0.2 µl/min into the hippocampi (coordinates from bregma: 1.98 mm posterior, 1.6 mm lateral, 2.00 mm ventral). 1D11 (10 ng/µl) was dissolved in PBS; TGFβ1 (1 ng/µl) was formulated in 20 mM Na_2PO_4, 130 mM NaCl, 15% propylene glycol, 20% polyethylene glycol 400, and 1% ethanol (pH 7.2) [13]. The same amount of mouse immunoglobulin G (IgG) was injected contralaterally as a control. For intraperitoneal injection, 10 ng of TGFβ1 was prepared in 100 μl of the above formulation and injected immediately.

Tissue Preparation

Mice were anesthetized with chloral hydrate 400 mg/kg (Sigma-Aldrich) and perfused transcardially with 0.9% saline. Brains were removed. Each hemi-brain was dissected into hippocampus, cortex, thalamus, brain stem, and cerebellum. Each subregion was weighed and lysed in 100 to 400 μl of 1 × Cell Culture Lysis Reagent (Promega, Madison, WI, USA). Luciferase activities from tissue homogenates were measured with a tube luminometer (Lumat LB 9507; Berthold Technologies, Oak Ridge, TN, USA) using a commercial luciferase assay kit (Promega) and normalized to weight.

Bioluminescence Imaging

Bioluminescence was detected with the *In Vivo* Imaging System (IVIS; Xenogen, Hopkinton, MA, USA) [11]. Mice were injected intraperitoneally with D-luciferin 150 mg/kg (Xenogen) 10 minutes before imaging and anesthetized with isofluorane during imaging. Photons emitted from living mice were acquired as photons/s/cm ^{2}/steradian (sr) using LivingImage software (Xenogen) and integrated over 5 minutes. For photon quantification, a region of interest was manually selected and kept constant during all experiments; the signal intensity was converted into photons/s/mm^2/sr. For longitudinal comparison of bioluminescence, measurements were expressed as x-fold induction over the baseline levels for each mouse obtained at the outset of an experiment. In addition, a background bioluminescent reading obtained in nontransgenic mice injected with luciferin was subtracted from all values.

Results and Discussion

SBE-luc Reporter Mice

In primary astrocytes derived from SBE-luc mice, the reporter gene is principally activated by TGFβ1 and to a lesser extent by TGFβ2, TGFβ3, and activin [11]. In agreement with these findings, mice injected peripherally with recombinant TGFβ1 showed a detectable increase in bioluminescence emitted from the brain (Fig. 1A). Similarly, recombinant TGFβ1 injected stereotactically into the hippocampus resulted in a twofold induction of luciferase activity on the ipsilateral side compared with the contralateral side (Fig. 1B) ($p = 0.023$). Furthermore, SBE-luc

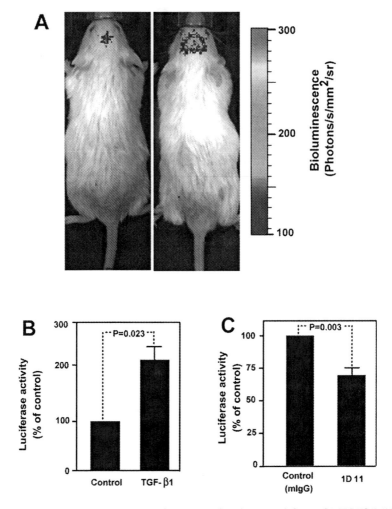

Fig. 1 Specificity of SBE-luc reporter mice. **A** Transforming growth factor-β1 (TGFβ1) (10 ng) was injected intraperitoneally into SBE-luc mice (T9-7F). The mouse shown was imaged 4 hours later (right panel) and compared with baseline photon emission (before injection, left panel). **B** TGFβ1 (10 ng) was stereotactically injected into the hippocampi of SBE mice (T9-55F). Brains were removed 4 hours after injection, and hippocampi were dissected. Luciferase activity in the tissue homogenate was measured and expressed as a percentage of the control ($n = 3$). **C** SBE-luc mice (T9-55F) were lesioned with lipopolysaccharaide (LPS) (2 mg/kg, i.p.). Five hours later, the neutralizing antibody 1D11 (10 ng in phosphate-buffered saline, or PBS) was injected stereotactically into one hippocampus, and mouse immunoglobu-lin G (IgG) into the contralateral hippocampus as a control. Brains were removed 24 hours after LPS injection, and hippocampi were dissected. Luciferase activity in the tissue homo-genate was measured and expressed as a percentage of the control ($n = 3$)

mice were injected intraperitoneally with LPS, which results in prominent activation of the reporter gene in the brain [11]. Stereotactic injection of antibody 1D11, which neutralizes the activity of TGFβs (TGFβ1, 2, and 3) 5 hours later into the same mice resulted in a 30% reduction of reporter activity compared with a control antibody (Fig. 1C) ($p = 0.003$). These results demonstrate again that the reporter responds, at least in part, to TGFβ-dependent signaling. Limited bioavailability due to neutralization or lack of accessibility may account for the relatively modest responses with recombinant TGFβ1 and 1D11. Activin, which also signals via Smad2/3 protein, may contribute to reporter gene activation in vivo.

Bioluminescent Imaging of Excitotoxic Brain Injury

Excitotoxic injury of neurons results in excessive activation of glutamate receptors and subsequent Ca^{2+} influx. It has been postulated to be responsible not only for neuronal loss associated with seizures, traumatic and ischemic brain injury, and hypoxia but also for neuronal damage in Alzheimer's disease and other neurodegenerative disorders [14]. We therefore used a model of excitotoxic injury to determine if it would activate the SBE-luc reporter gene and, more importantly, if SBE-luc mice could be used to monitor brain injury in such a model. SBE-luc mice were injured with KA, a potent inducer of excitotoxicity that causes epileptic seizures followed by inflammatory responses and oxidative stress, resulting in neuronal degeneration and death [15]. The KA injury paradigm shares many neurochemical and histopathological alterations with those observed in neurodegenerative disorders in humans and has been used widely as a model of excitotoxicity and neurodegeneration [15].

Bioluminescence emitted from the brain was monitored noninvasively in living mice using a bioluminescence imaging system. Intraperitoneal injection of luciferin into SBE-luc mice resulted in detectable emission of photons from the brain, and this signal was increased several-fold 24 hours after KA lesioning (Fig. 2). Time course studies of bioluminescence after KA injury showed a significantly increased bioluminescent signal for up to 5 days (Fig. 2). Consistent with these findings, coronal slices of brains from KA-lesioned or control mice showed increased emission of photons from the lesioned brain slices. Interestingly, photons were emitted mainly from the hippocampus and to a lesser extent from the cortex (data not shown). These data demonstrate that a KA lesion induces activation of the reporter gene and that such injury can be monitored noninvasively over time in live animals.

Noninvasive Imaging of Endotoxic Brain Injury In Vivo

Neuroinflammation is a hallmark of many neurodegenerative diseases [16]. LPS triggers a host of inflammatory responses and has been used to mimic inflammatory

Bioluminescence
(Fold induction)   2.97   1.64   1.96

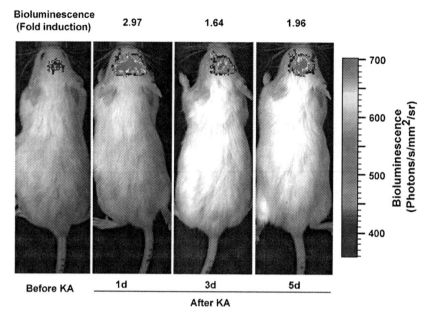

Before KA   ___1d___   3d   5d___
                          After KA

Fig. 2 Bioluminescent imaging of kainic acid (KA)-induced brain injury. Three-month-old SBE-luc mice (T9-7F) were imaged immediately before the KA lesion (30 mg/kg i.p., 0 hours) or at the indicated time points after injection. Bioluminescence was quantified in defined areas over the head and expressed as x-fold induction over baseline values obtained at 0 hours

conditions of human neurodegenerative diseases [17]. A significant increase in bioluminescence was previously observed in brains of SBE-luc mice injected peripherally with LPS [11]. To determine the temporal pattern of reporter gene activation after endotoxin challenge, photon emission in living SBE-luc mice was quantified at various time points after administration of LPS (Fig. 3A). Photon emission increased rapidly in the brain, peaking at around 5 hours and remaining at similar levels for up to 48 hours (Fig. 3A). Individual variations in this temporal response could be visualized for the first time (Fig. 3B).

Taken together, these results demonstrate that acute brain injury induced by excitotoxins or endotoxins can be monitored in living mice. More importantly, bioluminescence quantified in living mice was recently found to correlate with postmortem neurodegeneration and neuroinflammation (Luo, Lin, and Wyss-Coray, unpublished observation). Thus, the activity of a reporter gene may be used as a marker for brain injury and to monitor disease progression.

SBE-luc mice, described here, can be used for repeated and efficient imaging of brain injury and neurodegeneration in living animals. They could be particularly suited for the rapid testing of in vivo activity, blood-brain barrier permeability, and other pharmacological properties of chemical compounds identified in cell culture or other screens. Related models tailored to measure

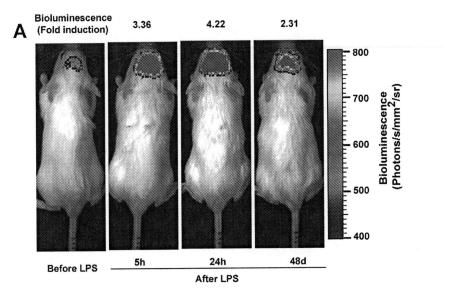

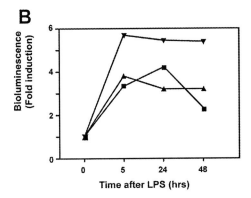

Fig. 3 Temporal activation of reporter gene by LPS in brains of living mice. Three-month-old SBE-luc mice (T9-7F) were imaged immediately before administration of LPS (2 mg/kg, 0 hours) or at the indicated time points after injection. **A** Bioluminescence at various time points after LPS administration in an individual mouse. **B** Bioluminescence expressed as x-fold induction in three individual mice

activation of specific signaling pathways may be used to monitor memory processes, behavioral traits, and other brain functions.

Acknowledgments This work was supported by the NIH (T.W.C.; AG20603, AG023708); the Veterans Administration Geriatric Research, Education and Clinical Center (GRECC); and the Mental Illness Research, Education and Clinical Center (MIRECC) services (T.W.C.). A.L. was supported by a National Research Service Award.

References

1. Contag CH, Bachmann H. Advances in in vivo bioluminescence imaging of gene expression. Annu Rev Biomed Eng 2002;4:235–260.
2. Greer LF III, Szalay AA. Imaging of light emission from the expression of luciferases in living cells and organisms: a review. Luminescence 2002;17(1):43–74.
3. Dennler S, Goumans MJ, ten Dijke P. Transforming growth factor b signal transduction. J Leukoc Biol 2002;71:731–740.
4. Massagué J, Blain SW, Lo RS. TGF-β signaling in growth control, cancer, and heritable disorders. Cell 2000;103:295–309.
5. Border WA, Ruoslahti E. Transforming growth factor-β in disease: the dark side of tissue repair. J Clin Invest 1992;90:1–7.
6. Flanders KC, Ren RF, Lippa CF. Transforming growth factor-βs in neurodegenerative disease. Prog Neurobiol 1998;54:71–85.
7. Derynck R, Zhang YE. Smad-dependent and Smad-independent pathways in TGF-beta family signaling. Nature 2003;425(6958):577–584.
8. Shi Y, Massague J. Mechanisms of TGF-beta signaling from cell membrane to the nucleus. Cell 2003;113(6):685–700.
9. Derynck R, Zhang Y, Feng XH. Smads: transcriptional activators of TGF-beta responses, Cell 1998;95(6):737–740.
10. Yang YC, Piek E, Zavadil J, et al. Hierarchical model of gene regulation by transforming growth factor beta. Proc Natl Acad Sci U S A 2003;100(18):10269–10274.
11. Lin AH, Luo J, Mondshein LH, et al. Global analysis of Smad2/3-dependent TGF-β signaling in living mice reveals prominent tissue-specific responses to injury. J Immunol 2005;175(1):547–554.
12. Schauwecker PE, Steward O. Genetic determinants of susceptibility to excitotoxic cell death: implications for gene targeting approaches. Proc Natl Acad Sci U S A 1997;94:4103–4108.
13. Ledbetter S, Kurtzberg L, Doyle S, Pratt BM. Renal fibrosis in mice treated with human recombinant transforming growth factor-beta2. Kidney Int 2000;58(6):2367–2376.
14. Rothman SM, Olney JW. Excitotoxicity and the MDA receptor-still lethal after eight years, Trends Neurosci 1995;18:57–58.
15. Wang Q, Yu S, Simonyi A, et al. Kainic acid-mediated excitotoxicity as a model for neurodegeneration. Mol Neurobiol 2005;31(1–3):3–16.
16. Akiyama H, Barger S, Barnum S, et al. Inflammation and Alzheimer's disease; Neuroinflammation Working Group. Neurobiol Aging 2000;21:383–421.
17. Hauss-Wegrzyniak B, Dobrzanski P, Stoehr JD, Wenk GL. Chronic neuroinflammation in rats reproduces components of the neurobiology of Alzheimer's disease. Brain Res 19988;780(2):294–303.

Chapter 18
Alzheimer's Disease Neuroimaging Initiative

Susanne G. Mueller, Michael W. Weiner, Leon J. Thal, Ronald C. Petersen,
Clifford Jack, William Jagust, John Q. Trojanowski, Arthur W Toga,
and Laurel Beckett

Introduction

Age is a major risk factor for almost all neurodegenerative diseases and
particularly for dementias such as Alzheimer's disease (AD). Owing to the
increasing life expectancy in developed countries, the frequency of AD and its
socioeconomic impact are gaining more and more importance. Intensive
research has led to the unraveling of a number of pathophysiological mechanisms involved in AD and thus to the development of new, promising AD
treatment strategies. However, the incorporation of these new drugs into
clinical treatment has been slow because the clinical trials necessary to establish
their efficacy require large patient samples and lengthy observation times. This
is mostly due to the fact that neuropsychological tests (e.g., the ADAS Cog [1]),
which are traditionally used to assess treatment efficacy in such trials, have a
relatively poor test–retest reliability [intraclass correlation coefficients (ICCs)
ca. 0.5–0.8] and thus reduced statistical power to detect differences between
treatment groups.

However, there is growing evidence that neuroimaging [2,3] and blood/
cerebrospinal fluid (CSF) biomarkers [4] provide valuable additional information to complement that gained from clinical and neuropsychological measures.
Neuroimaging, for example, has a much higher test–retest reliability than
cognitive measures (e.g., ICC > 0.95 for hippocampal volume [5]). This greatly
increases its power to detect longitudinal change and treatment effects. It is also
generally accepted that loss of synapses and neurons due to neurodegeneration
results in reduced brain activity/metabolism [i.e.,decreased fluorodeoxyglucose
positron emission tomography (FDG-PET) activity], and volume loss on structural magnetic resonanace imaging (MRI). Furthermore, both FDG-PET and
structural MRI have been quantitatively validated to some extent by

S.G. Mueller
University of California, San Francisco, California, USA

A. Fisher et al. (eds.), *Advances in Alzheimer's and Parkinson's Disease*,
© Springer 2008

correlation with cognitive/functional measures and correlation with neuro-pathology at autopsy [6,7].

However, atrophy and hypometabolism are not specific for AD. Biochemical biomarkers (e.g., isoprostanes, sulfatides, phosphorylated tau231) might allow better discrimination of AD from other forms of dementia [8]. Until now, it has not been determined which combination of these complementary measures is best suited to allow an accurate diagnosis in the early stages of AD and mild cognitive impairment (MCI) or to monitor the treatment efficacy of new drugs. To address these questions, a new, multicenter AD research project called the Alzheimer's Disease Neuroimaging Initiative (ADNI) was launched in October 2004.

Alzheimer' Disease Neuroimaging Initiative

The ADNI (Initiative PI: Dr. M.W. Weiner, Center for Imaging of Neurodegenerative Diseases) is funded by the National Institute on Aging (NIA) and the National Institute of Biomedical Imaging and Bioengineering (NIBIB) of the National Institutes of Health (NIH); several pharmaceutical companies (Pfizer, Wyeth, Eli Lilly, Merck, GlaxoSmithKline, AstraZeneca, Novartis, Eisai, Elan, Forest Laboratories, Bristol Meyers Squibb); and foundations (Alzheimers Association, Institute for Study of Aging) in conjunction with the NIH Foundation. It is to date the most comprehensive effort to identify neuroimaging measures and biomarkers associated with cognitive and functional changes in healthy elderly and those with MCI or AD, and it encompasses clinical sites in the United States and Canada (Fig. 1). The overall goals of the ADNI are the following.

1. Development of optimized methods leading to uniform standards for acquir-ing longitudinal, multicenter MRI and FDG-PET data on patients with AD and MCI as well as healthy elderly controls.
2. Use of these optimized methods for the acquisition of longitudinal structural and metabolic imaging data in a large cohort of healthy elderly and those with MCI and AD and validation of these imaging surrogates with parallel acquired biochemical biomarkers and clinical and cognitive measures.
3. Identification of those neuroimaging measures, cognitive measures, and biochemical biomarkers that provide maximum power for the diagnosis of MCI and AD and for the assessment of treatment effects in trials involving healthy elderly and those with MCI and AD.
4. Creation of a generally accessible imaging and clinical data repository that describes longitudinal changes in brain structure and metabolism, cognitive function, and biochemical biomarkers in healthy elderly and those with MCI and AD.

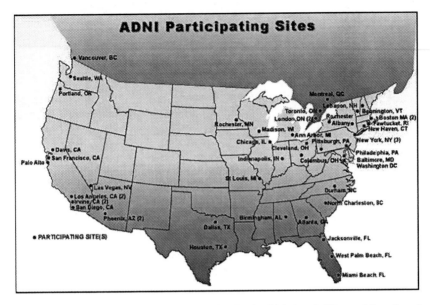

Fig. 1 Overview of the clinical site participating in the Alzheimer's Disease Neuroimaging Initiative

Study Outline

There are several ADNI cores (Fig. 2).

1. A clinical coordination center (PIs: Drs. L. Thal, University of California, San Diego and R. Peterson, Mayo Clinic, Rochester). It is responsible for subject recruitment and maintenance, uniform collection and quality control of clinical and neuropsychological data, deposition of the processed clinical data in a common database, and testing clinical hypotheses.
2. Two neuroimaging cores: (1) MRI core (PI: Dr. C. Jack, Mayo Clinic Rochester) and (2) PET core (PI: Dr. W. Jagust, University of California, Berkeley). They are responsible for determining the optimal imaging parameters, ensuring uniform collection and quality control of the neuroimaging data, and testing imaging hypotheses.
3. A biomarker core (PIs: Drs. J. Trojanowski and L. Shaw, University of Pennsylvania). It oversees the collection and storage of blood, urine, and CSF specimens; is responsible for the development of immortalized cell lines for genetic analyses; and performs all standard clinical laboratory tests and measurements of selected AD biomarkers (ApoE genotype, isoprostanes, tau, Aβ, sulfatides, homocysteine).
4. An informatics core (PI: Dr. A Toga, University of California, Los Angeles). It receives and stores all raw and processed MRI and PET images in a generally accessible data repository (http://www.loni.ucla.edu/ADNI/Data).

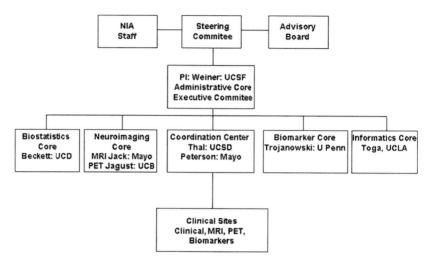

Fig. 2 Overview of the structural organization of the Alzheimer's Disease Neuroimaging Initiative

5. A biostatistics core (PI: Dr. L. Beckett, University of California, Davis). It is in charge of the development of statistical tools and all statistical analyses.

The total duration of the ADNI is 5 years. This time is divided into two main phases: a preparation phase of 6 months followed by an execution phase of 54 months. During an enrollment phase of the first 12 months of the execution phase (June 2005–June 2006), a total of 800 subjects were recruited, including 200 healthy elderly subjects, 400 MCI patients, and 200 mild AD patients. They are to be followed over a period of 2 years (AD patients) to 3 years (healthy elderly, MCI patients). All study participants undergo clinical/cognitive assessments and 1.5-tesla (T) structural MRI examinations at specified intervals. Approximately 50% of each group undergo, in addition, FDG-PET scans at the same time intervals, and 25% of each group (who have not undergone PET) have additional 3-T MRI examinations. Biomarkers in blood and urine are regularly assayed for all participants, and in 25% of the patients the biomarkers are also assayed in CSF.

Preparation Phase

Clinical Coordination Center

During the preparation phase, the clinical core establishes a network of clinical sites and develops a plan for the recruitment and retention of subjects. Furthermore, the clinical core prepares the final clinical protocol and an informed consent document, which is distributed to the sites for local institutional review board approval.

Neuroimaging Core

The neuroimaging core uses the preparation period to develop the final imaging protocols for MRI and PET acquisitions and to develop the procedures necessary for control of data quality and scan consistency.

For the MRI part, the following sequences are evaluated: several T1-weighted three-dimensional (3D) volume sequences, a multiecho 3D volume sequence, and an axial dual echo fast spin echo. A number of "test bed" sites, which operate one of the three MR platforms (Siemens, GE, and Philips) used in the Initiative optimize and standardize these sequences across platforms and test them in a small number of control and AD subjects. In addition, various methods for corrections of B0 inhomogeneity, gradient nonlinearity, and motion artifacts are evaluated. Several state-of-the-art processing methods—boundary shift integral (BSI), voxel-based morphometry (VBM), tensor-based morphometry (TBM), measurements of cortical thickness, semiautomated hippocampal volumetry— are tested regarding their ability to differentiate cross-sectionally between groups and to provide consistent measurements in the absence of change in short duration serial examinations. The sequences, parameters, and correction methods to be used for the execution phase are selected based on the outcome of these comparisons. A similar procedure is used to select and optimize sequences and parameters for 3T acquisitions. Finally, each site participating in the study has to undergo a qualification procedure to demonstrate the capability of perfect, consistent execution of these standardized imaging sequences.

Similar as the MRI part, the PET part of the neuroimaging core also develops a standardized protocol to which all participating centers must adhere for all studies acquired for the Initiative. The participating sites have to undergo a qualification process that involves the acquisition of phantom data using a protocol with specified radioactivity, imaging times, and geometry. These data are used to assess image quality and uniformity, effective image resolution, and the noise and sensitivity of each scanner and to calculate scanner-specific 3D correction images, which then are applied to all human PET studies.

Execution Phase

Clinical Core

All clinical data and results of the laboratory tests obtained during the execution phase are transmitted to the ADNI-CC at the University of California, San Diego and regularly entered into the clinical database of the Initiative. After removing all identifying information, the entire clinical database is placed on a public website so it can be accessed by the Initiative investigators, the pharmaceutical Industry, and the public.

Neuroimaging Core

After completing an examination, both MR and PET sites send the raw data to the imaging data repository at the Laboratory of Neuroimaging (LONI) of the University of California, Los Angeles. From there, the centers responsible for imaging quality control access the data.

Quality control for the MRI arm of the study is performed by the Department of Radiology (Dr. Jack) at the Mayo Clinic, Rochester using visual and computerized methods. After applying the corrections for gradient nonlinearity and so on and stripping all identifying information, the images are made accessible to the public on the imaging data repository at LONI. At about 2 to 2 years 1 month into the project, all data acquired so far (baseline, 6 months, 12 months) will be analyzed by a variety of state-of-the-art automated processing applications available at the participating centers (BSI, volumetric measurement of entorhinal cortex and hippocampus, tensor-based morphometry, voxel-based segmentation/brain parcellation, T1/T2 relaxometry, voxel-based morphometry, cortical time-lapse maps, cortical thinning, parametric 3D surface mesh modeling of subcortical structures, 4D tensor maps). This allows comparison of these methods regarding their immunity to artifacts and cerebrovascular disease, their ability to detect the expected cross-sectional and longitudinal changes, and their processing speed and cost efficiency. Based on those results, the most promising image processing methods will be identified and the remaining means channeled accordingly.

The PET arm of the study employs similar procedures. Quality control is overseen by the team at the University of Michigan. PET images are visually and quantitatively inspected and 3D correction images are applied. The fully corrected images and the various processing stages between the fully corrected image and the initial reconstruction are sent back to the neuroimaging data repository at LONI. As an additional quality control, all corrected PET data are analyzed continuously with three state-of-the-art processing methods—statistical parametric mapping (SPM), stereotactic surface projections (SSP), volume of interest analyses (VOI)—to ensure that the quality of the data is sufficient for the automated processing methods and for hypothesis testing.

Current State of the ADNI

At the time of this writing (April 2005), the coordination center has recruited a network of clinical study sites. The final clinical protocol has been established, as has the informed consent form to be distributed for approval by the local institutional review boards. The final decision on the MRI sequences to be used in the ADNI will be made in early May 2005. It is anticipated that in June 2005 approximately 20 performance sites will have IRB approval and will be able to begin enrollment. By September 2005, remainder of the selected sites should have IRB approval and will start enrollment.

Acknowledgments This work was supported by NIH grant U01AG024964-01.

References

1. Mohs RC, Marin D, Haroutunian V. Early clinical and biological manifestations of Alzheimer's disease: implications for screening and treatment. In: Fillit HM, O'Connell AW (eds) Drug Discovery and Development for Alzheimer's Disease 2000. New York: Springer, 2002, pp 57–63.
2. Kantarci K, Jack CR Jr. Neuroimaging in Alzheimer disease: an evidence-based review. Neuroimaging Clin N Am 2003;13(2):197–209.
3. Kantarci K, Jack CR. Quantitative magnetic resonance techniques as surrogate markers of Alzheimer's disease. NeuroRx 2004;1(2):196–205.
4. Irizarry MC. Biomarkers of Alzheimer disease in plasma. NeuroRx 2004;1(2):226–234.
5. Hsu Y, Schuff N, Du A, et al. Comparison of automated and manual hippocampal MR volumetry in aging and Alzheimer disease. J Magn Res Imaging 2002;16(3):305–310.
6. Davis PC, Gearing M, Gray L, et al. The CERAD experience. Part VIII. Neuroimaging-neuropathology correlates of temporal lobe changes in Alzheimer's disease. Neurology 1995;45(1):178–179.
7. Mielke R, Schroder R, Fink GR, et al. Regional cerebral glucose metabolism and post-mortem pathology in Alzheimer's disease. Acta Neuropathol (Berl) 1996;91(2):174–179.
8. De Leon MJ, DeSanti S, Zinkowski R, et al. MRI and CSF studies in the early diagnosis of Alzheimer's disease. J Intern Med 2004;256(3):205–223.

Chapter 19
Two Hits and You're Out? A Novel Mechanistic Hypothesis of Alzheimer Disease

Xiongwei Zhu, George Perry, and Mark A. Smith

Introduction

The pathological presentation of Alzheimer disease (AD), the leading cause of senile dementia, involves regionalized neuronal death and an accumulation of intraneuronal and extracellular filaments termed neurofibrillary tangles and senile plaques, respectively [reviewed in ref. 1]. A clearer understanding of the mechanism(s) responsible for neuronal death and dysfunction should not only lead to a greater understanding of the underlying pathophysiology of the disease but also unveil potential therapeutic opportunities. To date, despite intensive efforts, the mechanism(s) responsible for AD remain elusive. This incomplete understanding of disease pathogenesis has greatly affected the development of accurate animal and cellular models and, thereby, has retarded the development of therapeutic modalities. Even though several independent hypotheses have been proposed to link the pathological lesions and neuronal cytopathology with, among others, apolipoprotein E genotype [2,3], hyperphosphorylation of cytoskeletal proteins [4], and amyloid-β metabolism [5], not one of these theories alone is sufficient to explain the diversity of biochemical and pathological abnormalities of AD, which involves a multitude of cellular and biochemical changes. Furthermore, attempts to mimic the disease by a perturbation of one of these elements using cell or animal models, including transgenic animals, do not result in the same spectrum of pathological alterations. The most striking case is that although amyloid-β plaques are deposited in some transgenic rodent models overexpressing amyloid β protein precursor (AβPP) [6], there is no neuronal loss (a seminal feature of AD) and the behavioral changes poorly mimic the human disease.

What many of these theories have failed to incorporate is that AD is a disease of aging [7]. Importantly, this holds true even in individuals with a genetic predisposition (i.e., those individuals with an autosomal dominant inheritance

X. Zhu
Department of Pathology, Case Western Reserve University, Cleveland, OH 44106, USA

A. Fisher et al. (eds.), *Advances in Alzheimer's and Parkinson's Disease*,
© Springer 2008

of AD) or in those with Down's syndrome who develop the pathology of AD. Therefore, age is a clear contributor in 100% of AD cases, whatever the genetic background. The free radical theory of aging [8] posits that the aging process is associated with: (1) an increase in the adventitious production of oxygen-derived radicals, i.e., reactive oxygen species (ROS); together with (2) a concurrent decrease in the ability to defend against such ROS, which leads to the accumulation of oxidatively modified macromolecules. The decrease in ROS buffering capacity also leads to a compromised ability to deal with abnormal sources of ROS such as those associated with genetic predisposition and/or disease status. Studies during the past decade have established oxidative stress and damage not only in lesions of AD but also in neurons at risk of death [9–22]. Researchers are now establishing how oxidative stress is related to other possible causes of AD as well as whether oxidative stress is an initiator or is, instead, a result of the disease process. Notably, although oxidative stress is not unique to AD, it does represent one of the earliest pathological events in the disease. Therefore, although oxidative stress is a fundamental aspect of the disease, other factors, likely in synergy, also impinge on disease initiation and progression.

Given the postmitotic nature of adult neurons, it is somewhat surprising that in AD susceptible cortical neurons display an activated cell cycle phenotype normally only seen during developmental neurogenesis, in mitotically active cells, and in neoplastic cells [reviewed in ref. 23]. In neoplasia such ectopic mitogenicity is, by definition, due to a successfully dysregulated cell cycle, whereas in the vulnerable neurons of AD it is due to an emergence from terminal differentiation and attempted reentry into the cell cycle [23]. However, as yet there is no evidence suggesting a successful nuclear division in AD, implying that the neurons do not complete mitosis (M-phase) [reviewed in ref. 24]. In fact, terminally differentiated neurons may lack the ability to complete the cell cycle such that the mitotic alterations (i.e., reactivation of cell cycle machinery) may contribute to neuronal death [24]. Like oxidative stress, cell cycle alterations are extremely early events in disease pathogenesis and likely act in synergy to initiate and propagate disease.

Here, we propose a "two hit" hypothesis of AD, stating that: (1) susceptible neurons are subject to two independent insults—oxidative and mitotic; and (2) both are necessary and sufficient to lead to AD.

Oxidative Stress, Oxidative Stress Signaling, and Alzheimer's Disease

Free radical production occurs as a ubiquitous by-product of both oxidative phosphorylation and the myriad oxidases necessary to support aerobic metabolism. In addition to this background level of ROS, there are a number of other contributory sources in AD that are thought to play an important role in the disease process. They include but are not limited to the following: (1) Iron, in a redox-active state, is increased in neurofibrillary tangles and in amyloid-β deposits and is involved in ROS production [17,25]. Iron catalyzes the formation of •OH from H_2O_2 as well as the formation of advanced glycation end-products. Furthermore, iron-induced lipid peroxidation is potentiated by aluminum [26], which also accumulates in neurofibrillary tangle-containing neurons [27]. (2) Activated microglia, such as those that surround most senile plaques [28], are a source of NO and O_2^- [29], which can react to form peroxynitrite, leaving nitrotyrosine as an identifiable marker [16,30]. (3) Amyloid-β itself has been directly implicated in ROS formation through peptidyl radicals [31–34]. (4) Advanced glycation end-products [9] in the presence of transition metals [17] can undergo redox cycling with consequent ROS production [35–38]. Additionally, advanced glycation end-products and amyloid-β activate specific receptors such as the receptor for advanced glycation end-products (RAGE) and the class A scavenger-receptor to increase ROS production [39,40]. (5) Abnormalities in mitochondrial metabolism, such as deficiencies in key enzyme functions, resulting in part from detection of the mitochondrial genome, may be a major initiating source of ROS [41–46].

An exact determination of the contribution of each source of oxidative stress is complicated if for no other reason than that most sources involve positive feedback. Nonetheless, the overall result is oxidative damage including advanced glycation end-products [9], nitration [16,30,47,48], lipid peroxidation adduction products [18,49–54], and carbonyl-modified neurofilament protein and free carbonyls [9,11,14,18,37,55–60]. It is notable and of mechanistic importance that such oxidative modifications extend beyond the lesions to neurons that do not display obvious signs of degenerative change. Indeed, because oxidative crosslinking makes proteins not only insoluble [reviewed in refs. 12, 15] but also resistant to proteolytic removal [61] by competitively inhibiting the proteasome [62], oxidative crosslinking may be a significant, initiating factor in the formation of neurofibrillary tangles [63] in the face of the numerous proteolytic activities that are highly active against abnormal proteins [64]. Indeed, it may not be coincidental that similar fibrillary inclusions, found in neurodegenerative diseases other than AD, are also extensively ubiquinated (e.g., Lewy/Pick bodies and Rosenthal fibers [65,66]) and are also oxidatively modified [67–69]. Moreover, the induction of antioxidant enzymes such as heme oxygenase-1, Cu/Zn superoxide dismutase, catalase, GSHPx, GSSG-R, peroxiredoxins, and several heat shock proteins and their association with intracellular pathology [10,60,70–72] provide more credence to the fact that the vulnerable neuronal cells are mobilizing antioxidant defense in the face of increased oxidative stress.

As alluded to above, there is increasing evidence that the very earliest neuronal and pathological changes characteristic of AD show evidence of oxidative damage, and such a notion has considerable experimental support [20,21,73–75]. An early, contributing role for oxidative stress and damage is borne out by clinical management of oxidative stress, which appears to reduce the incidence and severity of AD [76,77]. Indeed, increased levels of isoprostane, a product of polyunsaturated fatty acid oxidation, in living patients with mild cognitive impairment (MCI) and probable AD suggests that lipid peroxidation is present at the very earliest stages of the disease [78–80]. That oxidative damage—marked by lipid peroxidation, nitration, reactive carbonyls, or nucleic acid oxidation—is increased in vulnerable neurons regardless of whether they contain neurofibrillary tangles suggests that increases in neuronal oxidative damage must precede neurofibrillary pathology formation [19,21]. Moreover, a marked accumulation of active oxidative modification products, such as 8-hydroxyguanosine (8OHG) and nitrotyrosine, temporally precedes amyloid-β deposition by decades in the cytoplasm of cerebral neurons from patients with Down's syndrome, who invariably develop AD symptoms in their teens and twenties [20,81]. That oxidative damage is the earliest event preceding the formation of tau and amyloid-β-containing pathologies is also confirmed in AD brains [21,74] and, compellingly, in AβPP transgenic mice models, where oxidative stress precedes amyloid-β deposition [75,82].

In sum, oxidative stress appears to play an early and chronic role in both the initiation and progression of AD.

Mitotic Abnormalities, Mitotic Signaling, and Alzheimer's Disease

The cell cycle is a highly regulated process with numerous checks and balances that ensures a homeostatic balance between cell proliferation and cell death in the presence of appropriate environmental signals. The cell cycle is typically divided into four phases: the S phase of DNA replication and the M phase of mitosis, separated by two gap phases called G_1 and G_2. It is the sequential expression and activation of cyclin/cyclin-dependent kinase (cdk) complexes, the main regulators of cell cycle progression, that orchestrate the transition from one phase to another [83]. Cells can exit the cell cycle to stay at a resting (G_0) phase, which is the case in terminally differentiated cells. Triggered by the presence of mitotic growth factors, the resting G_0 cells may reenter G_1 phase owing to the expression/activation of cyclin D/cdk 4,6 complex. Thereafter, the G_1/S transition is controlled by activation of the cyclin E/cdk2 complex [84] such that the absence of cyclin E and/or the inhibition of the cyclin E/cdk2 complex by p21, p27, and p53 causes the cell cycle to be arrested at the G_1 checkpoint. The subsequent fate of the G_1-arrested cells depends on the presence or absence of cyclin A [83] such that in the absence of cyclin A the cells return to G_0 and redifferentiate. However, in the presence of cyclin A, the cells become committed to division, lack the ability to redifferentiate, and if unable to complete the cell cycle die via an apoptotic pathway [85]. Therefore, once beyond late G_1 any arrest in the cell cycle leads to cell death. The DNA replication in the S phase and the transition to the G_2 phase is regulated by the activation of cyclin A/cdk2 complex and proliferating cell nuclear antigen (PCNA). The G_2/M phase transition is controlled by cyclin B/cdc 2 complex.

Any perturbation of these regulators results in the arrest of the cell cycle at G_2/M transition point and cell death.

Although the scheme depicted above is sufficient to describe the behavior of a continuously dividing cell, it fails to provide a mechanism for cells that remain as a steady-state population as is the case for terminally differentiated neurons.

However, in recent years, emerging evidence has shown that vulnerable neurons in AD exhibit phenotypic changes characteristic of mitotic cells, suggesting that these neurons, although not necessarily capable of completing the cell cycle, are capable of reentering the cell cycle [86–94]. In support of this notion, various components of the cell cycle machinery are activated in vulnerable neurons in AD [reviewed in refs. 23, 95]. For example, the presence of cyclin D, cdk4, and Ki67 in diseased neurons suggests that vulnerable neurons in AD are no longer in a quiescent (G_0) phase [86–89]. Moreover, the presence of cyclin E/cdk2 complex indicates that neurons have passed G_1 [88] and are therefore committed to division or death without the possibility of dedifferentiation. In support of this assertion, the presence of coordinated DNA replication suggests that the susceptible neurons may complete a nearly full S phase [91]. Moreover, the aberrant expression of cyclin B1/cdc2 complex indicates that degenerating neurons in AD may even, in some cases, reach G_2 phase [89,90,94,96]. However, the highly unorganized nature of the cell cycle in AD neurons [24] is evident from: (1) the concurrent expression and aberrant localization of PCNA and cyclin B [97]; (2) the concurrent appearance of cdk4 and p16 [87]; and (3) the presence of cyclin E and cyclin B but absence of cyclin D and cyclin A [89]. These abnormalities point to inadequate or failed control of the cell cycle in these neurons, which may contribute to their eventual death in AD patients [24]. Notably, like oxidative stress, mitotic abnormalities are among the very earliest neuronal changes to occur in the disease [87,88,97,98] and not end-stage epiphenomena of the neuropathology. Indeed, cell cycle markers occur prior to the appearance of gross cytopathological changes [99], and the proximal nature of mitotic events in the disease process is evident in pre-AD patients with mild cognitive impairment [100].

In sum, like oxidative stress, there is accumulating evidence that cell cycle alterations represent a very early and thereafter chronic contributor to disease initiation and progression.

Two-Hit Hypothesis

As detailed above, oxidative stress and aberrant mitotic signaling both play early roles in the pathogenesis of AD. However, the temporal relation between these two events was, until recently, unclear. Studies of oxidative stress signaling and mitotic signaling pathways reveal that oxidative stress and aberrant mitotic stimuli are both necessary to initiate and propagate AD [101]. In other words, "two hits" are necessary for the development of AD, whereas individuals subject to only "one hit" remain free of disease (Fig. 1). To illustrate this concept,

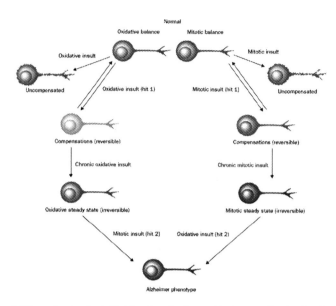

Fig. 1 Two-hit hypothesis. An initial insult, whether oxidative or mitotic, that is chronic and above threshold limits, leads to a new steady state (either oxidative steady state or mitotic steady state). It is in this new steady state where neurons are vulnerable to the subsequent "second hit," which causes the AD phenotype. (From *Lancet Neurology* 2004;3:219–226, reprinted with permission from Elsevier.)

although it is clear that oxidative stress is a pervasive feature in AD at all stages, it is apparent that few neurons (less than 1/10,000 at any given time) exhibit signs of apoptosis [102,103] as would be expected under conditions of acute, high-level oxidative stress. Therefore, AD is associated with lower, chronic levels of oxidative stress that in other situations induce an adaptive response rather than cell death [104–107]. Hence, we suspect that a uniquely chronic, tolerable exposure of neurons to oxidative stress provides an explanation for the low levels of neuronal apoptosis in AD as well as the abnormally sustained activation of SAPK pathways [108,109]. Tolerable levels of oxidative stress provoke compensatory changes that lead to a shift in neuronal homeostasis and, although initially reversible, become permanent adaptive changes under chronic oxidative stress. In this new steady state—"oxidative steady state"—neurons still function relatively normally perhaps for decades [108], and individuals remain relatively cognitively intact (Fig. 1). In fact, because oxidative stress is much higher in pre-AD and AD than with normal aging, it is likely that neurons at an oxidative steady state devote much of their compensatory potential to resisting oxidative stress. Unfortunately, such compensations make the neurons uniquely vulnerable to secondary insults that require other compensatory changes in other pathways, such as those that regulate cell size and growth. Normally, neurotrophic factors such as brain-derived neurotrophic factor (BDNF) and nerve growth factor (NGF) promote the survival, growth, and/or synaptogenesis of neurons [110]. However, the ectopic expression of, or increased sensitivity to, neurotrophic factors in response to cellular stress in an "oxidative steady state" may serve as the second hit, and may trigger a catastrophe in these neurons that leads to AD-type changes [111–113]. Conversely, neurons that have reentered what will become a futile attempt at division (i.e., mitotic steady state) are more vulnerable to changes in oxidative stress that requires further adaptation. In other words, the onset of AD, at least in the absence of genetic factors, is a stochastic process that, given the nature of the detrimental "hits," is age-related in penetrance (Fig. 1).

Genetic Factors and the Two-Hit Hypothesis

Mutations in at least three genes—AβPP and the two homologous presenilin genes PS1 and PS2—are associated with early-onset AD [114]. Although all these mutations inevitably lead to increased amyloid-β production, the exact mechanism(s) by which mutations in these genes are involved in AD pathogenesis remains elusive. However, it is notable that these proteins share a common function—namely, a role in cell cycle control that may be key to the "two hit hypothesis."

AβPP is a single-pass membrane protein expressed at the cell surface, whose cytoplasmic C-terminus interacts with several adaptor proteins, including Fe65 and AβPP-BP1, which function as regulators of the cell cycle [115–117]. For example, AβPP-BP1 is a cell cycle protein that normally negatively regulates the progression of cells into the S phase and positively regulates progression into mitosis [115,116]. The other adaptor, Fe65, is a nuclear protein that also negatively regulates G_1 to S phase cell cycle progression by inhibiting the key S phase enzymes [117]. It is therefore conceivable that AβPP may act as a cell surface receptor to relay cell cycle-related signals. Moreover, AβPP and its proteolytic fragments (i.e., amyloid-β peptide and sAβPP) are mitogenic [91,118–122], and amyloid-β itself can promote activation of the mitotic cycle in cultured differentiated neurons that enter the S phase and start the replication of DNA [120]. sAβPP has been shown to have epithelial growth factor activity, inducing two- to threefold increases in the rates of cell proliferation and cell migration [121,122]. It therefore follows that overexpression, or mutation of AβPP, may push neurons into an aberrant cell cycle. In support of this hypothesis, FAD mutants of AβPP have a greater capacity to drive DNA synthesis than expression of wild AβPP [115,116]. The compensatory changes in response to such genetic stress that serves as a first hit may leave neurons vulnerable to an additional hit. In this regard, it has been demonstrated that neuronal cells bearing AβPPsw mutants have significantly enhanced vulnerability to oxidative stress [123,124], reduced trophic factors [125], ultraviolet irradiation, and staurosporine [126]. AβPP transgenic mice also show increased vulnerability to oxidative stress-related conditions such as ischemia [127] and traumatic brain injury [128].

Presenilin (PS) 1 and 2 proteins also play a role in cell cycle control. For example, the overexpression of both PS1 and PS2 proteins resulted in G_1 phase arrest of the cell cycle [129,130], which may be due to the decrease in Cdk4 activity and phosphorylation of the retinoblastoma tumor suppressor protein [131]. Overexpression of FAD PS1/2 mutants further increases cell cycle arrest compared to wild-type PS1/2, and the degree to which the FAD PS1 mutants inhibit cell cycle progression correlates somewhat with the age of AD onset induced by the mutations [130]. Conversely, PS1 deficiency results in accelerated entry into the S phase and a prolonged S phase of the cell cycle [132,133]. Therefore, the disruption of PS1/2 function caused by FAD mutants could affect regulation of the cell cycle. Neurons under such mitotic stress, which we term the "mitotic steady-state," must devote much of their compensatory potential to fight against it and would be extremely vulnerable to an additional "hit." Indeed, multiple lines of evidence demonstrate that although expression of pathogenic PS mutants is not toxic it does enhance the susceptibility to apoptotic and necrotic insults both in vitro and in vivo.

In summary, both AβPP and PS1/2 play important roles in cell cycle control. Therefore, it is conceivable that the disruption caused by FAD mutants may impair the cell cycle control of susceptible neurons. Given the fact that massive neuronal loss occurs only relatively later in life, it is conceivable that the compensatory changes to such genetic stress lead to a steady state that we call "mitotic steady state" where susceptible neurons still function normally but are vulnerable to a second oxidative hit, as evidenced by their enhanced vulnerability to additional insults.

Conclusions

That AD, like cancer, is a disease of "two hits" explains not only why current therapeutic strategies are often found wanting with respect to efficacy but also why current models of disease pathogenesis fail to replicate the human condition. Models utilizing a two hit strategy are currently in development and should allow the development of pharmacological modalities for AD.

Acknowledgments   Work in the authors' laboratories is supported by funding from the National Institutes of Health, the Alzheimer's Association, Philip Morris USA Inc., and Philip Morris International. This work first appeared on October 23, 2004 at the Alzheimer Research Forum (http://www.alzforum.org/res/adh/cur/zhu/default.asp).

References

1. Smith MA. Alzheimer disease. Int Rev Neurobiol 1998;42:1–54.
2. Corder EH, Saunders AM, Strittmatter WJ, et al. Gene dose of apolipoprotein E type 4 allele and the risk of Alzheimer's disease in late onset families. Science 1993;261(5123):921–923.

3. Roses AD. Apolipoprotein E genotyping in the differential diagnosis, not prediction, of Alzheimer's disease. Ann Neurol 1995;38(1):6–14.
4. Trojanowski JQ, Schmidt ML, Shin RW, et al. Altered tau and neurofilament proteins in neuro-degenerative diseases: diagnostic implications for Alzheimer's disease and Lewy body dementias. Brain Pathol 1993;3(1):45–54.
5. Selkoe DJ. Alzheimer's disease: genotypes, phenotypes, and treatments. Science 1997;275(5300):630–631.
6. Hsiao K, Chapman P, Nilsen S, et al. Correlative memory deficits, Aβ elevation, and amyloid plaques in transgenic mice. Science 1996;274(5284):99–102.
7. Katzman R. Alzheimer's disease, N Engl J Med1986;314(15):964–973.
8. Harman D. Aging: a theory based on free radical and radiation chemistry. J Gerontol 1956;11(3):298–300.
9. Smith MA, Taneda S, Richey PL, et al. Advanced Maillard reaction end products are associated with Alzheimer disease pathology. Proc Natl Acad Sci USA 1994;91(12):5710–5714.
10. Smith MA, Kutty RK, Richey PL, et al. Heme oxygenase-1 is associated with the neurofibrillary pathology of Alzheimer's disease. Am J Pathol 1994;145(1):42–47.
11. Smith MA, Rudnicka-Nawrot M, Richey PL, et al. Carbonyl-related posttranslational modification of neurofilament protein in the neurofibrillary pathology of Alzheimer's disease. J Neurochem 1995;64(6):2660–2666.
12. Smith MA, M. Sayre LM, Monnier CM, Perry G. Radical AGEing in Alzheimer's disease. Trends Neurosci 18(4):172–176.
13. Smith MA, Sayre LM, Vitek MP, et al. Early AGEing and Alzheimer's. Nature 1995;374(6520):316.
14. Smith MA, Perry G, Richey PL, et al. Oxidative damage in Alzheimer's. Nature 1996;382(6587):120–121.
15. Smith MA, Siedlak SL, Richey PL, e al. Quantitative solubilization and analysis of insoluble paired helical filaments from Alzheimer disease. Brain Res 1996;717(1-2):99–108.
16. Smith MA, Richey Harris PL, Sayre LM, et al. Widespread peroxynitrite-mediated damage in Alzheimer's disease. J Neurosci 1997;17(8):2653–2657.
17. Smith MA, Harris PL, Sayre LM, Perry G. Iron accumulation in Alzheimer disease is a source of redox-generated free radicals. Proc Natl Acad Sci USA 1997;94(18):9866–9868.
18. Sayre LM, Zelasko DA, Harris PL, et al. 4-Hydroxynonenal-derived advanced lipid peroxidation end products are increased in Alzheimer's disease. J Neurochem 1997;68(5):2092–2097.
19. Nunomura A, Perry G, Pappolla MA, et al. RNA oxidation is a prominent feature of vulnerable neurons in Alzheimer's disease. J Neurosci 1999;19(6):1959–1964.
20. Nunomura A, Perry G, Pappolla MA, et al. Neuronal oxidative stress precedes amyloid-beta deposition in Down syndrome. J Neuropathol Exp Neurol 2000;59(11):1011–1017.
21. Nunomura A, Perry G, Aliev G, et al. Oxidative damage is the earliest event in Alzheimer disease. J Neuropathol Exp Neurol 2001;60(8):759–767.
22. Perry G, Castellani RJ, Smith MA, et al. Oxidative damage in the olfactory system in Alzheimer's disease. Acta Neuropathol (Berl) 2003;106(6):552–556.
23. Raina AK, Zhu X, Rottkamp CA, et al. Cyclin' toward dementia: cell cycle abnormalities and abortive oncogenesis in Alzheimer disease. J Neurosci Res 2000;61(2):128–133.
24. Bowser R, Smith MA. Cell cycle proteins in Alzheimer's disease: plenty of wheels but no cycle. J Alzheimers Dis 2002;4(3):249–254.
25. Sayre LM, Perry G, Harris PL, et al. In situ oxidative catalysis by neurofibrillary tangles and senile plaques in Alzheimer's disease: a central role for bound transition metals. J Neurochem 2000;74(1):270–279.
26. Oteiza PI. A mechanism for the stimulatory effect of aluminum on iron-induced lipid peroxidation. Arch Biochem Biophys 1994;308(2):374–379.

27. Good PF, Perl DP, Bierer LM, Schmeidler J. Selective accumulation of aluminum and iron in the neurofibrillary tangles of Alzheimer's disease: a laser microprobe (LAMMA) study. Ann Neurol 1992;31(3):286–292.

28. Cras P, Kawai M, Siedlak S, et al. Neuronal and microglial involvement in beta-amyloid protein deposition in Alzheimer's disease. Am J Pathol 1990;137(2):241–246.

29. Colton CA, Gilbert DL. Production of superoxide anions by a CNS macrophage, the microglia. FEBS Lett 1987;223(2):284–288.

30. Good PF, Werner P, Hsu A, et al. Evidence of neuronal oxidative damage in Alzheimer's disease. Am J Pathol 1996;149(1):21–28.

31. Butterfield DA, Hensley K, Harris M, et al. beta-Amyloid peptide free radical fragments initiate synaptosomal lipoperoxidation in a sequence-specific fashion: implications to Alzheimer's disease. Biochem Biophys Res Commun 1994;200(2):710–715.

32. Butterfield DA, Bush AI. Alzheimer's amyloid beta-peptide (1-42): involvement of methionine residue 35 in the oxidative stress and neurotoxicity properties of this peptide. Neurobiol Aging 2004;25(5):563–568.

33. Hensley K, Carney JM, Mattson MP, et al. A model for beta-amyloid aggregation and neurotoxicity based on free radical generation by the peptide: relevance to Alzheimer disease. Proc Natl Acad Sci U S A 1994;91(8):3270–3274.

34. Sayre LM, Zagorski MG, Surewicz WK, et al. Mechanisms of neurotoxicity associated with amyloid beta deposition and the role of free radicals in the pathogenesis of Alzheimer's disease: a critical appraisal. Chem Res Toxicol 1997;10(5):518-526.

35. Baynes JW. Role of oxidative stress in development of complications in diabetes. Diabetes 1991;40(4):405-412.

36. Yan SD, Yan SF, Chen X, et al. Non-enzymatically glycated tau in Alzheimer's disease induces neuronal oxidant stress resulting in cytokine gene expression and release of amyloid beta-peptide. Nat Med 1995;1(7):693–699.

37. Yan SD, Chen X, Schmidt AM, et al. Glycated tau protein in Alzheimer disease: a mechanism for induction of oxidant stress. Proc Natl Acad Sci USA 1994;91(16):7787–7791.

38. Munch G, Kuhla B, Luth HJ, et al. Anti-AGEing defences against Alzheimer's disease. Biochem Soc Trans 2003;31(Pt 6):1397–1399.

39. El Khoury J, Hickman SE, Thomas CA, et al. Scavenger receptor-mediated adhesion of microglia to beta-amyloid fibrils. Nature 1996;382(6593):716–719.

40. Yan SD, Chen X, Fu J, et al. RAGE and amyloid-beta peptide neurotoxicity in Alzheimer's disease. Nature 382(6593):685–691.

41. Davis RE, Miller S, Herrnstadt C, et al. Mutations in mitochondrial cytochrome c oxidase genes segregate with late-onset Alzheimer disease. Proc Natl Acad Sci USA 1997;94(9):4526–4531.

42. Hirai K, Aliev G, Nunomura A, et al. Mitochondrial abnormalities in Alzheimer's disease. J Neurosci 2001;21(9):3017–3023.

43. Coskun PE, Beal MF, Wallace DC. Alzheimer's brains harbor somatic mtDNA control-region mutations that suppress mitochondrial transcription and replication. Proc Natl Acad Sci U S A 2004;101(29):10726–10731.

44. Lustbader JW, Cirilli M, Lin C, et al. ABAD directly links Aβ to mitochondrial toxicity in Alzheimer's disease. Science 2004;304(5669):448-452.

45. Manczak M, Park BS, Jung Y, Reddy PH. Differential expression of oxidative phosphorylation genes in patients with Alzheimer's disease: implications for early mitochondrial dysfunction and oxidative damage. Neuromol Med 2004;5(2):147–162.

46. Trimmer PA, Keeney PM, Borland MK, et al. Mitochondrial abnormalities in cybrid cell models of sporadic Alzheimer's disease worsen with passage in culture. Neurobiol Dis 2004;15(1):29–39.

47. Williamson KS, Gabbita SP, Mou S, et al. The nitration product 5-nitro-gamma-tocopherol is increased in the Alzheimer brain. Nitric Oxide 2002;6(2):221–227.

48. Castegna A, Thongboonkerd V, Klein JB, et al. Proteomic identification of nitrated proteins in Alzheimer's disease brain. J Neurochem 2003;85(6):1394–1401.
49. Palmer AM, Burns MA, Selective increase in lipid peroxidation in the inferior temporal cortex in Alzheimer's disease. Brain Res 1994;645(1-2):338–342.
50. Butterfield DA, Drake J, Pocernich C, Castegna A. Evidence of oxidative damage in Alzheimer's disease brain: central role for amyloid beta-peptide. Trends Mol Med 2001;7(12):548–554.
51. Tamaoka A, Miyatake F, Matsuno S, et al. Apolipoprotein E allele-dependent antioxidant activity in brains with Alzheimer's disease. Neurology 2000;54(12):2319–2321.
52. Lovell MA, Ehmann WD, Butler SM, Markesbery WR. Elevated thiobarbituric acid-reactive substances and antioxidant enzyme activity in the brain in Alzheimer's disease. Neurology 1995;45(8):1594–1601.
53. Markesbery WR, Lovell MA, Four-hydroxynonenal, a product of lipid peroxidation, is increased in the brain in Alzheimer's disease. Neurobiol Aging 1998;19(1):33–36.
54. Guan Z, Wang Y, Cairns NJ, et al. Decrease and structural modifications of phosphatidylethanolamine plasmalogen in the brain with Alzheimer disease. J Neuropathol Exp Neurol 1999;58(7):740–747.
55. Wataya T, Nunomura A, Smith MA, et al. High molecular weight neurofilament proteins are physiological substrates of adduction by the lipid peroxidation product hydroxynonenal. J Biol Chem 2002;277(7):4644–4648.
56. Smith CD, Carney JM, Starke-Reed PE, et al. Excess brain protein oxidation and enzyme dysfunction in normal aging and in Alzheimer disease. Proc Natl Acad Sci USA 1991;88(23):10540–10543.
57. Ledesma MD, Bonay P, Colaco C, Avila J. Analysis of microtubule-associated protein tau glycation in paired helical filaments. J Biol Chem 1994;269(34):21614–21619.
58. Vitek MP, Bhattacharya K, Glendening JM, et al. Advanced glycation end products contribute to amyloidosis in Alzheimer disease. Proc Natl Acad Sci USA 1994;91(11):4766–4770.
59. Montine TJ, Amarnath V, Martin ME, et al. E-4-hydroxy-2-nonenal is cytotoxic and cross-links cytoskeletal proteins in P19 neuroglial cultures. Am J Pathol 1996;148(1):89–93.
60. Takeda A, Smith MA, Avila J, et al. In Alzheimer's disease, heme oxygenase is coincident with Alz50, an epitope of tau induced by 4-hydroxy-2-nonenal modification. J Neurochem 2000;75(3):1234–1241.
61. Cras P, Smith MA, Richey PL, et al. Extracellular neurofibrillary tangles reflect neuronal loss and provide further evidence of extensive protein cross-linking in Alzheimer disease. Acta Neuropathol (Berl) 1995;89(4):291–295.
62. Friguet B, Stadtman ER, Szweda LI, Modification of glucose-6-phosphate dehydrogenase by 4-hydroxy-2-nonenal: formation of cross-linked protein that inhibits the multicatalytic protease. J Biol Chem 1994;269(34):21639–21643.
63. Perry G, Mulvihill P, Manetto V, et al. Immunocytochemical properties of Alzheimer straight filaments. J Neurosci 1987;7(11):3736–3738.
64. Smith MA, Perry G. Alzheimer disease: an imbalance of proteolytic regulation? Med Hypotheses 1994;42(4):277–279.
65. Galloway PG, Grundke-Iqbal I, Iqbal K, Perry G. Lewy bodies contain epitopes both shared and distinct from Alzheimer neurofibrillary tangles. J Neuropathol Exp Neurol 1988;47(6):654–663.
66. Manetto V, Abdul-Karim FW, Perry G, et al. Selective presence of ubiquitin in intracellular inclusions. Am J Pathol 1989;134(3):505–513.
67. Castellani R, Smith MA, Richey PL, et al. Evidence for oxidative stress in Pick disease and corticobasal degeneration. Brain Res 1995;696(1-2):268–271.
68. Castellani R, Smith MA, Richey PL, Perry G. Glycoxidation and oxidative stress in Parkinson disease and diffuse Lewy body disease. Brain Res 1996;737(1-2):195–200.

69. Castellani RJ, Perry G, Harris PL, et al. Advanced glycation modification of Rosenthal fibers in patients with Alexander disease. Neurosci Lett 1997;231(2):79–82.
70. Pappolla MA, Omar RA, Kim KS, Robakis NK. Immunohistochemical evidence of oxidative [corrected] stress in Alzheimer's disease. Am J Pathol 1992;140(3):621–628.
71. Aksenov MY, Tucker HM, Nair P, et al. The expression of key oxidative stress-handling genes in different brain regions in Alzheimer's disease. J Mol Neurosci 1998;11(2):151–164.
72. Lee SC, Zhao ML, Hirano A, Dickson DW, Inducible nitric oxide synthase immunoreactivity in the Alzheimer disease hippocampus: association with Hirano bodies, neurofibrillary tangles, and senile plaques. J Neuropathol Exp Neurol 1999;58(11):1163–1169.
73. Perry G, Smith MA. Is oxidative damage central to the pathogenesis of Alzheimer disease? Acta Neurol Belg 1998;98(2):175–179.
74. Nunomura A, Chiba S, Lippa CF, et al. Neuronal RNA oxidation is a prominent feature of familial Alzheimer's disease. Neurobiol Dis 2004;17(1):108–113.
75. Pratico D, Uryu K, Leight S, et al. Increased lipid peroxidation precedes amyloid plaque formation in an animal model of Alzheimer amyloidosis. J Neurosci 2001;21(12):4183–4187.
76. Sano M, Ernesto C, Thomas RG, et al. A controlled trial of selegiline, alpha-tocopherol, or both as treatment for Alzheimer's disease: The Alzheimer's Disease Cooperative Study. N Engl J Med 1997;336(17):1216–1222.
77. Stewart WF, Kawas C, Corrada M, Metter EJ. Risk of Alzheimer's disease and duration of NSAID use. Neurology 1997;48(3):626–632.
78. Pratico D, Lee MY V, Trojanowski JQ, et al. Increased F2-isoprostanes in Alzheimer's disease: evidence for enhanced lipid peroxidation in vivo. FASEB J 1998;12(15):1777–1783.
79. Pratico D, Clark CM, Lee VM, et al. Increased 8,12-iso-iPF2alpha-VI in Alzheimer's disease: correlation of a noninvasive index of lipid peroxidation with disease severity. Ann Neurol 2000;48(5):809–812.
80. Pratico D, Clark CM, Liun F, et al. Increase of brain oxidative stress in mild cognitive impairment: a possible predictor of Alzheimer disease. Arch Neurol 2002;59(6):972–976.
81. Odetti P, Angelini G, Dapino D, et al. Early glycoxidation damage in brains from Down's syndrome. Biochem Biophys Res Commun 1998;243(3):849–851.
82. Smith MA, Hirai K, Hsiao K, et al. Amyloid-beta deposition in Alzheimer transgenic mice is associated with oxidative stress. J Neurochem 1998;70(5):2212–2215.
83. Grana X, Reddy EP. Cell cycle control in mammalian cells: role of cyclins, cyclin dependent kinases (CDKs), growth suppressor genes and cyclin-dependent kinase inhibitors (CKIs). Oncogene 1995;11(2):211–219.
84. Sherr CJ. G1 phase progression: cycling on cue. Cell 1994;79(4):551–555.
85. Meikrantz W, Schlegel R. Apoptosis and the cell cycle. J Cell Biochem 1995;58(2):160–174.
86. Smith TW, Lippa CF. Ki-67 immunoreactivity in Alzheimer's disease and other neurodegenerative disorders. J Neuropathol Exp Neurol 1995;54(3):297–303.
87. McShea A, Harris PL, Webster KE, et al. Abnormal expression of the cell cycle regulators P16 and CDK4 in Alzheimer's disease. Am J Pathol 1997;150(6):1933–1939.
88. Nagy Z, Esiri MM, Smith AD. Expression of cell division markers in the hippocampus in Alzheimer's disease and other neurodegenerative conditions. Acta Neuropathol (Berl) 1997;93(3):294–300.
89. Nagy Z, Esiri MM, Cato AM, Smith AD. Cell cycle markers in the hippocampus in Alzheimer's disease. Acta Neuropathol (Berl) 1997;94(1):6–15.
90. Harris PL, Zhu X, Pamies C, et al. Neuronal polo-like kinase in Alzheimer disease indicates cell cycle changes. Neurobiol Aging 2000;21(6):837–841.
91. Yang Y, Geldmacher DS, Herrup K. DNA replication precedes neuronal cell death in Alzheimer's disease. J Neurosci 2001;21(8):2661–2668.
92. Ogawa O, Lee HG, Zhu X, et al. Increased p27, an essential component of cell cycle control, in Alzheimer's disease. Aging Cell 2003;2(2):105–110.
93. Ogawa O, Zhu X, H. G. Lee HG, et al. Ectopic localization of phosphorylated histone H3 in Alzheimer's disease: a mitotic catastrophe? Acta Neuropathol (Berl) 2003;105(5):524–528.

94. Zhu X, McShea A, Harris PL, et al. Elevated expression of a regulator of the G_2/M phase of the cell cycle, neuronal CIP-1-associated regulator of cyclin B, in Alzheimer's disease. J Neurosci Res 2004;75(5):698–703.

95. Zhu X, Raina AK, Smith MA. Cell cycle events in neurons: proliferation or death? Am J Pathol 1999;155(2):327–329.

96. Vincent I, Jicha G, Rosado M, Dickson DW. Aberrant expression of mitotic cdc2/cyclin B1 kinase in degenerating neurons of Alzheimer's disease brain. J Neurosci 1997;17(10):3588–3598.

97. Busser J, Geldmacher DS, Herrup K. Ectopic cell cycle proteins predict the sites of neuronal cell death in Alzheimer's disease brain. J Neurosci 1998;18(8):2801–2807.

98. Zhu X, Rottkamp CA, Raina AK, et al. Neuronal CDK7 in hippocampus is related to aging and Alzheimer disease. Neurobiol Aging 2000;21(6):807–813.

99. Vincent I, Zheng JH, Dickson DW, et al. Mitotic phosphoepitopes precede paired helical filaments in Alzheimer's disease. Neurobiol Aging 1998;19(4):287–296.

100. Yang Y, Mufson EJ, Herrup K. Neuronal cell death is preceded by cell cycle events at all stages of Alzheimer's disease. J Neurosci 2003;23(7):2557–2563.

101. Zhu X, Raina AK, Perry G, Smith MA. Alzheimer's disease: the two-hit hypothesis. Lancet Neurol 2004;3(4):219–226.

102. Perry G, Nunomura A, Smith MA. A suicide note from Alzheimer disease neurons? Nat Med 1998;4(8):897–898.

103. Perry G, Zhu X, Smith MA. Do neurons have a choice in death? Am J Pathol 2001;158(1):1–2.

104. Keyse SM, Tyrrell RM. Heme oxygenase is the major 32-kDa stress protein induced in human skin fibroblasts by UVA radiation, hydrogen peroxide, and sodium arsenite. Proc Natl Acad Sci U S A 1989;86(1):99–103.

105. Rushmore TH, King RG, Paulson KE, Pickett DB. Regulation of glutathione S-transferase Ya subunit gene expression: identification of a unique xenobiotic-responsive element controlling inducible expression by planar aromatic compounds. Proc Natl Acad Sci U S A 1990;87(10):3826–3830.

106. Davies JM, Lowry CV, Davies KJ. Transient adaptation to oxidative stress in yeast. Arch Biochem Biophys 1995;317(1):1–6.

107. Wiese AG, Pacifici RE, Davies KJ. Transient adaptation of oxidative stress in mammalian cells. Arch Biochem Biophys 1995;318(1):231–240.

108. LeBel CP, Bondy SC. Oxidative damage and cerebral aging. Prog Neurobiol 1992;38(6):601–609.

109. Chao M, Zhu X, Raina AK, et al. Sources contributing to the initiation and propagation of oxidative stress in Alzheimer disease. Proc Indian Natl Sci Acad Part B 2003;69:251–260.

110. Mattson MP, Chan SL, Duan W. Modification of brain aging and neurodegenerative disorders by genes, diet, and behavior. Physiol Rev 2002;82(3):637–672.

111. Allen SJ, MacGowan SH, Treanor JJ, et al. Normal beta-NGF content in Alzheimer's disease cerebral cortex and hippocampus. Neurosci Lett 1991;131(1):135–139.

112. Crutcher KA, Scott SA, Liang S, et al. Detection of NGF-like activity in human brain tissue: increased levels in Alzheimer's disease. J Neurosci 1993;13(6):2540–2550.

113. Connor B, Young D, Lawlor P, et al. Trk receptor alterations in Alzheimer's disease. Brain Res Mol Brain Res 1996;42(1):1–17.

114. Hardy J. Amyloid, the presenilins and Alzheimer's disease. Trends Neurosci 1997;20(4):154–159.

115. Chen Y, McPhie DL, Hirschberg L, Neve RL. The amyloid precursor protein-binding protein APP-BP1 drives the cell cycle through the S-M checkpoint and causes apoptosis in neurons. J Biol Chem 2000;275(12):8929–8935.

116. Neve RL, McPhie DL, Chen Y. Alzheimer's disease: a dysfunction of the amyloid precursor protein(1). Brain Res 2000;886(1-2):54–66.

117. Bruni P, Minopoli G, Brancaccio Y, et al. Fe65, a ligand of the Alzheimer's beta-amyloid precursor protein, blocks cell cycle progression by down-regulating thymidylate synthase expression. J Biol Chem 2002;277(38):35481–35488.

118. Schubert D, Cole G, Saitoh Y, Oltersdorf T. Amyloid beta protein precursor is a mitogen. Biochem Biophys Res Commun 1989;162(1):83–88.

119. Milward EA, Papadopoulos R, Fuller SJ, et al. The amyloid protein precursor of Alzheimer's disease is a mediator of the effects of nerve growth factor on neurite outgrowth. Neuron 1992;9(1):129–137.

120. Copani A, Condorelli F, Caruso A, et al. Mitotic signaling by beta-amyloid causes neuronal death. FASEB J 1999;13(15):2225–2234.

121. Hoffmann J, Twiesselmann C, Kummer MP, et al. A possible role for the Alzheimer amyloid precursor protein in the regulation of epidermal basal cell proliferation. Eur J Cell Biol 2000;79(12):905–914.

122. Schmitz A, Tikkanen R, Kirfel G, Herzog V. The biological role of the Alzheimer amyloid precursor protein in epithelial cells. Histochem Cell Biol 2002;117(2):171–180.

123. Eckert A, Steiner B, Marques C, et al. Elevated vulnerability to oxidative stress-induced cell death and activation of caspase-3 by the Swedish amyloid precursor protein mutation. J Neurosci Res 2001;64(2):183–192.

124. Marques CA, Keil U, Bonert A, et al. Neurotoxic mechanisms caused by the Alzheimer's disease-linked Swedish amyloid precursor protein mutation: oxidative stress, caspases, and the JNK pathway. J Biol Chem 2003;278(30):28294–28302.

125. Leutz S, Steiner B, Marques CA, et al. Reduction of trophic support enhances apoptosis in PC12 cells expressing Alzheimer's APP mutation and sensitizes cells to staurosporine-induced cell death. J Mol Neurosci 2002;18(3):189–201.

126. Xu X, Yang D, Wyss-Coray T, et al. Wild-type but not Alzheimer-mutant amyloid precursor protein confers resistance against p53-mediated apoptosis. Proc Natl Acad Sci U S A 1999;96(13):7547–7552.

127. Koistinaho M, Kettunen MI, Goldsteins G, et al. Beta-amyloid precursor protein transgenic mice that harbor diffuse A beta deposits but do not form plaques show increased ischemic vulnerability: role of inflammation. Proc Natl Acad Sci USA 2002;99(3):1610–1615.

128. Nakagawa Y, Nakamura M, McIntosh TK, et al. Traumatic brain injury in young, amyloid-beta peptide overexpressing transgenic mice induces marked ipsilateral hippocampal atrophy and diminished Aβ deposition during aging. J Comp Neurol 1999;411(3):390–398.

129. Janicki SM, Monteiro MJ. Presenilin overexpression arrests cells in the G_1 phase of the cell cycle: arrest potentiated by the Alzheimer's disease PS2(N141I) mutant. Am J Pathol 1999;155(1):135–144.

130. Janicki SM, Stabler SM, Monteiro MJ. Familial Alzheimer's disease presenilin-1 mutants potentiate cell cycle arrest. Neurobiol Aging 2000;21(6):829–836.

131. Prat MI, Adamo AM, Gonzalez SA, et al. Presenilin 1 overexpressions in Chinese hamster ovary (CHO) cells decreases the phosphorylation of retinoblastoma protein: relevance for neurodegeneration. Neurosci Lett 2002;326(1):9–12.

132. Soriani M, Pietraforte D, Minetti M. Antioxidant potential of anaerobic human plasma: role of serum albumin and thiols as scavengers of carbon radicals. Arch Biochem Biophys 1994;312(1):180–188.

133. Yuasa S, Nakajima M, Aizawa H, et al. Impaired cell cycle control of neuronal precursor cells in the neocortical primordium of presenilin-1-deficient mice. J Neurosci Res 2002;70(3):501–513.

Chapter 20
Vascular Risk Factors and Risk for Alzheimer's Disease and Mild Cognitive Impairment: Population-Based Studies

Hilkka Soininen, Miia Kivipelto, and Aulikki Nissinen

Introduction

Dementia is a common, disabling disorder in the elderly. Because of the worldwide aging phenomenon of the population, dementia has increasing public health relevance. The major cause of dementia is Alzheimer's disease (AD). It has been estimated that the number of AD patients will quadruple by 2050 unless a means of prevention or cure is found. From the same projections, it has been proposed that interventions that could postpone disease onset by 5 years would decrease the projected prevalence of AD by 50% [1]. Until recently, age and family history were the only well established risk factors for AD, and no strategies have been available for preventing AD.

Alzheimer's disease is a multifactorial disease resulting from an interaction between genetic susceptibility and environmental factors. Thus, its prevention is likely to be at least partly possible. To determine interventions that would prevent or delay the onset of AD, modifiable risk factors for the disease must first be identified. These risk factors could then be used as targets for intervention, both pharmacological and nonpharmacological; moreover, population-based health education and intervention programs could be developed. Because the process leading to the clinical manifestation of AD takes decades, long-term prospective studies are necessary to identify its environmental risk factors. This is a short review of the recent findings of a Finnish population-based cohort study concerning vascular- and lifestyle-related factors for dementia/AD and for mild cognitive impairment (MCI), which is considered a high risk state for the development of AD.

H. Soininen
Department of Neurology, University of Kuopio, Kuopio 70211, Finland

A. Fisher et al. (eds.), *Advances in Alzheimer's and Parkinson's Disease*,
© Springer 2008

Subjects and Methods

CAIDE Study

The Cardiovascular Risk Factors, Aging, and Dementia (CAIDE) study is a large, longitudinal, population-based study in Finland investigating the role of vascular, lifestyle, and genetic factors in the development of dementia/AD and MCI. Participants were derived from four separate and independent population-based random samples studied within the framework of the North Karelia Project and the FINMONICA study in 1972, 1977, 1982, or 1987 (baseline, midlife visit). Vascular risk factors and health behavior were assessed during these visits. Those individuals still alive, aged 65 to 79 at the end of 1997, and living in two geographically defined areas in or close to the towns of Kuopio and Joensuu (n = 2293) were the target of this study. From among these subjects, a random sample of 2000 persons was invited to the reexamination carried out in 1998 (first follow-up). Altogether, 1449 (73%) individuals participated.

Dementia and MCI were diagnosed in a three-phase study design: a screening phase, a clinical phase, and a differential diagnostic phase [2]. A total of 61 participants were diagnosed as having dementia according to DSM-IV criteria, and 48 of them fulfilled the diagnostic criteria of AD according to NINCDS-ADRDA criteria. The dementia diagnoses of the nonparticipants were derived from medical records of hospitals and health care centers. The total number of dementia cases increased to 117 (5.9% of the population) when these diagnoses were also taken into account. The diagnostic criteria proposed by the Mayo Clinic AD Research Center [3] were applied for diagnosing MCI. The second follow-up of this cohort is now ongoing.

Kuopio MCI Study

In another cohort [4] we estimated the incidence of MCI among cognitively healthy elderly subjects during a 3-year follow-up and evaluated the impact of demographic and vascular factors as well as the APOE ε4 allele on conversion to MCI. At baseline, the cognitive abilities of 806 of 1150 eligible subjects (aged 60–76 years) from a population-based sample were examined. Cognitively intact subjects (n = 747) were followed for an average of 3 years.

Results

Vascular Risk Factors in AD and MCI

We have shown in the CAIDE cohort that high systolic blood pressure (BP) ($\geq$ 160 mmHg) and high serum total cholesterol ($\geq$ 6.5 mmol/L) at midlife

Table 1 Main Findings from the CAIDE Study: Midlife
Risk Factors for Dementia and AD Later in Life

Vascular
High systolic BP (≥ 160 mmHg)
High serum cholesterol (≥ 6.5 mmol/L)
Obesity (BMI ≥ 30 kg/m^2)
Lifestyle related
Frequent alcohol drinking
Moderate use of saturated fatty acids/lack of polyunsaturated fatty acids
Physical inactivity

BP, blood pressure; BMI, body mass index.

(on average 21 years earlier) are significant risk factors for late-life AD (Table 1) [5]. Midlife high cholesterol also increased the risk of late-life MCI, and the effect of high systolic BP approached significance [2]. The APOE ε4 allele was an independent risk factor for AD, even after adjustment for midlife vascular risk factors and other confounders [odds ratio (OR) 2.1, 95% confidence interval (CI) 1.1–4.1]. Similarly, elevated midlife cholesterol level (OR 2.8, 95% CI 1.2–6.7) and systolic BP (OR 2.6, 95% CI 1.1–6.6) were independent risk factors for AD, even after adjustment for ApoE genotype and other confounding factors [6]. These results suggest that the risk of AD related to elevated cholesterol and BP is independent of and greater than that related to the ApoE ε4 allele. There is also evidence that the risk of dementia related to the ApoE ε4 allele may vary by sex and could be reduced or modified by interventions targeted to treatable risk factors. For instance, our results have indicated that the ApoE ε4 allele may be a significant risk factor for AD only among persons not using antihypertensive drugs [7].

We have also found that systolic BP decreased during the 21-year follow-up in individuals diagnosed with AD, whereas values increased in nondemented subjects. Also, serum cholesterol values decreased significantly during the follow-up among subjects with AD and MCI; and at the time of diagnosis, there were no longer any differences in BP or cholesterol levels between the groups. These changes may explain at least partly the inconsistent/negative results from the earlier cross-sectional or short-term follow-up studies on this issue. These findings give further support to the view that the temporal relation between risk factors and AD cannot be determined unequivocally in studies with relatively short follow-up times.

Our recent findings indicate that obesity at midlife [body mass index (BMI) > 30 kg/m^2] is a significant risk factor for late-life dementia, even after adjusting for other vascular risk factors and disorders [8]. Obesity, high systolic BP, and high serum cholesterol at midlife were all independent risk factors for late-life dementia/AD with odds ratios of around 2 for each factor. Clustering of these midlife vascular risk factors increased the risk of dementia in an additive manner: The odds ratio of dementia in participants who had all three risk factors compared with participants who had none of these was 6.2. In the

Kuopio MCI study, 66 (8.8%) of 747 cognitively intact elderly subjects had converted to MCI during the 3-year follow-up period. The global incidence rate of MCI was 25.94/1000 person-years. Persons with higher age (OR 1.08, 95% CI 1.01–1.16), who were ApoE ε4 allele carriers (OR 2.04, 95% CI 1.15–3.64), or who had medicated hypertension (OR 1.86, 95% CI 1.05–3.29) were more likely to convert to MCI than those individuals of lower age and without an ApoE ε4 allele or medicated hypertension. Persons with high education (OR 0.79, 95% CI 0.70–0.89) were less likely to convert to MCI than persons with low or no education. In subjects with both the ApoE ε4 allele and medicated hypertension, the crude OR for conversion was 3.92 (95% CI 1.81–8.49). In subjects with cardiovascular disease, the crude OR for conversion was 2.13 (95% CI 1.26–3.60). Sex, measured BP, diabetes, or cerebrovascular disease at baseline had no significant effect on the conversion to MCI.

Lifestyle-Related Factors and AD

We have recently found in our population that midlife alcohol drinking is related to the risk of late-life MCI in a U-shaped manner with both never-drinkers and frequent drinkers having a higher risk than infrequent alcohol drinkers [9]. However, this was not the case for dementia. The ApoE genotype seemed to modify the association between midlife alcohol drinking and subsequent dementia as a direct association was detected only among the ApoE ε4 allele carriers. Based on these findings we suggest that the path from normal cognitive function to MCI is influenced by environmental factors (e.g., frequent alcohol drinking). However, a combined effect of detrimental environmental factors (frequent alcohol drinking in this case) and the genetic susceptibility may be required for cognitive impairment to progress further to dementia. Part of the effect of heavy alcohol drinking on the risk of AD may be mediated through the increased risk of cerebrovascular disease in drinkers.

Our recent findings indicate that moderate intake of unsaturated fats at midlife is protective, whereas a moderate intake of saturated fats may increase the risk of dementia and AD, especially among ApoE ε4 carriers. Regular midlife physical activity seems to be protective against late-life dementia and AD. Also, here the association was more pronounced among the ApoE ε4 carriers.

Conclusion

There is increasing evidence that AD and MCI have many modifiable risk factors. Vascular risk factors, especially elevated BP and serum cholesterol values earlier in life, seem to play an important role in the development of AD, as suggested by both epidemiological studies and studies that have reported a decreased occurrence of AD in persons receiving antihypertensive or lipid-lowering drug treatments [10]. These findings are also supported by experimental studies. It is important to note that many of the other lifestyle-related risk factors for AD (e.g., alcohol drinking, dietary fat intake, physical inactivity) are vascular-related.

Based on the available data, it seems possible that early interventions aimed at reducing these cardiovascular risk factors may have a profound impact on the future occurrence of dementia and cognitive impairment. Lifestyle intervention may be of even greater significance for persons with genetic susceptibility for AD. These findings should encourage people to adopt positive lifestyle options in their youth and at midlife to increase their probability of enjoying vital years later in life not only physically but also cognitively. The role of lifestyle and pharmacological interventions for the prevention of AD need to be studied further. The putative interactions between genetic and environmental factors especially merit further investigation.

References

1. Brookmeyer R, Gray S, Kawas C (1998) Projections of Alzheimer's disease in the United States and the public health impact of delaying disease onset. Am J Public Health 88:1337–1342.
2. Kivipelto M, Helkala EL, Hänninen T, et al (2001a) Midlife vascular risk factors and late-life mild cognitive impairment: a population based study. Neurology 56:1683–1689.
3. Petersen RC, Smith GE, Ivnik RJ, et al (1995) Apolipoprotein E status as a predictor of the development of Alzheimer's disease in memory-impaired individuals. JAMA 273:1274–1278.
4. Tervo S, Kivipelto M, Hänninen T, et al (2004) Incidence and risk factors for mild cognitive impairment: a population-based three year follow-up study of cognitively healthy elderly subjects. Dement Geriatr Cogn Disord 17:96–203.
5. Kivipelto M, Helkala EL, Laakso MP, et al. (2001b) Midlife vascular risk factors and Alzheimer's disease in later life: longitudinal, population-based study. BMJ 322:1447–1451.
6. Kivipelto M, Helkala EL, Laakso MP, et al (2002a) Apolipoprotein E ε4 allele, elevated midlife cholesterol and systolic blood pressure are independent risk factors for late-life Alzheimer's disease. Ann Intern Med 137:149–155.
7. Kivipelto M, Helkala EL, Nissinen A, et al (2002b) Vascular risk factors, ApoE ε4 allele and gender and the risk of Alzheimer's disease: perspectives on prevention. Drug Dev Res 56:85–94.
8. Kivipelto M, Anttila T, Fratiglioni L, et al (2005) Obesity and vascular risk factors at midlife and the risk of dementia and Alzheimer's disease. Arch Neurol 62:1556–1560.
9. Anttila T, Helkala EL, Viitanen M, et al. (2004) Alcohol drinking at midlife and the risk of mild cognitive impairment and dementia in late-life: a population-based study. BMJ 329:539–542.
10. Kivipelto M, Laakso MP, Tuomilehto J, et al (2002c) Hypertension and hypercholesterolemia as risk factors for Alzheimer's disease: potential for pharmacological intervention. CNS Drugs 16:435–444.

Chapter 21
Cholesterol Transport and Production in Alzheimer's Disease

Judes Poirier, Louise Lamarre-Théroux, Doris Dea, Nicole Aumont, and Jean Francois Blain

Introduction

Apolipoproteins, lipid carrier molecules, are instrumental in the coordination of the metabolism of lipid following peripheral and central nervous system (CNS) injuries. Apolipoprotein E (ApoE) in particular is unique among apolipoproteins in that it has a special relevance to nervous tissue. It was shown to coordinate the mobilization and redistribution of cholesterol and phospholipids for the repair, growth, and maintenance of myelin and neuronal membranes, during development or after injury, in the peripheral nervous system. In the brain, ApoE plays a pivotal role in the mobilization and redistribution of cholesterol and phospholipids during membrane remodeling associated with synaptic plasticity [1]. The low concentrations of other key plasma apolipoproteins such as ApoA1, C1, and B in the brain further emphasize the critical role of ApoE in this particular tissue [2].

Cholesterol Homeostasis in the Brain

Cholesterol and other lipids are used for membrane synthesis and for many other anabolic or catabolic activities by cells throughout the body, including those of the CNS, a site of high lipid turnover. Although cells composing the nervous tissue are capable of de novo synthesis of lipid molecules, they can also bind and take up lipoproteins that are made available in the local environment for their lipid requirements.

Cholesterol requirements of most mammalian cells are met by two separate, yet interrelated processes [3]. One process is the endogenous synthesis of cholesterol. This synthesis pathway, which involves more than 20 reactions, is regulated primarily by the activity of the 3-hydroxy-3-methylglutaryl coenzyme

J. Poirier
McGill Centre for Studies in Aging, McGill University, Montreal, Quebec, Canada

A. Fisher et al. (eds.), *Advances in Alzheimer's and Parkinson's Disease*,
© Springer 2008

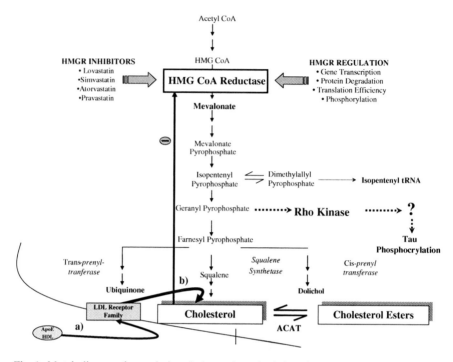

Fig. 1 Metabolic cascade regulating cholesterol synthesis in eukaryotic cells

A reductase (HMGR), which catalyzes the formation of mevalonate, the key precursor molecule in the synthesis of cholesterol (Fig. 1). The other process involves the utilization of lipoprotein-derived cholesterol following internalization of the lipoprotein bound to its surface receptor (ApoE-rich lipoprotein complexes). Brain cells, particularly astrocytes and neurons, cultured in vitro synthesize cholesterol at a rate that is inversely proportional to the cholesterol content in the growth environment. According to this scheme, more cholesterol is imported through the ApoE-ApoE (low density lipoprotein, or LDL) receptor system than is synthesized de novo by neurons. In contrast, oligodendrocytes and Schwann cells, which are involved in axonal myelination, synthesize de novo cholesterol rather than importing it through the ApoE-ApoE receptor pathway. Inhibition of cholesterol synthesis in oligodendrocytes in response to phenylketonuria [4,5] or tellurium [6] salts exposure causes drastic demyelization in both animal models and humans. Cholesterol homeostasis in brain cells is thus maintained by the perfect balance between cholesterol influx through the ApoE receptor family and cholesterol synthesis via the HMGR pathway, the rate-limiting step in cholesterol biosynthesis.

Because the blood-brain barrier prevents the passage of whole macromolecular complexes, lipoprotein particles are assembled locally in the CNS using components imported from the plasma or originating from local synthesis and secretion. Lipoprotein particles the size of plasma high density lipoprotein (HDL) are produced and released in the cerebrospinal fluid (CSF) [7]. ApoE-HDL in the CSF represents a potential and apparently unique source of lipids for cells of the nervous parenchyma as well. They are also responsible for the extracellular transport and distribution of beta-amyloid peptides in the brain and in the plasma. In addition to lipids, as their name implies, lipoproteins contain a protein moiety referred to as apolipoproteins (Apos), some of which are regulators for extracellular enzymatic reactions involved in lipid metabolism. Other Apos are ligands for cell surface receptors that mediate the influx of lipoprotein particles and their subsequent intracellular metabolism.

Apolipoprotein E and Cholesterol Transport in the CNS: Backbone for the Maintenance of Synaptic Integrity and Plasticity

The brain is a major site of ApoE mRNA expression in humans, marmosets, rats, and mice, ranking second only to the liver in humans [2]. Transcripts for ApoE are distributed throughout all regions of the brain and have been localized to astrocytes and microglia by in situ hybridization and site mutagenesis analyses. Accordingly, ApoE was shown to be synthesized and secreted mostly by glial cells (> 95%) and to serve as a ligand for the members of the LDL receptor family in the brain. Primary cultures of hippocampal neurons from rat embryos and prosimians have the capacity to internalize ApoE-containing lipoproteins; intraneuronal localization of ApoE has been observed in several other studies. Abnormal neurons containing neurofibrillary tangles in brains of individuals with Alzheimer's disease (AD) were also shown to contain ApoE. An important biochemical characteristic of human apoE stems from a genetic polymorphism, first established by Utermann et al. [8] using isoelectric focusing. Polymorphisms within the human ApoE gene (located on chromosome 19) account for the three major ApoE isoforms, designated ApoE2, ApoE3, and ApoE4, arising from respective alleles $\varepsilon 2$, $\varepsilon 3$, and $\varepsilon 4$. The result of this polymorphism is three homozygous genotypes ($\varepsilon 2/\varepsilon 2$, $\varepsilon 3/\varepsilon 3$, $\varepsilon 4/\varepsilon 4$) and three heterozygous genotypes ($\varepsilon 2/\varepsilon 3$, $\varepsilon 2/\varepsilon 4$, $\varepsilon 3/\varepsilon 4$). The most common isoform, apoE3, differs from ApoE2 and ApoE4 by amino acid substitutions at residues 112 and 158 (apoE is 299 amino acids long). The ApoE2 isoform has cysteine residues at sites 112 and 158, ApoE3 has cysteine at site 112 and arginine at site 158, and ApoE4 has arginine at both sites [9].

In the nervous system, the importance of the polymorphic nature of apoE has recently been revealed with regard to function in neuronal plasticity and with respect to other pathologies, such as AD [10–13]. Apolipoprotein $\varepsilon 4$ allele was shown to be strongly associated with the familial and sporadic forms of Alzheimer disease [14,15]. The Apo $\varepsilon 4$ allele affects in a gene dose manner the rate of progression of the disease, the extent of the neuronal cell loss, cholinergic activity, accumulation of amyloid plaques in hippocampal and cortical areas, and total beta-amyloid production and beta-amyloid deposition in the brains of AD subjects. ApoE4 carriers were also shown to exhibit poor synaptic remodeling and defective compensatory plasticity in vulnerable brain areas in AD [10–13]. Indeed, the role of ApoE in the maintenance of synaptic integrity and plasticity is so central to brain physiology that the ability of a subject to recover from traumatic brain injuries is highly dependent on Apo ε4 allele content.

Cholesterol Homeostasis and Cholesterol-Lowering Drugs

Cholesterol homeostasis in brain cells is maintained by the perfect balance between cholesterol influx through the ApoE/ApoE receptor family pathway and cholesterol synthesis via the HMGR pathway, the rate-limiting step in cholesterol biosynthesis. Under normal circumstances, cholesterol synthesis via the HMGR pathway (Fig. 1) is required only when lipoprotein internalization by the ApoE/ApoE receptor pathway is insufficient to meet the cholesterol requirement of the cell. The endoplasmic reticulum (ER)-bound HMGR is regarded as the rate-limiting enzyme in the synthesis of cholesterol, a critical membrane lipid, precursor of steroid hormones (glucocorticoids and estrogen), and signaling molecules involved in embryogenesis. The other, shorter form of HMGR, localized in the peroxisomal compartment of embryonic cells does not appear to play an important role in cholesterol homeostasis. The peroxisomal form is far more resistant to usual HMGR inhibitors, such as simvastatin, than is the ER form. Its body distribution has not yet been examined. In cells grown in excess of cholesterol-rich lipoproteins, the HMGR activity is down-regulated in favor of cholesterol reuptake via the ApoE receptor pathway [16]. A similar process was described in the peripheral nervous system (PNS) and the CNS during the acute phase of regeneration that ensues as a result of degradation of dead cells following experimental injury or a chronic neurodegenerative process [17–19]. To maintain cellular cholesterol homeostasis, there is a rather potent negative feedback system regarding HMGR activity and gene expression that results in a decrease in the synthesis of cholesterol in response to excess intracellular sterol internalization via the ApoE receptor family. This first and most important feedback regulation of HMGR activity is through a decrease in gene transcription. The factor that has been shown to regulate the expression of the reductase is the controlled degradation of the HMGR protein. Lastly, there is evidence from hamsters for a modulation in translation efficiency of mRNA for HMGR, resulting in decreased or increased reductase protein and activity.

In the AD brain, ApoE (cholesterol transport) and HMG CoA reductase (cholesterol synthesis) activity and mRNA prevalence display a rather striking interdependence. Figure 2A illustrates the HMG CoA reductase activity measured in cortical areas of a large group of autopsy-confirmed AD patients and age-matched control subjects. Postmortem delays and sexes were matched.

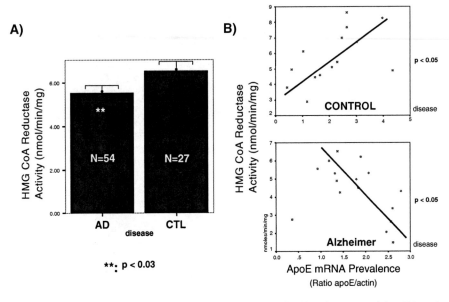

Fig. 2 3-Hydroxy-3-methylglutaryl coenzyme A (HMG CoA) reductase activity (**A**) and mRNA prevalence (**B**) in cortical areas in autopsy-confirmed sporadic cases with Alzheimer's disease (AD) and in age-matched and sex-matched control subjects. **B** Actin was used as a positive control mRNA in our real-time quantitative polymerase chain reaction (PCR) assays, as described by Aleong et al. [20]

A modest but significant reduction of the HMG CoA reductase activity was measured in AD subjects when compared to healthy control cases ($p < 0.03$). Figure 2B illustrates the relation between ApoE mRNA prevalence and HMG CoA reductase activity, using real-time quantitative reverse transcription-polymerase chain reaction (RT-PCR), in the cortical areas of autopsy-confirmed subjects with and without definitive sporadic AD diagnosis. The analysis indicates a significant positive correlation between the ApoE mRNA concentration and HMG CoA reductase in healthy control subjects, but an inverse correlation between the ApoE and HMG CoA reductase in autopsy-confirmed AD cases ($p < 0.05$).

The inverse relation between cholesterol synthesis and cholesterol transport via the ApoE transport system is strikingly similar to observations in deafferented hippocampus of rodents (Fig. 3) [19]. In this experimental model of deafferentation and reinnervation, ApoE synthesis (mRNA and protein) [18] is markedly up-regulated in response to ongoing deafferentation, whereas HMG CoA reductase activity [19] becomes progressively repressed during the terminal proliferation and synaptic remodeling (Fig. 3). During the active phase of reinnervation, the uptake of cholesterol esters via the ApoE (LDL) receptor pathway is so efficient that neurons progressively reduce cholesterol synthesis to maintain intracellular cholesterol homeostasis. The biochemical similarities between the

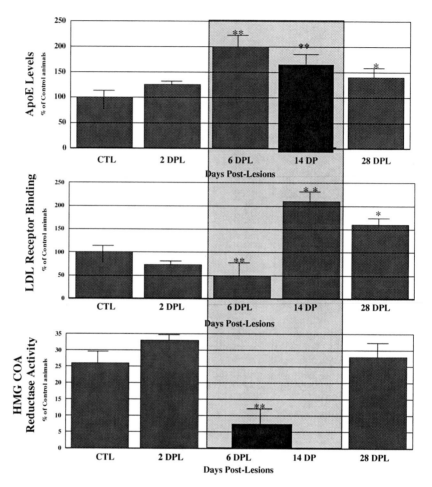

Fig. 3 Time course analysis of the alterations in apolipoprotein E (ApoE) levels, low density lipoprotein (LDL) receptor density, and HMG CoA reductase activity in entorhinal cortex-lesioned rats during: (**a**) the deafferentation phase (0–6 days after lesioning) and (**b**) the reinnervation phase (7–40 days after lesioning) in the adult animals. (Adapted from Poirier et al. [14,19])

hippocampal deafferentation model and the deafferented cortical areas of sporadic AD subjects supports the notion that the AD brain is actively resisting the deafferentation process by promoting a compensatory response designed to maintain cholesterol homeostasis during active synaptic remodeling.

Evidence obtained from cross-sectional epidemiological studies indicates that the utilization of HMGR inhibitors (statins) in middle-age humans reduces circulating cholesterol levels in the blood and confers some protection against sporadic AD [21–23]. However, recent longitudinal prospective epidemiological studies do not support the original findings on risk reduction [24,25].

It is interesting to note that statins, by way of selective inhibition of cholesterol synthesis in liver, cause marked up-regulation of the LDL receptors, which then interact with blood-derived lipoproteins. Blood lipoproteins, cholesterol, and triglycerides are thus captured by these cell surface mediators of cholesterol internalization. This, in turn, leads to a marked reduction of blood cholesterol concentration and a compensatory HDL-cholesterol induction. It should be noted that the whole process results in transient but marked increases in cholesterol concentrations in the liver. The effects of statins on brain cell biology remain the subject of speculation, as studies claim that lipophilic statins cross the blood-brain barrier poorly, whereas others can cross the blood-brain barrier at a low rate.

However, if one assumes that a very small portion of statins does penetrate the brain-blood barrier and is transported to the neurons of the CNS, it is quite conceivable that a mechanism similar to the one in place in the liver is being activated in neurons, inhibiting HMG CoA reductase activity and massively up-regulating LDL receptor expression. This would, in turn, enhance ApoE-HDL internalization and cholesterol uptake. In this scenario, the increased availability of cholesterol would facilitate terminal proliferation and synaptic remodeling, two processes that require huge amounts of lipid.

Conclusion

The very nature of the various cholesterol homeostatic processes in the CNS suggests that the AD brain, although seriously impaired by the neurodegenerative disease process, is actively engaged in synaptic remodeling and compensatory sprouting, particularly in ApoE3 and ApoE2 subjects. The use of pharmacological facilitators of cholesterol transport, such as probucol, a potent ApoE transporter, or statins, selective ApoE (LDL) receptor inducers, should enhanced local cholesterol availability and promote compensatory plastic responses in subjects who exhibit symptoms of the disease or who are nonsymptomatic carriers of a genetic predisposition to AD.

References

1. Poirier J (1994) Apolipoprotein E in animal models of CNS injury and in Alzheimer's disease. TINS 17:525–530.
2. Elshourbagy NA, Liao WS, Mahley RW, Taylor JM (1985) Apolipoprotein E mRNA is abundant in the brain and adrenals, as well as in the liver, and is present in other peripheral tissues of rats and marmosets. Proc Natl Acad Sci U S A 82:203–207.
3. Poirier J (2003) Apolipoprotein E and cholesterol metabolism in the pathogenesis and treatment of Alzheimer's disease. Trends Mol Med 9:94–101.

4. Castillo M, Zafra MF, Garcia-Peregrin E (1988) Inhibition of brain and liver 3-hydroxy-3-methylglutaryl-CoA reductase and mevalonate-5-pyrophosphate decarboxylase in experimental hyperphenylalaninemia. Neurochem Res 13:551–555.

5. Shefer S, Tint GS, Jean-Guillaume D, et al (2000) Is there a relationship between 3-hydroxy-3-methylglutaryl coenzyme A reductase activity and forebrain pathology in the PKU mouse? J Neurosci Res 61:549–563.

6. Toews AD, Roe EB, Goodrum JF, et al (1997) Tellurium causes dose-dependent coordinate down-regulation of myelin gene expression. Mol Brain Res 49:113–119.

7. LaDu MJ, Gilligan SM, Lukens JR, et al (1998) Nascent astrocyte particles differ from lipoproteins in CSF. J Neurochem 70:2070–2081.

8. Utermann G, Langenbeck U, Beisiegel U, Weber W (1980) Genetics of the apolipoprotein E system in man. Am J Hum Genet 32:339–347.

9. Davignon J, Gregg RE, Sing CF (1988) Apolipoprotein E polymorphism and atherosclerosis. Arteriosclerosis 8:1–21.

10. Arendt T, Schindler C, Bruckner MK, et al (1997) Plastic neuronal remodeling is impaired in patients with Alzheimer's disease carrying apolipoprotein epsilon 4 allele. J Neurosci 17:516–529.

11. Ji Y, Gong Y, Gan W, et al (2003a) Apolipoprotein E isoform-specific regulation of dendritic spine morphology in apolipoprotein E transgenic mice and Alzheimer's disease patients. Neuroscience 122:305–315.

12. Ji Y, Gong Y, Gan W, et al (2003b) Apolipoprotein E isoform-specific regulation of dendritic spine morphology in apolipoprotein E transgenic mice and Alzheimer's disease patients. Neuroscience 122:305–315.

13. Poirier J, Delisle MC, Quirion R, et al (1995) Apolipoprotein E4 allele as a predictor of cholinergic deficits and treatment outcome in Alzheimer disease. Proc Natl Acad Sci USA 92:12260–12264.

14. Poirier J, Davignon J, Bouthillier D, et al (1993b) Apolipoprotein E polymorphism and Alzheimer's disease. Lancet 342:697–699.

15. Strittmatter WJ, Saunders AM, Schmechel D, et al (1993) Apolipoprotein E: high-avidity binding to beta-amyloid and increased frequency of type 4 allele in late-onset familial Alzheimer disease. Proc Natl Acad Sci U S A 90:1977–1981.

16. Brown MS, Goldstein JL, Siperstein MD (1973) Regulation of cholesterol synthesis in normal and malignant tissue. Fed Proc 32:2168–2173.

17. Boyles JK, Zoellner CD, Anderson LJ, et al (1989) A role for apolipoprotein E, apolipoprotein A-I, and low density lipoprotein receptors in cholesterol transport during regeneration and remyelination of the rat sciatic nerve. J Clin Invest 83:1015–1031.

18. Poirier J, Hess M, May PC, Finch CE (1991) Astrocytic apolipoprotein E messenger-RNA and GFAP messenger-RRNA in hippocampus after entorhinal cortex lesioning. Mol Brain Res 11:97–106.

19. Poirier J, Baccichet A, Dea D, Gauthier S (1993a) Cholesterol synthesis and lipoprotein reuptake during synaptic remodelling in hippocampus in adult rats. Neuroscience 55:81–90.

20. Aleong R, Aumont N, Dea D, Poirier J (2003) Non-steroidal anti-inflammatory drugs mediate increased in vitro glial expression of apolipoprotein E protein. Eur J Neurosci 18:1428–1438.

21. Jick H, Zornberg GL, Jick SS, et al (2000) Statins and the risk of dementia. Lancet 356:1627–1631.

22. Rockwood K, Kirkland S, Hogan SB, et al (2002) Use of lipid-lowering agents, indication bias, and the risk of dementia in community-dwelling elderly people. Arch Neurol 59:223–227.

23. Wolozin B, Kellman W, Ruosseau P, et al (2000) Decreased prevalence of Alzheimer disease associated with 3-hydroxy-3-methyglutaryl coenzyme A reductase inhibitors. Arch Neurol 57:1439–1443.

24. Wagstaff LR, Mitton MW, Arvik BM,. Doraiswamy PM (2003) Statin-associated memory loss: analysis of 60 case reports and review of the literature. Pharmacotherapy 23:871–880.
25. Zamrini E, McGwin G, Roseman JM (2004) Association between statin use and Alzheimer's disease. Neuroepidemiology 23:94–98.

Chapter 22
Cholesterol and Aβ Production: Methods for Analysis of Altered Cholesterol De Novo Synthesis

Jakob A. Tschäpe, Marcus O.W. Grimm, Heike S. Grimm, and Tobias Hartmann

Introduction

The main characteristics of Alzheimer's disease (AD) are massive cerebral accumulation of amyloid composed of fibrillary aggregates of the amyloid beta peptide (Aβ) and intracellular accumulation of abnormally phosphorylated tau protein, associated with widespread neurodegeneration. The clinical picture is characterized by progressive and irreversible dementia, which is eventually fatal (for reviews see refs. [1–4]).

Up to now four genes have been identified, harboring point mutations that significantly affect AD pathogenesis [5]. All of these mutations result in increased β-amyloid 42 (Aβ42) levels [6]. Three of these genes-the amyloid precursor protein (APP), presenilin 1 (PS1), and presenilin 2 (PS2)—are involved in the molecular pathway of AD. The fourth gene, apolipoprotein E (ApoE) suggests a possible link to lipid pathways. ApoE encodes a lipid-binding protein that transports lipids between cells and is therefore an important factor in lipid homeostasis [7]. Human ApoE exists in three major alleles. In epidemiological studies the ε4 allele of ApoE increases the risk for hypercholesterolemia and decreases the age of onset of AD [8]. Moreover, ApoE transgenic and knockout mice show altered Aβ deposition, indicating that ApoE and lipids might play a significant role in Aβ pathology [9]. Alterations in lipid homeostasis have long been recognized to interfere severely with neuronal and glial functions and to cause neurodegenerative diseases [10].

Studies on the infrequent familial forms of the disease (FAD), especially with presenilin FAD, indicate increased production of Aβ42 to be associated invariably with FAD and to cause AD [1,2,5,11,12]. Aβ is a proteolytic fragment of APP, a protein of unclear function. The BACE I-catalyzed cleavage of APP initiates Aβ generation; and the resulting C-terminal fragment C99—but not full-length APP—is a substrate for γ-secretase, a ubiquitous multimeric

J.A. Tschäpe
Center for Molecular Biology Heidelberg (ZMBH), University of Heidelberg, Heidelberg D-69120, Germany

A. Fisher et al. (eds.), *Advances in Alzheimer's and Parkinson's Disease*,
© Springer 2008

protease. The γ-secretase cleavage results mainly in Aβ40 and Aβ42 peptides. The active center of the γ-secretase complex is formed by presenilin 1 or presenilin 2 (PS1 or PS2); and double knockout of PS1 and PS2 results in complete loss of γ-secretase activity and Aβ production [13]. Importantly, the cleavage site of C99 is located at the center of the transmembrane domain of APP/C99, possibly causing sensitivity to alterations in membrane composition [14].

Lipids and Alzheimer's Disease

Until recently little indication existed suggesting a link between AD, the proteases that process APP, and homeostasis of cholesterol and other lipids. However, this picture has changed dramatically over the last several years [15–18]. Cellular and biochemical studies showed that APP processing and the proteases involved are sensitive to cholesterol and cholesterol trafficking [19–27]; in vivo studies revealed that cholesterol feeding and cholesterol lowering by medication affects Aβ42 production and the amyloid burden [22,28–30]. Likewise, the relevance of cholesterol homeostasis in AD was further substantiated in epidemiological studies, revealing that hypercholesterolemia is a risk factor in AD [31,32]. This led to preliminary clinical trials with cholesterol- lowering drugs [33–37]. Recently, beneficial effects have been described for AD patients treated with atorvastatin [38].

The correlation between apolipoprotein E (ApoE) allele frequency and AD risk [8,39], the early increase in brain cholesterol levels during AD progression [40], and studies on lipid composition in AD brains showing age-dependent changes in cholesterol levels [41] further support the importance of cholesterol and lipids in AD.

It is now well documented that cholesterol and ganglioside GM1 enhance γ-secretase-mediated Aβ production in vitro [19,21,22,26,42], whereas inhibition of 3-hydroxy-3-methylglutaryl-CoA (HMG-CoA) reductase with statins reduces Aβ levels in cerebrospinal fluid (CSF) and brain tissue [22]. Similar results were found using the cholesterol synthesis inhibitor BM15.766. BM15.766 inhibits Δ7-reductase, which catalyzes the final step of cholesterol biosynthesis Treatment with this substance decreased the amyloid load in a transgenic mouse model [30]. In cell culture, depletion of cholesterol with cyclodextrin leads to reduced Aβ generation [19], and cholesteryl esters stimulate nonamyloidogenic APP degradation [43]. Furthermore, it has been shown recently that certain steps of APP processing are sensitive to alterations in sphingolipid levels. Inhibition of serin palmitoyltransferase (SPT) causes increased Aβ42 secretion along with an increased release of other cleavage products [44].

Thus, the Aβ–generating machinery is highly responsive to small alterations in lipid composition [19,21,22,26]. Even small changes that are achieved by altering the standard mouse diet to a cholesterol-enriched diet causes increased cerebral amyloid load [28,29] and increases accumulation of Aβ in animal brains [45].

Cholesterol, sphingomyelin (SM), and gangliosides play an important part in raft biology, lipid microdomains, which introduces lateral heterogeneity into the membrane layer and accumulates specific proteins. Rafts are involved in modulating cellular function and cellular structures [46–49] including APP processing [42,50–55]. BACE I and γ-secretase or PS, respectively, are present in rafts [50,51,53–56].

Although several aspects remain unclear, the accumulating evidence points strongly toward a link between lipids and AD etiopathology that is substantial and comprehensive. Missing in this scheme, however, is a molecular explanation for the specific lipid sensitivity of this proteolytic system.

To investigate the mechanisms underlying these aspects of AD pathology, new methods are needed to establish that identify of the enzymes targeted by products of APP processing.

Experiments and Methods

Cholesterol De Novo Synthesis Assay

The biochemistry of cholesterol biosynthesis, including all metabolic steps, is well understood. The key enzyme of this synthetic pathway is HMG-CoA reductase (HMGR), which is used as a therapeutic target with statins when lowering cholesterol blood levels is desirable (Fig. 1).

To identify the influence of certain pathways and proteins relevant in AD on cholesterol metabolism, we established a new method to analyze cholesterol de novo synthesis and validated it in comparison to other methods and control experiments. We used mouse embryonic fibroblast (MEF) cell lines with differing cholesterol synthesis rates (Fig. 2).

For cholesterol de novo synthesis, cells were incubated for 6 hours with 0.2 μCi 3-hydroxy-3-methyl[3-^{14}C]glutaryl-coenzyme-A, ^{14}C-Mevalonate (1 μCi) or ^{14}C-acetate (1 μCi). The lipid extraction was done according to Bligh and Dyer [57] with some modifications, which are briefly described. All steps were performed at room temperature. The samples were vortexed in the glass tubes for 60 minutes. After 1 ml CHCl$_3$ was added the samples were vortexed for 10 minutes. Phase separation occurred by adding 1 ml ddH$_2$O. The resulting hydrophilic and lipophilic phases were thoroughly mixed by vortexing for 30 minutes. Afterward the phases were separated by centrifugation for 10 minutes at 3000×g. The lower lipophilic phase was transferred into a new glass tube. The transfer of the lower lipophilic phase has to be quantitative without contaminating the organic phase with the interphase or hydrophilic phase. The interphase and the hydrophilic phase contain nucleic acids, carbohydrates, and proteins, whereas the lipophilic phase contains the lipids, especially cholesterol, sphingolipids, and

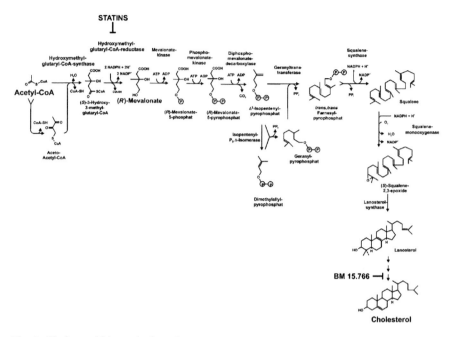

Fig. 1 Cholesterol biosynthetic pathway

phospholipids. The procedure was carried out two more times, starting with the addition of CHCl₃/MeOH/ HCl (5:10:0.075). After the last extraction, the lipophilic phases containing the extracted lipids were combined and then evaporated under slight continuous nitrogen flow. The samples were resolved again in 200 μl CHCl₃/MeOH/HCl (5:10:0.075) by vortexing for 60 minutes. An amount of 50 μl

Fig. 2 Comparison of cholesterol de novo synthesis measured by different methods. Use of 3-hydroxy-3-methyl[3-1⁴C]-glutaryl-coenzyme A as a substrate (instead of ¹⁴C-acetate in other studies) does not influence results

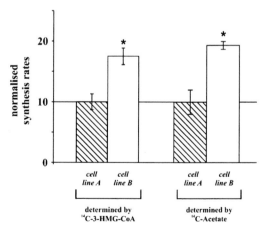

is needed for the phosphorus determination, and the remaining 150 μl was applied to scintillation counting. This was added to 2.5ml scintillation solution Ultima Gold (Packard Instruments) and thoroughly vortexed for 30 minutes. Incorporation of the radioactive precursors into cholesterol was determined using an LS 6000IC scintillation counter (Beckman).

Samples were normalized to the phospholipid content, measured indirectly by phosphorus determination of the organic phase, as briefly described. To quantify the inorganic phosphorus concentration, the phospholipids were digested by refluxing in perchloric acid to release inorganic phosphate. Afterward the inorganic phosphate was converted in phosphomolybdic acid, which is reduced to a blue compound for spectrometric determination [58]. To prepare the chromogenic solution, 16 g ammonium molybdate were dissolved in 120 ml water (reagent A). Hydrochloric acid (40 ml) and mercury (10 ml) were added to 80 ml of reagent A and thoroughly mixed. The supernatant was used as reagent B. Then, 40 ml of reagent A was combined with 200 ml concentrated sulfuric acid and added to reagent B. This results in reagent C. Subsequently, 25 ml of reagent C was added to 45 ml methanol, 5 ml chloroform, and 20 ml water. This chromogenic solution can be stored at 4°C for several weeks. The chromogenic solution (13 μl) was added to 50 μl of the sample, and the solution was heated at 100°C for 75 seconds. After the samples were thoroughly mixed and cooled for 5 minutes to room temperature, 500 ml nonane was added. The samples were mixed briefly and incubated for 15 minutes. The tubes were centrifuged for 3 minutes at 3000 × g before the absorbance of the supernatant at 730 nm was compared with the blank, where, instead of the sample, 50 μl CHCl$_3$/MeOH/HCl (5:10:0.075) was used.

To validate the results for cholesterol de novo synthesis, we determined the total cholesterol and cholesterol ester contents of the cells used. The same difference in cholesterol synthesis was observed (Fig. 3).

Cholesterol levels (steady state) were determined using the Amplex-Red cholesterol assay (Molecular Probes). To analyze the cholesterol content in cells, 0.2 mg protein was used.

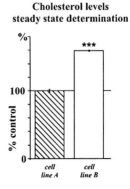

Fig. 3 Cholesterol levels determined by steady-state determination. The measurement of cholesterol at steady state resembles the results obtained with the de novo synthesis method, thus validating this assay

Fig. 4 Identifying
HMG-CoA reductase
(HMGR) as target of
regulation. The enhanced
cholesterol de novo
synthesis in line B was only
observed upon incubation
with 14C-HMG-CoA,
indicating that in this cell
line the cholesterol de novo
synthesis is enhanced via
regulation of HMGR

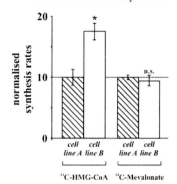

Identifying the Target Enzyme

To prove whether HMGR is the enzyme targeted in the cell line with altered cholesterol synthesis, cells were incubated with either ^{14}C-HMG-CoA (as the substrate of HMGR) or ^{14}C-mevalonate (as the product of HMGR).

The enhanced cholesterol de novo synthesis in line B was observed only upon incubation with ^{14}C-HMG-CoA, indicating that in this cell line the cholesterol de novo synthesis is enhanced via regulation of HMGR (Fig. 4).

To further validate this assay, cells were incubated with a specific inhibitor of HMGR lovastatin [22]. Treatment with lovastatin (5 µM), was performed for 24 hours followed by 6 hours of conditioning in the presence of the inhibitor.

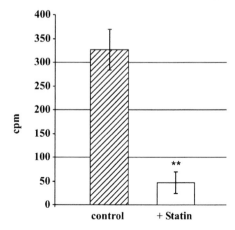

Fig. 5 Validation of the
^{14}C-HMG-CoA cholesterol
de novo synthesis assay.
Lovastatin, a specific
HMGR inhibitor,
suppresses the cholesterol de
novo synthesis to
approximately 15% in
COS7 cells, thus again
validating the assay

Lovastatin suppressed the cholesterol de novo synthesis to approximately 15% in COS7 cells, validating the assay (Fig. 5).

Conclusions

Regarding the close link between cholesterol homeostasis and AD pathology provided by clinical studies (see above), a role for the regulation of cholesterol metabolism in AD pathology can be assumed. As already shown, proteins with functions in lipid metabolism—particularly enzymes of the sphingolipid and ganglioside pathway—seem to be involved frequently in neurodegenerative disorders (Table 1). For example, in Niemann-Pick disease the sphingomyelin level is increased drastically by a lack of the sphingomyelin-degrading enzyme, the acid sphingomyelinase (Table 1). However, it should be noted that these diseases are often fatal during childhood or early adulthood, and therefore little is known about their relevance to AD.

Table 1 Lipid-Related Neuronal Disorders with the Affected Enzyme or Lipid

Disease	Affected lipids	Enzymatic defects
Faber disease	Ceramide (Cer)	Acid ceramidase
Niemann-Pick disease	Sphingomyelin	Sphingomyelinase
Krabbe disease	Galactosylceramide (GalCer) Galactosylsphingosine	Galactosylceramidase
Gaucher disease	Glucosylceramide (GlcCer) Glucosylsphingosine	Glucosylceramidase
Fabry disease	Digalactosylceramide (DiGalCer)	α-Galactosidase A
Tay-Sachs disease	GM ganglioside	β-Hexosaminidase A
Sandhoff disease	GM ganglioside	β-Hexosaminidase A & B
Metachoromatic leukodystrophy	Sulfatide	Arylsulfatase A (Sulfatidase)
Multiple sulfatase deficiency	Sulfatide	Arylsulfatase A, B, C
Sulfatidase activator deficiency (sap-B deficiency)	Sulfatide Globotriaosylceramide Digalactosylceramide (DiGalCer) GM3 ganglioside	Sulfatidase activator (SAP-1, sap-B)
SAP-2 deficiency	Glucosylceramide	SAP-2 (sap-C)
SAP precursor deficiency	All glycolipids with short sugar chains (e.g., Cer, GlcCer, LacCer, GalCer, DigalCer, sulfatide)	sap precursor sap-A, B, C, D
GM1-gangliosidosis	GM1 ganglioside	GM1 ganglioside Galactosidase
GM2-gangliosidosis (B1 variant)	GM2 ganglioside	β-Hexosaminidase A
GM2-gangliosidosis (AB variant)	GM2 ganglioside	β-Hexosaminidase A

It seems promising to investigate the metabolic pathways of these lipids, including cholesterol, in respect to their putative roles in AD pathology. Methods to investigate lipid metabolism and alterations herein may contribute to better understanding of the cellular processes underlying AD pathology.

Considering the established connection between cholesterol and APP processing or Aβ production, respectively, the question arises whether APP itself or its cleavage products may influence lipid homeostasis in return. Furthermore, the APP cleaving machinery could be part of such regulatory cycles. As a first step, knockout cell lines and tissue probes from either APP or presenilin knockout mice could be assayed for cholesterol anabolism and other lipid levels and/or syntheses.

Investigating the connections between lipid metabolism and proteins involved in AD may help us understand the clinical and epidemiological observations linking these lipids with AD. Understanding this process may facilitate new access to the malregulated molecular processes that underlie the early steps in the pathology of Alzheimer's disease.

Acknowledgments   This work was funded by the EU via QLK-172-2002 Lipidiet, DFG, and BMBF.

References

1. Selkoe DJ. Alzheimer's disease: genes, proteins, and therapy. Physiol Rev 2001; 81:741–766.
2. Sisodia SS, St George-Hyslop PH. γ-Secretase, Notch, Aβ and Alzheimer's disease: where do the presenilins fit in? Nat Rev Neurosci 2002;3:281–290.
3. Hardy JA, Higgins GA. Alzheimer's disease: the amyloid cascade hypothesis. Science 1992;256:184–185.
4. Masters CL, Beyreuther KT. The pathology of the amyloid A4 precursor of Alzheimer's disease. Ann Med 19889;21:89–90.
5. De Strooper B, Annaert W. Proteolytic processing and cell biological functions of the amyloid precursor protein. J Cell Sci 2000;113(Pt 11):1857–1870.
6. Sinha S, Lieberburg I. Cellular mechanisms of beta-amyloid production and secretion. Proc Natl Acad Sci U S A 1999;96:11049–11053.
7. Mahley RW, Rall SCJ. Apolipoprotein E: far more than a lipid transport protein. Annu Rev Genomics Hum Genet 2000;1:507–537.
8. Corder EH, Saunders AM, Strittmatter WJ, et al. Gene dose of apolipoprotein E type 4 allele and the risk of Alzheimer's disease in late onset families. Science 261:921–923.
9. Bales KR, Verina T, Dodel RC, et al. Lack of apolipoprotein E dramatically reduces amyloid beta-peptide deposition. Nat Genet 1997;17:263–264.
10. Nathan BP, Bellosta S, Aanan DA, et al. Differential effects of apolipoproteins E3 and E4 on neuronal growth in vitro. Science 1994;264:850–852.
11. Tanzi RE. A genetic dichotomy model for the inheritance of Alzheimer's disease and common age-related disorders. J Clin Invest 1999;104:1175–1179.
12. Younkin SG. Evidence that A beta 42 is the real culprit in Alzheimer's disease. Ann Neurol 1995;37:287–288.
13. De Strooper B. Aph-1, Pen-2, and Nicastrin with presenilin generate an active gamma-secretase complex. Neuron 2003;38:9–12.

14. Grziwa B, Grimm MO, Masters CL, et al. The transmembrane domain of the amyloid precursor protein in microsomal membranes is on both sides shorter than predicted. J Biol Chem 2003;278:6803–6808.
15. Casserly I, Topol E. Convergence of atherosclerosis and Alzheimer's disease: inflammation, cholesterol, and misfolded proteins. Lancet 2004;363:1139–1146.
16. Puglielli L, Tanzi RE, Kovacs DM. Alzheimer's disease: the cholesterol connection. Nat Neurosci 2003;6:345–351.
17. Simons K, Ehehalt R. Cholesterol, lipid rafts, and disease. J Clin Invest 2002;110:597–603.
18. Hartmann T. Cholesterol, Abeta and Alzheimer's disease. TINS 2001;24:45–48.
19. Simons M, Keller P, De Strooper B, et al. Cholesterol depletion inhibits the generation of beta-amyloid in hippocampal neurons. Proc Natl Acad Sci U S A 1998;95:6460–6464.
20. Mizuno T, Nakata M, Naiki H, et al. Cholesterol-dependent generation of a seeding amyloid beta-protein in cell culture. J Biol Chem 1999;274:15110–15114.
21. Wahrle S, Das P, Nyborg AC, et al. Cholesterol-dependent gamma-secretase activity in buoyant cholesterol-rich membrane microdomains. Neurobiol Dis 2002;9:11–23.
22. Fassbender K, Simons M, Bergmann C, et al. Simvastatin strongly reduces levels of Alzheimer's disease beta-amyloid peptides Abeta 42 and Abeta 40 in vitro and in vivo. Proc Natl Acad Sci U S A 2001;98:5856–5861.
23. Burns M, Gaynor K, Olm V, et al. Presenilin redistribution associated with aberrant cholesterol transport enhances beta-amyloid production in vivo. J Neurosci 2003;23:5645–5649.
24. Runz H, Rietdorf J, Tomic I, et al. Inhibition of intracellular cholesterol transport alters presenilin localization and amyloid precursor protein processing in neuronal cells. J Neurosci 2002;22:1679–1689.
25. Tschäpe JA, Hammerschmied C, Mühlig-Versen M, et al. The neurodegeneration mutant lochrig interferes with cholesterol homeostasis and Appl processing. EMBO J 21:6367–6376.
26. Yamazaki T, Chang TY, Haass C, Ihara Y. Accumulation and aggregation of amyloid beta-protein in late endosomes of Niemann-pick type C cells. J Biol Chem 2001;276:4454–4460.
27. Kojro E, Gimpl G, Lammich S, et al. Low cholesterol stimulates the nonamyloidogenic pathway by its effect on the alpha-secretase ADAM 10. Proc Natl Acad Sci USA 2001;98:5815–5820.
28. Shie FS, Jin LW, Cook DG, et al. Diet-induced hypercholesterolemia enhances brain A beta accumulation in transgenic mice. Neuroreport 2002;13:455–459.
29. Refolo LM, Malester B, LaFrancois J, et al. Hypercholesterolemia accelerates the Alzheimer's amyloid pathology in a transgenic mouse model. Neurobiol Dis 2000;7:321–331.
30. Refolo LM, Pappolla MA, LoFrancois J, et al. A cholesterol-lowering drug reduces beta-amyloid pathology in a transgenic mouse model of Alzheimer's disease. Neurobiol Dis 2001;8:890–899.
31. Kivipelto M, Helkala EL, Laakso MP, et al. Midlife vascular risk factors and Alzheimer's disease in later life: longitudinal, population based study. BMJ 2001;322:1447–1451.
32. Pappolla MA, Bryant-Thomas TK, Herbert D, et al. Mild hypercholesterolemia is an early risk factor for the development of Alzheimer amyloid pathology. Neurology 2003;61:199–205.
33. Rockwood K, Kirkland S, Hogan DB, et al. Use of lipid-lowering agents, indication bias, and the risk of dementia in community-dwelling elderly people. Arch Neurol 2002;59:223–227.
34. Buxbaum JD, Cullen EI, Friedhoff LT. Pharmacological concentrations of the HMG-CoA reductase inhibitor lovastatin decrease the formation of the Alzheimer beta-amyloid peptide in vitro and in patients. Front Biosci 2002;7:a50–a59.
35. Simons M, Schwärzler F, Lütjohann D, et al. Treatment with simvastatin in normocholesterolemic patients with Alzheimer's disease: a 26-week randomized, placebo-controlled, double-blind trial. Ann Neurol 2002;52:346–350.
36. Jick H, Zornberg GL, Jick SS, et al. Statins and the risk of dementia. Lancet 2000;356:1627–1631.

37. Wolozin B, Kellman W, Ruosseau P, et al. Decreased prevalence of Alzheimer disease associated with 3-hydroxy-3-methyglutaryl coenzyme A reductase inhibitors. Arch Neurol 2000;57:1439–1443.
38. Sabbagh M, et al. Benefits of atorvastatin in subjects with Alzheimer's disease. Arch Neurol (in press).
39. Strittmatter WJ, Roses AD. Apolipoprotein E and Alzheimer disease. Proc Natl Acad Sci U S A 1995;92:4725–4727.
40. Cutler RG, Kelly J, Storie K, et al. Involvement of oxidative stress-induced abnormalities in ceramide and cholesterol metabolism in brain aging and Alzheimer's disease. Proc Natl Acad Sci U S A 2004;101:2070–2075.
41. Wood WG, Schroeder F, Igbavboa U, et al. Brain membrane cholesterol domains, aging and amyloid beta-peptides. Neurobiol Aging 2002;23:685–694.
42. Zha Q, Ruan Y, Hartmann T, et al. GM1 ganglioside regulates the proteolysis of amyloid precursor protein. Mol Psychiatry 2004;9:946–952.
43. Puglielli L, Konopka G, Pack-Chung E, et al. Acyl-coenzyme A: cholesterol acyltransferase modulates the generation of the amyloid beta-peptide. Nat Cell Biol 2001;3:905–912.
44. Sawamura N, Ko M, Yu W, et al. Modulation of amyloid precursor protein cleavage by cellular sphingolipids. J Biol Chem 2004;279:11984–11991.
45. Sparks DL, Scheff SW, Hunsaker JC 3rd, et al. Induction of Alzheimer-like beta-amyloid immunoreactivity in the brains of rabbits with dietary cholesterol. Exp Neurol 1994;126:88–94.
46. Simons K, Vaz WL. Model systems, lipid rafts, and cell membranes. Annu Rev Biophys Biomol Struct 2004;33:269–295.
47. Simons K, Toomre D. Lipid rafts and signal transduction. Nat Rev Mol Cell Biol 2000;1:31–39.
48. Simons K, Ikonen E. How cells handle cholesterol. Science 2000;290:1721–1726.
49. Hakomori S, Handa K. Glycosphingolipid-dependent cross-talk between glycosynapses interfacing tumor cells with their host cells: essential basis to define tumor malignancy. FEBS Lett 2002;531:88–92.
50. Vetrivel KS, Cheng H, Lin W, et al. Association of gamma-secretase with lipid rafts in post-Golgi and endosome membranes. J Biol Chem 2004;279:44945–44954.
51. Cordy JM, Hussain I, Dingwall C, et al. Exclusively targeting beta-secretase to lipid rafts by GPI-anchor addition up-regulates beta-site processing of the amyloid precursor protein. Proc Natl Acad Sci U S A 2003;100:11735–11740.
52. Marlow L, Cain M, Pappolla MA, Sambamurti K. Beta-secretase processing of the Alzheimer's amyloid protein precursor (APP). J Mol Neurosci 2003;20:233–239.
53. Ehehalt R, Keller P, Haass C, et al. Amyloidogenic processing of the Alzheimer beta-amyloid precursor protein depends on lipid rafts. J Cell Biol 2003;160:113–123.
54. Riddell DR, Christie G, Hussain I, Dingwall C. Compartmentalization of beta-secretase (Asp2) into low-buoyant density, noncaveolar lipid rafts. Curr Biol 2001;11:1288–1293.
55. Parkin ET, Turner AJ, Hooper NM. Amyloid precursor protein, although partially detergent-insoluble in mouse cerebral cortex, behaves as an atypical lipid raft protein. Biochem 1999;J 344(Pt 1):23–30.
56. Lee SJ, Liyanage U, Bickel PE, et al. A detergent-insoluble membrane compartment contains A beta in vivo. Nat Med 1998;4:730–734.
57. Bligh EG, Dyer WJ. A rapid method of total lipid extraction and purification. Can J Med Sci 1959;37:911–917.
58. Hundrieser KE, Clark RM, Jensen RG. Total phospholipid analysis in human milk without acid digestion. Am J Clin Nutr 1985;41:988–993.

Chapter 23
Glycosaminoglycans and Analogs in Neurodegenerative Disorders

Lucilla Parnetti and Umberto Cornelli

Introduction

A variety of etiologically diverse neurodegenerative disorders, which include Alzheimer's disease (AD), spongiform encephalopathies, tauopathies, and synucleinopathies are characterized by abnormal protein deposition. In these diseases, protein aggregates constituted by fibrils with a high percentage of ß-pleated sheet secondary structure appear as dense filamentous inclusions [1,2]. Some of them have received the name of "amyloid," a term first coined by Virchow in 1851 with the meaning of starch-like substance.

Alzheimer's disease is histopathologically characterized by the presence of extracellular amyloid plaques, also named senile plaques, and intracellular neurofibrillary tangles (NFT). Senile plaques are mainly constituted by the ß-amyloid peptide (Aß), a 39- to 43-amino-acid peptide that results from the proteolytic processing of an integral membrane protein referred to as the amyloid precursor protein (APP). Aß aggregates consist of a mixture of fibrils, protofibrils, and low-molecular-weight oligomeric intermediates, which have consistently been shown to be toxic to neurons in culture as well as in vivo after injection into the brains of aged monkeys [3]. NFTs are formed by hyperphosphorylated forms of tau protein. These fibrillar aggregates of tau are not an exclusive hallmark of AD, as they are also present in the brains of patients with tauopathies. These neurodegenerative diseases include progressive supranuclear palsy, amyotrophic lateral sclerosis/parkinsonism-dementia complex of Guam, Pick's disease, corticobasal degeneration, sporadic frontotemporal dementias, and familial frontotemporal dementia with parkinsonism-linked to chromosome 17. Although full-length tau protein is the major component of NFTs, a shorter tau peptide is also able to form fibrillar polymers [4].

Spongiform encephalopathies, also called prion diseases, include scrapie in sheep and Creutzfeldt-Jakob disease, kuru, fatal familial insomnia, and

L. Parnetti
Neurology SectionUniversity of Perugia, Italy

A. Fisher et al. (eds.), *Advances in Alzheimer's and Parkinson's Disease*,
© Springer 2008

Gerstmann-Straussler-Scheinker disease in humans. Spongiform encephalopathies are characterized by the accumulation of a conformationally modified form of the prion protein (PrP^{sc}), which may also aggregate into amyloid plaques. The PrP^{sc} prion is not only neurotoxic but may also be a transmissible agent, as it may induce conversion of the normal PrPc into a pathogenic conformation [5]. Synthetic peptides comprising residues 106 to 126 of PrP have been demonstrated to form fibrils and be toxic to cultured neuronal cells [6].

α-Synuclein, a protein of 140 amino acids found in abundance in the presynaptic regions of neurons, is the principal component of Lewy bodies (LBs). α-Synuclein also is the major component of the glial cytoplasmic inclusions of multiple system atrophy, and a fragment of α-Synuclein has been identified as the non-Aß component of Alzheimer's plaques. α-Synucleinopathies include Parkinson's disease, Lewy body variant of Alzheimer's disease, and dementia with Lewy bodies.

Because aberrant processing of aggregate-forming proteins is a crucial event leading to neurodegeneration in many neurological disorders, great attention has been paid to the molecular mechanisms underlying the formation of protein aggregates, with a special emphasis on the implication of proteases, chaperone-like factors, and other interacting proteins. Sulfated glycosaminoglycans (sGAGs) may play a role as modulators of the aggregation and toxicity properties of the Aß, prion, tau, and α-Synuclein peptides. The minimal core sequences responsible for the polymerization of these peptides into fibrils contain similarly cationic motifs (HHQK in Aß, KTNMKH in PrP, NIHHK in tau, KTKEGV in α-Synuclein), which might contribute to sGAG binding. The in vitro interactions between sGAGs and Aß, prion, tau and α-Synuclein peptides have been demonstrated, and some evidence suggests that sGAGs are associated with these protein inclusions in vivo.

Synthesis and Physiological Functions of GAG

From a chemical point of view, GAGs are unbranched linear polymers of repeated disaccharide units, which are usually composed by an amino sugar and an uronic acid. Hyaluronic acid is a nonsulfated GAG composed of glucuronic acid and N-acetylglucosamine. The other types of GAG are sulfated (sGAG). The hypersulfated form of heparan sulfate is referred to as heparin. Chondroitin and heparan sulfate are the main sGAGs present in rat brain, although heparin has also been found in glial cells. The biosynthesis of sGAG polymers takes place on a core protein that is first covalently bound to a sugar; then the remaining sugar moieties are added, generating a GAG bound to a core protein. The sulfation of sugar residues occurs within the Golgi apparatus. The resulting sGAG-bearing protein is called proteoglycan (PG). The sizes of the core proteins are variable, and there are also variations in the number and types of attached GAG. Often a core protein bears different types of sGAG, and core proteins may remain permanently or transitorily bound to sGAG. PGs are among the most abundant components of the extracellular matrix of many tissues including the brain. Glial cells, including both astrocytes and oligodendrocytes, are responsible for the biosynthesis of some of these PGs. With regard to the regulation of PG biosynthesis, a possible role of microtubules as modulators of PG biosynthesis in nonneural cells, and secretion in neural cells, has been proposed [7].

Proteoglycans can interact with other proteins through their GAG in a noncovalent way. These interactions are probably ionic in nature, in which anionic sites of sGAG may bind to basic residues in proteins. It is still unclear under which conditions PGs are degraded to yield the core protein and free sGAG. PG may be internalized into the cell by endocytosis and transported to endosomes and lysosomes, where the core protein is hydrolyzed and the GAG component could be either transported to the nucleus or recycled back to the cell surface through the Golgi apparatus. A deficit in GAG degradation resulting in GAG accumulation and neurological dysfunction occurs in several lysosomal enzyme deficiencies. It is not known if in pathological situations free sGAG are released to the cytoplasm as a result of leakage from membrane-bound endosomal or lysosomal compartments [8].

The interaction of sGAG with various proteins, including extracellular signaling polypeptides, could regulate their function [9]. This function may be related to signal transduction mechanisms in which changes in the phosphorylation state of many proteins may be induced [10–12]. There is a wide divergence between what might be expected to be the function of PGs in the nervous system, based on studies in vitro, and the lack of an obvious phenotype in knockout animals lacking the proteoglycan. Regarding redundancy of function, PGs represent quite a complicated case, as their function may be localized in the GAG chain, in the protein, or in both. Thus, if the function is due to the GAG chain, any other PG with a corresponding GAG may compensate so long as it is expressed in the same spatiotemporal manner. It remains to be clarified to what extent a given GAG structure with all its modifications can exist or be mimicked on another protein core. It is reasonable to assume that the lack of neural PG expression may induce behavioral defects in learning or memory that are revealed only by behavioral tests [13].

Interactions of sGAG with ß-Amyloid Peptide

ß-Amyloid peptide (Aß), the principal component of the senile plaque (SP) found in AD, is a peptide usually containing 40 to 42 residues generated as a minor cleavage product from APP by α- and γ-secretases. Aß is able to polymerize in vitro, yielding fibrils similar to those found in SPs; sGAG including heparan, keratan, dermatan, and chondroitin sulfates strongly favor Aß polymerization in vitro, and the binding of sGAG to Aß has been found to decrease Aß degradation [14]. Thus, sGAG might enhance polymerization of Aß into fibrils, which would become resistant to proteolytic cleavage, thus stimulating the deposition of Aß into SPs in patient brains in vivo. A heparin-binding site on proteins with the consensus motif BBXB (B for basic amino acid) has been proposed by Cardin and Weintraub [15]. Aß contains the sequence HHQK, which may serve as a suitable binding site for sGAG [16]. This domain, and specifically region 1 to 11, is critical for the proinflammatory activity of Aß [17,18], whereas residues 13 to 16 are responsible for microglia activation [19]. Heparin, the most sulfated GAG, has been shown to bind to the region 12 to 17 of Aß [20], near the complement and contact system activating residues.

Detailed structural studies on the substitution of histidine or lysine residues by alanine in this motif have supported the importance of HHQK for the ionic interaction of Aß with sGAG [16]; this is in agreement with a previous study by the same group demonstrating the importance of the sulfate moieties of sGAG for the formation of amyloid fibrils [21]. sGAG may associate with Aß not only in vitro but also in vivo; thus, both heparan and chondroitin sulfate PGs are localized at sites of Aß deposition in the brains of patients with AD [22,23]. According to these observations, sGAG may be considered important cofactors for the formation of amyloid plaques in AD.

Cell culture studies have indicated a possibly neuroprotective role of sGAG, which might be relevant in AD. Thus, heparan and chondroitin sulfate GAG attenuate the neurotoxic effect of Aß in primary neuronal cultures [24] and in neuron-like cell lines [25]. Because sGAG can bind Aß, sGAG-mediated neuroprotection may be due to the sequestering of Aß. For example, Aß oligomeric and protofibrillar intermediates, which are highly neurotoxic, may preferentially associate with sGAG, resulting in the formation of dense filamentous inclusions in which Aß molecules are prevented from interacting with neuronal membranes. According to this hypothesis, the generation of SPs in AD would be a partially protective response aimed at reducing Aß neurotoxicity. According to this hypothesis, it has been shown that neuronal oxidative stress, which is a marker of neuronal dysfunction, precedes the formation of Aß amyloid plaques, and in some cases the extent of amyloid deposits is associated with a decreased level of neuronal oxidative stress [26]. This observation is in good agreement with the weak correlation between fibrillar amyloid load and the degree of cognitive impairment in patients with AD [27]. Supportive of this view is the fact that the chondroitin sulfate content is inversely correlated with the amount of hyperphosphorylated tau in cortical areas of patients with AD. In particular, neurons ensheathed by chondroitin sulfate-rich perineuronal nets seem healthier and contain less phosphorylated tau than the others [28].

Tau protein hyperphosphorylation, a hallmark of AD, is also observed in neuronal cultures treated with Aß [29] in transgenic mice that overexpress mutant APP and accumulate Aß [30] and in aged primates after injection of Aß into the brain [3]. Thus, the level of hyperphosphorylated tau can be considered as a good marker for Aß-induced neuronal dysfunction [31].

Assuming that sGAG may inhibit Aß neurotoxicity, those cortical neurons surrounded by a chondroitin sulfate-rich extracellular matrix would be less vulnerable to Aß and would therefore contain less hyperphosphorylated tau, as it is indeed found in AD brains. In addition to their interaction with Aß, sGAG (particularly heparin) may also bind to another site in the APP molecule, modulating the growth of neurites [32]. It is thus conceivable that sGAG may not only attenuate Aß neurotoxicity but also modulate APP effects on neuronal plasticity.

Interaction of sGAG with Tau Protein and Peptides

Neurofibrillary tangles (NFTs) are aggregates of paired helical filaments (PHFs), and the main component of PHFs is the microtubule-associated protein known as tau. Tau protein is able to self-assemble in vitro, yielding fibrillar polymers similar to PHFs. Interestingly, in the brains of patients with AD, sGAG-bearing PGs co-localize not only with Aß in SPs but also with tau protein in NFTs [33]. The in vitro polymerization of tau is facilitated by sGAGs, which also determine PHF helicity; again, nonsulfated GAGs do not have this role [34]. sGAG are present both in situ in NFTs and in PHFs isolated from AD patient brains. NFTs (and PHFs) are located both extracellularly and inside neurons. Tau protein of extracellular NFTs may easily interact, after neuronal cell lysis, with sGAGs, but it is more difficult to explain how tau and sGAGs may associate with each other in neuronal cell bodies.

Leakage of sGAGs from endosomes and lysosomes might facilitate their association with tau under certain pathological conditions. In frontotemporal dementia with parkinsonism-linked to chromosome 17 (FTDP-17), mutated tau proteins show a higher capacity for self-assembly in vitro in the presence of sGAGs when compared to the wild type tau [35]. The minimal tau sequence that is required for self-assembly into filaments in the presence of sGAGs corresponds to an 18-amino-acid peptide comprising residues 317 to 335. No FTDP-17 mutations have been found in this tau peptide, indicating that tau polymerization is also significantly affected by other regions of the molecule in addition to the minimal core peptide able to form fibrils. Residues 317 to 335 contain the sequence KCGSLGNIHHKPGGG, which might be involved in the interaction of tau protein with sGAGs. The motif HHK is important for the interaction with sGAG, as the substitution of one histidine and one lysine for other residues prevents the interaction of tau peptide with sGAG.

Interaction of sGAG with Prion Protein and Peptides

Prion is an acronymon for proteinaceous infectious particle and refers to the putative agent responsible for transmissible spongiform encephalopathies. The prion is devoid of any nucleic acid and is composed only of a conformationally modified protein. In the brain, the normal prion protein (PrP) results from cleavage of a precursor protein at its N- and C-terminal regions, yielding a mature protein with a conformation characterized by high α-helical content. This conformation changes by a partial loss of α-helical structure and a large increase of ß-sheet content, resulting in the toxic and infective PrP^{sc} protein. The region of the PrP protein comprising residues 109 to 122 is essential for the conformational change from PrP to PrP^{sc} [36]. A synthetic PrP peptide comprising residues 106 to 126 is able to polymerize in vitro into fibrils, which are toxic for cultured neurons. This peptide contains a hydrophobic core with the sequence AGAAAAGA as well as a motif with the sequence KTNMKH, which might interact with sGAGs. In fact, GAGs bind in vitro to PrP peptide 106–126 and modulate its ability to polymerize into fibrils [37].

Thus, sGAGs diminish the polymerization of PrP peptide 106–126 and result in the formation of filaments with a different morphology, whereas hyaluronic acid does not inhibit PrP peptide polymerization; it is therefore possible that the interaction between the basic motif KTNMKH of the PrP peptide and the negatively charged sulfate moieties of sGAGs may decrease the efficiency of peptide polymerization, whereas other interactions between GAGs (both sulfated and nonsulfated) and PrP peptide may affect fibril morphologies. In addition to their binding to the PrP peptide 106–126, sGAG have also been found to interact with the full-length prion protein [38] and modulate its intracellular trafficking [39].

Heparan sulfate and pentosan polysulfate can directly stimulate the cell-free conversion of PrP to PrP^{sc}, and pentosan polysulfate binds to PrP, triggering a conformational change that may potentiate its PrP^{sc}-induced conversion [40]. Consequently, some membrane-bound sGAGs, or similar polyanionic molecules, may be essential cofactors for the conversion of PrPc into PrP^{sc} in vivo. In agreement with this hypothesis, it has been shown that the generation of prion infectivity in vitro requires the presence of both protease-resistant PrP^{sc} and of nonprotein components of prion amyloid-like fibrils, which are possibly sGAG [41]. Thus, purified and solubilized protease-resistant PrP^{sc} isolated from scrapie-infected hamster brain exhibits low infectivity unless it is mixed with the nonprotein fraction of prion fibrils. Interestingly, the ability of this nonprotein fraction to facilitate the reconstitution of infectivity is considerably reduced after digestion with heparanase. Furthermore, the addition of purified heparan sulfate to purified and solubilized protease-resistant PrP^{sc} results in augmented infectivity. This means that sGAG-like molecules are important in the production of prion infectivity.

Because sGAGs stimulate the in vitro conversion of PrP to PrP^{sc}, whereas they partially inhibit polymerization of the prion peptide 106–126 into amyloid-like fibrils, conversion and polymerization are probably distinct processes mediated by different conformational changes of the implicated peptides. The prion peptide 88–153 contains the minimal sequence required for disease transmissibility, but it does not efficiently interact with PrP to favor its conversion to PrP^{sc} because of its insoluble ß Sheeted structure and aggregation state;

probably larger and less fibrillogenic peptides can guarantee more efficient interaction with PrP. Similar to what has been found for Aß, sGAG appears to play a neuroprotective role in prion diseases. Heparan sulfate (and other sGAGs) significantly block PrP^{sc} formation and infective prion propagation in both cultured cells [42] and experimental animals [43]. Furthermore, sGAG significantly inhibit the toxicity of PrP peptide 106-126 on cultured neuron-like cells [37]. A possible explanation of sGAG's paradoxical pathogenetic role in vitro, in contrast to a protective role in vivo, is that free exogenous sGAG may compete for, and interfere with, the association of PrP with an endogenous membrane-bound sGAG (or other similar polyanionic cofactor) that might be required for efficient conversion to PrP^{sc} [44].

Interaction of SGAG with α-Synuclein

The function of α-Synuclein is not completely known. The isolated protein lacks any organized secondary or tertiary structure, acquiring a high content of α-helix upon binding to membranes and phospholipids vesicles. Therefore, it may have a role in regulating synaptic vesicle formation. α-Synuclein has also been shown to have protein chaperone activity, to bind to and inhibit phospholipase D, and to bind to microtubule-associated proteins tau (i.e., synphilin-1). Incubation of recombinant α-Synuclein for prolonged periods at 37°C leads to the formation of typical amyloid fibrils with a morphology similar to that of fibrils isolated from Lewy bodies; the kinetics of fibril formation is highly dependent on pH and ionic strength. The extent and rate of formation of amyloid fibrils from α-Synuclein are greatly enhanced by GAGs—and specifically by heparin; fibrils formed from wild-type α-Synuclein in the presence of heparin have a different morphology with respect to those formed in its absence. It is well documented that Lewy bodies contain heparan sulfate PGs [37].

Based on these premises, it is reasonable to assume that GAGs/PGs play a role in the formation of the amyloid deposits of α-Synuclein found in Lewy bodies [45]. This process might be triggered by the leakage of heparan sulfate PGs into the cytoplasm of neurons during the early stages of cell degeneration, analogous to the case of NFT formation (i.e., interaction between heparan sulfate and protein tau) in Alzheimer's disease. Therefore, PG or GAG could be directly involved in the etiology of Parkinson's disease and related α-Synucleinopathies.

Conclusions and Therapeutic Implications

As reported before, in vitro work has shown that PGs and GAGs directly bind ßamyloid and synthetic ß-amyloid peptides [46–50], stimulating fibril nucleation, growth, and stability [51,52]. These data suggest that PGs and their GAG side chains may act as "seed molecules" [51] for ß-amyloid fibril formation, thereby promoting the senile plaque formation seen in Alzheimer's disease. Much of the research suggests that the charge, size, and degree of sulfation of these PGs/GAGs influence their ability to interact with ß-amyloid [51,53,54]

The minimal core sequences responsible for the polymerization of Aß, prion peptide, tau peptide, and α-Synuclein contain, similarly, cationic motifs that may be implicated in the interaction with sGAGs. Thus, these peptides share two important features: the abilities to polymerize into filaments and to associate with sGAGs. The interaction of sGAGs with these peptides, especially Aß-peptide, may be of great relevance in vivo.

Studies performed on cell culture models suggest that sGAGs may have significant neuroprotective properties. One possibility is that sGAG-induced neuroprotection results from the sequestering of toxic peptides or proteins into dense filamentous inclusions, which become inert as they are rendered unable to interact with neuronal membranes. Of course, modulation of the activity of secreted proteins with neurotrophic properties such as fibroblast growth factor (FGF) and APP may also contribute to sGAG-induced neuroprotection.

In the brains of patients with AD, it may be hypothesized that the association of highly toxic Aß oligomeric and protofibrillar aggregates with sGAG-bearing PGs might both reduce neurotoxicity and favor the formation of senile plaques. A similar situation may also apply to prion diseases. In this case, an endogenous membrane-bound sGAG (or a related polyanionic molecule) possibly enhances the conversion of PrP to pathogenic PrP^{sc}. In contrast, the presence of high concentrations of extracellular sGAG may sequester PrP (and PrP^{sc}), rendering them unable to continue to stimulate the conversion of PrP to PrP^{sc}. Whereas Aß and PrP are secreted peptides, thus interacting with both extracellular and membrane-bound sGAG-bearing PGs, tau is a cytoplasmic protein. Accordingly, the relevance of the interaction between tau and sGAGs remains to be firmly established, although some studies have suggested that sGAGs are indeed associated with intracellular NFTs. Thus, under certain pathological conditions, sGAGs might be released from membrane organelles (endosomes and lysosomes) into the cytoplasm, where sGAG would favor tau polymerization into filaments. In this case, intracellular sGAG may contribute to the pathogenic process in tauopathies. If this hypothesis is correct, the prevention of sGAG leakage into the cytoplasm might be useful to inhibit the progression of tauopathies and AD. Curiously, chondroitin sulfate PGs have been found to be associated with intraneuronal protein inclusions and reactive astrocytes in a variety of neurodegenerative diseases.

Thus, the regulation of the biosynthesis, trafficking, and degradation of brain sGAGs may be a target for novel therapeutic interventions in some neurodegenerative diseases. In particular, enhanced secretion of sGAGs into the extracellular medium may have a significant neuroprotective effect in AD, prion diseases, tauopathies, and synucleinopathies. Of relevance in this regard is the existence of specialized extracellular matrix microenvironments in brain tissue. Thus, several types of neuron are ensheathed by chondroitin sulfate

PG-rich, hyaluronic acid-containing, lattice-like coatings referred to as perineuronal nets [55]. Precisely those neurons ensheathed by perineuronal nets are the least vulnerable to tau hyperphosphorylation and aggregation in AD and aging brains [56]. Further research into the formation of sGAG-rich perineuronal nets may provide important clues to understanding sGAG-mediated neuroprotection.

In vivo studies have shown that GAGs and other synthetic polysulfated compounds can attenuate the neurotoxic effects of ß -amyloid in cell culture [25,57,58] by inhibiting ß-amyloid's interaction with cells [58]. Snow et al. [59] have shown that when ß-amyloid, along with heparan sulfate proteoglycan, is infused into the hippocampus of rats, there is an increased fibrillary ß-amyloid deposition in the neuropil. However, when ß-amyloid and only the heparan sulfate GAG side chains were co-infused, no Congo red-positive deposit was detected, suggesting alternative roles for PGs and GAGs in ß-amyloid aggregation in vivo. Castillo et al. [53] found the presence of GAG sulfate moieties to be critical in the enhancement of ß-amyloid fibril formation, as unsulfated heparin GAG did not lead to ß-amyloid fibrillogenesis, thus demonstrating the importance of these GAG sulfate moieties in vivo.

Heparin has been shown to reduce cytotoxic and inflammatory activity of Aß in vitro [60]. More recently, it was shown that long-term, peripheral treatment with enoxaparin, a low-molecular-weight heparin, in transgenic mice overexpressing human amyloid precurson protein 751 significantly lowered the number and decreased the area occupied by cortical ß-amyloid deposits and the total ß-amyloid cortical concentration, also reducing the number of activated astrocytes surrounding ß-amyloid deposits [61]. In further experiments in vitro, the same authors found that the drug: (1) dose-dependently attenuated the toxic effect of ß-amyloid on neuronal cells; and (2) reduced the ability of ß-amyloid to activate complement and contact systems. These results clearly showed the potential role of sGAG, and specifically heparin, as anti-amyloid agents.

Consistent observations show that chondroitin sulfate-derived monosaccharides represent the minimal GAG subunit required for Aß binding and that lateral aggregation between Aß fibers into mature amyloid fibers requires a sulfated GAG disaccharide [48]. This means that the size constraints of the monosaccharide are insufficient to facilitate the association of fibers but sufficient to bind Aß. Therefore, these monosaccharide compounds may represent a potential tool for stabilizing toxic Aß intermediates. Alternatively, GAG-derived disaccharides may represent a template in which to develop drugs (GAG mimetics) that decrease available monomer by accelerating precipitation of Aß fibers, thus acting as antiamyloid agents.

In vitro and in vivo studies have shown that small polysulfated compounds are protective against Aß-induced effects. Several underlying mechanisms can cause the protective actions of these sulfated GAG mimetics. First, they may inhibit binding of heparan sulfate to Aß and may thus interfere with fibrillogenesis. Second, GAG mimetics may inhibit formation of ß-pleated sheets. Finally, these highly sulfated compounds may block adherence of Aß to the cell surface. Any of these effects could result in protection of neurons and vascular cells against Aß-mediated cellular toxicity. For application in a therapeutic setting, these compounds must meet strict criteria, such as good bioavailability (including the ability to cross the blood–brain barrier efficiently) and a good safety profile. It is encouraging to note that even after chronic exposure to GAG mimetics no intrinsic cellular toxicity was observed. Furthermore, such a therapeutic strategy has already been shown to be effective in a mouse model, where these molecules clearly reduced the progression of inflammation-associated amyloid. Clinical trials with therapeutic inhibitors of amyloid formation are underway, an approach that may be rewarding not only in the treatment of AD but also in other amyloid disorders.

Currently, small organic molecules that mimic sulfated GAGs are being developed. These mimetics compete with natural sulfated molecules and may prevent formation of amyloid fibrils and binding of amyloidogenic proteins to the cell surface. Phase II trials with this new class of drugs in humans are underway. Previous clinical studies have shown a therapeutic effect of Ateroid (a compound consisting of both high- and low-molecular-weight glycans) on Alzheimer's disease and multiinfarct dementia in elderly patients [62–66]. This compound has also been shown to alleviate both behavioral and neurochemical impairments seen in aged F344 rats [67], which along with the clinical studies suggests a role for GAGs in treating not only AD but also other age-related dementias. The active components of this heterogeneous mixture of both high- and low-molecular-weight glycans, might be the latter, as the blood–brain barrier likely prevents the high-molecular-weight glycan component from entering the central nervous system. Studies show that low-molecular-weight GAGs are indeed capable of passing the blood–brain barrier if given to rats intravenously, subcutaneously, or orally [68].

Low-molecular-weight GAGs are able to prevent amyloid-induced tau conformational changes and reactive astrocytosis if given before or after amyloid deposition has occurred [69]. Additionally, the effects of low-molecular-weight GAGs are independent of amyloid deposit formation, as these compounds do not break up the injected amyloid deposit in this model. Previous studies have shown that amyloid can induce tau phosphorylation and conformation changes [70–75], suggesting a link between amyloid and altered tau pathology. Gotz et al. [76] showed that an intrahippocampal amyloid ß 1–42 injection could significantly accelerate NFT formation and reactive astrocytosis in P301L mutant tau transgenic mice. Interestingly, these studies showed that ß-amyloid induced neuropathology at sites distant from the ß-amyloid injection site, supporting the hypothesis that damage to presynaptic terminals projecting to the amyloid injection site can lead to altered tau pathology. The ability of low-molecular-weight GAGs to prevent these changes provides compelling evidence of their potential therapeutic effects in AD.

Currently, it is unknown what molecular weight glycosaminoglycan fraction of these heterogeneous compounds crosses the blood–brain barrier. However, it is reasonable to accept that there is a molecular weight range of GAGss that block the effects of ß-amyloid. Likely, GAGs are able to prevent cellular responses to amyloid without altering amyloid deposition. Kisilevsky et al [77]. found that small-molecule disulfates (molecular weight 900–1000) could inhibit in vitro acceleration of amyloid fibril formation by heparan sulfate. These compounds, slightly smaller in molecular mass than low-molecular-weight GAGs, have a similar amount of sulfate residues, substantiating the importance of sulfates in inhibiting amyloid-induced neuropathology and/or fibril formation.

Because there is no measurable change in the Congo red-stained positive deposit after low-molecular-weight GAG treatment, it is unlikely that low-molecular-weight GAGs are acting as amyloid fibrillogenesis inhibitors. Alternatively, in vitro studies have shown that GAGs may act by coating amyloid, thus preventing ß-amyloid from interacting with neurons [58]. This mechanism would explain why low-molecular-weight GAG treatment leads to decreased amyloid-induced neuropathology without altering the Congo red-stained positive deposits. One possible mechanism to explain the effects of low-molecular-weight GAGs in this model is their role as neurotrophic modulators. Heparin, heparin sulfate, and heparin-derived GAG oligosaccharides have been shown to bind and potentiate the effects of FGF [78–82], also leading to an increase in both dendritic branching and dendritic spine densities in rat hippocampus [83]. It is possible that low-molecular-weight glycosaminoglycans may act as modulators of FGF, potentiating FGF-induced mitogenic and injury recovery processes, effectively allowing cells to better recover from the amyloid insult.

In conclusion, low-molecular-weight GAGs, and specifically heparin-derived oligosaccharides, are heparin fragments with no anticoagulant activity and a strong binding affinity to amyloid. They represent a potential therapeutic treatment for amyloid-induced neuropathology, specifically that of Alzheimer's disease.

References

1. Goedert M, Spillantini MG, Davies SW. Filamentous nerve cell inclusions in neurodegenerative diseases. Curr Opin Neurobiol 1998;8:619–632.
2. Kaytor MD, Warren ST. Aberrant protein deposition and neurological disease. J Biol Chem 1999;274:37507–37510.
3. Geula C, Wu CK, Saroff D, et al. Aging renders the brain vulnerable to amyloid beta-protein neurotoxicity. Nat Med 1998;4:827–831.
4. Pérez M, Valpuesta JM, Medina M, et al. Polymerization of tau into filaments in the presence of heparin: the minimal sequence required for tau–tau interaction. J Neurochem 1996;67:1183–1190.
5. Prusiner SB. Prion diseases and the BSE crisis. Science 1997;278:245–251.
6. Thellung S, Florio T, Corsaro A, et al. Intracellular mechanisms mediating the neuronal death and astrogliosis induced by the prion protein fragment 106–126. Int J Dev Neurosci 2000;18:481–492.

7. Jortikka MO, Parkkinen JJ, Inkinen RI, et al. The role of microtubules in the regulation of proteoglycan synthesis in chondrocytes under hydrostatic pressure. Arch Biochem Biophys 2000;374:172–180.

8. Goedert M, Crowther RA, Jakes R, et al. Filamentous tau protein and alpha-synuclein deposits in neurodegenerative diseases. In: Iqbal K, Swaab DF, Winblad B, Wisniewski HM (eds) Alzheimer's Disease and Related Disorders. Chichester, UK: Wiley, 1999, pp 245–258.

9. Park PW, Reizes O, Bernfield M. Cell surface heparan sulfate proteoglycans: selective regulators of ligand–receptor encounters. J Biol Chem 2000;275:29923–29926.

10. Maccarana M, Casu B, Lindahl U. Minimal sequence in heparin/heparan sulfate required for binding of basic fibroblast growth factor. J Biol Chem 1993;268:23898–23905.

11. Lopez-Casillas F, Payne HM, Andres JL, Massague J. Betaglycan can act as a dual modulator of TGF-beta access to signalling receptors: mapping of ligand binding and GAG attachment sites. J Cell Biol 1994;124:557–568.

12. Tatebayashi Y, Iqbal K, Grundke-Iqbal I. Dynamic regulation of expression and phosphorylation of tau by fibroblast growth factor-2 in neural progenitor cells from adult rat hippocampus. J Neurosci 1999;19:5245–5254.

13. artmann U, Maurer P. Proteoglycans in the nervous system–the quest for functional roles in vivo. Matrix Biol 2001;20:23–35.

14. Gupta-Bansal R, Frederickson RC, Brunden KR. Proteoglycan-mediated inhibition of A beta proteolysis: a potential cause of senile plaque accumulation. J Biol Chem 1995;270:18666–18671.

15. Cardin AD, Weintraub HJ. Molecular modeling of protein–glycosaminoglycan interactions. Arteriosclerosis 1989;9:21–32.

16. McLaurin J, Fraser PE. Effect of amino-acid substitutions on Alzheimer's amyloid-beta peptide–glycosaminoglycan interactions. Eur J Biochem 2000;267:6353–6361.

17. Velazquez P, Cribbs DH, Poulos TL, Tenner AJ. Aspartate residue 7 in amyloid beta protein is critical for classical complement pathway activation: implications for Alzheimer's disease pathogenesis. Nat Med 1997;3:77–79.

18. Bergamaschini L, Donarini C, Foddi C, et al. The region 1-11 of Alzheimer amyloid-ß is critical for activation of contact-kinin system. Neurobiol Aging 2001;22;63–69.

19. Giulian D, Haverkamp LJ, Yu J, et al. The HHQK domain of beta-amyloid provides a structural basis for the immunopathology of Alzheimer's disease. J Biol Chem 1998;273;29719–29726.

20. Watson DJ, Lander AD, Selkoe DJ. Heparin-binding properties of the amyloidogenic peptides Abeta and amylin: dependence on aggregation state and inhibition by congo red. J Biol Chem 1997;272:31617–31624.

21. Fraser PE, Nguyen JT, Chin DT, Kirschner DA. Effects of sulfate ions on Alzheimer ß/A4 peptide assemblies: implications for amyloid fibril-proteoglycan interactions. J Neurochem 1992;59:1531–1540.

22. Snow AD, Mar H, Nochlin D, et al. The presence of heparan sulfate proteoglycans in the neuritic plaques and congophilic angiopathy in Alzheimer's disease. Am J Pathol 1988;33:456–463.

23. De Witt DA, Silver J, Canning DR, Perry G. Chondroitin sulphate proteoglycans are associated with the lesions of Alzheimer's disease. Exp Neurol 1993;121:149–152.

24. Woods AG, Cribbs DH, Whittemore ER, Cotman CW. Heparan sulfate and chondroitin sulfate glycosaminoglycan attenuate beta-amyloid (25–35) induced neurodegeneration in cultured hippocampal neurons. Brain Res 1995;697:53–62.

25. Pollack SJ, Sadler II, Hawtin SR, et al. Sulfonated dyes attenuate the toxic effects of beta-amyloid in a structure-specific fashion. Neurosci Lett 1995;197:211–214.

26. Nunomura APG, Pappolla MA, Friedland RP, et al. Neuronal oxidative stress precedes amyloid-beta deposition in Down syndrome. J Neuropathol Exp Neurol 2000;59:1011–1017.

27. Klein WL, Krafft GA, Finch CE. Targeting small A beta oligomers: the solution to an Alzheimer's disease conundrum? Trends Neurosci 2001;24:219–224.
28. Brückner G, Hausen D, Härtig W, et al. Cortical areas abundant in extracellular matrix chondroitin sulphate proteoglycans are less affected by cytoskeletal changes in Alzheimer's disease. Neuroscience 1999;92:791–805.
29. Busciglio J, Lorenzo A, Yeh J, Yankner BA. ß-Amyloid fibrils induce tau phosphorylation and loss of microtubule binding. Neuron 1995;14:879–888.
30. Sturchler-Pierrat C, Abramowski D, Duke M, et al. Two amyloid precursor protein transgenic mouse models with Alzheimer disease-like pathology. Proc Natl Acad Sci U S A 1997;94:13287–13292.
31. Yankner BA.Mechanisms of neuronal degeneration in Alzheimer's disease. Neuron 1996;16:921–932.
32. Small DH, Nurcombe V, Reed G, et al. A heparin-binding domain in the amyloid protein precursor of Alzheimer's disease is involved in the regulation of neurite outgrowth. J Neurosci 1994;14:2117–2127.
33. Perry G, Kawai M, Tabaton M, et al. Neuropil threads of Alzheimer's disease show a marked alteration of the normal cytoskeleton. J Neurosci 1991;11:1748–1755.
34. Arrasate M, Pérez M, Valpuesta JM, Avila J. Role of glycosaminoglycans in determining the helicity of paired helical filaments. Am J Pathol 1997;151:1115–1122.
35. Arrasate M, Pérez M, Armas-Portela R, Avila J. Polymerization of tau peptides into fibrillar structures: the effect of FTDP-17 mutations. FEBS Lett 1999;446:199–202.
36. Forloni G, Angeretti N, Chiesa R, et al. Neurotoxicity of a prion protein fragment. Nature 1993;362:543–546.
37. Pérez M, Wandosell F, Colaço C, Avila J. Sulphated glycosaminoglycans prevent the neurotoxicity of a human prion protein fragment. Biochem J 1998;335:369–374.
38. Gabizon R, Meiner Z, Halimi M, Ben-Sasson SA. Heparin-like molecules bind differentially to prion-proteins and change their intracellular metabolic fate. J Cell Physiol 1993;157:319–325.
39. Shyng SL, Lehmann S, Moulder KL, Harris DA. Sulfated glycans stimulate endocytosis of the cellular isoform of the prion protein, PrPc, in cultured cells. J Biol Chem 1995;270:30221–30229.
40. Wong C, Xiong LW, Horiuchi M, et al. Sulfated glycans and elevated temperature stimulate PrP(sc)-dependent cell-free formation of protease-resistant prion protein. EMBO J 2001;20:377–386.
41. Shaked GM, Meiner Z, Avraham I, et al. Reconstitution of prion infectivity from solubilized protease-resistant PrP and non-protein components of prion rods. J Biol Chem 2001;276:14324–14328.
42. Caughey B, Raymond GJ. Sulfated polyanion inhibition of scrapie-associated PrP accumulation in cultured cells. J Virol 1993;67:643–650.
43. Kimberlin RH, Walker CA. Suppression of scrapie infection in mice by heteropolyanion 23, dextran sulfate, and some other polyanions. Antimicrob Agents Chemother 1986;30:409–413.
44. Perry G, Richey P, Siedlak SL, et al. Basic fibroblast growth factor binds to filametous inclusions of neurodegenerative diseases. Brain Res 1992;579:350–352.
45. Cohlberg JA, Li J, Uversky VN, Fink AL. Heparin and other glycosaminoglycans stimulate the formation of amyloid fibrils from alpha-synuclein in vitro. Biochemistry 2002;41:1502–1511.
46. Brunden KR, Richter-Cook NJ, Chaturvedi N, Frederickson RC. pH-dependent binding of synthetic beta-amyloid peptides to glycosaminoglycans. J Neurochem 1993; 61:2147–2154.
47. Buee L, Ding W, Anderson JP, et al. Binding of vascular heparan sulfate proteoglycan to Alzheimer's amyloid precursor protein is mediated in part by the N-terminal region of A4 peptide. Brain Res 1993;627:199–204.

48. Fraser PE, Darabie AA, McLaurin JA. Amyloid-beta interactions with chondroitin sulfate-derived monosaccharides and disaccharides: implications for drug development. J Biol Chem 2001;76:6412–6419.
49. Leveugle B, Scanameo A, Ding W, Fillit H. Binding of heparan sulfate glycosaminoglycan to beta-amyloid peptide: inhibition by potentially therapeutic polysulfated compounds. Neuroreport 1994;5:1389–1392.
50. McLaurin J., Franklin T, Zhang X, et al. Interactions of Alzheimer amyloid-beta peptides with glycosaminoglycans effects on fibril nucleation and growth. Eur J Biochem 1999;266:1101–1110.
51. Castillo GM, Ngo C, Cummings J, et al. Perlecan binds to the beta-amyloid proteins (A beta) of Alzheimer's disease, accelerates A beta fibril formation, and maintains A beta fibril stability. J Neurochem 1997;69:2452–2465.
52. McLaurin J, Franklin T, Kuhns WJ, Fraser PE. A sulfated proteoglycan aggregation factor mediates amyloid-beta peptide fibril formation and neurotoxicity. Amyloid 1999;6:233–243.
53. Castillo GM, Lukito W, Wigh, TN, Snow AD. The sulfate moieties of glycosaminoglycans are critical for the enhancement of beta-amyloid protein fibril formation. J Neurochem 1999;72:1681–1687.
54. Fukuchi K, Hart M, Li L. Alzheimer's disease and heparan sulfate proteoglycan. Front Biosci 1998;3:d327–d337.
55. Celio MR, Spreafico R, De Biasi S, Vitellaro-Zuccarello L. Perineuronal nets: past and present. Trends Neurosci 1998;21:510–515.
56. Härtig W, Klein C, Brauer K, et al.Hyperphosphorylated protein tau is restricted to neurons devoid of perineuronal nets in the cortex of aged bison. Neurobiol Aging 2001;22:25–33.
57. Pollack SJ, Sadler II, Hawtin SR, et al. Sulfated glycosaminoglycans and dyes attenuate the neurotoxic effects of beta-amyloid in rat PC12 cells. Neurosci Lett 1995a;184:113–116.
58. Sadler II, Smith DW, Shearman MS, et al. Sulfated compounds attenuate beta-amyloid toxicity by inhibiting its association with cells. Neuroreport 1995;7:49–53.
59. Snow AD, Sekiguchi R, Nochlin D, et al. An important role of heparan sulfate proteoglycan (Perlecan) in a model system for the deposition and persistence of fibrillar A beta-amyloid in rat brain. Neuron 1994;12:219–234.
60. Bergamaschini L, Donarini C, Rossi E, et al. Heparin attenuates cytotoxic and inflammatory activity of Alzheimer amyloid-beta in vitro. Neurobiol Aging 2002; 23:531–536.
61. Bergamaschini L, Rossi E., Storini C, et al. Peripheral treatment with enoxaparin, a low molecular weight heparin, reduces plaques and beta-amyloid accumulation in a mouse model of Alzheimer's disease. J Neurosci 2004;24:4181–4186.
62. Ban TA, Morey LC, Santini V. Clinical investigations with Ateroid in old-age dementias. Semin Thromb Hemost 1991;17:161–163.
63. Conti L, Placidi GF, Cassano GB. Ateroid in the treatment of dementia: results of a clinical trial. In: Ban TA, Lehmann HE (eds) Diagnosis and Treatment of Old Age Dementias. Basel: Karger, 1989,pp 76–84.
64. Conti L, Re F, Lazzerini F, et al. Glycosaminoglycan polysulfate (Ateroid) in old-age dementias: effects upon depressive symptomatology in geriatric patients. Prog Neuropsychopharmacol Biol Psychiatry 1989;13:977–981.
65. Parnetti L, Ban TA, Senin U. Glycosaminoglycan polysulfate in primary degenerative dementia-pilot study of biologic and clinical effects. Neuropsychobiology 1995; 31:76–80.
66. Cornelli U. The therapeutical approach to Alzheimer's disease. In: Casu JHAB (ed) Non-Anticoagulant Actions of Glycosaminoglycans (GAGs). New York: Plenum, 1996, pp 249–279.

67. Lorens SA, Guschwan M, Hata N, et al. Behavioral, endocrine, and neurochemical effects of sulfomucopolysaccharide treatment in the aged Fischer 344 male rat. Semin Thromb Hemost 1991;17:164–173.
68. Ma Q, Dudas B, Hejna M, et al. The blood-brain barrier accessibility of a heparin-derived oligosaccharides C3. Thromb Res 2002;105:447–453.
69. Walzer M, Lorens S, Hejna M, et al. Low molecular weight glycosaminoglycan blockade of beta amyloid induced neuropathology. Eur J Pharmacol 2002;445:211–220.
70. Chambers CB, Sigurdsson EM, Hejna MJ, et al. Amyloid-beta injection in rat amygdala alters tau protein but not mRNA expression. Exp Neurol 2000;162:158–170.
71. Kowall NW, McKee AC, Yankner BA, Beal MF. In vivo neurotoxicity of beta-amyloid [beta(1–40)] and the beta(25–35) fragment. Neurobiol Aging 1992;13:537–542.
72. Sigurdsson EM, Lorens SA, Hejna MJ, et al. Local and distant histopathological effects of unilateral amyloid-beta 25–35 injections into the amygdala of young F344 rats. Neurobiol Aging 1996;17:893–901.
73. Sigurdsson EM, Lee JM, Dong XW, et al. Bilateral injections of amyloid-beta 25–35 into the amygdale of young Fischer rats: behavioral, neurochemical, and time dependent histopathological effects. Neurobiol Aging 1997;18:591–608.
74. Sigurdsson EM, Lee JM, Dong XW, et al. Laterality in the histological effects of injections of amyloid beta 25–35 into the amygdala of young Fischer rats. J Neuropathol Exp Neurol 1997;56:714–725.
75. Takashima A, Honda T, Yasutake K, et al. Activation of tau protein kinase I/glycogen synthase kinase-3beta by amyloid beta peptide (25–35) enhances phosphorylation of tau in hippocampal neurons. Neurosci Res 1998;31:317–323.
76. Gotz J, Chen F, van Dorpe J, Nitsch RM. Formation of neurofibrillary tangles in P301 tau transgenic mice induced by Abeta 42 fibrils. Science 2001;293:1491–1495.
77. Kisilevsky R, Lemieux LJ, Fraser PE, et al. Arresting amyloidosis in vivo using small-molecule anionic sulphonates or sulphates: implications for Alzheimer's disease. Nat Med 1995;1:143–148.
78. Damon DH, D'Amore PA, Wagner JA. Sulfated glycosaminoglycans modify growth factor-induced neurite outgrowth in PC12 cells. J Cell Physiol 1988;135:293–300.
79. Damon DH, Lobb RR, D'Amore PA, Wagner JA. Heparin potentiates the action of acidic fibroblast growth factor by prolonging its biological half-life. J Cell Physiol 1989;138:221–226.
80. Neufeld G, Gospodarowicz D, Dodge L, Fujii DK. Heparin modulation of the neurotropic effects of acidic and basic fibroblast growth factors and nerve growth factor on PC12 cells. J Cell Physiol 1987;131:131–140.
81. Walicke PA. Interactions between basic fibroblast growth factor (FGF) and glycosoaminoglycans in promoting neurite outgrowth. Exp Neurol 1988;102:144–148.
82. Zhou FY, Kan M, Owens RT, et al. Heparin-dependent fibroblast growth factor activities: effects of defined heparin oligosaccharides. Eur J Cell Biol 1997;73:71–80.
83. Mervis RF, McKean J, Zats S, et al. Neurotrophic effects of the glycosaminoglycan C3 on dendritic arborization and spines in the adult rat hippocampus: a quantitative Golgi study. CNS Drug Rev 2000;6:44–46.

Chapter 24
Prospective Role of Glycosaminoglycans in Apoptosis Associated with Neurodegenerative Disorders

Bertalan Dudas, Amira Lemes, Umberto Cornelli, and Israel Hanin

Introduction

Damaged cells in vertebrates are commonly removed by apoptosis (programmed cell death) that involves cell packaging to prevent inflammation associated with necrotic processes. The extrinsic route of apoptosis is triggered by stimuli from the external environment via membrane-bound "death receptors," whereas the internal pathway is initiated usually by mitochondrial damage involving cytochrome-c release from damaged mitochondria (Fig. 1). Caspase-3 activation is the result of activation of the extrinsic and/or intrinsic pathways associated with cell shrinkage and DNA degradation. Because caspase-3 plays a pivotal role in both of the intrinsic and extrinsic routes of apoptosis, it is commonly used as a marker of programmed cell death.

Apoptotic processes are believed to play a crucial role in cell death associated with various neurodegenerative disorders. Dysregulation of the proteins of the apoptotic cascade has been reported in patients with Alzheimer's disease (AD) [1,2]. Moreover, caspases are commonly believed to be involved in generating toxic protein fragments that may participate in the pathological changes characteristic for AD [3]. Thus, attenuation of the abnormal activation of apoptotic processes may be a valuable therapeutic tool in the treatment of neurodegenerative conditions including AD.

Neuroprotective Role of Glycosaminoglycans

Previous studies revealed that glycosaminoglycans (GAGs) play a pivotal role in the pathogenesis of AD. GAGs bind to amyloid-β (Aβ) and attenuate the proteoglycan-mediated protection of Aβ against proteolysis.

B. Dudas
Neuroendocrine Organization Laboratory (NEO), Lake Erie College of Osteopathic Medicine, Erie, PA 16509, USA

A. Fisher et al. (eds.), *Advances in Alzheimer's and Parkinson's Disease*,
© Springer 2008

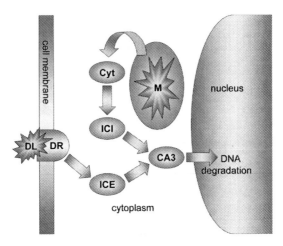

Fig. 1 Major elements of the apoptotic pathway. During apoptosis the damaged cells are packaged and removed entirely by the surrounding cells to prevent the inflammation characteristically seen in necrotic processes. Apoptosis is generated by external stimuli (extrinsic pathway) and internal events (intrinsic pathway). In the extrinsic pathway, "death ligands" (DL) are bound to specific "death receptors" (DR) on the cell membrane. The receptor–ligand interaction activates the initiator caspases of the external pathway (ICE). The intrinsic route of the apoptotic pathway is a result of an intracellular lesion that typically involves mitochondrial damage (M). This lesion leads to cytochrome-c (Cyt) release from the damaged mitochondria, activating the initiator caspases of the internal pathway (ICI). The extrinsic and intrinsic pathways unite at caspase-3 (CA3), which is activated by initiator caspases (ICE and ICI). Consequent processes result in degradation of DNA in the nucleus and packaging of the cell into small units that are easily cleared up by the adjacent cells

Stereotactic injection of Aβ fragments (Aβ 25–35) into the amygdala increases tau-2 immunoreactivity in the rat hippocampus [4]. This phenomenon was blocked by administration of low-molecular-weight GAGs [5]. Moreover, intraventricular administration of a specific cholinotoxin, ethylcholine aziridinium (AF64A), resulted in cholinergic lesions in the septum and cingulum bundle. Simultaneous administration of low-molecular-weight GAGs attenuated the AF64A-induced cholinergic damage [6]. It has been also shown that GAGs inhibit the aggregation and toxicity of Aβ itself [7] and increase axonal growth and arborization in the rat hippocampus [8]. Recent studies suggested that GAGs may act via modulating the release of nerve growth factor (NGF) by stimulating the effect of growth factors in cell culture [9,10] and/or reducing the expression of cholinotoxin-stimulated growth factor receptor expression in the rat septum [11]. However, the exact mechanism of the neuroprotective/neurorepair effect of GAGs is not known.

Role of GAGS in Apoptosis Associated with Neurodegeneration

Because GAGs were shown to express neuroprotective properties against various lesions in cell culture [4] and in vivo [1,2], it is conceivable that GAGs may protect against neuronal lesions via influencing stressor-induced apoptosis (programmed cell death). Indeed, heparin inhibits glomerular cell apoptosis in cell culture [9] and attenuates trophoblast apoptosis, explaining the clinical benefits of heparin on recurrent pregnancy loss [12]. Hyaluronan, a major GAG constituent of the extracellular matrix of mammalian bone marrow, also attenuates apoptosis induced by dexamethasone in malignant multiplex myeloma (MM) cells [13]. This finding may explain why MM cells in the bone marrow can escape conventional chemotherapy. Moreover, chondroitin sulfate and heparan sulfate attenuate apoptosis in fetal lung fibroblasts [14]. Because low-molecular-weight GAGs derived from heparin have been shown to penetrate the blood-brain barrier [15], it is plausible that these molecules can protect against neuronal apoptotic processes that are major hallmarks of neurodegenerative disorders.

In the present study we examined the effect of C3, a low-molecular-weight GAG, on AF64A-mediated apoptosis using immunohistochemistry of the caspase-3 apoptotic marker. To reveal whether the immunohistochemical pattern of apoptotic cells correlates with the cholinergic lesions, we also visualized the cholinergic system, using choline acetyltransferase (ChAT) immunohistochemistry.

Methods

Animals

Male Fischer 344 (Harlan) rats ($n = 6$/group) weighing approximately 250 to 300 g were used. The animals were group-housed and maintained in a colony room, with water and rat chow readily available at all times. A 12-hour light/dark cycle was maintained for the duration of the experiment.

Drugs and Chemicals

AF64A was prepared as described previously by Fisher and Hanin [16] Briefly, an aqueous solution of acetylethylcholine mustard HCl (Research Biochemicals, Wayland, MA, USA) was brought to pH 11.3 with NaOH. After stirring for 30 minutes at room temperature, the pH was lowered to 7.4 with HCl and stirred for 60 minutes. The amount of AF64A was then adjusted to 1 nmol/2 µl. Control animals received distilled water 2 µl/side treated in the same way as the AF64A solution.

C3 was obtained from unfractionated porcine mucosal heparin by means of controlled depolymerization induced by γ-radiation (US patent 4,987,222, LDO, Trino Vercellese, Italy). Unfractioned heparin in the solid state or in solution is treated with a rectilinear γ-ray beam at doses of 2.5 to 20.0 Mrad. This irradiation is supplied in successive stages with cooling intervals between the various stages. The degraded material is then fractionated to remove high-molecular-weight (MW) fractions and to obtain C3 with a narrow MW range around ∼2 kDa. The MW profile of C3 was determined using gel permeation chromatography/high performance liquid chromatography (GPC-HPLC). It was found that C3 is a mixture of oligosaccharides (average MW ∼2.4 kDa) and a composition of oligosaccharides containing primarily four to eight dextrose units.

Administration of drugs and compounds

All animals were given C3 (25 mg/kg) or saline once daily by oral gavage for 7 days before AF64A administration and 7 days thereafter. Either AF64A or vehicle was infused bilaterally, in a sequential manner, intracerebroventricularly (i.c.v.) with a 30-gauge needle inserted through a burr hole drilled into the skull in both the right and left ventricles. Stereotactic coordinates were (from the bregma): posterior 0.8 mm, lateral ± 1.5 mm, ventral (from the dura) 3.6 mm. The rate of infusion was 1.0 (l/min; the needle was left in place for 5 minutes after infusion and then withdrawn slowly. The vehicle of AF64A was distilled water prepared in the same manner as the AF64A. To anesthetize the animals, sodium pentobarbital 60 mg/kg i.p. was administered.

Animal Sacrifice and Tissue Preparation for Histochemical Assays

The animals were decapitated 7 days after AF64A administration. Following sacrifice, the brains were placed in an ice-cold bath of saline and then on an ice-cold surface to dissect the brain. The brains were stored in 4% paraformaldehyde at 4°C for subsequent immunohistochemistry. Coronal sections (40 mm) were cut on a freezing microtome (American Optical, Buffalo, NY, USA), placed in 0.2% sodium azide in phosphate-buffered saline (PBS), and stored at 4°C until used for immunohistochemistry.

Immunohistochemistry

The sections were removed from the buffer and subsequently placed in 0.3% H_2O_2 in PBS for 10 minutes and then in 0.3% Triton X-100 in PBS for 30 minutes. The sections were then washed in PBS and incubated in anti-caspase rabbit antibody 1:1000 (Chemicon, Temecula, CA, USA) or anti-ChAT rabbit antibody 1:2000 (Chemicon) for 24 hours at room temperature. Subsequently, the tissue was washed in PBS and then incubated for 1.0 hour in biotinylated anti-rabbit immunoglobulin (IgG) secondary antibody 1:1000 (Vectastain ABC Elite kit; Vector Laboratories, Burlingame, CA, USA). Following two washes in PBS, the tissue was incubated for 1 hour in a solution of avidin horseradish peroxidase 1:1000 (Vector). The sections were next washed in PBS, immersed for 10 minutes in Tris-buffered saline pH 7.6 (TBS), and reacted with $3'-3'$ diaminobenzidine tetrahydrochloride (DAB) solution containing nickel ions. Tissue sections were finally washed in TBS, slide-mounted, dried, defatted, and cover-slipped.

Results

Effect of Intracerebroventricular AF64A Administration on Caspase-3 Immunoreactivity

AF64A/Saline Administration

Compared to the control animals (Fig. 2A), intraventricular injection of AF64A (1 nmol/2 µl/side) induced the appearance of caspase-3 immunoreactivity in the septum 7 days after the stereotactic injection (Fig. 2C). The largest number of caspase-3-immunoreactive perikarya was detected in the lateral septal nucleus.

AF64A/C3 Administration

Orally administered C3 significantly attenuated the appearance of the caspase-3 immunoreactive structures in the septum (Fig. 2D) compared to the AF64A-treated rats (Fig. 2C).

Vehicle of AF64A/C3 Administration

Orally administered C3 did not show any effect on the septal caspase-3 expression (Fig. 2B) compared to the control animals (Fig. 2A).

Effect of Intracerebroventricular AF64A Administration on ChAT Immunoreactivity

AF64A/Saline Administration

Compared to the control animals (Fig. 2E), AF64A resulted in two major histological changes regarding the cholinergic cell population:

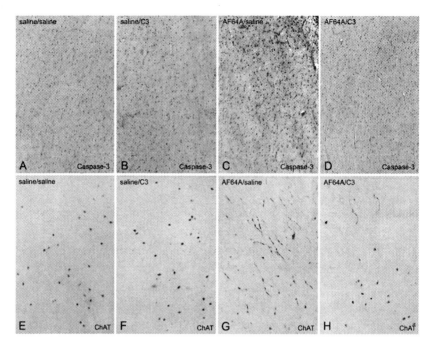

Fig. 2 Neuroprotective role of C3 against AF64A-induced caspase-3 activity (**A–D**) and cholinergic (**E–H**) deficit. Compared to the control animals (**A, E**), intraventricular injection of AF64A (1 nmol/2 μl/side) induced the appearance of caspase-3 immunoreactivity (**C**) and damaged ChAT-IR fiber varicosities (**G**) primarily in the lateral septum, at 7 days after the stereotactic injection. Orally administered C3 significantly attenuated this phenomenon (**D, H**). Orally administered C3 alone did not have any detectable effect on septal caspase-3 or ChAT expression (**B, F**)

Intraventricular injection of AF64A (1 nmol/2 μl/side) (1) induced the appearance of ChAT immunoreactive (IR), abnormal fiber varicosities in the lateral septal nucleus, and (2) reduced the number of ChAT-IR perikarya in the medial septal nucleus at 7 days after the stereotactic AF64A injection (Fig. 2G). These histological features were in accord with our previous findings [6,17].

AF64A/C3 Administration

Orally administered C3 attenuated the appearance of the damaged ChAT-IR fiber varicosities in the septum (Fig. 2H) compared to the AF64A-treated rats (Fig. 2G). Also, C3 administration resulted in increased cholinergic perikarya in the medial septal nucleus (Fig. 2H) when compared to the AF64A-treated animals (Fig. 2G).

Vehicle of AF64A/C3 Administration

No significant effect on septal ChAT expression was observed after oral C3 treatment alone (Fig. 2F) compared to the control animals (Fig. 2E).

Discussion and Conclusion

In our experiments, i.c.v. administration of AF64A increased septal caspase-3 expression primarily in the lateral septal nucleus. This histological change correlated with the AF64A-induced decrease of the number of cholinergic perikarya in the medial septum and with the appearance of abnormal cholinergic axon varicosities in the lateral septum. These findings suggest that AF64A exerts its cholinotoxic activity, at least partially, via apoptotic pathways. Subsequent administration of C3, a low-molecular-weight GAG, reduced the amount of AF64A-induced caspase-3-IR structures in the lateral septum. Because C3 also reduced the number of abnormal, septal cholinergic fiber varicosities, it is conceivable that some of the caspase-3-IR structures are indeed cholinergic.

The ability of C3 to attenuate AF64A-induced caspase-3 expression indicates that GAGs may have a direct modulatory role on apoptotic processes by affecting caspase-3 expression. Although GAGs are known to induce apoptosis in cell culture, our findings suggest that C3 may involve inhibition of apoptotic processes in its protection against neurodegeneration in vivo. As previous studies have shown that C3 is capable of crossing the blood-brain barrier [15], C3 treatment may represent a promising novel approach for the therapy of Alzheimer's disease and other neurodegenerative disorders. Indeed, C3 has been accepted as a low-risk compound based on the results of toxicology tests, and it is at present in a Phase I clinical trial.

Acknowledgment The present study was supported by NIA/STTR grant 1-R41-AG15740–02.

References

1. Engidawork E, Gulesserian T, Seidl R, et al. Expression of apoptosis related proteins in brains of patients with Alzheimer's disease. Neurosci Lett 2001;303:79–82.
2. Engidawork E, Gulesserian T, Yoo BC, et al. Alteration of caspases and apoptosis-related proteins in brains of patients with Alzheimer's disease. Biochem Biophys Res Commun 2001;281:84–93.
3. Rideout HJ, Stefanis L. Caspase inhibition: a potential therapeutic strategy in neurological diseases. Histol Histopathol 2001;16:895–908.

4. Sigurdsson EM, Lorens SA, Hejna MJ, et al. Local and distant histopathological effects of unilateral amyloid-beta 25-35 injections into the amygdala of young F344 rats. Neurobiol Aging 1996;17:893–901.

5. Dudas B, Cornelli U, Lee JM, et al. Oral and subcutaneous administration of the glycosaminoglycan C3 attenuates Aβ (25-35)-induced abnormal tau protein immunoreactivity in rat brain. Neurobiol Aging 2002;23:97–104.

6. Rose M, Dudas B, Cornelli U, Hanin I. Protective effect of the heparin-derived oligosaccharide C3, on AF64A-induced cholinergic lesion in rats. Neurobiol Aging 2003;24:481–490.

7. Kisilevsky R, Lemieux LJ, Fraser PE, et al. Arresting amyloidosis in vivo using small-molecule anionic sulphonates or sulphates: implications for Alzheimer's disease. Nat Med 1995;1:143–148.

8. Mervis RF, McKean J, Zats S, et al. Neurotrophic effects of the glycosaminoglycan C3 on dendritic arborization and spines in the adult rat hippocampus: a quantitative golgi study. CNS Drug Reviews 2000;44–46.

9. Damon DH, D'Amore PA, Wagner JA. Sulfated glycosaminoglycans modify growth factor-induced neurite outgrowth in PC12 cells. J Cell Physiol 1988;135:293–300.

10. Lesma E, Di Giulio AM, Ferro L, et al. Glycosaminoglycans in nerve injury. 1. Low doses of glycosaminoglycans promote neurite formation. J Neurosci Res 1996;46:565–571.

11. Dudas B, Rose M, Cornelli U, Hanin I. Low molecular weight glycosaminoglycan C3 attenuates AF64A-stimulated, low-affinity nerve growth factor receptor-immunoreactive axonal varicosities in the rat septum. Brain Res 2005;1033:34–40.

12. Bose P, Black S, Kadyrov M, et al. Heparin and aspirin attenuate placental apoptosis in vitro: implications for early pregnancy failure. Am J Obstet Gynecol 2005;192:23–30.

13. Vincent T, Molina L, Espert L, Mechti N. Hyaluronan, a major non-protein glycosaminoglycan component of the extracellular matrix in human bone marrow, mediates dexamethasone resistance in multiple myeloma. Br J Haematol 2003;121:259–269.

14. Cartel NJ, Post M. Abrogation of apoptosis through PDGF-BB-induced sulfated glycosaminoglycan synthesis and secretion. Am J Physiol Lung Cell Mol Physiol 2005;288:L285–L293.

15. Ma Q, Dudas B, Hejna M, et al. The blood-brain barrier accessibility of a heparin-derived oligosaccharide, C3. Thromb Res 2002;105:447–453.

16. Fisher A, Hanin I. Potential animal models for senile dementia of Alzheimer's type, with emphasis on AF64A-induced cholinotoxicity. Annu Rev Pharmacol Toxicol 1986;26:161–181.

17. Rose M, Dudas B, Cornelli U, Hanin I. Glycosaminoglycan C3 protects against AF64A-induced cholinotoxicity in a dose-dependent and time-dependent manner. Brain Res 2004;1015:96–102.

Chapter 25
Stem Cell Therapy in Alzheimer's Disease

Kiminobu Sugaya, Young-Don Kwak, and Angel Alvarez

Introduction

Neural tissue transplantation therapy for neurodegenerative diseases is not a novel idea. Parkinson's disease has been a good target for transplantation therapy because of its specific loss of large neurons in the substantia nigra that send dopaminergic projections to the striatum. Hence a conventional treatment for Parkinson's disease is augmentation of the dopamine content in the striatum with L-dopa. Based on experience with L-dopa treatment, transplantation of fetal tissue producing dopamine to the striatum of Parkinson's disease has been tested for many years. Although they are not neuroreplacement therapies for degenerating neurons in the substantia nigra, many of the clinical trials significantly ameliorated the behavioral deficit. Thus, the possibilities of this approach for Alzheimer's disease have been discussed. However, the use of human fetal neuronal tissue not only raises ethical concerns but also is impractical because neural tissue from multiple fetuses is required for each patient. Thus, we must seek alternatives to fetal tissue.

Recent advances in stem cell technologies are expanding our ability to replace many tissues throughout the body. Cognitive impairment caused by degeneration of neuronal cells has been considered incurable because of a long-held "truism" claiming that neurons do not regenerate during adulthood. This statement has now been challenged, and we have found new evidence that neurons do indeed have the potential to be renewed after maturation. The discovery of multipotent neural stem cells (NSCs) in the adult brain [1,2] has brought revolutionary changes in the theory of neurogenesis, which currently posits that regeneration of neurons can occur throughout life, thus opening a door for the development of novel therapies to treat neurodegenerative diseases, including Alzheimer's disease, by neuronal regeneration using stem cell transplantation.

K. Sugaya
Biomolecular Sciences Center, Burnett College of Biomedical Sciences, University of Central Florida, Orlando, FL 32816-2364, USA

A. Fisher et al. (eds.), *Advances in Alzheimer's and Parkinson's Disease,*
© Springer 2008

Selection of Stem Cells

Human NSCs may be the most promising candidate for neuroreplacement therapy. However, the ethical issues and risk of immunological rejection limit their value. Although tissue rejection may not be particularly problematic for use in neuroreplacement strategies because the brain does not produce an immune response unless traumatic damage has occurred in the tissue, the ideal biological source of cells for replacement therapies would be autologous transplantation of stem cells derived from the patient's own tissues. It is also not known whether a large volume of heterologous neuronal transplantation could change the character or personality of an individual, and some patients might have psychological difficulties accepting brain tissue from outside sources.

Some researchers are trying to find autologous transplantable cell sources in embryonic stem (ES) cells that extensively proliferate and theoretically differentiate into any type of somatic cells. ES cells could also be modified, by cloning, into cells that possess the same genetic material as that of the patient. On the other hand, we must develop methods of enriching the cell of interest because the ES cell does not have specific information to become a certain type of cell. In other words, ES cells are not committed to become neural cells like NSCs. Recently, McKay's group reported that a highly enriched population of midbrain NSCs can be derived from mouse ES cells. In their report dopamine neurons generated by these stem cells showed electrophysiological and behavioral properties expected of neurons from the midbrain, thus encouraging the use of ES cells in cell replacement therapy for Parkinson's disease.

Even after this success, though, we still have to jump a hurdle in addition to the ethical issue of using human embryos before we can rely on the autologous cell therapy using ES cells: It is related to tissue-specific epigenetic modification. Cloning by nuclear transfer is an inefficient process in which most clones die before birth and survivors often display growth abnormalities [3]. This may be due to the tissue-specific DNA methylation pattern from somatic nuclei used in cloning [4]. Thus, cloned ES cells may also receive the tissue-specific epigenetic modification and may not be fully functional as neural cells.

Transdifferentiation of Adult Stem Cells

Bone marrow contains stem-like cells used not only for hematopoiesis but also for production of a variety of nonhematopoietic tissues. A subset of stromal cells in bone marrow, referred to as mesenchymal stem cells (MeSCs), are capable of producing multiple mesenchymal cell lineages, including bone, cartilage, fat, tendons, and other connective tissues [5–8]. Recent reports have shown that human MeSCs (HMeSCs) also have the ability to differentiate into a diverse family of cell types that may be unrelated to their phenotypical embryonic origin, including muscle and heptocytes [9–14]. Although adult stem cells continue to possess some multipotency, cell types produced from adult stem cells are limited by their tissue-specific character. To overcome this barrier, lineage alterations of stem cell are necessary. However, the regulation mechanisms of tissue-specific stem cell fate remain unclear. Thus, to differentiate MeSCs into neural cells, alteration of their epigenetic information before transplantation may be necessary. Nonetheless, neuroreplacement therapy by MeSCs transplantation must clear some hurdles before it can be considered for clinical use.

The potential therapeutic use of HMeSCs in the central nervous system has been discussed [15,16], and several in vivo transplantation studies showed neural and glial differentiation of HmeSCs [17–21]. However, technologies to induce neural lineage from HMeSCs are not fully established. Verfaillie's group (Jiang et al. [22]) recently identified multipotent progenitor cells that co-purify with MeSCs in the adult bone marrow [22]. The authors claimed that these cells contribute to most, if not all, somatic cell types. Thus, this subpopulation of HMeSCs may be mainly responsible for the neural differentiation.

To investigate neural differentiation of HMeSCs in vivo, we injected HMeSCs into the lateral ventricle of mature mice; the HMeSCs had been expanded without differentiation and labeled by the incorporation of BrdU into nuclear DNA. Four to six weeks after transplantation, the mouse brains were analyzed by immunohistochemistry for human specific βIII-tubulin and GFAP, markers for neurons and astrocytes, respectively. the migration and differentiation patterns of the transplanted HMeSCs were quite similar to our previous results with HNSCs transplanted into rats. The main difference was the size of the donor cells compared to the host cells, which were in the order of humans $>$ rats $>$ mice. Intensely and extensively stained with βIII-tubulin neurons, BrdU-positive nuclei were found in the bilateral cingulate and parietal cortices and in the hippocampus. The βIII-tubulin-positive neurons found in the cerebral cortex typically demonstrated a dendrite pointing to the edge of the cortex. In the hippocampus, donor-derived neurons exhibited multiple morphologies and varied in cellular size and shape, with one or more processes and branching.

Recently, two groups reported spontaneous fusion of stem cells [23,24]. In these reports, the authors found that stem cells acquired phenotypes from other cells by fusion, which may occur when these stem cells directly touch other cells after transplantation. To investigate the possibility of the neural differentiation of HMeSCs without fusion, we co-cultured BrdU-labeled HMeSCs with differentiated HNSCs. The HNSCs were differentiated in 12-well tissue culture plates under basal media conditions. The HMeSCs were then transferred onto a tissue culture 0.4 μm membrane insert and placed on top of the differentiated HNSCs under basal media conditions.

Immunocytochemical examination 7 days after co-culture revealed that HMeSCs differentiated into βIII-tubulin-immunopositive, small bipolar and unipolar cells (approximately 40% of the total), and GFAP-immunopositive large flattened multipolar cells (approximately 60% of the total). Thus, most

HMeSCs are converted to neural cells, indicating that not only one subset of the HMeSCs is responsible for neural differentiation. The general cell morphology in both HNSC- and HMeSC-differentiated cultures were similar. This result indicates that HMeSCs are capable of becoming neurons and astrocytes when co-cultured with differentiated NSCs. Because no exogenous differentiation factors, such as retinoic acid or BDNF, were added to these cultures, and no cell-to-cell contact existed in this co-culture system, it is reasonable to hypothesize that the membrane-permeable endogenous factor(s) released from differentiating HNSCs altered the cell fate decisions of HMeSCs. In our in vitro experiment, HMeSCs were cultured on the membrane insert and were kept totally separated from the HNSCs. Thus, the possibility of fusion between HMeSCs and HNSCs can be excluded.

These results indicate that the brain environment may produce factor(s) that allow the differentiation of not only NSCs but also MeSCs into neurons. They also suggest that HMeSCs may serve as an alternative to HNSCs for potential therapeutic use in neuroreplacement.

Dedifferentiation of Adult Stem Cells

Nanog, also referred to as early embryo-specific NK (*ENK*) [25], is a recently discovered gene responsible for maintaining pluripotency in embryonic stem cells [26,27]. This unique gene and its cousin, Nanog2, are genetically distinct members of the ANTP class of homeo-domain proteins and have at least 12 identified pseudogenes [28,29]. Structurally, Nanog contains three α helixes encoded within the homeodomain portion and can be divided into three regions with respect to the central homeodomain sequence [30]. The N-terminal region is rich in serine and threonine residues, indicating phosphate-regulated transactivation [30], possibly through SMAD4 interactions [29]; the C-terminal domain is seven times as active, with an unusual motif of equally spaced tryptophans separated by four amino acids, each flanked with serine or threonine residues [26,27,29,30]. Gene expression studies have shown Nanog to be active in embryonic stem cells, tumors, and some adult tissue. Nanog expression precipitously decreases with differentiation [31] and maintains self-renewal in embryonic stem cells by gene transfection [26]. In culture, Nanog guards against differentiation and acts concomitantly with Oct-4, Wnt [32], and BMP-4 [33], yet it utilizes a STAT-3-independent mechanism to maintain an undifferentiated state [26,27]. The role of Nanog in regulating pluripotency makes this gene a potential candidate for increasing the potency of adult stem cells.

To determine if Nanog could restore pluripotency in adult stem cells rather than simply maintain the state in embryonic cells, we developed a two-step process of dedifferentiation and development along an alternative lineage. Initially, we cultured human mesenchymal stem cells in six-well culture plates and allowed them to adhere and grow for at least 48 hours to achieve approximately 75% confluence. Cells were subsequently transfected with a mammalian cell vector or control vehicle, cultured, and examined. Cells transfected with Nanog became nonadherent and proliferated in the presence of the remaining adherent cells acting as a feeder layer. Control samples receiving empty transfection did not show any proliferation and decreased dramatically, likely due to the toxicity of the transfection. When the nonadherent cells were transferred to wells without a feeder layer, they tended to die by apoptosis. This may be caused by either an absence of feeder cell proteins or decreased cell density. Cellular transformation occurred in a pattern of transfection, nonadherence, survival, and proliferation. Transformed cells proliferated in three-dimensional clusters for months in culture. These characteristics are similar to those of embryonic stem cells.

To test the hypothesis that cells can be dedifferentiated using Nanog and committed to an alternate lineage, we utilized a co-culture system of differentiated human neural stem cells and transformed mesenchymal cells. Neural stem cell spheres were placed in 12-well plates and differentiated using serum-free basal medium, as previously described. Neuronal stem cells began to differentiate by becoming adherent and migrating radially outward from the original neural sphere. Following neural stem cell differentiation, these cells were utilized as feeder cells in our co-culture system by placing modified cells inside co-culture chambers that separated modified stem cells from the feeder layer with a 0.2-μm semipermeable membrane.

Co-culturing experiments showed that embryoid body-like clusters began differentiation within 48 hours. Control cells with our empty vector treatments failed to show any signs of neural differentiation. Embryoid-like bodies adhered to membranes, and differentiation occurred as neural cells migrated radially outward. These migrating cells were immunopositive for βIII-tubulin and GFAP, indicating neurons and astrocytes, respectively.

Here we demonstrate a novel method to improve transdifferentiation of mesenchymal stem cells into neural cells by gene expression. This study, to our knowledge, represents the first report of increasing pluripotency in adult cells by means of gene expression. The ability to dedifferentiate cells by overexpression of pluripotency genes is a simple and exciting prospect for cellular engineering. This novel technology may allow us to develop autologous stem cell therapies.

APP Function in Stem Cell Biology

We found evidence that APP fragments are secreted from apoptotic HNSCs, and induce differentiation of other HNSCs in vitro. We also observed that exogenously added secreted-type APP (sAPP) induces the differentiation of HNSCs, whereas antibody recognizing the N'-terminal of APP prevents the differentiation of HNSCs. These findings indicate that APP signaling is one of the regulatory systems involved in the differentiation of NSCs. We also found that HNSCs transplanted into the APP-knockout mouse brain could not migrate properly and failed to repair brain lesions, whereas HNSCs transplanted into wild-type mice successfully migrated into the proper position and differentiated into the right kind of cells. This result not only is the first finding of a phenotypical change in APP-knockout mice but also indicates a physiological role for APP in the regeneration of adult brain cells. Furthermore, we found that the addition of a higher concentration of sAPP or the overexpression of APP by transgenes to HNSC cultures caused glial, rather than neural, differentiation of these cells. These findings indicate that the pathological alteration of APP metabolism in AD induces glial differentiation of neural stem cells and leads to exhaustion of the stem cell population, which may be important for ongoing neurogenesis in the adult brain.

Although many factors are released following apoptotic cell death, several studies point to an important correlation between apoptosis and the APP. Damaged neurons and neurons committed to apoptosis demonstrate signals strongly immunopositive for APP [34,35]. Moreover, amyloidogenic fragments produced from APP are reported to be released into the extracellular space from neuronal cells under a serum-deprived condition [36]. The expression of APP is also reported to increase during retinoic acid-induced neuronal differentiation [37]. The mRNA expression of β-amyloid precursor-like proteins (APLP-1 and APLP-2) is also up-regulated during retinoic acid-induced differentiation of human SH-SY5Y neuroblastoma cells. The increase in APP expression levels during neuronal differentiation in various cell culture systems suggests an important cellular function for APP during the differentiation process. From these observations, we hypothesized that under serum-free differentiation conditions, APP fragments released from apoptotic cells serve as regulation and differentiation factors for neighboring HNSCs.

Amyloid precursor protein is also known to be up-regulated during development and after brain damage [38]—events that involve migration and differentiation of NSCs. Secreted APP (sAPP) has also been reported to produce protein kinase C and synaptogenesis in cultured neurons [39], in addition to significantly enhancing the proliferation and growth of neural stem cells [40]. Moreover, it has been shown that sAPP is able to activate MAPK (ERK) in PC12 cells via the Ras pathway [41]. Because MAPK activation can induce proliferation or differentiation, it is possible that sAPP activates this pathway in HNSCs and induces cell differentiation. These facts, together with our findings, indicate that one of APP's physiological functions may be the regulation of NSC biology to allow for successful formation and replacement of proper structures and neuronal circuits. A possible scenario for reconstructing neuronal circuits with guidance by NSCs might be that sAPP released from damaged or dying cells preferentially induces glial differentiation of a population of NSCs. These NSC-derived glial cells can then produce factors that may support surrounding damaged cells [42] and promote neuronal migration and differentiation of other NSCs in this area. This scenario fits nicely with our in vitro observations that the initial apoptotic cell death-induced glial differentiation was followed by neuronal differentiation [43]. Thus, under normal physiological conditions, APP may be necessary to recover from brain damage. In the case of familial AD or Down Syndrome (DS), the increased levels of APP fragments produced in the brains of these patients may modify the biological equilibrium of HNSCs in such a way that a pathological shift toward premature differentiation of HNSCs occurs, thereby exhausting the HNSC population. Because the effective natural replacement of degenerating neurons in the adult brain during aging or disease processes may be important for maintaining normal brain function, the HNSC population exhaustion would pose serious problems.

In our study, the addition of recombinant sAPP to the cell culture medium dose-dependently differentiated HNSCs under serum-free differentiation conditions. We also characterized the cell population of sAPP-treated HNSCs at 5 days in vitro (DIV) under serum-free differentiation conditions by double immunofluorescence labeling of GFAP and βIII-tubulin. Treatment with sAPP dose-dependently (25, 50, 100 ng/ml) increased the population of GFAP-positive cells from an average of 45% in controls (no sAPP) to an average of 83% using the highest concentration of sAPP (100 ng/ml at 5 DIV). Interestingly, it was observed that the lowest dose of sAPP treatment (25 ng/ml) also increased neuronal differentiation. However, higher doses of sAPP (50 and 100 ng/ml) dose-dependently decreased βIII-tubulin-positive neurons in the total population of differentiated HNSCs. These results indicate that sAPP released from dying cells promotes differentiation of HNSCs while causing gliogenesis at higher doses.

To confirm the glial differentiation-promoting effect of APP, HNSCs were transfected with mammalian expression vectors containing genes for either wild-type APP or APP differentiated under serum-free unsupplemented conditions. HNSCs transfected with APP revealed a significantly higher level of glial differentiation than did HNSCs transfected with the vector alone at 5 DIV. These results indicate that APP overexpression can also induce glial differentiation of HNSCs, possibly contributing to gliogenesis seen in AD. Bahn et al., reported that stem cells from people with Down's syndrome differentiated into astrocytes rather than neurons [44]. Because Downs' patients have inherited three copies of APP (which resides on chromosome 21), this abnormal differentiation may result from an overdose of APP [45]. In addition to characteristic physical manifestations, Down's syndrome patients often exhibit early-onset AD. Arai et al. suggested that APP plays a role in neuronal development and that the earlier appearance of AD in adult DS patients is associated with an abnormal regeneration process related to aging [46]. Thus, we speculate that transplantation therapy of AD with HNSCs may not be effective in an environment where APP metabolism is altered as it might lead to excessive gliogenesis.

It is not clear whether adult neurogenesis is essential for normal cognitive function during aging. Aged transgenic APP mice exhibit neuronal loss and extensive gliogenesis in the neocortex [47]. Nonetheless, it is tempting to hypothesize that pathologically altered APP metabolism could impair NSC migration and differentiation into the proper ratio of neurons and glia in AD. This pathological APP effect on AD brain may also prevent successful neuro-replacement therapy for AD using NSCs by shifting the differentiation pattern of the transplanted cells to glial cells rather than neurons. APP signaling may be one of the regulatory systems involved in the differentiation of NSCs. More-over, the pathological alteration of APP metabolism in AD may induce glial differentiation of NSCs and lead to exhaustion of the stem cell population. These events may be an important function in the ongoing neurogenesis of the adult brain. Incidentally, this possibility raises the question of whether Aβ immunization, which may also reduce APP fragments, is helpful for maintain-ing stem cell function in AD. Our opinion is that it may be not helpful because HNSCs transplanted into APP knockout mice do not migrate or differentiate effectively into neurons in the cerebral cortex; in contras, in our studies we have seen excellent neural differentiation of transplanted HNSCs in the cerebral cortex of wild-type mice. HNSCs may play an important role in neuroregenera-tion; and if APP is indeed involved in the regulation of HNSCs as we propose, destruction of the APP system may jeopardize the maintenance of brain func-tion. Although the rate of endogenous neuroregeneration in the adult brain may be minimal, in the long run a defect in this process might significantly harm normal brain function.

Conclusion

The first clinical trials of NSC transplantation for AD may be imminent, but this approach will take time to be established as a therapy for AD. We need to know the effects stem cell treatment will have on the AD brain and how AD brain environment affects stem cell biology. Our efforts in stem cell and AD research may well reduce the time needed to develop better treatments for AD.

Acknowledgments These studies were supported by the National Institutes of Health (R01 AG 23472) and the Alzheimer Association (IIRG-03-5577).

References

1. Alvarez-Buylla A, Kirn, Birth JR. migration, incorporation, and death of vocal control neurons in adult songbirds. J Neurobiol 1997;33(5):585–601.
2. Gould E, Reeves AJ, Fallah M, et al. Hippocampal neurogenesis in adult Old World primates. Proc Natl Acad Sci U S A 1999;96(9):5263–5267.
3. Rideout WM 3rd, Eggan K, Jaenisch R. Nuclear cloning and epigenetic reprogramming of the genome. Science 2001;293(5532):1093–1098.
4. Humpherys D, Eggan K, Akutsu H, et al. Epigenetic instability in ES cells and cloned mice. Science 2001;293(5527):95–97.

5. Majumdar MK, Thiede MA, Mosca JD, et al. Phenotypic and functional comparison of cultures of marrow-derived mesenchymal stem cells (MSCs) and stromal cells. J Cell Physiol 1998;176(1):57–66.

6. Pereira RF, Halford KW, O'Hara MD, et al. Cultured adherent cells from marrow can serve as long-lasting precursor cells for bone, cartilage, and lung in irradiated mice. Proc Natl Acad Sci U S A 1995;92(11):4857–4861.

7. Prockop DJ. Marrow stromal cells as stem cells for nonhematopoietic tissues. Science 1997;276(5309):71–74.

8. Pittenger MF, Mackay AM, Beck SC, et al. Multilineage potential of adult human mesenchymal stem cells. Science 1999;284(5411):143–147.

9. Ferrari G, Cusella-De Angelis G, Coletta M, et al. Muscle regeneration by bone marrow-derived myogenic progenitors. Science 1998;279(5356):1528–1530.

10. Makino S, Fukuda K, Miyoshi S, et al. Cardiomyocytes can be generated from marrow stromal cells in vitro. J Clin Invest 1999;103(5):697–705.

11. Petersen BE, Bowen WC, Patrene KD, et al. Bone marrow as a potential source of hepatic oval cells. Science 1999;284(5417):1168–1170.

12. Mackenzie TC, Flake AW. Human mesenchymal stem cells persist, demonstrate site-specific multipotential differentiation, and are present in sites of wound healing and tissue regeneration after transplantation into fetal sheep. Blood Cells Mol Dis 2001;27(3):601–604.

13. Imasawa T, Utsunamiya Y, Kawamura T, et al. The potential of bone marrow-derived cells to differentiate to glomerular mesangial cells. J Am Soc Nephrol 2001;12(7):1401–1409.

14. Liechty KW, MacKenzie TC, Shaaban AF, et al. Human mesenchymal stem cells engraft and demonstrate site-specific differentiation after in utero transplantation in sheep. Nat Med 2000;6(11):1282–1286.

15. Prockop DJ, Azizi SA, Phinney DG, et al. Potential use of marrow stromal cells as therapeutic vectors for diseases of the central nervous system. Prog Brain Res 2000;128:293–297.

16. Bianco P, Riminucci M, Gronthos S, Robey PG. Bone marrow stromal stem cells: nature, biology, and potential applications. Stem Cells 2001;19(3):180–192.

17. Schwarz EJ, Alexander GM, Prockop DJ, Azizi SA. Multipotential marrow stromal cells transduced to produce L-dopa: engraftment in a rat model of Parkinson disease. Hum Gene Ther 1999;10(15):2539–2549.

18. Chopp M, Zhang XH, Li Y, et al. Spinal cord injury in rat: treatment with bone marrow stromal cell transplantation. Neuroreport 2000;11(13):3001–3005.

19. Chen J, Li Y, Chopp M. Intracerebral transplantation of bone marrow with BDNF after MCAo in rat. Neuropharmacology 2000;39(5):711–716.

20. Li Y, Chopp M, Chen J, et al. Intrastriatal transplantation of bone marrow nonhematopoietic cells improves functional recovery after stroke in adult mice. J Cereb Blood Flow Metab 2000;20(9):1311–1319.

21. Kopen GC, Prockop DJ, Phinney DG. Marrow stromal cells migrate throughout forebrain and cerebellum, and they differentiate into astrocytes after injection into neonatal mouse brains. Proc Natl Acad Sci U S A 1999;96(19):10711–10716.

22. Jiang Y, Jahagirdar BN, Reinhardt RL, et al. Pluripotency of mesenchymal stem cells derived from adult marrow. Nature 2002;418(6893):41–49.

23. Terada N, Hamazaki T, Oka M, et al. Bone marrow cells adopt the phenotype of other cells by spontaneous cell fusion. Nature 2002;416(6880):542–545.

24. Ying QL, Nichols J, Evans EP, Smith AG. Changing potency by spontaneous fusion. Nature 2002;416(6880):545–548.

25. Wang SH, Tsai MS, Chang MF, Li H. A novel NK-type homeobox gene, ENK (early embryo specific NK), preferentially expressed in embryonic stem cells. Gene Exp Patterns 2003;3(1):99–103.

26. Chambers I, Colby D, Robertson M, et al. Functional expression cloning of Nanog, a pluripotency sustaining factor in embryonic stem cells. Cell 2003;113(5):643–655.

27. Mitsui K, Tokuzawa Y, Itoh H, et al. The homeoprotein Nanog is required for maintenance of pluripotency in mouse epiblast and ES cells. Cell 2003;113(5):631–642.
28. Booth HA, Holland PW. Eleven daughters of NANOG. Genomics 2004;84(2):229–238.
29. Hart AH, Hartley L, Ibrahim M, Robb L. Identification, cloning and expression analysis of the pluripotency promoting Nanog genes in mouse and human. Dev Dyn 2004;230(1):187–198.
30. Pan GJ, Pei DQ. Identification of two distinct transactivation domains in the pluripotency sustaining factor nanog. Cell Res 2003;13(6):499–502.
31. Richards M, Tan SP, Tan JH, et al. The transcriptome profile of human embryonic stem cells as defined by SAGE. Stem Cells 2004;22(1):51–64.
32. Sato N, Meijer L, Skaltsounis L, et al. Maintenance of pluripotency in human and mouse embryonic stem cells through activation of Wnt signaling by a pharmacological GSK-3-specific inhibitor. Nat Med 2004;10(1):55–63.
33. Ying QL, Nichols J, Chambers I, Smith A. BMP induction of Id proteins suppresses differentiation and sustains embryonic stem cell self-renewal in collaboration with STAT3. Cell 2003;115(3):281–292.
34. LeBlanc A, Liu H, Goodyer C, et al. Caspase-6 role in apoptosis of human neurons, amyloidogenesis, and Alzheimer's disease. J Biol Chem 1999;274(33):23426–23436.
35. Piccini A, Ciotti MT, Vitolo OV, et al. Endogenous APP derivatives oppositely modulate apoptosis through an autocrine loop. Neuroreport 2000;11(7):1375–1379.
36. Hugon J, Esclaire F, Lesort M, et al. Toxic neuronal apoptosis and modifications of tau and APP gene and protein expressions. Drug Metab Rev 1999;31(3):635–647.
37. Hung AY, Koo EH, Haass C, et al. Increased expression of beta-amyloid precursor protein during neuronal differentiation is not accompanied by secretory cleavage. Proc Natl Acad Sci U S A 1992;89(20):9439–9443.
38. Murakami N, Yamaki T, Iwamoto Y, et al. Experimental brain injury induces expression of amyloid precursor protein, which may be related to neuronal loss in the hippocampus. J Neurotrauma 1998;15(11):993–1003.
39. Ishiguro M, Ohsawa I, Takamura C, et al. Secreted form of beta-amyloid precursor protein activates protein kinase C and phospholipase Cgamma1 in cultured embryonic rat neocortical cells. Brain Res Mol Brain Res 1998;53(1-2):24–32.
40. Ohsawa I, Takamura C, Kohsaka S. Fibulin-1 binds the amino-terminal head of beta-amyloid precursor protein and modulates its physiological function. J Neurochem 2001;76(5):1411–1420.
41. Greenberg SM, Koo EH, Selkoe DJ, et al. Secreted beta-amyloid precursor protein stimulates mitogen-activated protein kinase and enhances tau phosphorylation. Proc Natl Acad Sci U S A 1994;91(15):7104–7108.
42. Miyachi T, Asai K, Tsuiki H, et al. Interleukin-1beta induces the expression of lipocortin 1 mRNA in cultured rat cortical astrocytes. Neurosci Res 2001;40(1):53–60.
43. Brannen CL, Sugaya K. In vitro differentiation of multipotent human neural progenitors in serum-free medium. Neuroreport 2000;11(5):1123–1128.
44. Bahn S, Mimmack M, Ryan M, et al. Neuronal target genes of the neuron-restrictive silencer factor in neurospheres derived from fetuses with Down's syndrome: a gene expression study. Lancet 2002;359(9303):310–315.
45. Sawa A. Neuronal cell death in Down's syndrome. J Neural Transm Suppl 1999;57:87–97.
46. Arai Y, Suzuki A, Mizuguchi M, et al. Developmental and aging changes in the expression of amyloid precursor protein in Down syndrome brains. Brain Dev 1997;19(4):290–294.
47. Bondolfi L, Calhoun M, Ermini F, et al. Amyloid-associated neuron loss and gliogenesis in the neocortex of amyloid precursor protein transgenic mice. J Neurosci 2002;22(2):515–522.

Chapter 26
Oral Aβ Vaccine Using a Recombinant Adeno-Associated Virus Vector in an Alzheimer's Disease Mouse Model

Takeshi Tabira and Hideo Hara

Introduction

Alzheimer's disease (AD) is characterized by progressive loss of cognitive function due to β-amyloid deposits in the central nervous system [1]. Immunization of amyloid precursor protein (APP)-transgenic mice with synthetic Aβ in complete and, subsequently, incomplete Freund's adjuvant showed a marked reduction of amyloid burden in the brain [2]. Repetitive passive transfer of Aβ antibodies [3] was also effective for reducing the amyloid deposits. Although Aβ is not an infectious agent, this treatment is now widely accepted as a "vaccination" from its analogous mechanism. Because vaccinated mice showed diminished memory [4,5], clinical trials were performed in humans in the United States and Europe. The Phase II trial of AN-1792 vaccine composed of synthetic Aβ1-42 and adjuvant QS21 was halted because of the complication of subacute meningoencephalitis, which appeared in 6% of patients [6]. However, autopsy cases with or without the complication suggested effective clearance of Aβ following its vaccination [7–9], and patients who produced antibodies against senile plaque amyloid showed better cognitive function, or less cognitive decline, than those who did not produce such antibodies [10,11]. Therefore, Aβ vaccination seems to be a promising way to delay the onset or to slow the progression of AD if the complication is minimized. We have developed an oral Aβ vaccine using the recombinant adeno-associated virus vector (AAV), which successfully reduced the amyloid burden in tg2576 APP transgenic mice without any complications [12].

T. Tabira
National Institute for Longevity Sciences NCGG, Morioka, Obu City,
Aichi 474-8522, Japan

A. Fisher et al. (eds.), *Advances in Alzheimer's and Parkinson's Disease*,
© Springer 2008

Results

Production of AAV/Aβ Vaccine

The AAV vector carrying human Aβ 1-43 or Aβ ~1-21 was constructed using plasmid DNA pTRUF2, and the secreted form of Aβ was made by linking the APP signal sequence (SS) to the Aβ sequence (Fig. 1a). Human embryonic kidney (HEK), 293 cells, were co-transfected with SS-Aβ pTRUF2 and plasmid pXX2 and pXX6 as described [13] (Fig. 1b). Recombinant AAV titers were in the range of 1×10^{13} to 2×10^{13} viral genomes/ml.

To confirm the secretion of Aβ, we transfected HEK293 cells with the SS-Aβ 1-43 pTRUF2 expression vector. An immunoprecipitation and Western blot method revealed Aβ monomer in the cell lysate and Aβ oligomers in the conditioned medium (Fig. 2). When AAV/Aβ was given to mice, Aβ expression was observed primarily in the lamina propria of the upper part of the small intestine. There was no increase of Aβ 1-43 or Aβ 1-21 in the serum. Transduction of AAV was confirmed by the polymerase chain reaction (PCR) in intestinal cells but not in the liver, spleen, heart, lung, or kidney 4, 11, and 21 weeks after treatment, suggesting an absence of widespread infection with the virus vector.

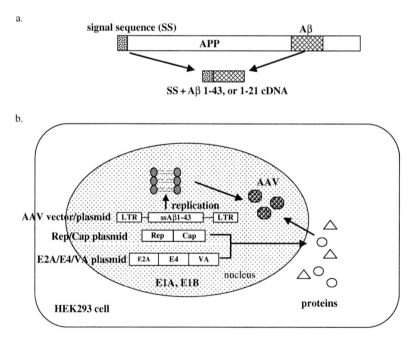

Fig. 1 cDNA construct of Aβ vaccine and its production. **a** To make a secreted form of Aβ, a signal sequence of APP is ligated to Aβ 1–43 or Aβ 1–21 cDNA. **b** AAV vaccine carrying Aβ cDNA was produced in HEK293 cells transfected with three plasmids, shown in the figure

Fig. 2 Western blot of Aβ
vaccine-infected cells.
HEK293 cells were infected
with AAV/Aβ, and a
Western blot was conducted
with the cell lysate and
supernatant. The cell lysate
shows 4 kDa Aβ monomer,
and the supernatant shows
an aggregated form of Aβ.
Antibody: 4G8

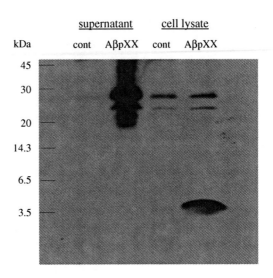

Vaccination and Tissue Examinations

AAV/Aβ 43 or AAV/Aβ 21 diluted with phosphate-buffered saline (PBS) and 5×10^{11} genome in a final volume of 0.1 ml were administered once to tg2576 mice (Taconic) using, in our preliminary study, an orogastric tube in the treated group; control mice received 0.1 ml of PBS once. Because immunization with vector alone did not show any significant difference, we used PBS as the control. Mice were randomized to four groups: a group treated at age 15 weeks; a group treated at age 30 weeks; a group treated at age 45 weeks; and control groups treated with PBS at each age. Each group consisted of four to six mice. At the age of 56 weeks all mice were anesthetized with Nembutal, and their brains were extirpated and fixed in 4% paraformaldehyde with 0.1 M phosphate buffer, pH 7.6, for immunohistochemical analysis.

Oral vaccination with AAV/Aβ 43 or AAV/Aβ 21 resulted, at age 56 weeks, in a marked reduction of Aβ deposition in all treated groups compared to the controls examined (Fig. 3). Quantitative image analyses in three regions of the brain showed a significant decrease of Aβ burden in all vaccinated mice compared to control mice.

Hematoxylin and eosin (H&E) staining of the brain sections of the treated mice showed no lymphocytic infiltration in either leptomeninges or cerebral white matter, and there was no hemorrhagic lesion in the brain. Immunohistochemical studies did not reveal any cellular infiltration positive for CD3, CD4, CD86, CD19, or CD11b in brain sections. Iba-1-positive activated microglia were more numerous in vaccinated mice, and some microglia cells containing phagocytosed Aβ were observed. In contrast, GFAP-positive cells were less frequent in vaccinated mice.

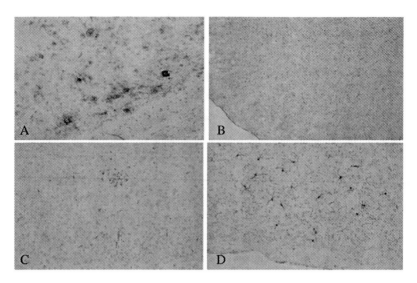

Fig. 3 Aβ burden in the brain of control and vaccinated mice. Aβ burden was significantly reduced in mice that had received AAV/Aβ vaccine at 15 weeks (**B**), 30 weeks (**C**), and 45 weeks (**D**) older than the controls (**A**). A disrupted plaque showing a group of activated microglia positive for Aβ staining is seen in C

Immune Responses to Aβ

In the treated tg2576 mice, IgG antibodies were detected in the serum at 4 weeks, and they remained elevated for more than 6 months (Fig. 4a). The antibody isotypes were mainly IgG1 and to a lesser extent IgG2b; but IgG2a was not detected, and IgA was low. The immune sera from vaccinated mice stained the amyloid plaques in the brain (Fig. 4c). The proliferative response of spleen T cells against Aβ peptide was not detected in the vaccinated mice or in control mice.

Aβ Vaccine-Mediated Meningoencephalitis

The exact mechanism of meningoencephalitis in patients who received AN-1792 vaccine is not known yet. The autopsied brain showed cellular infil-trates composed mainly of CD4+ T cells and CD8+ T cells. The magnetic resonance imaging (MRI) findings were similar to postvaccinal encephalomye-litis, although gray matter lesions were more pronounced in AN-1792-related meningoencephalitis. It has been reported that there are Aβ-reactive T cells in the human peripheral blood, with the frequency higher in the elderly [14]. Thus, it is highly probable that the AN-1792-related meningoencephalitis is autoim-mune encephalitis, probably due to Th1 immune responses to Aβ. If this is the case, it may be possible to induce similar conditions experimentally in animals.

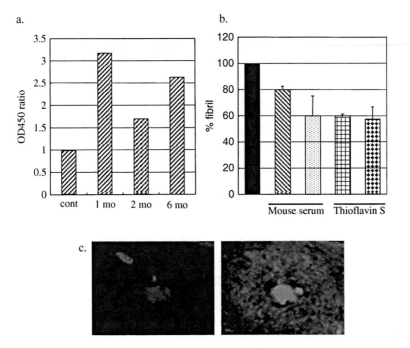

Fig. 4 Antibody responses in AAV/Aβ vaccine-treated mice. Oral AAV/Aβ vaccine was given once, and serum antibodies were measured by an enzyme-linked immunosorbent assay (ELISA). **a** The antibody to Aβ was well elevated 1 month after vaccination, and the levels were still high 6 months after vaccination. **b** The serum from vaccinated mice inhibited Aβ aggregation as strongly as thioflavin S. **c** The serum from vaccinated mice stained senile plaque amyloid in an Alzheimer patient's brain. *Left*, mouse serum. *Right*, thioflavin S

Although there is a report showing experimental meningoencephalitis in B6 mice immunized with Aβ [15], we and others could not confirm that observation. However, Monsonego et al. could induce encephalitis in APP mice crossed with interferon-γ transgenic mice, which have augmented Th1 immune responses (A. Monsonego, personal communication). Thus, it is reasonable to assume that the meningoencephalitis is mediated by autoimmune Th1 T cells reactive to Aβ.

Gut Immune System and Advantage of AAV/Aβ Vaccine

It is well known that the gut immune system is strongly shifted to Th2 (Fig. 5). There are two types of T-helper cells, Th1 and Th2. Th1 cells mainly help cellular immune responses and suppress Th2 cells. It is known that effector T cells for autoimmune encephalomyelitis are of the Th1 type. In contrast, Th2 T cells mainly help humoral immune responses and suppress Th1 cells. Thus, the use of the gut immune system has a big advantage for inducing antibodies and suppressing adverse T-cell immune responses. Because anti-Aβ

Fig. 5 Gut immune system. The gut immune system is strongly shifted to Th2, which helps the humoral immune response and suppressed Th1 response

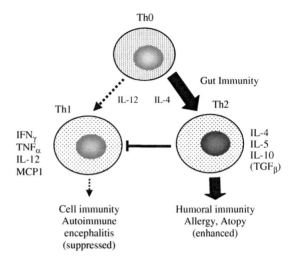

antibodies were continuously elevated for more than 6 months in mice that received our oral Aβ vaccine, it might be sufficient for patients to take the oral vaccine once or twice a year. In addition, AAV was detected only in the gut, without spreading to other organs, including germ cells. Adeno-associated viral DNA normally does not integrate into the cellular genome; instead, it remains in the episomal region. Moreover, because the turnover of epithelial cells of the gastrointestinal tract is relatively quick, the recombinant AAV is eliminated during the course of renewal of the epithelial cells, suggesting lower risk in case of an unexpected event. It is interesting that most epithelial cells in the murine and probably the human gut are exfoliated into the gut lumen, whereas those of the guinea pig and monkey gut are engulfed by macrophages in the lamina propria [16] (Fig. 6). Thus, most Aβ cDNA in epithelial cells seems to be deleted along with exfoliation of gut epithelial cells in humans.

Fig. 6 Renewal of gut epithelium. Gut epithelial cells are renewed in a few days by exfoliation in mice, rats, and humans, whereas they are engulfed by macrophages in guinea pigs and monkeys. (From Iwanaga [16], with permission)

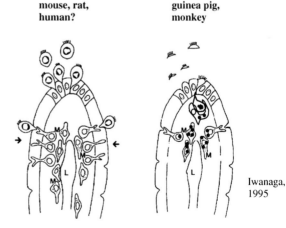

Mechanism of Aβ Vaccine

The mechanism by which Aβ vaccine clears β-amyloid from the brain tissue is still unknown. There are several hypotheses: (1) There is Fc receptor-mediated uptake of Aβ-antibody complexes by activated microglia [4]. (2) Antibody-mediated disaggregation of amyloid fibrils occurs [17]. Several reports have indicated that therapeutically active antibodies recognize mainly the residue 4-10 of Aβ peptide, and that these antibodies inhibit Aβ fibrillogenesis and cytotoxicity [18–20]. (3) DeMattos et al [21]. hypothesized that injected antibodies sequester Aβ from the peripheral blood and eventually pull Aγ out of the brain. In our vaccinated mice, some activated microglia contained Aβ, and sera from vaccinated mice showed inhibition of Aβ aggregation (Fig. 4b). Thus, all three mechanisms seem to be likely with our vaccine.

Conclusion

The oral Aβ vaccine described herein appears to be safe and beneficial for AD. We are currently testing this vaccine in monkeys.

Acknowledgments This work was supported partially by grants from the Ministry of Health, Welfare, and Labor (Medical Frontier) and the Organization for Pharmaceutical Safety and Research (MF-3). The authors are grateful to Dr. Xiao Xiao at the University of Pittsburgh for providing us with the AAV vector.

References

1. Selkoe DJ. Alzheimer's disease: genes, proteins and therapies. Physiol Rev 2001;81:742–761.
2. Schenk D, Barbour R, Dunn W, et al. Immunization with amyloid-β attenuates Alzheimer-disease-like pathology in the PDAPP mouse. Nature 1999;400:173–177.
3. Bard F, Cannon C, Barbour R, et al. Peripherally administered antibodies against amyloid beta-peptide enter the central nervous system and reduce pathology in a mouse model of Alzheimer disease. Nat Med 2000;6:916–919.
4. Janus C, Pearson J, McLaurin J, et al. Ab peptide immunization reduces behavioral impairment and plaques in a model of Alzheimer's disease. Nature 2000;408:979–982.
5. Morgan D, Diamond DM, Gottschall PE, et al. Aβ peptide vaccination prevents memory loss in an animal model of Alzheimer's disease. Nature 2000;408:982–985.
6. Orgogozo JM, Gilman S, Dartigues JF, et al. Subacute meningoencephalitis in a subset of patients with AD after Ab42 immunization. Neurology 2003;61:46–54.
7. Nicoll JA, Wilkinson D, Holmes C, et al. Neuropathology of human Alzheimer disease after immunization with amyloid-β peptide: a case report. Nat Med 2003;9:448–452.
8. Ferrer I, Boada Rovira M, Sanchez Guerra ML, et al. Neuropathology and pathogenesis of encephalitis following amyloid-beta immunization in Alzheimer's disease. Brain Pathol 2004;14:11–20.

9. Masliah E, Hansen L, Adame A, et al. Aß vaccination effects on plaque pathology in the absence of encephalitis in Alzheimer disease. Neurology 2005;64:129–131.
10. Hock C, Konietzlo U, Streffer JR, et al. Antibodies against beta-amyloid slow cognitive decline in Alzheimer's disease. Neuron 2003;38:547–554.
11. Gilman S, Koller M, Black RS, et al. Clinical effects of Ab immunization (AN1792) in patients with AD in an interrupted trial. Neurology 2005;64:1553–1562.
12. Hara H, Monsonego A, Yuasa K, et al. Development of a safe oral Abeta vaccine using recombinant adeno-associated virus vector for Alzheimer's disease. J Alzheim Dis 2004;6:483–488.
13. Xiao X, Li J, Samulski RJ. Production of high-titer recombinant adeno-associated virus vectors in the absence of helper adenovirus. J Virol 1998;72:2224–2232.
14. Monsonego A, Zota V, Kami A, et al. Increased T cell reactivity to amyloid beta protein in older humans and patients with Alzheimer disease. J Clin Invest 2003;112:415–422.
15. Furlan R, Brambilla E, Sanvito F, et al. Vaccination with amyloid-beta peptide induces autoimmune encephalomyelitis in C57/BL6 mice. Brain 2003;126:285–291.
16. Iwanaga T. The involvement of macrophages and lymphocytes in the apoptosis of enterocytes. Arch Histol Cytol 1995;58:151–159.
17. McLaurin J, Cecal R, Kierstead ME, et al. Therapeutically effective antibodies against amyloid-beta peptide target amyloid-beta residues 4-10 and inhibit cytotoxicity and fibrillogenesis. Nat Med 2002;8:1263–1269.
18. Frenkel D, Katz O, Solomon B. Immunization against Alzheimer's beta-amyloid plaques via EFRH phage administration. Proc Natl Acad Sci USA 2000;97:11455–11459.
19. Bard F, Barbour R, Cannon C, et al. Epitope and isotype specificities of antibodies to β-amyloid peptide for protection against Alzheimer's disease-like neuropathology. Proc Natl Acad Sci U S A 2003;100:2023–2028.
20. DeMattos RB, Bales KR, Cummins DJ, et al. Peripheral anti-Aβ antibody alters CNS and plasma Aβ clearance and decreases brain Aβ burden in a mouse model of Alzheimer's disease. Proc Natl Acad Sci U S A 2001;98:8850–8855.
21. DeMattos RB, Bales KR, Cummins DJ, et al. Brain to plasma amyloid-β efflux: a measure of brain amyloid burden in a mouse model of Alzheimer's disease. Science 2002;295:2264–2267.

Chapter 27
In Vivo Targeting of Amyloid Plaques Via Intranasal Administration of Phage Anti-β-Amyloid Antibodies

Beka Solomon

Introduction

Alzheimer's disease (AD) is characterized neuropathologically by progressive deposition of β-amyloid protein (AβP) 40–42 residues in senile plaques. This is one of the main hallmarks of the disease, occurring in brain parenchyma and cerebral blood vessels [1].

In vitro performance of AβP-specific antiaggregating monoclonal antibodies suggests their application in AD therapy [2,3]. Delivery difficulties of antibodies as drugs are accentuated when the brain is the target organ because access for macromolecules from the circulation is generally prevented by the presence of the blood-brain barrier (BBB) and the blood-cerebrospinal fluid (CSF) barrier [4]. Techniques of genetic engineering have been applied to minimize the size of the antibodies from 135–900 to 25 kDa[5] while maintaining their biological activity. These single-chain antibodies (ScFvs) can be displayed on the surface of a phage for further manipulation or can be used as soluble ScFv molecules. Anti-AβP ScFv antibodies devoid of the Fc region maintain the properties of the whole antibodies, recognize AβP in vivo and in vitro, and preserve the antiaggregating properties toward β-amyloid (Aβ).

We have proposed that phages can serve as antibody brain delivery systems [6]. Because of its linear structure, the filamentous phage has high permeability to various membranes [7] and may reach the affected sites in the brain via the olfactory tract.

Phages have distinct advantages over animal viruses as delivery vehicles. They are simple systems whose large-scale production and purification is highly efficient and less expensive than that of animal viral vectors. Having evolved for prokaryotic infection, assembly, and replication, the bacteriophage cannot replicate in mammalian cells.

B. Solomon
Department of Molecular Microbiology and Biotechnology, George S. Wise Faculty of Life Sciences, Tel Aviv University, Ramat Aviv, Tel Aviv 69978, Israel

A. Fisher et al. (eds.), *Advances in Alzheimer's and Parkinson's Disease*,
© Springer 2008

Here we show that repeated intranasal administration of filamentous phage anti-Aβ antibody fragments into Alzheimer's APP transgenic mice enables Aβ in the olfactory bulb, the hippocampus region, and various brain regions to dissolve.

In Vivo Targeting of Amyloid Plaques

Intranasal administration of the phage as a delivery vector of anti-Aβ antibody fragment into Alzheimer's APP transgenic mice enabled in vivo targeting of the plaques. Similar to the parent whole antibody, the ability of the ScFv antibody to prevent AβP toxicity toward PC12 cells is dose-dependent [8].

Aβ deposits were visualized in the olfactory bulb and hippocampus region [6]. The hippocampus is one of the first affected regions during AD, and Aβ plaques can be detected in this locus before any clinical signs are observed [9].

Aβ plaques were visualized by both Tioflavin-S (ThS) and fluorescence-labeled anti-phage antibodies. After intranasal phage administration of ScFv in these mice, Aβ brain plaques were specifically labeled in the two brain sections (olfactory and hippocampus) where most of the early phages are located. The stained plaques, detected in vitro by immunofluorescence techniques, show high sensitivity, similar to those from human AD patients and transgenic mice. The filamentous phage maintains the biological activity of a displayed foreign molecule of anti-AβP ScFv and efficiently penetrates biological membranes. No evidence of the phage was found in those specific brain regions after phage immunization via intraperitoneal administration, strongly indicating that the olfactory neuron route may target the plaques to specific regions.

The intranasal route of administration may provide a simple, practical, rapid method of delivery of therapeutic agents into the central nervous system (CNS) because of the unique connection between the nose and the brain. Intranasal delivery of agents to the CSF is not surprising, as CSF normally drains along the olfactory axon bundles as they traverse the cribriform plate of the skull and approach the olfactory submucosa in the roof of the nasal cavity, where the CSF is then diverted into the nasal lymphatics [10]. The olfactory neural pathway provides both intraneuronal and extraneuronal pathways into the brain.

Passive Immunization of Transgenic Mice with Phage-ScFv Against β-Amyloid Peptide

To demonstrate the ability of the anti-Aβ ScFv-phage to dissolve Aβ in vivo, hAPP transgenic mice were exposed to intranasal administration of the filamentous phage with an anti-Aβ antibody fragment.

The experiments were performed on transgenic (Tg) mice carrying the hAPP gene with both London (717) and Swedish (670/671) mutations. The 9- to 10-month-old Tg mice were treated with ScFv-phage. Administration of 50 μl containing 1011 phages per mouse took place every 3 weeks for a total of 6 months. Following the passive immunization protocol, the mice were subjected to training for the Morris water maze test for 4 days. The cognitive average of the treated animals was close to that of the non-Tg animals, indicating a healthy pattern of learning and memorizing the learned information [11].

Targeting β-Amyloid Plaques

Intranasal treatment for 6 months with phage anti-AβP ScFv of Tg mice over-expressing hAPP resulted in reduction and/or total elimination of the plaque load and considerably reduced brain inflammation (Fig. 1). Thioflavin-S staining of mouse brain sections showed that the plaque load (defined as the area occupied by amyloid plaques divided by the total brain-section area) in the treated mice was an average 50% of the plaque load in the control Tg mice.

Disappearance of filamentous phage from the brain without inducing adverse effects, as shown by histological studies, as well as considerable improvement in cognitive function suggest that the intranasal route of administration may provide a simple, practical, rapid method for delivering therapeutic agents into the CNS while avoiding the BBB.

Here we demonstrate the feasibility of a novel delivery approach of antibodies raised against the AβP displayed on filamentous bacteriophage, to the CNS via intranasal administration. Intranasal administration was chosen as a direct delivery route of vectors to the CNS via the olfactory system or by close neuronal tissue. Olfactory receptor neurons are bipolar cells that reside in the epithelial lining of the nose, high in the nasal cavity. Their axons traverse

AMYLOID BURDEN (%)

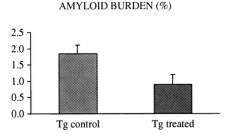

Fig. 1 Anti-β-amyloid ScFv-treated mice showed reduced plaque load. Brain sections of the phage ScFv-treated mice stained with thioflavin-S were examined. Mice in the control group contained plaque load values ranging from 1.55% to 2.20%. The treatment resulted in a reduction of about 50% of the plaque load compared to that in untreated transgenic (Tg) animals

the cribriform plate and project to the first synapse of the olfactory pathway in the olfactory bulb of the brain. They appear to form a "highway" by which viruses or other transported substances may gain access to the CNS [12].

Intranasal application presents several important advantages. It provides a simple, rapid delivery route for therapeutic agents into the CNS because of the unique connection between the nose and brain. The olfactory system has been shown to be one of the first brain parts affected by AD; thus, it is targeted first. Following the olfactory tract, phage-displayed antibodies may directly reach the hippocampal area, another region affected during the early stages of the disease. Substances delivered by intranasal administration are likely to have a longer brain residence time than those delivered by intravenous administration and to exert their effect on the brain rather than on peripheral tissues.

In experiments reported previously [13], filamentous phages were injected intravenously into mice and were subsequently rescued from the different organs, showing that their integrity was not affected during membrane penetration into the organs. Filamentous bacteriophages offer an obvious advantage over other vectors. The filamentous phages M13, f1, and fd are well understood at both structural and genetic levels [14]. They were genetically engineered to display antigen and/or antibody and were used in various biological systems to present foreign proteins on their surfaces [7,15]. We showed a direct correlation between the number of applications and the amount of phage detected in the brain in both regions. The linear structure of the phage is suggested to confer penetration properties via various membranes. There was no evidence of filamentous phage spreading to other brain sections, which strongly emphasizes the olfactory tract as the most probable path in this case. In a control experiment, we performed intranasal administration of chloroform-treated spheroid phages in mice under the same experimental conditions, and no presence of phages was detected [9].

Clearance of phage was evaluated by immunohistochemistry (IHC) and titered assays. Most organs were IHC-negative for phage within 3 days. Virtually all were negative by 3 weeks.16 The half-life of phages in plasma was 3.6 hours. After 72 hours phage was cleared from the blood, mainly through hepatic and renal excretion.

Several hypotheses may be considered regarding the disappearance of filamentous phage from the brain without inducing a toxic effect, as shown in histology studies, as well as the long life-span of challenged animals. As in other reported cases, immune mechanisms may be involved that activate scavenger cells as microglia [17,18].

The genetically engineered filamentous bacteriophage proved to be an efficient, nontoxic viral delivery vector to the brain, offering an obvious advantage over other mammalian vectors. The bacteriophage lacks the ability to infect mammalian cells unless designed to do so. Due to its structure, the filamentous phage is highly permeable to various membranes; and following the olfactory tract, it may directly reach the hippocampal area via the limbic system to target affected sites. To our knowledge, this is the first demonstration that filamentous bacteriophage exhibits an ability to penetrate the CNS while preserving both the inert properties of the vector and the ability to carry foreign molecules.

Six months' treatment with phage ScFv of transgenic mice overexpressing hAPP resulted in a reduction and/or total elimination of the plaque load as well as considerable reduction of brain inflammation. Disappearance of filamentous phage from the brain without inducing adverse effects, as shown by histological studies, and considerable improvement in cognitive function suggests that the intranasal route of administration may provide a simple, practical, rapid way to deliver therapeutic agents into the CNS, thereby avoiding the BBB. The results described above demonstrate that entry of anti-AβP antibody-phage into the brain dissolves amyloid plaques in the olfactory and other brain regions, suggesting that an immunotherapeutic approach with phage-displaying recombinant anti-AβP antibodies may serve as a safe and potent tool for reducing disease symptoms.

Acknowledgments The author thanks Rachel Cohen-Kupiec for phage anti-AβP ScFv; Vered Lavie, Maria Becker, and Orna Goren for neuropathological staining; Rela Koppel for technical assistance; and Faybia Margolin for editing.

References

1. Selkoe DJ. Toward a comprehensive theory for Alzheimer's disease: hypothesis—Alzheimer's disease is caused by the cerebral accumulation and cytotoxicity of amyloid beta-protein. Ann NY Acad Sci 2000;924:17.
2. Solomon B, Koppel R, Hanan E, Katzav T. Monoclonal antibodies inhibit in vitro fibrillar aggregation of the Alzheimer β-amyloid peptide. Proc Natl Acad Sci U S A 1996;93:452.
3. Solomon B, Koppel R, Frenkel D, Hanan-Aharon E. Disaggregation of Alzheimer β-amyloid by site-directed mAb. Proc Natl Acad Sci U S A 1997;94:4109.
4. Brightman MW, Ishihara S, Chang L. Penetration of solutes, viruses and cells across the blood brain barrier. Curr Top Microbiol Immunol 1995;202:63.
5. Winter G, Griffiths AD, Hawkins RE, Hoogenboom HR. Making antibody by phage display technology. Annu Rev Immunol 1994;12:433.
6. Frenkel D, Solomon B. Filamentous phage as vector-mediated antibody delivery to the brain. Proc Natl Acad Sci U S A 2002;16:99(8):5675.
7. McCafferty J, Griffith AD, Winter G, Chiswell DJ. Phage antibodies: filamentous phage displaying antibody variable domains. Nature 1990;348:552.
8. Frenkel D, Solomon B, Benhar I. Modulation of Alzheimer's β-amyloid neurotoxicity by site-directed single-chain antibody. J Neuroimmunol 2000;106:23.
9. Naslund J, Haroutunian V, Mohs R, et al. Correlation between elevated levels of amyloid beta-peptide in the brain and cognitive decline. JAMA 2000;283:1571.
10. Thorne RG, Emory CR, Ala TA, Frey WH Jr. Quantitative analysis of the olfactory pathway for drug delivery to the brain. Brain Res 1995;692:278.
11. Lavie V, Becker M, Cohen-Kupiec R, et al. EFRH-phage immunization of Alzheimer's disease animal model improves behavioral performance in Morris water maze trials, J Mol Neurosci 2004;24(1):105.
12. Draghila R, Caillaud C, Manicom R, et al. Gene delivery into the central nervous system by nasal instillation in rats. Gene Ther 1995;2:418.

13. Pasqualini R, Ruoslahti E. Organ targeting in vivo using phage display peptide libraries. Nature (Lond) 1996;380:364.
14. Greenwood J, Willis EA, Perham NR. Multiple display of foreign peptides on a filamentous bacteriophage: peptides from Plasmodium falciparum circumsporozoite protein as antigens. J Mol Biol 1991;220:821.
15. Scott JK, Smith GP. Searching for peptide ligands with an epitope library. Science 1990;249:386.
16. Zou J, Dickerson MT, Owen NK, et al. Biodistribution of filamentous phage peptide libraries in mice. Mol Biol Rep 2004;31:121.
17. Kreutzberg GW. Microglia: a sensor for pathological events in the CNS. Trends Neurosci 1996;19:312.
18. Chan A, Magnus T, Gold R. Phagocytosis of apoptotic inflammatory cells by microglia and modulation by different cytokines: mechanism for removal of apoptotic cells in the inflamed nervous system. Glia 2001;33:87.

Chapter 28
Decreased ProBDNF: The Cause of Alzheimer's-Associated Neurodegeneration and Cognitive Decline?

Margaret Fahnestock, S. Peng, D.J. Garzon, and Elliott J. Mufson

Introduction

Brain-derived neurotrophic factor (BDNF) is synthesized widely throughout the brain but in particularly high amounts in the cortex and hippocampus [1]. BDNF supports the survival and function of the neurons that are severely affected in Alzheimer's disease (AD). This includes basal forebrain cholinergic and entorhinal cortex neurons, circuits critical for learning and memory [2]. BDNF synthesized and secreted by neurons in the cortex and hippocampus is captured by high-affinity TrkB receptors on the axon terminals of basal forebrain and entorhinal cortex neurons and is transported in a retrograde manner back to the cell bodies [3]. A small but continuous supply of BDNF is required by these neurons to maintain survival and differentiation [4,5]. BDNF is also required for the survival and function of hippocampal and cortical neurons [6,7]. It can be transported anterogradely by these neurons and secreted in an activity-dependent fashion [8], and has been shown to provide support to hippocampal neurons via both autocrine and paracrine interactions [9]. BDNF influences excitability of hippocampal and cortical neurons and has been shown to be important for synaptic plasticity and for learning and memory [10–14].

BDNF is synthesized as a pre-proprotein. Following cleavage of the signal sequence in the endoplasmic reticulum, proBDNF, with a molecular weight of approximately 36 kDa, is secreted in both basal and activity-dependent fashion and can be processed both intra- and extracellularly to produce mature BDNF [15–19]. ProBDNF is biologically active, as it has been shown to bind and activate TrkB [15,20]. Both pro and mature forms of BDNF are present in human brain [21].

M. Fahnestock
Department of Psychiatry and Behavioural Neurosciences, McMaster University,
Hamilton, Canada

A. Fisher et al. (eds.), *Advances in Alzheimer's and Parkinson's Disease*, 279
© Springer 2008

In AD, the loss of synapses correlates with the degree of dementia [22–24]. The cause of selective loss of synapses in AD is not understood, although there are a variety of theories, including oxidative damage, mitochondrial damage, and down-regulation of survival signaling pathways. The studies presented in this chapter explore the hypothesis that decreased proBDNF and BDNF cause loss of synaptic and neuronal function in AD. To support this hypothesis it is necessary to demonstrate that: (1) BDNF is decreased in AD; (2) AD patho-physiology could cause BDNF down-regulation; (3) BDNF down-regulation occurs early enough to be responsible for functional deficits in AD; and (4) decreased BDNF correlates with cognitive decline.

Results

Decreased BDNF in AD

We have previously shown that BDNF mRNA is decreased 3.4-fold in AD parietal cortex compared to control subjects [25]. A comparable decrease has also been demonstrated in AD hippocampus [26]. Consequently, proBDNF is decreased 40% in Alzheimer's disease parietal cortex [21,27], and reductions in mature BDNF of 23% to 62% have been reported in cortical and hippocampal AD tissues [28,29].

The decrease in BDNF mRNA, however, does not represent global down-regulation of BDNF. There are at least seven BDNF transcripts expressed in human brain tissue and more if splice variants are taken into account [30–32]. However, only three of these—specifically, transcripts 1, 2, and 4 (the latter previously known as transcript 3)—are down-regulated in AD brain compared to controls [31]. What biochemical change in the AD brain is responsible for specific down-regulation of BDNF transcripts 1, 2, and 4?

AD Pathophysiology and BDNF Down-regulation

According to the amyloid cascade hypothesis [33], aggregated β-amyloid (Aβ) is thought to be responsible for the neuronal and synaptic loss and cognitive dysfunction in AD. However, the mechanism by which Aβ might cause these deficits is unclear. If decreased BDNF is the mechanism, Aβ should cause down-regulation of BDNF. We tested this in a cell culture model using the human neuroblastoma cell line SH-SY5Y differentiated with retinoic acid. We found that 2 days of exposure to a nontoxic dose of Aβ 1–42 induced significant down-regulation of BDNF in these cells. We have also demonstrated specific down-regulation of BDNF transcript 4 by Aβ [34].

Early BDNF Down-regulation and Functional Deficits in AD

For down-regulation of BDNF to be a cause rather than an effect of neuronal dysfunction and neurodegeneration, down-regulation of BDNF must occur early in AD. To test this, we quantified proBDNF and BDNF protein in parietal cortex of subjects classified as normal (NCI), mildly cognitively impaired (MCI), or AD. We found that proBDNF and BDNF were both significantly reduced even in the MCI subjects [29].

Decreased BDNF and Cognitive Decline

Interestingly, transgenic mice with 50% decreased levels of BDNF exhibit defects in long-term potentiation (LTP), a model of memory formation [11,35,36]. These data demonstrate that the similarly reduced levels of BDNF expression in AD are sufficient to compromise cognitive function. In support of this, we have shown that proBDNF and BDNF protein levels are positively correlated with cognitive scores, Mini-Mental State Examination (MMSE) scores, and Global Cognitive Score (GCS) in normal, MCI, and AD subjects [29].

Conclusions

We have shown that: (1) BDNF mRNA transcripts 1, 2, and 4 are selectively decreased in AD, indicating targeted down-regulation; (2) BDNF mRNA is down-regulated by Aβ 1–42 in human neuroblastoma cells; (3) BDNF transcript 4, the predominant transcript in the human brain, is reduced in AD and is down-regulated by Aβ; (4) proBDNF and BDNF levels decrease during the preclinical stages of AD; and (5) proBDNF and BDNF levels correlate with cognitive decline. Because Aβ down-regulates the major BDNF transcript affected in vivo, these data suggest that Aβ is likely to be at least partially responsible for decreased BDNF expression in AD. Furthermore, because decreased BDNF synthesis occurs early in the disease and is correlated with cognitive dysfunction, our data strongly suggest that a lack of proBDNF and BDNF contributes to synaptic loss and cognitive dysfunction in AD.

Acknowledgments This work was supported by awards from The Scottish Rite Charitable Foundation of Canada to M.F. and D.J.G.; by grant MOP-64382 from the CIHR to M.F.; by a studentship from the Alzheimer Society of Canada to D.G.; and by NIH grants AG14449, AG10161, AG10688, and AG09466 to E.J.M.

References

1. Hofer M, Pagliusi SR, Hohn A, et al. Regional distribution of brain-derived neurotrophic factor mRNA in the adult mouse brain. EMBO J 1990;9(8):2459–2464 .
2. Hyman BT, Van Hoesen GW, Damasio AR, Barnes CL. Alzheimer's disease: cell-specific pathology isolates the hippocampal formation. Science 1984;225(4667):1168–1170.
3. DiStefano PS, Friedman B, Radziejewski C, et al. The neurotrophins BDNF, NT-3, and NGF display distinct patterns of retrograde axonal transport in peripheral and central neurons. Neuron 1992;8(5):983–993.
4. Knusel B, Beck KD, Winslow JW, et al. Brain-derived neurotrophic factor administration protects basal forebrain cholinergic but not nigral dopaminergic neurons from degenerative changes after axotomy in the adult rat brain. J Neurosci 1992;12(11):4391–4402.
5. Ando S, Kobayashi S, Waki H, et al. Animal model of dementia induced by entorhinal synaptic damage and partial restoration of cognitive deficits by BDNF and carnitine. J Neurosci Res 2002;70(3):519–527.
6. Ghosh A, Carnahan J, Greenberg ME. Requirement for BDNF in activity-dependent survival of cortical neurons. Science 1994;263(5153):1618–1623.
7. Lowenstein DH, Arsenault L. The effects of growth factors on the survival and differentiation of cultured dentate gyrus neurons. J Neurosci 1996;16(5):1759–1769.
8. Kohara K, Kitamura A, Morishima M, Tsumoto T. Activity-dependent transfer of brain-derived neurotrophic factor to postsynaptic neurons. Science 2001;291(5512):2419–2423.
9. Lindholm D, Carroll P, Tzimagiogis G, Thoenen H. Autocrine-paracrine regulation of hippocampal neuron survival by IGF-1 and the neurotrophins BDNF, NT-3 and NT-4. Eur J Neurosci 1996;8(7):1452–460.
10. Kang H, Schuman EM. Long-lasting neurotrophin-induced enhancement of synaptic transmission in the adult hippocampus. Science 1995;267(5204):1658–1662.
11. Korte M, Kang H, Bonhoeffer T, Schuman E. A role for BDNF in the late-phase of hippocampal long-term potentiation. Neuropharmacology 1998;37(4-5):553–559.
12. Mcallister AK, Katz LC, Lo DC. Neurotrophins and synaptic plasticity. Annu Rev Neurosci 1999;22:295–318.
13. Broad KD, Mimmack ML, Keverne EB, Kendrick KM. Increased BDNF and trk-B mRNA expression in cortical and limbic regions following formation of a social recognition memory. Eur J Neurosci 2002;16(11):2166–2174.
14. Hariri AR, Goldberg TE, Mattay VS, et al. Brain-derived neurotrophic factor val66met polymorphism affects human memory-related hippocampal activity and predicts memory performance. J Neurosci 2003;23(17):6690–6694.
15. Mowla SJ, Farhadi HF, Pareek S, et al. Biosynthesis and post-translational processing of the precursor to brain-derived neurotrophic factor. J Biol Chem 2001;276(16):12660–12666.
16. Lee R, Kermani P, Teng KK, Hempstead BL. Regulation of cell survival by secreted proneurotrophins. Science 2001;294(5548):1945–1948.
17. Chen ZY, Patel PD, Sant G, et al. Variant brain-derived neurotrophic factor (BDNF) (Met66) alters the intracellular trafficking and activity-dependent secretion of wild-type BDNF in neurosecretory cells and cortical neurons. J Neurosci 2004;24(18):4401–4411.
18. Pang PT, Teng HK, Zaitsev E, et al. Cleavage of proBDNF by tPA/plasmin is essential for long-term hippocampal plasticity. Science 2004;306(5695):487–491.
19. Seidah NG, Benjannet S, Pareek S, et al. Cellular processing of the neurotrophin precursors of NT3 and BDNF by the mammalian proprotein convertases. FEBS Lett 1996;379(3):247–250.
20. Fayard B, Loeffler S, Weis J, et al. The secreted brain-derived neurotrophic factor precursor pro-BDNF binds to TrkB and p75NTR but not to TrkA or TrkC. J Neurosci Res 2005;80(1):18–28.
21. Michalski B, Fahnestock M. Pro-brain-derived neurotrophic factor is decreased in parietal cortex in Alzheimer's disease. Brain Res Mol Brain Res 2003;111(1-2):148–154.

22. DeKosky ST, Scheff SW. Synapse loss in frontal cortex biopsies in Alzheimer's disease: correlation with cognitive severity. Ann Neurol 1990;27(5):457–464.
23. Terry RD, Masliah E, Salmon DP, et al. Physical basis of cognitive alterations in Alzheimer's disease: synapse loss is the major correlate of cognitive impairment. Ann Neurol 1991;30(4):572–580.
24. Coleman P, Federoff H, Kurlan R. A focus on the synapse for neuroprotection in Alzheimer disease and other dementias. Neurology 2004;63(7):1155–1162.
25. Holsinger RM, Schnarr J, Henry P, et al. Quantitation of BDNF mRNA in human parietal cortex by competitive reverse transcription-polymerase chain reaction: decreased levels in Alzheimer's disease. Brain Res Mol Brain Res 2000;76(2):347–354.
26. Phillips HS, Hains JM, Armanini M, et al., BDNF mRNA is decreased in the hippocampus of individuals with Alzheimer's disease. Neuron (1991)7(5), 695–702.
27. Fahnestock M, Garzon D, Holsinger RM, Michalski B. Neurotrophic factors and Alzheimer's disease: are we focusing on the wrong molecule? J Neural Transm Suppl 2002;62:241–252.
28. Ferrer I, Marin C, Rey MJ, et al. BDNF and full-length and truncated TrkB expression in Alzheimer disease: implications in therapeutic strategies. J Neuropathol Exp Neurol 1999;58(7):729–739.
29. Peng S, Wuu J, Mufson EJ, Fahnestock M. ProBDNF and mature BDNF are decreased in the preclinical stages of Alzheimer's disease. J Neurochem 2005;93(6):1412–1421.
30. Aoyama M, Asai K, Shishikura T, et al. Human neuroblastomas with unfavorable biologies express high levels of brain-derived neurotrophic factor mRNA and a variety of its variants. Cancer Lett 2001;164(1):51–60.
31. Garzon D, Yu G, Fahnestock M. A new brain-derived neurotrophic factor transcript and decrease in brain-derived neurotrophic factor transcripts 1, 2 and 3 in Alzheimer's disease parietal cortex. J Neurochem 2002;82(5):1058–1064.
32. Liu QR, Walther D, Drgon T, et al. Human brain derived neurotrophic factor (BDNF) genes, splicing patterns, and assessments of associations with substance abuse and Parkinson's disease. Am J Med Genet B Neuropsychiatr Genet 2005;134(1):93–103.
33. Sommer B. Alzheimer's disease and the amyloid cascade hypothesis: ten years on. Curr Opin Pharmacol 2002;.2(1):87–92.
34. Garzon D, Fahnestock M. Transcriptional down-regulation of BDNF by oligomeric amyloid-beta application to human neuroblastoma cells. Abstract Viewer/Itinerary Planner. Washington, DC: Society for Neuroscience, 2005. Online.
35. Patterson SL, Abel T, Deuel TA, et al., Recombinant BDNF rescues deficits in basal synaptic transmission and hippocampal LTP in BDNF knockout mice. Neuron 1996;16(6):1137–1145.
36. Bartoletti A, Cancedda L, Reid SW, et al. Heterozygous knock-out mice for brain-derived neurotrophic factor show a pathway-specific impairment of long-term potentiation but normal critical period for monocular deprivation. J Neurosci 2002;22(23):10072–10077.

Chapter 29
Shift in the Balance of TRKA and ProNGF in Prodromal Alzheimer's Disease

Elliott J. Mufson, Scott E. Counts, S. Peng, and Margaret Fahnestock

Introduction

Cholinergic basal forebrain (CBF) neurons in the nucleus basalis (NB) provide the primary source of cholinergic innervation to the cerebral cortex [1,2]. These cortical projection neurons undergo extensive degeneration in late-stage Alzheimer's disease (AD), which correlates with clinical severity and disease duration [3–5]. CBF neurons require nerve growth factor (NGF) for their survival and biological activity [6–10]. NGF is derived from its precursor molecule, proNGF [11,12], which is the predominant form found in the central nervous system (CNS) and may play a role in cell survival. Cellular responses to NGF are initiated by the binding and activation of its cognate receptors TrkA and p75NTR [6–13], which are produced within CBF neurons and transported in a retrograde manner to the cortex and hippocampus [13]. More than two decades have passed since it was hypothesized that degeneration of CBF neurons was due to a loss of neurotrophic support from target sites that produce NGF in AD [14–16]. These observations were based solely on studies that examined autopsy material harvested from late-stage AD patients.

However, to better understand the roles of NGF, proNGF, the common neurotrophin receptor p75[NTR], and the high-affinity trkA receptor in basal cortical dysfunction and their relation to cognitive impairment, we examined postmortem brain tissue from subjects during the prodromal stages of AD derived from the Religious Orders Study (ROS), a longitudinal clinicopathological study of aging and dementia in retired Catholic clergy [17–22]. Each ROS participant underwent an annual detailed clinical evaluation, including a battery of tests for function in five cognitive domains (orientation, attention, memory, language, perception), and agreed to brain autopsy and neuropathological analysis. These individuals were categorized as having no cognitive impairment (NCI), mild cognitive impairment (MCI), or AD. Here, we describe

E.J. Mufson
Rush University Medical School, Department of Neurological Sciences, Chicago, USA

A. Fisher et al. (eds.), *Advances in Alzheimer's and Parkinson's Disease*,
© Springer 2008

the findings generated from our clinical molecular pathological studies of the cholinotrophic basal forebrain system derived from the ROS cohort. These studies have provided unique insights into the role that changes in the NGF system play in CBF neuronal dysfunction during the progression of AD.

Results

Preservation of Basocortical Cholinergic Markers in MCI and AD

Quantitative unbiased stereological counting studies revealed that the number of cholinergic NB perikarya expressing either choline acetyltransferase (ChAT), the synthetic enzyme for acetylcholine (ACh), or the vesicular ACh transporter (VAChT) remain unchanged in people clinically characterized with MCI or mild AD compared to those diagnosed as NCI [19]. Other studies demonstrated that ChAT activity in NB cortical projection sites was stable during the early stages of AD or increased in the superior frontal cortex in the hippocampus of MCI subjects [23]. Taken together, these observations suggest that basocortical cholinergic tone is preserved in MCI and early AD, supporting the notion that cholinergic NB neurotransmission dysfunction is a late-stage event in AD [24].

Cholinotrophic Alterations in MCI and AD

Stereological counting experiments revealed that the number of NB perikarya expressing either TrkA or $p75^{NTR}$ was reduced $\sim$50% in MCI and mild AD compared to NCI [21,22]. Interestingly, there was a significant positive correlation between the number of TrkA-immunoreactive (-ir) NB neurons and performance on the Boston Naming Test and Global Cognitive Score measures of cognitive function [21]. In a similar vein, the number of $p75^{NTR}$-ir NB neurons was significantly correlated with performance on the Mini-Mental State Examination (MMSE) and the Global Cognitive Score [22]. These chemoanatomical findings indicate that during the early stages of AD many cholinergic NB neurons appear to undergo a phenotypic silencing of NGF receptor expression that correlates with cognitive impairment in the absence of frank cell loss or cholinergic deficits. The pronounced early defects in NGF receptor expression may be a precursor to the eventual extensive cell loss seen in the CBF in end-stage AD.

NGF Receptor Levels in MCI and AD

NGF receptors are synthesized in NB perikarya and anterogradely transported to the cortex. Quantitative immunoblotting experiments to measure TrkA and p75^{NTR} protein levels in five NB cortical projection zones (anterior cingulate, superior frontal, superior temporal, inferior parietal, visual cortex) revealed that the cortical levels of p75^{NTR} protein were stable across the diagnostic groups, whereas TrkA protein in the cortex was reduced ~50% (Fig. 1) in mild AD compared to NCI and MCI. Furthermore, cortical TrkA levels were positively correlated with MMSE performance [18]. There is also a decrease in trkA mRNA in end-stage AD. Because TrkA signaling is associated with neuronal survival [7,9,10,23], this specific reduction of cortical TrkA receptor protein may play a key role in the loss of cholinergic NB neurons observed in later stages of AD [24].

Pro-NGF Increases in MCI and AD

Interestingly, proNGF levels increase about 40% to 60% in the inferior parietal cortex of subjects diagnosed with MCI, mild AD [25], or severe AD [12] compared to aged controls (Fig. 2). Quantitative Western blotting demonstrated that

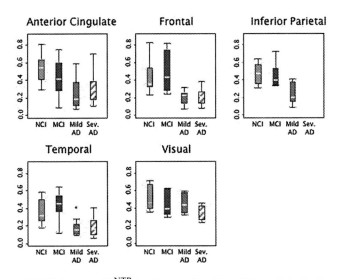

Fig. 1 Cortical TrkA but not p75^{NTR} levels are reduced in subjects clinically diagnosed with mild/moderate or severe AD. **A** Representative immunoblot of detergent lysates from superior temporal cortex were separated by sodium dodecyl sulfate polyacrylamide gel electrophoresis (SDS-PAGE) and immunoblotted for TrkA, p75^{NTR}, and β-tubulin. **B** Cortical TrkA was reduced about 50% in the anterior cingulate, superior frontal, superior temporal, and inferior parietal cortex in Alzheimer's disease (AD) subjects compared to subjects diagnosed with no cognitive impairment (NCI) or mild cognitive impairment (MCI): $p < 0.01$ via one-way analysis of variance (ANOVA) with post hoc Tukey's studentized range test for multiple comparisons. Densitometry was performed by normalizing TrkA or p75^{NTR} immunoreactive signals to β-tubulin signals on the same blots

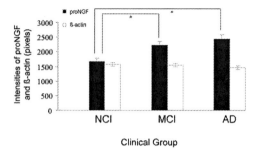

Fig. 2 Histogram showing that pro-nerve growth factor (proNGF) protein expression increases about 40% in the inferior parietal cortex in MCI patients and about 60% in AD patients. Densitometry was performed by normalizing proNGF immunoreactive signals to β-actin signals on the same Western blots. $*p < 0.001$ via one-way ANOVA with post hoc Tukey's studentized range test for multiple comparisons. (From Peng et al., [25] with permission.)

proNGF is the predominant form of NGF present in the cortex of aged intact humans. This finding is intriguing given the previous demonstration that levels of total NGF, as detected by enzyme-linked immunosorbent assay (ELISA) were unchanged in the cortex in MCI and mild AD [26]. This discrepancy may be due to technical differences in the studies (e.g., immunoblotting versus ELISA) or may indicate that if total NGF levels are unchanged the proNGF/mature NGF ratio increases during disease progression. The increase in proNGF was negatively correlated with cognitive scores during the progression of AD [25].

Conclusions

The biological consequences of the reduction of trkA protein and proNGF accumulation in the cortex during the prodromal stages of AD have yet to be determined. Recent findings indicate that recombinant proNGF binds trkA and promotes neuronal survival and neurite outgrowth similar to mature NGF but is approximately fivefold less active than mature NGF [11,12]. Although trkA-mediated proNGF retrograde transport has not been demonstrated, proNGF accumulation in NB target sites may be associated with reduced cortical trkA levels and/or defective retrograde transport to cholinotrophic basal forebrain neurons. Whereas trkA levels in the cortex were positively associated with MMSE scores [18], cortical proNGF levels were negatively correlated with MMSE performance [27]. Thus, the concomitant reduction of TrkA and accumulation of proNGF in the cortex may be an early pathobiological marker for the onset of AD.

On the other hand, data indicate that increases in cortical proNGF may result in pro-apoptotic signaling through binding to the p75^{NTR} receptor [28]. It is well known that NGF binding to p75^{NTR} in the absence of TrkA induces apoptosis in glia and some neurons [13]. Supporting this hypothesis are findings using a different form of recombinant proNGF that binds p75^{NTR} with high affinity and promotes neuronal apoptosis [28]. Recent investigations suggest that the pro-apoptotic effect(s) of p75^{NTR}-mediated proNGF signaling depends on interactions between p75^{NTR} and the neurotensin receptor sortilin [29]. Sortilin expression appears to mediate p75^{NTR}-induced apoptosis following proNGF treatments [28], suggesting that sortilin is a p75^{NTR} binding partner associated with the activation of cell death mechanisms [28–31].

Acknowledgments This work was supported by grants AG000257, AG14449, AG16668, AG09446, and AG10161 (E.J.M.), The Scottish Rite Charitable Foundation of Canada (MF) and grant MOP-64382 from the Canadian Institutes of Health Research (M.F.). A list of participating groups can be found at the website: http://www.rush.edu/rumc/page-R12394.html.

References

1. Mesulam MM, Geula C. Nucleus basalis (Ch4) and cortical cholinergic innervation in the human brain: observations based on the distribution of acetylcholinesterase and choline acetyltransferase. J Comp Neurol 1988;275(2):216–240.
2. Mesulam MM, Mufson EJ, Levey AI, Wainer BH. Cholinergic innervation of cortex by the basal forebrain: cytochemistry and cortical connections of the septal area, diagonal band nuclei, nucleus basalis (substantia innominata), and hypothalamus in the rhesus monkey. J Comp Neurol 1983;214(2):170–197.
3. Bartus RT, Dean RL 3rd, Beer B, Lippa AS. The cholinergic hypothesis of geriatric memory dysfunction. Science 1982;217(4558):408–414.
4. Mufson EJ, Bothwell M, Kordower JH. Loss of nerve growth factor receptor-containing neurons in Alzheimer's disease: a quantitative analysis across subregions of the basal forebrain. Exp Neurol 1989;105(3):221–232.
5. Whitehouse PJ, Price DL, Clark AW, et al. Alzheimer disease: evidence for selective loss of cholinergic neurons in the nucleus basalis. Ann Neurol 1981;10(2):122–126.
6. Kaplan DR, Miller FD, Signal transduction by the neurotrophin receptors. Curr Opin Cell Biol 1997;9(2):213–221.
7. Lad SP, Neet KE, Mufson EJ, Nerve growth factor: structure, function and therapeutic implications for Alzheimer's disease. Curr Drug Target CNS Neurol Disord 2003;2(5):315–334.
8. Sofroniew MV, Howe CL, Mobley WC. Nerve growth factor signaling, neuroprotection, and neural repair. Annu Rev Neurosci 2001;24:1217–1281.
9. Chao MV, Hempstead BL. p75 and Trk: a two-receptor system. Trends Neurosci 1995;18(7):321–326.
10. Bothwell M. Functional interactions of neurotrophins and neurotrophin receptors. Annu Rev Neurosci 1995;18:223–253.
11. Fahnestock M, Yu G, Coughlin MD. ProNGF: a neurotrophic or an apoptotic molecule? Prog Brain Res 2004;146:107–110.
12. Fahnestock M, Michalski B, Xu B, Coughlin MD. The precursor pro-nerve growth factor is the predominant form of nerve growth factor in brain and is increased in Alzheimer's disease. Mol Cell Neurosci 2001;18(3):210–220.

13. Sobreviela T, Clary DO, Reichardt LF, et al. TrkA-immunoreactive profiles in the central nervous system: co-localization with neurons containing p75 nerve growth factor receptor, choline acetyltransferase, and serotonin. J Comp Neurol 1994;350(4):587–611.

14. Appel SH. A unifying hypothesis for the cause of amyotrophic lateral sclerosis, parkinsonism, and Alzheimer disease. Ann Neurol 1981;10(6):499–505.

15. Hefti F, Hartikka J, Knusel B, Function of neurotrophic factors in the adult and aging brain and their possible use in the treatment of neurodegenerative diseases. Neurobiol Aging 1989;10(2):515–533.

16. Kordower JH, Mufson EF. NGF and Alzheimer's disease: unfulfilled promise and untapped potential. Neurobiol Aging 1989;10(5):543–544.

17. Bennett DA, Wilson RS, Schneider JA, et al. Natural history of mild cognitive impairment in older persons. Neurology 2002;59(2):198–205.

18. Counts SE, Nadeem M, Wuu J, et al. Reduction of cortical TrkA but not p75$^{(NTR)}$ protein in early-stage Alzheimer's disease. Ann Neurol 2004;56(4):520–531.

19. Gilmor ML, Erickson JD, Varoqui H, et al. Preservation of nucleus basalis neurons containing choline acetyltransferase and the vesicular acetylcholine transporter in the elderly with mild cognitive impairment and early Alzheimer's disease. J Comp Neurol 1999;411(4):693–704.

20. Kordower JH, Chu Y, Stebbins GT, et al. Loss and atrophy of layer II entorhinal cortex neurons in elderly people with mild cognitive impairment. Ann Neurol 2001;49(2):202–213.

21. Mufson, Ma SY, Cochran EJ, et al. Loss of nucleus basalis neurons containing trkA immunoreactivity in individuals with mild cognitive impairment and early Alzheimer's disease. J Comp Neurol 2000;427(1):19–30.

22. Mufson EJ, Ma SY, Dills J, et al. Loss of basal forebrain p75$^{(NTR)}$ immunoreactivity in subjects with mild cognitive impairment and Alzheimer's disease, J Comp Neurol 2002;443(2):136–153.

23. DeKosky ST, Ikonomovic M, Styren SD, et al. Upregulation of choline acetyltransferase activity in hippocampus and frontal cortex of elderly subjects with mild cognitive impairment. Ann Neurol 2002;51(2):145–155.

24. Davies P, Malone AJ. Selective loss of central cholinergic neurons in Alzheimer's disease. Lancet 1976;2(8000):1403.

25. Peng S, Wuu J, Mufson EJ, Fahnestock M. Increased proNGF Levels in subjects with mild cognitive impairment and mild Alzheimer's disease. J Neuropathol Exp Neurol 2004;63(6):641–649.

26. Mufson EJ, Ikonomovic SD, Styren SE, et al. Preservation of brain nerve growth factor in mild cognitive impairment and Alzheimer disease. Arch Neurol 2003;60(8):1143–1148.

27. Lee R, Kermani P, Teng KK, Hempstead BL. Regulation of cell survival by secreted proneurotrophins. Science 2001;294(5548):1945–1948.

28. Barrett GL. The p75 neurotrophin receptor and neuronal apoptosis. Prog Neurobiol 2000;61(2):205–229.

29. Nykjaer A, Lee R, Teng KK, et al. Sortilin is essential for proNGF-induced neuronal cell death. Nature 2004;427(6977):843–848.

30. Roux PP, Barker PA. Neurotrophin signaling through the p75 neurotrophin receptor. Prog Neurobiol 2002;67(3):203–233.

31. Mamidipudi V, Wooten MW. Dual role for p75(NTR) signaling in survival and cell death: can intracellular mediators provide an explanation? J Neurosci Res 2002;68(4):373–384.

Chapter 30
Neuroprotective Effects of Trophic Factors and Natural Products: Involvement of Multiple Intracellular Kinases

Stéphane Bastianetto, Wen-Hua Zheng, Yingshan Han,
Lixia Gan, and Rémi Quirion

Introduction

Various studies have shown that growth factors, such as insulin-like trophic factor-1 (IGF-1), can promote neuronal survival in various models of toxicity. For example, IGF-1 was shown to be able to protect and even rescue hippocampal neurons exposed to ß-amyloid peptides (Aß), which likely play a role in Alzheimer's disease (AD) [1,2]. The neuroprotective effects of IGF-1 are sometimes shared by brain-derived neurotrophic factor (BDNF), albeit with lower potency for the neurotrophin [3]. The phosphatidylinositol 3-kinase (PI3K)/Akt kinase/FOXO pathway is a key molecular target involved in the neuroprotective and neurorescuing effects of IGF-1 [3–6]. However, the detailed intracellular mechanism(s) underlying the neuroprotective effect of IGF-1 remains to be established.

Epidemiological studies have suggested that high intake of polyphenols derived from fruits and vegetables may reduce the risk of dementia or Parkinson's disease [7]. Most in vitro and in vivo animal studies have been targeted at the possible involvement of polyphenols found in red wine and teas (i.e., resveratrol and catechins) as they display strong neuroprotective properties in various models of toxicity [8–16]. For example, we previously reported that resveratrol and catechins protected hippocampal neurons against toxicity induced by the nitric oxide (NO) donor sodium nitroprusside (SNP) [9] and Aß peptides [13]. Our studies and those of others indicated that mechanisms underlying the neuroprotective effects of these polyphenols did not involve solely their antioxidant properties but also their modulatory actions on intracellular effectors such as protein kinase C (PKC) [12,13,15,17]. We briefly review here intracellular signaling pathways that are modulated by various growth factors and polyphenols and their possible role in neuronal survival.

S. Bastianetto
Douglas Hospital Research Centre, Department of Psychiatry, McGill University, 6875 Blvd LaSalle, Montréal, Québec, H4H 1R3, Canada

A. Fisher et al. (eds.), *Advances in Alzheimer's and Parkinson's Disease*, 291
© Springer 2008

Neuroprotective Effects of IGF-1 Involved the PI3/AKT Pathway and FOXO3a

Accumulated evidence indicates that IGF-I, apart from regulating growth and development, protects neurons against various toxic agents such as Aß peptides, glucose, or serum deprivation [2]. These effects are likely mediated by the activation of intracellular pathways involving phosphatidylinositide 3-kinase (PI3K)/Akt kinase and FOXO3a, a member of the Forkhead family of transcription factors that acts as a substrate for Akt kinases [4,5]. We reported earlier that IGF-1 (10–100 nM) rapidly induced the phosphorylation/ inactivation of endogenous FOXO3a in both PC12 cells and primary cultured neurons [4,5]. The PI3K/Akt kinase pathway mediates this action as the PI3-kinase inhibitors wortmannin (0.5 μM) and LY294002 (50 μM) inhibited the phosphorylation of both Akt and FOXO3a produced by IGF-1, whereas PD98059 (50 μM), a MEK kinase inhibitor, and rapamycin, a p70S6 pathway inhibitor, were ineffective [4]. Moreover, IGF-1 (100 nM) blocked the nuclear translocation of FOXO3a in hippocampal neurons and promoted survival in parallel with the phosphorylation of Akt and FOXO3a, suggesting that these events mediated the survival/rescuing properties of IGF-1 in neurons [5].

Effects of Trophic Factors on Other Transcription Factors of the FoxO Family

FoxO1 is another member of the FoxO subfamily of forkhead transcription factors that is apparently targeted by insulin and growth factors in the regulation of metabolism and survival in peripheral tissues [5,18]. Accordingly, we examined next the effects of various growth factors on the nuclear/cytoplasmic shuttling of FoxO1 in PC12 cells. IGF-1 and nerve growth factor (NGF) potently induced the nuclear exclusion of FoxO-green fluorescent protein, whereas NT-3 and NT-4 exerted much weaker effects [19]. In contrast, BDNF failed to induce FoxO1 translocation. The translocation of FoxO1 was inhibited by LY294002, a well established PI3K/Akt kinase inhibitor. Moreover, FoxO1 was phosphorylated by all trophic factors tested here, with the exception of BDNF. These findings indicate that the PI3kinase/Akt pathway is involved in the regulation of nuclear/cytoplasmic shuttling of FoxO1 by various trophic factors in neuronal cells [19]. Hence, both FOXO3a and FoxO1 can be regulated by trophic factors, in particular IGF-1, via the activation of the PI3K/Akt cascade.

Neuroprotective Effects of Resveratrol Against Aß-Induced Toxicity

We recently reported that resveratrol (20 µM) protected hippocampal neurons before, during, after exposure to Aß 25–35. Similar effects were observed when cells were exposed to Aß 1-40 and Aß 1–42, and were shared by phorbol 12-myristate-13-acetate [13]. Pretreatment with the PKC inhibitor GF 109203X (1 µM) blocked the effect of resveratrol, whereas inhibitors of PI3K or MAP kinases (i.e., LY294002 and PD98059, respectively) were ineffective [13]. Moreover, treatment of hippocampal cells with resveratrol (20 µM) induced the phosphorylation of various isoforms of PKC and blocked the inhibitory effects of Aß 25–35 on the phophorylation of PKC, indicating that this enzyme (particularly the PKC-δ isoform) mediated the protective effects of resveratrol in our model [13]. In contrast, resveratrol failed to modulate the phosphorylation of either Akt kinase or MAP kinases in this model [13].

Tea Extracts Protect Hippocampal Neurons Through Their Catechins Gallate Esters

Based on the purported protective effects of tea in various diseases including cardiovascular and neurodegenerative disorders [20,21], we investigated the potential effectiveness of a green and a black tea extract against Aβ toxicity in rat primary hippocampal cultured cells (Bastianetto and Quirion, unpublished data). Our results demonstrated that both total extracts (5–25 µg/ml) displayed neuroprotective action against toxicity induced by Aβ peptides (Aβ 25–35, Aβ 1–40, and Aβ 1–42). These effects were shared by epicatechin gallate (ECG) (1–20 µM) and the most abundant green tea catechin, epigallocatechin gallate (EGCG) (1–10 µM), the former being the most potent flavanol. In contrast, the nongallate moiety, epicatechin and epigallocatechin, failed to protect cells in the same range of concentrations. Similarly, the green tea extract (25 µg/ml) and EGCG (10 µM) shared with the Ginkgo biloba extract EGb 761 (100 µg/ml). a well known extract that is prescribed in Europe for the treatment of cognitive deficits [22]) the ability to block the toxic effects of Aß 1–42. Interestingly, thioflavin and Western blot assays indicated that EGCG (10 µM) inhibited the fibrillization of Aß as well as Aß oligomers (known as Aß-derived diffusible neurotoxin ligands, or ADDLs) that have been reported to mediate Aß 1–42-induced toxicity [23]. Taken together, these results indicate that catechin gallates, particularly EGCG, likely mediate the neuroprotective effects of green tea extract that may be associated, at least in part, with their inhibitory action on the formation of Aß fibrils/oligomers.

Conclusion

In this brief review, we have provided evidence that IGF-1 and neurotrophins exerted neuroprotective/neurorescuing effects in various models of toxicity, with IGF-1 apparently being one of the most effective trophic factors. In addition, we have demonstrated that the survival-promoting effects of both IGF-1 and BDNF mostly depend on activation of the PI3K/Akt pathway, and that the mitogen-activated protein kinase (MAPK) pathway likely plays a minor role (Table 1). The exact mechanism(s) by which Akt mediates the survival effects of IGF-1 and BDNF in cultured neurons remains to be fully established. However, it is well known that activated Akt can phosphorylate (this event leading to their inactivation) proapoptotic proteins such as Bad [24], caspase-9 [25], and the FOXO family of transcription factors [4,5]. It is therefore likely that the activation of Akt induced the inactivation of various proapoptotic molecules, leading to the inhibition of apoptosis. Akt may also indirectly regulate major death survival pathways such as p53, NF-κB, or even CREB under certain conditions [26,27]. Hence, activation of the PI3K/Akt cascade by trophic factors such as IGF-1 resulted in a variety of downstream intracellular events (focused here on FOXOs) involved in their survival promoting properties.

The neuroprotective effects of trophic factors are shared by the polyphenols resveratrol and catechin gallate esters that are present in high amounts in red wine and teas, respectively. The effect of resveratrol is mediated by the activation of PKC, supporting the hypothesis that the modulation of kinases plays an important role in the neuroprotective abilities of polyphenols, as previously reported for EGCG [12]. Moreover, EGCG appeared to be the most potent green tea catechin to protect against Aß-induced toxicity. This effect did not seem to involve the antioxidant properties of this catechin, but its inhibition of the formation of Aß oligomers and fibrils (Table 1). Finally, in addition to a role for PKC in the neuroprotective effects of resveratrol, it has recently been shown

Table 1 Involvement of kinases and Aß formation in the purported neuroprotective effects of trophic factors, resveratrol, and epigallocatechin gallate

Factor	PI3K/Akt	MAPK	PKC	Aß formation
Trophic factors				
IGF-1	Yes	Modest	Yes (PKC?)	ND
NGF	Yes	Yes	Yes	ND
BDNF	Yes	Yes	Yes	ND
Polyphenols				
Resveratrol	No	No	Yes	ND
EGCG	Yes	ND	Yes	Inhibition

Aß, β-amyloid; MAPK, mitogen-activated protein kinase; PKC, protein kinase C; IGF-1, insulin-like growth factor; NGF, nerve growth factor; BDNF, brain-derived neurotrophic factor; EGCG, epigallocatechin gallate; ND, not determined

that this molecule can increase the expression of the transcription factor egr1 [28] and reverse the phosphorylation of stress-activated protein kinase/c-Jun N-terminal kinase (SAPK/JNK) in a model of paclitaxel-induced apoptosis [29]. Hence, polyphenols likely modulate various intracellular pathways. Hence, future studies using genomic and proteomic approaches are required to obtain a more precise understanding of the cellular mechanisms involved in their neuroprotective effects.

Acknowledgments This work was supported by CIHR grants to R. Quirion.

References

1. Selkoe DJ. The molecular pathology of Alzheimer's disease. Neuron 1991;6:487.
2. Dore S, Kar S, Quirion R. Rediscovering an old friend, IGF-I: potential use in the treatment of neurodegenerative diseases. Trends Neurosci 1997;20:326.
3. Zheng WH, Quirion R. Comparative signalling pathways of insulin-like growth factor-1 and brain-derived neurotrophic factor in hippocampal neurons and the role of the PI3 kinase pathway in cell survival. J Neurochem 2004;89:844.
4. Zheng WH, Kar S, Quirion R. Insulin-like growth factor-1-induced phosphorylation of the forkhead family transcription factor FKHRL1 is mediated by Akt kinase in PC12 cells. J Biol Chem 2000;275:39152.
5. Zheng WH, Kar S, Quirion R. Insulin-like growth factor-1-induced phosphorylation of transcription factor FKHRL1 is mediated by phosphatidylinositol 3-kinase/Akt kinase and role of this pathway in insulin-like growth factor-1-induced survival of cultured hippocampal neurons. Mol Pharmacol 2002;62:225.
6. Zheng WH, Kar S, Quirion R. FKHRL1 and its homologs are new targets of nerve growth factor Trk receptor signaling. J Neurochem 2002;80:1049.
7. Bastianetto S, Quirion R. Natural antioxidants and neurodegenerative diseases. Front Biosci 2004;9:3447.
8. Moosmann B, Behl C. The antioxidant neuroprotective effects of estrogens and phenolic compounds are independent from their estrogenic properties. Proc Natl Acad Sci U S A 1999,96:8867.
9. Bastianetto S, Zheng WH, Quirion R. Neuroprotective abilities of resveratrol and other red wine constituents against nitric oxide-related toxicity in cultured hippocampal neurons. Br J Pharmacol 2000;131:711.
10. Bastianetto S, Quirion, R. Resveratrol and red wine constituents: evaluation of their neuroprotective properties. Pharmacol News 2001;8:33–38.
11. Nagai K, Jiang MH, Hada J, et al. (–)-Epigallocatechin gallate protects against NO stress-induced neuronal damage after ischemia by acting as an anti-oxidant. Brain Res 2002;956:319.
12. Levites Y, Amit T, Mandel S, Youdim MB. Neuroprotection and neurorescue against Abeta toxicity and PKC-dependent release of nonamyloidogenic soluble precursor protein by green tea polyphenol (–)-epigallocatechin-3-gallate. FASEB J 2003;17:952.
13. Han YS, Zheng WH, Bastianetto S, et al. Neuroprotective effects of resveratrol against beta-amyloid-induced neurotoxicity in rat hippocampal neurons: involvement of protein kinase C. Br J Pharmacol 2004;141:997.
14. Mandel S, Weinreb O, Amit T, Youdim MB. Cell signaling pathways in the neuroprotective actions of the green tea polyphenol (–)-epigallocatechin-3-gallate: implications for neurodegenerative diseases. J Neurochem 2004;88:1555.

15. Mandel SA, Avramovich-Tirosh Y, Reznichenko L, et al. Multifunctional activities of green tea catechins in neuroprotection. Neurosignals 2005;14:46.
16. Choi YB, Kim YI, Lee KS, et al. Protective effect of epigallocatechin gallate on brain damage after transient middle cerebral artery occlusion in rats. Brain Res 2004;1019:47.
17. Levites Y, Amit T, Youdim MB, Mandel S. Involvement of protein kinase C activation and cell survival/cell cycle genes in green tea polyphenol (–)-epigallocatechin 3-gallate neuroprotective action. J Biol Chem 2002;277:30574.
18. Arden KC, Fox O. linking new signaling pathways. Mol Cell 2004;14:416.
19. Gan L, Zheng WH, Chabot JG, et al. Nuclear/cytoplasmic shuttling of the transcription factor FoxO1 is regulated by neurotrophic factors. J Neurochem 2005;93:1209.
20. Pan T, Jankovic J, Le W. Potential therapeutic properties of green tea polyphenols in Parkinson's disease. Drugs Aging 2003;20:711.
21. Weinreb O, Mandel S, Amit T, Youdim MB. Neurological mechanisms of green tea polyphenols in Alzheimer's and Parkinson's diseases. J Nutr Biochem 2004;15:506.
22. Kanowski S, Hoerr R. Ginkgo biloba extract EGb 761 in dementia: intent-to-treat analyses of a 24-week, multi-center, double-blind, placebo-controlled, randomized trial. Pharmacopsychiatry 2003;36:297.
23. Klein WL. Abeta toxicity in Alzheimer's disease: globular oligomers (ADDLs) as new vaccine and drug targets. Neurochem Int 2002;41:345.
24. Del Peso L, Gonzalez-Garcia M, Page C, et al. Interleukin-3-induced phosphorylation of BAD through the protein kinase Akt. Science 1997;278:687.
25. Cardone MH, Roy N, Stennicke HR, et al. Regulation of cell death protease caspase-9 by phosphorylation. Science 1998;282:1318.
26. Datta SR, Brunet A, Greenberg ME. , Cellular survival: a play in three Akts. Genes Dev 1999;13:2905.
27. Vivanco I, Sawyers CL. The phosphatidylinositol 3-kinase AKT pathway in human cancer. Nat Rev Cancer 2002;2:489.
28. Della Ragione F, Cucciolla V, Criniti V, et al. Antioxidants induce different phenotypes by a distinct modulation of signal transduction. FEBS Lett 2002;532:289.
29. Nicolini G, Rigolio R, Scuteri A, et al. Effect of trans-resveratrol on signal transduction pathways involved in paclitaxel-induced apoptosis in human neuroblastoma SH-SY5Y cells. Neurochem Int 2003;42:419.

Chapter 31
Intraneuronal Aβ and Alzheimer's Disease

Lauren M. Billings and Frank M. LaFerla

Introduction

Benign memory loss is an inevitable consequence of healthy aging. However, a significant percentage of the elderly experience an abnormal loss of intellectual functioning that manifests with extreme cognitive impairments that can be accompanied by aphasia, apraxia, and/or agnosia, thereby meeting the criteria for dementia [1]. Alzheimer's disease (AD) accounts for approximately two-thirds of senile dementia cases, affecting approximately 1 in 10 individuals over age 70 and 1 in 3 in individuals over age 85 (www.nih.gov). AD is commonly characterized by an insidious onset, with a gradual progression of symptoms that can greatly vary in time among individuals. The affected cognitive domains in AD are initially limited to memory systems, where the ability to retain and retrieve new information is impaired. Eventually, many other domains become affected, resulting in mood lability, anxiety, sleep disturbance, altered visuospatial perception, and impaired speech and judgment [2–4].

Our inability to diagnose early-stage AD accurately and reliably has considerably delayed efforts to identify a reliable biomarker in humans that could facilitate a clinical diagnosis. This early pre-AD stage, usually referred to as mild cognitive impairment (MCI), is generally considered a transitional stage prior to the development of AD but during which memory impairments occur with the preservation of other cognitive domains [4]. Following a battery of physical and neuropsychological tests, a patient can be given a diagnosis ranging from possible, to probable, to definite AD [3]. However, it is also plausible that these early symptoms are caused by a dementia that is unrelated neuropathologically to AD. In any event, the need to identify a biomarker for early AD, even MCI, is evident; multiple studies indicate that early-stage AD may be much more easily treated than the later stages [5,6]. In fact, findings from transgenic mouse studies indicate that a point-of-no-return may exist in

L.M. Billings
Department of Neurobiology and Behavior, University of California, Irvine, CA 92612, USA,

A. Fisher et al. (eds.), *Advances in Alzheimer's and Parkinson's Disease*,
© Springer 2008

late-stage, tangle-ridden AD brains [6]. The identification of a biomarker that signals the onset of AD symptoms has, until recently, remained elusive.

AD Pathology and Cognition

Pathologically, AD is characterized by cortical and hippocampal atrophy, intraneuronal accumulation of neurofibrillary tangles (NFTs) comprised of hyperphosphorylated and ubiquitinated microtubule-associated protein tau, and extracellular aggregates known as senile plaques. Plaques are comprised primarily of the small β-amyloid (Aß) peptide and are surrounded by dystrophic neurites, reactive astrocytes, and activated microglia.

Unlike amyotrophic lateral sclerosis (ALS) or Parkinson's disease, where the destruction of specific neuronal populations results in a clear phenotype, the relation between AD neuropathology and the clinical presentation of the disease has been the subject of a long-standing debate. A particularly central debate in AD research focuses on whether the hallmark lesions, Aβ plaques and tau tangles, are a cause or consequence. After decades of intensive research, the preponderance of evidence argues that they both may play an active role in the generation and progression of AD symptoms. Cytoskeletal pathology, particularly NFT formation, may cause impairments in neuronal trafficking, resulting in altered cellular transport or neuronal viability, or both [7]. However, findings from 3xTg-AD mice indicate that NFT formation develops after Aβ pathology and cognitive deficits emerge [8]. Furthermore, cerebrospinal fluid (CSF) tau levels do not correlate with the cognitive impairment, arguing against their use as a reliable peripheral biomarker for the onset of cognitive deterioration [9].

Synaptic dysfunction, a loss of synapses or cholinergic neurons particularly in brain regions involved in cognition and memory, may also mediate the profound deterioration of these faculties in AD [10–12]. However, a decline in cholinergic function appears to be associated with late stages of the disease and is also not likely an initiating factor [13,14]. Increasing evidence points to a role for pathological, early forms of Aβ, such as intraneuronal Aβ, in the pathogenesis of AD.

Amyloid Cascade Hypothesis

Over the past two decades, a growing body of evidence supports a central role for Aβ in AD. The strongest support comes from genetic studies that have identified mutations in both amyloid precursor protein (APP) and the presenilins (PS1 and PS2), which are involved in cleavage of the APP holoprotein to liberate Aβ [15–20]. In addition, the polymorphic allele of apolipoprotein ε4 (Apo ε4) was identified as a major susceptibility gene in late-onset AD [21]. All mutations identified so far, as well as the Apo ε4 polymorphism, modulate some aspect of APP metabolism, either by increasing total Aß levels, selectively increasing Aß$_{42}$ levels, or leading to enhanced stabilization of Aß. Although it is clear that mutations in tau, associated with frontotemporal dementia with parkinsonism, are sufficient to induce tau pathology and neuronal loss [22,23], the conspicuous presence of mutations in genes associated with Aß formation/processing in FAD cases has led to the widespread inference that Aβ is central to the genesis of AD. This is the fundamental tenet of the amyloid cascade hypothesis, which posits that all aspects of AD, including plaques, tangles, neuronal loss, and all resulting cognitive symptoms, are caused by aberrant Aβ processing [24].

Following the identification of mutant genes that underlie the genetic forms of AD, researchers were poised to determine the contribution of the hallmark AD neuropathology to the progression of the clinical phenotype through the generation of transgenic animal models harboring clinically relevant AD mutations. These APP or APP+PS mutant animal models develop plaque pathology and deficits in learning and/or memory and thus have been crucial in elucidating various mechanisms of plaque-dependent and plaque-independent changes in cognition [25–28].

Indeed, the amyloid-cascade hypothesis has gained substantial momentum with the generation of transgenic AD mouse models as well as cell lines transfected with AD-relevant mutations. There is a great deal of evidence to suggest that Aβ-mediated toxicity may trigger most of the pathological changes in AD brains. For example, it is well established that many domains of cellular functioning are affected by Aβ, including calcium homeostasis, mitochondrial function, and oxidative metabolism, which may all contribute directly to neurodegeneration (for a detailed review, see ref. [29]). In addition, multiple in vivo findings indicate that Aβ pathology results in the accumulation (and subsequent hyperphosphorylation) of tau, most likely through its inhibition of the proteasome [6,30,31].

Intraneuronal Aβ

Although the amyloid cascade hypothesis has garnered great support from molecular and histological studies using in vitro and in vivo models of AD, it has been difficult to directly connect Aβ to the clinical AD phenotype. In part, this is due to the consistently poor relation between Aβ plaque load and scores of cognitive performance in postmortem studies of AD patients: Some individuals can carry a large plaque load and still be relatively cognitively normal. In contrast, other patients may have severely impaired cognition but relatively few plaques [32–35].

The poor correlation between plaques and cognition is not unique to human AD studies and is a central finding in several studies of transgenic models of AD as well. For example, most AD models, harboring clinical mutations in APP or

PS (or both), tend to develop cognitive impairments prior to the emergence of overt plaque pathology [8,25–28]. Immunization studies in PDAPP and APP+PS1 mice have demonstrated that anti-Aβ treatment can rescue memory deficits without marked reductions in the Aβ plaque burden [36–38], which implies that the extensive extracellular Aβ burden in these mice is not solely responsible for the expression of cognitive deficits. Indeed, although some studies have found a relation between Aβ plaque load and learning [39] or memory [40], neither of these studies have been able to determine the biomarker that triggers the *onset* of early memory deficits.

Although the amyloid cascade hypothesis has generally been interpreted to mean that aggregated, extracellular Aβ pathology causes neuronal and synaptic dysfunction, the hypothesis does not rule out other forms of Aβ. Indeed, it is increasingly apparent that earlier forms of Aβ pathology may contribute to the initial memory loss in MCI or early AD. Intraneuronal Aβ (*i*Aβ) is an early pathological event not only in the human muscle disorder inclusion body myositis but also in brains of individuals with Down's syndrome, who invariably develop AD-like neuropathology [41,42], and is also seen in transgenic animal models of AD [8,43–50]. Furthermore, in brains from Down's syndrome individuals, *i*Aβ appears to precede the formation of amyloid plaques. Work by Gouras and colleagues [51] has demonstrated the existence of Aβ oligomers in AD brains from individuals with clinical diagnoses ranging from mild to severe cognitive dysfunction (CDR0.5-CDR1, respectively). These authors have also reported that *i*Aβ may mediate synaptic dysfunction [46]. Interestingly, in the 3xTg-AD mouse model of AD, synaptic and behavioral deficits are early events that develop coincidentally with *i*Aβ but prior to extracellular plaque formation [8,48]. Notably, *i*Aβ staining appears to be lower in more severe cases of AD (CDR2) when extracellular amyloid plaques are abundant, indicating that the *i*Aβ pathology is an early event and perhaps even a partial source for extracellular, aggregated Aβ [51].

Recent work from our laboratory demonstrated that retention deficits in the 3xTg-AD mice emerge at 4 months of age, coincidentally with the accumulation of Aβ in neurons of the hippocampus, cortex, and amygdala [8]. Immunotherapy, which effectively clears *i*Aβ, as well as plaque pathology [6] rescued the retention impairments on a hippocampus-dependent task where the pathology had been cleared but not on an amygdala-dependent task where the pathology had not been successfully lessened [8]. These findings indicated that intraneuronal Aβ triggers the onset of subtle changes in long-term retention for new information. These findings are supported by research by Cuello's group [52], who found that the expression of Aβ in neurons of the hippocampus and cortex in transgenic rats resulted in a subtle learning deficit. Previous studies using transgenic overexpression or inducible expression of Aβ in vivo demonstrated that Aβ42 is cytotoxic [53,54]. Furthermore, Zhang et al [55]. reported that Aβ 1–42 was cytotoxic when injected into human primary neurons but not when applied extracellularly. Taken with the behavioral findings, it is clear that early intraneuronal Aβ accumulation may initiate a cascade of cytotoxic events that ultimately interrupts vital learning and memory pathways.

Conclusion

The identification of a suitable biomarker that predicts the onset of AD symptoms will no doubt continue to be aided by studies using transgenic models of AD. Indeed, these models allow researchers to observe changes in behavior throughout the life-span of a single animal and to link these changes in cognition directly to coincidental changes in neuropathology. In particular, a longitudinal assessment of behavioral changes in transgenic mice will more accurately identify stepwise changes in cognition and how they relate to behavioral alterations than would a few cross-sectional time points chosen for their association with the existence of well established pathology. A particularly notable example of this strategy is the identification of early, subtle alterations in behavior in 3xTg-AD mice that are associated with early forms of *i*Aβ neuropathology.

A longitudinal assessment in a mouse harboring both plaques and tangles, such as the 3xTg-AD mice, will also help illuminate the relation between *i*Aβ, plaques, and tangles. For example, how the emergence of plaques and subsequently tangles contribute to the cognitive phenotype remains to be determined. Furthermore, it remains unknown whether tau is independently associated with late-stage cognitive deficits or if the clearance of plaques is sufficient to ameliorate the cognitive phenotype. Immunotherapeutic studies in 3xTg-AD mice at different ages and with different degrees of neuropathology will help clarify the contribution of each type of pathology to early and late changes in memory. Establishing a clear relation between cognition and the emergence of the hallmark pathologies of AD is a vital step in the process of identifying novel therapeutic targets and developing effective treatment for AD.

References

1. Ritchie K, Lovestone S. The dementias. Lancet 2002;360:1759–1766.
2. Albert MS. Cognitive and neurobiologic markers of early Alzheimer disease. Proc Natl Acad Sci U S A 1996;93:13547–13551.
3. Grossberg GT. Diagnosis and treatment of Alzheimer's disease. J Clin Psychiatry 2003;64(suppl 9):3–6.
4. Grundman M, Petersen RC, Ferris SH, et al. Mild cognitive impairment can be distinguished from Alzheimer disease and normal aging for clinical trials. Arch Neurol 2004;61:59–66.
5. Nicoll JA, Wilkinson D, Holmes C, et al. Neuropathology of human Alzheimer disease after immunization with amyloid-beta peptide: a case report. Nat Med 2003;9:448–452.

6. Oddo S, Billings L, Kesslak JP, et al. Abeta immunotherapy leads to clearance of early, but not late, hyperphosphorylated tau aggregates via the proteasome. Neuron 2004;43:321–332.

7. Goedert M, Jakes R, Spillantini MG, et al. Assembly of microtubule-associated protein tau into Alzheimer-like filaments induced by sulphated glycosaminoglycans. Nature 1996;383:550–553.

8. Billings LM, Oddo S, Green KN, et al. Intraneuronal Abeta causes the onset of early Alzheimer's disease-related cognitive deficits in transgenic mice. Neuron 2005;45:675–688.

9. Clark CM, Xie S, Chittams J, et al. Cerebrospinal fluid tau and beta-amyloid: how well do these biomarkers reflect autopsy-confirmed dementia diagnoses? Arch Neurol 2003;60:1696–1702.

10. Whitehouse PJ, Price DL, Struble RG, et al. Alzheimer's disease and senile dementia: loss of neurons in the basal forebrain. Science 1982;215:1237–1239.

11. Masliah E. Mechanisms of synaptic pathology in Alzheimer's disease. J Neural Transm Suppl 1998;53:147–158.

12. Masliah E, Ellisman M, Carragher B, et al. Three-dimensional analysis of the relationship between synaptic pathology and neuropil threads in Alzheimer disease. J Neuropathol Exp Neurol 1992;51:404–414.

13. Davis KL, Mohs RC, Marin D, et al. Cholinergic markers in elderly patients with early signs of Alzheimer disease. JAMA 1999;281:1401–1406.

14. DeKosky ST, Ikonomovic MD, Styren SD, et al. Upregulation of choline acetyltransferase activity in hippocampus and frontal cortex of elderly subjects with mild cognitive impairment. Ann Neurol 2002;51:145–155.

15. Goate A, Chartier-Harlin MC, Mullan M, et al. Segregation of a missense mutation in the amyloid precursor protein gene with familial Alzheimer's disease. Nature 1991;349:704–706.

16. Mullan M, Crawford F, Axelman K, et al. A pathogenic mutation for probable Alzheimer's disease in the APP gene at the N-terminus of beta-amyloid. Nat Genet 1992;1:345–347.

17. Sherrington R, Rogaev EI, Liang Y, et al. Cloning of a gene bearing missense mutations in early-onset familial Alzheimer's disease. Nature 1995;375:754–760.

18. Levy-Lahad E, Wijsman EM, Nemens E, et al. A familial Alzheimer's disease locus on chromosome 1. Science 1995;269:970–973.

19. Rogaev EI, Sherrington R, Rogaeva EA, et al. Familial Alzheimer's disease in kindreds with missense mutations in a gene on chromosome 1 related to the Alzheimer's disease type 3 gene. Nature 1995;376:775–778.

20. De Strooper B, Saftig P, Craessaerts K, et al. Deficiency of presenilin-1 inhibits the normal cleavage of amyloid precursor protein. Nature 1998;391:387–390.

21. Corder EH, Saunders AM, Strittmatter WJ, et al. Gene dose of apolipoprotein E type 4 allele and the risk of Alzheimer's disease in late onset families. Science 1993;261:921–923.

22. Poorkaj P, Bird TD, Wijsman E, et al. Tau is a candidate gene for chromosome 17 frontotemporal dementia. Ann Neurol 1998;43:815–825.

23. Hutton M, Lendon CL, Rizzu P, et al. Association of missense and 5'-splice-site mutations in tau with the inherited dementia FTDP-17. Nature 1998;393:702–705.

24. Hardy J, Selkoe DJ. The amyloid hypothesis of Alzheimer's disease: progress and problems on the road to therapeutics. Science 2002;297:353–356.

25. Mucke L, Masliah E, Johnson WB, et al. Synaptotrophic effects of human amyloid beta protein precursors in the cortex of transgenic mice. Brain Res 1994;666:151–167.

26. Hsiao KK. Understanding the biology of beta-amyloid precursor proteins in transgenic mice. Neurobiol Aging 1995;16:705–706.

27. Moechars D, Dewachter I, Lorent K, et al. Early phenotypic changes in transgenic mice that overexpress different mutants of amyloid precursor protein in brain. J Biol Chem 1999;274:6483–6492.

28. Van Dam D, D'Hooge R, Staufenbiel M, et al. Age-dependent cognitive decline in the APP23 model precedes amyloid deposition. Eur J Neurosci 2003;17:388–396.
29. LaFerla FM. Calcium dyshomeostasis and intracellular signalling in Alzheimer's disease. Nat Rev Neurosci 2002;3:862–872.
30. Gotz J, Chen F, van Dorpe J, Nitsch RM. Formation of neurofibrillary tangles in P3011 tau transgenic mice induced by Abeta 42 fibrils. Science 2001;293:1491–1495.
31. Lewis J, Dickson DW, Lin WL, et al. Enhanced neurofibrillary degeneration in transgenic mice expressing mutant tau and APP. Science 2001;293:1487–1491.
32. McKee AC, Kosik KS, Kowall NW. Neuritic pathology and dementia in Alzheimer's disease. Ann Neurol 1991;30:156–165.
33. Terry RD, Masliah E, Salmon DP, et al. Physical basis of cognitive alterations in Alzheimer's disease: synapse loss is the major correlate of cognitive impairment. Ann Neurol 1991;30:572–580.
34. Arriagada PV, Growdon JH, Hedley-Whyte ET, Hyman BT. Neurofibrillary tangles but not senile plaques parallel duration and severity of Alzheimer's disease. Neurology 1992;42:631–639.
35. Samuel W, Terry RD, DeTeresa R, et al. Clinical correlates of cortical and nucleus basalis pathology in Alzheimer dementia. Arch Neurol 1994;51:772–778.
36. Janus C, Pearson J, McLaurin J, et al. A beta peptide immunization reduces behavioural impairment and plaques in a model of Alzheimer's disease. Nature 2000;408:979–982.
37. Morgan D, Diamond DM, Gottschall PE, et al. A beta peptide vaccination prevents memory loss in an animal model of Alzheimer's disease. Nature 2000;408:982–985.
38. Dodart JC, Bales KR, Gannon KS, et al. Immunization reverses memory deficits without reducing brain Abeta burden in Alzheimer's disease model. Nat Neurosci 2002;5:452–457.
39. Chen G, Chen KS, Knox J, et al. A learning deficit related to age and beta-amyloid plaques in a mouse model of Alzheimer's disease. Nature 2000;408:975–979.
40. Hsiao K, Chapman P, Nilsen S, et al. Correlative memory deficits, Abeta elevation, and amyloid plaques in transgenic mice. Science 1996;274:99–102.
41. Gyure KA, Durham R, Stewart WF, et al. Intraneuronal abeta-amyloid precedes development of amyloid plaques in Down syndrome. Arch Pathol Lab Med 2001;125:489–492.
42. Mor C, Spooner ET, Wisniewsk KE, et al. Intraneuronal Abeta42 accumulation in Down syndrome brain. Amyloid 2002;9:88–102.
43. Wirths O, Multhaup G, Czech C, et al. Intraneuronal Abeta accumulation precedes plaque formation in beta-amyloid precursor protein and presenilin-1 double-transgenic mice. Neurosci Lett 2001;306:116–120.
44. Wirths O, Multhaup G, Czech C, et al. Intraneuronal APP/A beta trafficking and plaque formation in beta-amyloid precursor protein and presenilin-1 transgenic mice. Brain Pathol 2002;12:275–286.
45. Tabira T, Chui DH, Kuroda S. Significance of intracellular Abeta42 accumulation in Alzheimer's disease. Front Biosci 2002;7:a44–a49.
46. Takahashi RH, Almeida CG, Kearney PF, et al. Oligomerization of Alzheimer's beta-amyloid within processes and synapses of cultured neurons and brain. J Neurosci 2004;24:3592–3599.
47. Takahashi RH, Milner TA, Li F, et al. Intraneuronal Alzheimer abeta42 accumulates in multivesicular bodies and is associated with synaptic pathology. Am J Pathol 2002;161:1869–1879.
48. Oddo S, Caccamo A, Shepherd JD, et al. Triple-transgenic model of Alzheimer's disease with plaques and tangles: intracellular Abeta and synaptic dysfunction. Neuron 2003;39:409–421.
49. Shie FS, LeBoeuf RC, Jin LW. Early intraneuronal Abeta deposition in the hippocampus of APP transgenic mice. Neuroreport 2003;14:123–129.

50. Casas C, Sergeant N, Itier JM, et al. Massive CA1/2 neuronal loss with intraneuronal and N-terminal truncated Abeta42 accumulation in a novel Alzheimer transgenic model. Am J Pathol 2004;165:1289–1300.
51. Gouras GK, Tsai J, Naslund J, et al. Intraneuronal Abeta42 accumulation in human brain. Am J Pathol 2000;156:15–20.
52. Echeverria V, Ducatenzeiler A, Dowd E, et al. Altered mitogen-activated protein kinase signaling, tau hyperphosphorylation and mild spatial learning dysfunction in transgenic rats expressing the beta-amyloid peptide intracellularly in hippocampal and cortical neurons. Neuroscience 2004;129:583–592.
53. LaFerl, FM, Tinkle BT, Bieberich CJ, et al. The Alzheimer's A beta peptide induces neurodegeneration and apoptotic cell death in transgenic mice. Nat Genet 1995;9:21–30.
54. Magrane J, Smith RC, Walsh K, Querfurth HW. Heat shock protein 70 participates in the neuroprotective response to intracellularly expressed beta-amyloid in neurons. J Neurosci 2004;24:1700–1706.
55. Zhang Y, McLaughlin R, Goodyer C, LeBlanc A. Selective cytotoxicity of intracellular amyloid beta peptide1-42 through p53 and Bax in cultured primary human neurons. J Cell Biol 2002;156:519–529.

Chapter 32
Physiological Processing of the Cellular Prion Protein and βAPP: Enzymes and Regulation

Bruno Vincent, Moustapha Alfa Cisse, and Frédéric Checler

Introduction

Transmissible spongiform encephalopathies (TSEs) as well as Alzheimer's disease (AD) are neurodegenerative syndromes against which no efficient treatments are currently available. Although these pathologies are distinctive in regard to several aspects (incidence, infectious character of prions), they display common features such as the deposition of insoluble proteinaceous aggregates in the brain. Interestingly, the β-amyloid precursor protein (βAPP) and the cellular prion protein (PrP^c), the two culprit membrane proteins in AD and TSEs, respectively, are similarly processed in the middle of their toxic sequence. Indeed, so-called β-secretase cleavage precludes the integrity of the amyloid-β peptide and the 106 to 126 sequence of PrP^c and is up-regulated by protein kinase C (PKC) agonists. The recent identification of two members of the disintegrin family of proteases, ADAM10 and ADAM17, as constitutive and PKC-regulated β-secretase activities cleaving both βAPP and PrP^c further reinforced the link between Alzheimer's and prion diseases. Furthermore, a third ADAM protease, ADAM9, also participates indirectly to the processing of βAPP and PrP^c by modulating ADAM10 activity.

α-Secretase Processing of βAPP and PRP^C

Identification of the amyloid-β peptide (Aβ) in 19841 and the subsequent cloning of its precursor (βAPP) in 1987 [2] marked the beginning of intensive research in the field of AD. Because extracellular deposits in the brain of AD patients are invariably associated with high levels of Aβ, tremendous efforts have been made to elucidate the exact nature of β- and γ-secretases, the enzymes

B. Vincent
Institut de Pharmacologie Moléculaire et Cellulaire, CNRS UMR6097, Valbonne 06560, France

A. Fisher et al. (eds.), *Advances in Alzheimer's and Parkinson's Disease*,
© Springer 2008

responsible for Aβ production [3] (Fig. 1A). In addition to these two "amyloi-dogenic" entities, a third activity, by α-secretase, deserves particular attention because this cleavage occurs in the middle of the Aβ sequence (between lysine687 and leucine688 of αAPP or between residues 16 and 17 of Aβ) thus precluding or at least limiting its production (Fig. 1A). Moreover, α-secretase gives rise to the release in the extracellular space of a large ectodomain named sAPPα (for soluble APP α-cleaved) (Fig. 1A), which has both neurotrophic and neuroprotective functions [4]. Before any identification of genuine α-secretase candidates, several characteristics had been described for this cleavage. First, this cut mainly occurs at the plasma membrane, although part of it could also take place in intracellular compartments along the secretory pathway [5].

Second, it appears to display relaxed specificity, independent of the nature of the amino acid sequence at the targeted site but to be strongly driven by an α-helical conformation, systematically raising cleavages at the 12 and 13 resi-dues from the external leaflet of the plasma membrane [5]. Finally, in vitro

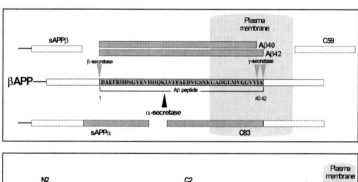

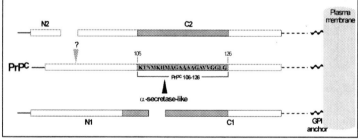

Fig. 1 Processing of βAPP and PrP^c. The amino acid human sequences of Aβ (**A**) and PrP106–126 (**B**) are indicated (gray bars) and aligned according to their α-secretase cleaving sites (black arrowheads). The derived products (sAPPα and C59 for βAPP; N1 and C1 for PrP^c) are mentioned. Sites targeted by β- and γ-secretases on βAPP and by the "disease-specific" unknown activity on PrP^c (gray arrowhead)s. Note that whereas β- and γ-secretase are genuine Aβ-forming enzymes (with the simultaneous production of secreted sAPPβ and intracellular C59), the yet unidentified "pathogenic" PrP-cleaving activity preserves the integrity only of the toxic 106–126 sequence and gives rise to the secreted N2 fragment and the membrane-associated C2 counterpart

experiments using a broad range of protease inhibitors suggested that metallo-proteases were likely responsible for the release of sAPPα from its precursor [6].

The interest in prion processing emerged more recently from work showing that both cellular and scrapie prion protein undergo hydrolysis in the brain in vivo [7]. Importantly, the profile of the proteolytic breakdown varied slightly regardless of whether one considered healthy or Creutzfeldt-Jakob disease (CJD) brains. Indeed, in normal brain, a unique cleavage occurs between amino acids 110 and 111 of PrP^c according to the human sequence numbering [7] and gives rise to the secreted N-terminal fragment named N1 and its C-terminal membrane-associated counterpart C1 (Fig. 1B). It is noteworthy that the 106 to 126 sequence of PrP^c had been shown to convey neurotoxic effects.[8] Thus, this cleavage could represent a putative mean to prevent the integrity of the 106 to 126 sequence similarly to the α-secretase-mediated breakdown of Aβ described above. Interestingly, an additional cleavage specifically operated in CJD brains takes place at the 90/91 site (Fig. 1B). This leftward shift preserves the integrity and toxic potential of the 106 to 126 sequence and could potentially promote the pathogenicity associated with prion diseases.

βAPP and PRP^C Physiological Processing: Up-regulation by PKC Agonists

A well documented aspect of the regulation of the β-secretase pathway concerns its activation by some kinases (Fig. 2). Indeed, several pharmacolo-gical agents targeting PKC, such as phorbol esters and the phosphatase inhibitor okadaic acid, strongly enhance the release of both N1 and sAPPα products [17,53]. In addition, it has been shown that some but not all PKC isozymes contribute to this pathway. Thus, the conventional α and β isoforms as well as the novel β isoform are involved in the regulated release of both sAPPα [30] and N1 (Alfa Cissé et al., unpublished data), whereas a fourth member, the novel PKCδ seems to participate only in N1 production (Alfa Cissé et al., unpublished data). This tends to prove that the molecular mecan-isms governing the PKC-mediated production of sAPPα and N1, although similar, display subtle differences. Noteworthy is the fact that PKCε is speci-fically expressed in the brain, is reduced in the brain of AD patients, and is able to reduce Aβ levels. These findings argue strongly in favor of its parti-cipation in the regulation of βAPP processing in the brain. Of course, it would be of interest to assess whether PKCβ levels are also modified in the brain of individuals with prion diseases.

The role of protein kinase A (PKA) in the regulated metabolism of βAPP and PrP^c is far more elusive. The initial demonstration that purified PKA as well as PKA agonists were able to stimulate the release of sAPPα independently of the PKC pathway [18,19] was challenged by a studies showing that cAMP inhibited sAPPα secretion [20] and that PKA inhibitors displayed no effect in this

βAPP   PrPC

βAPP	Refs		PrPC	Refs
sAPPα		Secreted product	N1	
C83		Membrane-associated product	C1	
	Refs	α-secretase candidates		Refs
ADAM9	9		ADAM9	50
ADAM10	10		ADAM10	51
ADAM17	11		ADAM17	51
Plasmin	12		Plasmin	52
Cathepsin B	13			
Gelatinase A	14			
Yapsins	15			
Calpains	16			
PKC	17	Kinases-mediated up-regulation of sAPPα secretion	PKC	53
PKA	18-19		No effect of PKA	53
No effect of PKA	20-21		?	
MAPK	22		?	
PI3K	23		?	
GSK3α	24		?	
JNK	25		?	
Cdk5	26		?	
Erk 1/2	27		?	
TrkA	28		?	
Rap1/Rac	29		?	
α, β, ε	30	PKC isoforms involved	α, β, δ, ε	Alfa-Cisse et al unpublished
M1/M3 muscarinic	31	Receptors involved	M1/M3 muscarinic	Alfa-Cisse et al unpublished
Thrombin	32		?	
mGluR1α	33		?	
EGF/NGF	34		?	
5-HT4	35	Agents up-regulating sAPPα secretion	?	
HGF	27		?	
FGF	36		?	
PLA2	37		?	
IL-1β	38		?	
Rab6	39		?	
Alkalizing agents	40		?	
Copper	41		?	
NSAIDs	42		?	
Ginkgo biloba	43		?	
17β-estradiol	44		?	
Testosterone	45		?	
Statins	46		?	
		ADAMs up-regulating agents	Refs	
		Acetylcholinesterase inhibitors	47-48	
		Statins	49	

Fig. 2 Comparison of βAPP- and PrPc-α-secretase processing pathways

paradigm [21]. Interestingly, PKA agonists such as forskolin or isoproterenol failed to induce N1 production in cells [53]. Along with PKC, several other kinases (listed in Fig. 2) have been shown to promote the α-secretase cleavage of βAPP. Whether they also play a role in the α-secretase-like processing of PrPc remains to be established and awaits future investigation.

During the search for receptors and ligands acting upstream of PKC activation, which ultimately leads to α-secretase up-regulation, it appeared that the M1/M3 subclass of muscarinic receptors coupled to PLC/PKC, but not the M2/M4 class (linked to adenylate cyclase/PKA), was a key component of this signaling cascade [31]. Indeed, whereas overexpression of M1/M3 receptors strongly increased sAPPα release [31], cells treated with M1/M3-specific agonists significantly induced N1 secretion (Alfa Cissé et al., unpublished data). Because cholinergic transmission is specifically affected in AD, one could postulate that such a defect would severely disturb the β-secretase cleavage of βAPP and progressively lead to aberrant Aβ production and senile plaque formation. This hypothesis is supported by the recent finding that sAPPα is reduced in platelets from AD patients and that normal sAPPα levels can be restored by treatment with acetylcholinesterase inhibitors [48]. Several other receptors, such as thrombin, EGF/NGF, and metabotropic receptors (Fig. 2), have also been proposed to participate in the regulated cleavage of βAPP.

In addition to the in-depth studied PKC-dependent regulation of α-secretase cleavage mentioned above, various other candidates able to stimulate sAPPα secretion have emerged during the past decade (Fig. 2), including hormones, cytokines, cations, cholesterol-lowering drugs, and even plant extracts. Overall, the obvious complexity of the events taking place upstream of the ultimate α-secretase cleavage suggests that identification and targeting of the last link in the chain (i.e., α-secretase per se) represents the simplest way to proceed as it would allow bypassing any putative deleterious side effects.

ADAM10 and ADAM17: Responsible for the Physiological Processing of βAPP and PRP^C?

Several plasma membrane-associated or intracellular α-secretase candidates have been proposed based on their ability to cleave synthetic peptides encompassing the α-site of βAPP and/or on their sensitivity to protease inhibitors. Thus, cathepsin B, an acetylcholinesterase-associated factor, the proteasome gelatinase A, plasmin, calpains, and yapsin have emerged as putative α-secretase activities (Fig. 2). Although one could reasonably postulate that more than a single protease processes βAPP or PrP^c at its α-site, none of the above cited candidates fulfills entirely the above cited characteristics assigned to α-secretase (i.e., predominant localization at the plasma membrane, sensitivity to metalloprotease inhibitors, up-regulation by phorbol esters).

The search for such enzymes has recently led to the identification of members of the ADAM (a disintegrin and metalloprotease) group as genuine α-secretases [54,55]. Structurally, ADAMs are type 1 integral membrane proteins with both a disintegrin domain and a metalloprotease catalytic site [56]. The first evidence that ADAM members could act as genuine α-secretase was provided when TACE [tumor necrosis factor-α (TNFα-converting enzyme,

ADAM17] was shown to contribute to the PKC-regulated secretion of sAPPα [11] Indeed, mouse embryonic fibroblasts that have been invalidated for TACE lost their ability to secrete increased amounts of sAPPα when stimulated with phorbol esters. The same was true when TACE-deficient cells were examined for their PDBu-stimulated N1 secretion [51]. In parallel, in human cells overexpressing TACE, PDBu-induced N1 production is drastically enhanced [51]. It is important to note that in both studies the basal amounts of sAPPα and N1 remained unaffected by TACE invalidation. These data imply that TACE is involved in PKC-mediated α-secretase activity but does not participate in the constitutive secretion of N1 and sAPPα. Two other important characteristics of TACE further reinforced its α-secretase nature: (1) TACE does not require any particular amino acid around its cleavage sites; and (2) recombinant TACE cleaves βAPP-derived synthetic peptides encompassing the α-secretase site at the expected peptide bond (Lys16–Leu17). Moreover, its implication in the regulated α-secretase pathway is supported by the ability of purified PKC to phosphorylate in vitro a construct containing the last 129 C-terminus amino acids of TACE (with two intracellular PKC phosphorylation sites on serines 803 and 808) fused to glutathione-S-transferase (GST). Whether these phosphorylation events have a direct impact on TACE activity remains to be elucidated.

The first demonstration that ADAM10 participates in the α-secretase processing of βAPP came from a study showing that overexpression of this enzyme in human cells significantly triggered both constitutive and PKC-dependent sAPPα secretion [10]. However, in additional experiments performed in cells devoid of PKC-regulated sAPPα production (Lovo cells), overexpression of ADAM10, but not TACE, still induced sAPPα release in the extracellular space [57]. Thus, ADAM10 seems to be mainly involved in the constitutive shedding of βAPP. Other studies supported the α-secretase function of ADAM10. First, like TACE, recombinant soluble ADAM10 cleaves βAPP-derived peptides at the expected Lys16–Leu17 peptidyl bond. Second, the ubiquitous expression of ADAM10 mRNA throughout the brain compared to the restricted pattern of TACE expression suggests a predominant role of ADAM10 as an in vivo α-secretase and places in question the physiological relevance of TACE. However, it has been postulated that embryonic fibroblasts derived from mice invalidated for ADAM10 were not altered in their ability to secrete sAPPα [58] This does not necessarily challenge the α-secretase role of ADAM10 and more likely suggests that compensatory mechanisms take place, as is often the case in knockout animals. In addition, we recently showed, in the same cell system, that ADAM10 invalidation indeed led to a significant loss of sAPPβ recovery in the extracellular space [50]. Nevertheless, the most compelling evidence of the physiological role played by ADAM10 as a βAPP-cleaving α-secretase was provided by the use of transgenic mice overexpressing this enzyme. Indeed, overexpression of ADAM10 in vivo not only was associated with increased production of sAPPα but also reduced Aβ levels; most importantly, it prevented plaque formation and alleviated the associated neurological defects [59].

Whether ADAM10 was also a key element in the processing of PrP^c inside its 106 to 126 neurotoxic sequence was investigated in our laboratory by means of both overexpresssion and invalidation approaches. ADAM10 overexpression specifically promoted the constitutive production of the N1 fragment in human HEK293 cells, whereas ADAM10–deficient embryonic fibroblasts reduced by 50% the constitutive formation of N1 without altering the PKC-dependent counterpart [51]. Additional support for ADAM10 being involved in the physiological processing of PrP^c has been brought by a recent study showing that, in vivo, the presence of the active form of the enzyme in human brain correlates with high levels of the C1 fragment [60] (the C-terminal membrane-associated counterpart of secreted N1) (Figs. 1, 3). The question immediately arising from such data is whether ADAM10-overexpressing cells or mice could be protected against PrP^{sc} infection. Answering this question is a priority of our laboratory. Finally, shedding enzymes have been generally described as proteolytic activities responsible for the release of extracellular domains from transmembrane proteins. Thus, identification of ADAM proteases as PrP^c-cleaving activities is the first demonstration that disintegrins are also involved in the shedding of glycosylphosphatidylinositol-anchored proteins.

ADAM9: Indirect Contribution to βAPP and PRP^C Physiological Processing by Modulating ADAM10 Activity

Among the numerous ADAM proteases identified so far, ADAM9 has been proposed to contribute to the β-secretase cleavage of βAPP. An early study showed that when co-expressed with βAPP ADAM9 exclusively produced sAPPα after treatment with phorbol ester [9]. The same researchers demonstrated that an alternatively spliced secreted form of the enzyme was also displaying an α-secretase activity toward βAPP [61]. Finally, a double-stranded RNA interference approach targeting ADAM9, ADAM10, and TACE in COS-7 cells suggested that endogenous α-secretase has, at least partially, these three activities [62]. Nevertheless, the observation that, unlike ADAM10 and TACE, ADAM9 does not cleave βAPP-related synthetic peptides at the expected Lys16–Leu17 site but, rather, at the His14-Gln15 bond [63], argues against a direct role of ADAM9 in the β-secretase processing of βAPP.

Part of this issue was clarified while we were investigating a putative contribution of ADAM9 to the physiological processing of PrP^c, as we had done previously for ADAM10 and TACE [50]. We first showed, in various cell types, that both transient and stable overexpression of ADAM9 increased N1 and sAPPα secreted products. Conversely, decreasing endogenous ADAM9 expression by an antisense approach drastically affected N1 and sAPPα secretions. One of the original findings of this study was that ADAM9 was totally unable to modulate PrP^c and βAPP α-secretase processing in cells invalidated for ADAM10. This somehow reconciled the apparently paradoxical findings mentioned above and was further reinforced by the fact that ADAM9-over-expressing intact cells, in the absence of ADAM10, were totally unable to cleave an α-secretase quenched fluorimetric substrate. We further showed that ADAM9 indirectly regulates α-secretase cleavage of PrP^c and βAPP most likely by contributing to the shedding of ADAM10 at the plasma membrane. Altogether, these results, as well as previous findings, suggest that ADAM9 would act upstream of ADAM10 which would likely represent the genuine βAPP- and PrP^c-cleaving enzyme.

Conclusion

The similar α-secretase-mediated processing of βAPP and PrP^c is illustrated in Fig. 3. The recent identification of members of the ADAM proteases as having α-secretase activity in regard to βAPP and PrP^c opened a new area of investigation that might prove useful for the treatment of Alzheimer's and prion diseases. Thus, an interesting common feature to the processing of these two proteins is that cleavage by ADAM10, TACE, and ADAM9 (in an indirect manner) occurs in the middle of their toxic core (Figs. 1, 3). It has long been hypothesized

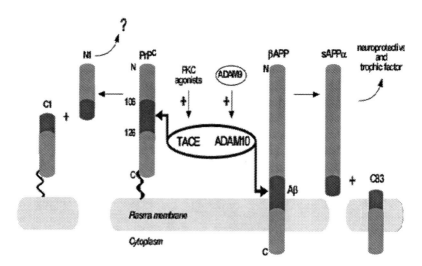

Fig. 3 α–Secretase cleavage of βAPP and PrP^c. Both βAPP and PrP^c are cleaved inside their "toxic" sequences Aβ and PrP106–126, respectively (dark gray segments). The constitutive cleavage is operated by ADAM10, which can be further enhanced by ADAM9, whereas PKC–regulated cleavage is performed by TACE. The secreted βAPP-derived product sAPPα has neurotrophic and neuroprotective functions. Whether such properties are also inherent to the N1 fragment is currently being investigated in our laboratory

that up-regulation of the α-secretase pathway would reduce Aβ formation. This theory has been recently supported by the elegant ADAM10 transgenic model in which sAPPα is increased, Aβ is reduced, and plaque formation and hippocampal deficits are delayed [59].

Although the sAPPα/N1 secretion can be up-regulated by several agents (Fig. 2), little is known about the underlying cellular mechanisms. Thus, it is of utmost importance to determine which of these pathways selectively promote ADAM activity and/or expression. Interestingly, it has emerged from recent work that acetylcholinesterase inhibitors and statins indeed increase ADAM10 expression and activity [48,49]. Therefore, medications displaying direct effects on α-secretase might hold promise as treatment of AD or prion diseases, although one must keep in mind that side effects, due to the large number of other substrates for ADAMs, have not been excluded.

References

1. Glenner GG, Wong CW. Alzheimer's disease: initial report of purification and characterization of a novel cerebrovascular amyloid protein. Biochem Biophys Res Commun 1984;120:885–890.
2. Kang J, Lemaire HG, Unterbeck A, et al. The precursor of Alzheimer's disease amyloid A4 protein resembles a cell-surface receptor. Nature 1987;325:733–736.
3. Wilquet V, De Strooper B. Amyloid-beta precursor protein processing in neurodegeneration. Curr Opin Neurobiol 2004;14:82–88.
4. Saitoh T, Sundsmo M, Roch JM, et al. Secreted form of amyloid β protein precursor is involved in the growth regulation of fibroblasts. Cell 1989;58:615–622.
5. Sisodia SS. β-Amyloid precursor protein cleavage by a membrane-bound protease. Proc Natl Acad Sci U S A 1992;89:6075–6079.
6. Roberts SB, Ripellino JA, Ingalls KM, et al. Non-amyloidogenic cleavage of the β-amyloid precursor protein by an integral membrane metalloendopeptidase. J Biol Chem 1994;269:3111–3116.
7. Chen SG, Teplow DB, Pachi P, et al. Truncated forms of the human prion protein in normal brain and in prion diseases. J Biol Chem 1995;270:19173–19180.
8. Forloni G, Angeretti N, Chiesa R, et al. Neurotoxicity of a prion protein fragment. Nature 1993;362:543–546.
9. Koike H, Tomioka S, Sorimachi H, et al. Membrane-anchored metalloprotease MDC9 has an α-secretase activity responsible for processing the amyloid precursor protein. Biochem J 1999;343:371–375.
10. Lammich S, Kojro E, Postina R, et al. Constitutive and regulated α-secretase cleavage of Alzheimer's amyloid precursor protein by a disintegrin metalloprotease. Proc Natl Acad Sci U S A 1999;96:3922–3927.
11. Buxbaum JD, Liu KN, Luo Y, et al. Evidence that tumor necrosis factor α-converting enzyme is involved in regulated β-secretase cleavage of the Alzheimer amyloid protein precursor. J Biol Chem 1998;273:27765–27767.
12. Ledesma MD, Da Silva JS, Crassaerts K, et al. Brain plasmin enhances APP α-cleavage and Aβ degradation and is reduced in Alzheimer's disease brains. EMBO Rep 2000;1:530–535.

13. Tagawa K, Kunishita T, Maruyama K, et al. Alzheimer's disease amyloid β-clipping enzyme (APP secretase): identification, purification, and characterization of the enzyme. Biochem Biophys Res Commun 1991;177:377–387.

14. Miyazaki K, Hasegawa M, Funahashi K, Umeda M. A metalloprotease inhibitor domain in Alzheimer amyloid protein precursor. Nature 1993;362:839–841.

15. Komano H, Seeger M, Gandy SE, et al. Involvement of cell surface glycosyl-phosphatidylinositol-linked aspartyl proteases in α-secretase-type cleavage and ectodomain solubilization of human Alzheimer β-amyloid precursor protein in yeast. J Biol Chem 1998;273:31648–31651.

16. Chen M, Durr J, Fernandez HL. Possible role of calpain in normal processing of β-amyloid precursor protein in human platelets. Biochem Biophys Res Commun 2000;273:170–175.

17. Buxbaum JD, Gandy SE, Cicchetti P, et al. Processing of Alzheimer β/A4 amyloid precursor protein: modulation by agents that regulate protein phosphorylation. Proc Natl Acad Sci U S A 1990;87:6003–6006.

18. Xu H, Sweeney D, Greengard P, Gandy SE. Metabolism of Alzheimer β-amyloid precursor protein: regulation by protein kinase A in intact cells and in a cell-free system. Proc Natl Acad Sci U S A 1996;93:4081–4084.

19. Marambaud P, Wilk S, Checler F. Protein kinase A phosphorylation of the proteasome: a contribution to the α-secretase pathway in human cells. J Neurochem 1996;67:2616–2619.

20. Efthimiopoulos S, Punj S, Manolopoulos V, et al. Intracellular cyclic AMP inhibits constitutive and phorbol ester-stimulated cleavage of amyloid precursor protein. J Neurochem 1996;67:872–875.

21. Marambaud P, Ancolio K, Alves da Costa C, Checler F. Effects of protein kinase A inhibitors on the production of Aβ40 and Aβ42 by human cells expressing normal and Alzheimer's disease-linked mutated βAPP and presenilin 1. Br J Pharmacol 1999;126:1186–1190.

22. Mills J, Charest DL, Lam F, et al. Regulation of amyloid precursor protein catabolism involves the mitogen-activated protein kinase signal transduction pathway. J Neurosci 1997;17:9415–9422.

23. Petanceska SS, Gandy SE. The phosphatidylinositol 3-kinase inhibitor wortmannin alters the metabolism of the Alzheimer's amyloid precursor protein. J Neurochem 1999;73:2316–2320.

24. Phiel CJ, Wilson CA, Lee VMY, Klein PS. GSK-3α regulates production of Alzheimer's disease amyloid-β peptides. Nature 2003;423:435–439.

25. Mudher A, Chapman S, Richardson J, et al. Dishevelled regulates the metabolism of amyloid precursor protein via protein kinase C/mitogen-activated protein kinase and c-jun terminal kinase. J Neurosci 2001;21:4987–4995.

26. Liu F, Su Y, Li B, Zhou Y, et al. Regulation of amyloid precursor protein (APP) phosphorylation and processing by p35/Cdk5 and p25/Cdk5. FEBS Lett 2003;547:193–196.

27. Liu F, Su Y, Li B, Ni B. Regulation of amyloid precursor protein expression and secretion via activation of ERK1/2 by hepatocyte growth factor in HEK293 cells transfected with APP751. Exp Cell Res 2003;287:387–396.

28. Rossner S, Ueberham U, Schliebs R, et al. Regulated secretion of amyloid precursor protein by TrkA receptor stimulation in rat pheochromocytoma-12 cells is mitogen activated protein kinase sensitive. Neurosci Lett 1999;271:97–100.

29. Maillet M, Robert SJ, Cacquevel M, et al. Crosstalk between Rap1 and Rac regulates secretion of sAPPββ Nat Cell Biol 2003;5:633–639.

30. Kinouchi T, Sorimachi H, Maruyama K, et al. Conventional protein kinase C (PKC)-α and novel PKCε, but not –δ, increase the secretion of an N-terminal fragment of Alzheimer's disease amyloid precursor protein from PKC cDNA transfected 3Y1 cells. FEBS Lett 1995;364:203–206.

31. Nitsch RM, Slack BE, Wurtman RJ, Growdon JH. Release of Alzheimer amyloid precursor derivatives stimulated by activation of muscarinic acetylcholine receptors. Science 1992;258:304–307.

32. Davis-Salinas J, Saporito-Irwin SM, Donovan FM, et al. Thrombin receptor activation induces secretion and nonamyloidogenic processing of amyloid β-protein precursor. J Biol Chem 1994;269:22623–22627.

33. Lee RKK, Wurtman RJ, Cox AJ, Nitsch RM. Amyloid precursor protein processing is stimulated by metabotropic glutamate receptors. Proc Natl Acad Sci U S A 1995;92:8083–8087.

34. Refolo LM, Salton SRJ, Anderson JP, et al. Nerve and epidermal growth factors induce the release of the Alzheimer amyloid precursor from PC12 cell cultures. Biochem Biophys Res Commun 1989;164:664–670.

35. Robert SJ, Zugaza JL, Fischmeister R, et al. The human serotonin 5-HT4 receptor regulates secretion of non-amyloidogenic precursor protein. J Biol Chem 2001;276:44881–44888.

36. Schubert D, Jin LW, Saitoh T, Cole G. The regulation of amyloid β protein precursor secretion and its modulatory role in cell adhesion. Neuron 1989;3:689–694.

37. Emmerling MR, Moore CJ, Doyle PD, et al. Phospholipase A_2 activation influences the processing and secretion of the amyloid precursor protein. Biochem Biophys Res Commun 1993;197:292–297.

38. Dash PK, Moore AN. Enhanced processing of APP induced by IL-1β can be reduced by indomethacin and nordihydroguaiaretic acid. Biochem Biophys Res Commun 1995;208:542–548.

39. McConlogue L, Castellano F, deWit C, et al. Differential effects of a rab6 mutant on secretory versus amyloidogenic processing of Alzheimer's β-amyloid precursor protein. J Biol Chem 1996;271:1343–1348.

40. Schrader-Fischer G, Paganetti PA. Effect of alkalizing agents on the processing of the β-amyloid precursor protein. Brain Res 1996;716:91–100.

41. Borchardt T, Camakaris J, Cappai R, et al. Copper inhibits β-amyloid production and stimulates the non-amyloidogenic pathway of amyloid precursor protein secretion. Biochem J 1999;344:461–467.

42. Avramovich Y, Amit T, Youdim MBH. Non-steroidal anti-inflammatory drugs stimulate secretion of non-amyloidogenic precursor protein. J Biol Chem 2002;277:31466–31473.

43. Colciaghi F, Borroni B, Zimmermann M, et al. Amyloid precursor protein metabolism is regulated toward alpha-secretase pathway by ginkgo biloba extracts. Neurobiol Dis 2004;16:454–460.

44. Xu H, Gouras GK, Greenfield JP, et al. Estrogen reduces neuronal generation of Alzheimer β-amyloid peptides. Nat Med 1998;4:447–451.

45. Gouras GK, Xu H, Gross RS, et al. Testosterone reduces neuronal secretion of Alzheimer's β-amyloid peptides. Proc Natl Acad Sci U S A 2000;97:1202–1205.

46. Refolo LM, Pappolla MA, LaFrancois J, et al. A cholesterol-lowering drug reduces β-amyloid pathology in a transgenic mouse model of Alzheimer's disease. Neurobiol Dis 2001;8:890–899.

47. Zimmermann M, Gardoni F, Marcello E, et al. Acetylcholinesterase inhibitors increase ADAM10 activity by promoting its trafficking in neuroblastoma cell lines. J Neurochem 2004;90:1489–1499.

48. Zimmermann M, Borroni B, Cattabeni F, et al. Cholinesterase inhibitors influence APP metabolism in Alzheimer diseases patients. Neurobiol Dis 2005;19:237–242.

49. Kojro E, Gimpl G, Lammich S, et al. Low cholesterol stimulates the nonamyloidogenic pathway by its effect on the α-secretase ADAM10. Proc Natl Acad Sci U S A 2001;98:5815–5820.

50. Alfa Cissé M, Sunyach C, Lefranc-Jullien S, et al. The disintegrin ADAM9 indirectly contributes to the physiological processing of cellular prion by modulating ADAM10 activity. J Biol Chem 2005;280:40624–40631.

51. Vincent B, Paitel E, Saftig P, et al. The disintegrins ADAM10 and TACE contribute to the constitutive and phorbol ester-regulated normal cleavage of the cellular prion protein. J Biol Chem 2001;276:37743–37746.

52. Praus M, Kettelgerdes G, Baier M, et al. Stimulation of plasminogen activation by recombinant cellular prion protein is conserved in the NH2-terminal fragment PrP23–110. Thromb Haemost 2003;89:812–819.

53. Vincent B, Paitel E, Frobert Y, et al. Phorbol ester-regulated cleavage of normal prion protein in HEK293 human cells and murine neurons. J Biol Chem 2000;275:35612–35616.

54. Allinson TMJ, Parkin ET, Turner AJ, Hooper NM. ADAMs family members as amyloid precursor protein α-secretases. J Neurosci Res 2003;74:342–352.

55. Vincent B. ADAM proteases: protective role in Alzheimer's and prion diseases? Curr Alz Res 2004;1:165–174.

56. Seals DF, Courtneidge SA. The ADAMs family of metalloproteases: multidomain proteins with multiple functions. Genes Dev 2003;17:7–30.

57. Lopez-Perez E, Zhang Y, Frank SLJ, et al. Constitutive α-secretase cleavage of the β-amyloid precursor protein in the furin-deficient LoVo cell line: involvement of the pro-hormone convertase 7 and the disintegrin ADAM10. J Neurochem 2001;76:1532–1539.

58. Hartmann D, De Strooper B, Serneels L, et al. The disintegrin/metalloprotease ADAM10 is essential for notch signalling but not for β-secretase activity in fibroblasts. Hum Mol Genet 2002;11:2615–2624.

59. Postina R, Schroeder A, Dewachter I, et al. Disintegrin-metalloproteinase prevents amyloid plaque formation and hippocampal defects in an Alzheimer disease mouse model. J Clin Invest 2004;113:1456–1464.

60. Laffont-Proust I, Faucheux BA, Hässig R, et al. The N-terminal cleavage of cellular prion protein in the human brain. FEBS Lett 2005;579:6333–6337.

61. Hotoda N, Koike H, Sasagawa N, Ishiura S. A secreted form of human ADAM9 has an α-secretase activity for APP. Biochem Biophys Res Commun 2002;293:800–805.

62. Asai M, Hattori C, Szabo B, et al. Putative function of ADAM9, ADAM10, and ADAM17 as APP α-secretase. Biochem Biophys Res Commun 2003;301:231–235.

63. Roghani M, Becherer JD, Moss ML, et al. Metalloprotease-disintegrin MDC9: intracellular maturation and catalytic activity. J Biol Chem 1999;274:3531–3540.

Chapter 33
Expression of Wnt Receptors, Frizzled, in Rat Neuronal Cells

Marcelo A. Chacón, Marcela Columbres, and Nibaldo C. Inestrosa

Introduction

Recent data suggest a role for the Wnt signaling pathway on neuronal physiology, including regulation of neuronal connectivity, axon remodeling, and synapse formation. On the other hand, we have demonstrated that activation of Wnt signaling can rescue neurons from amyloid-β (Aβ)-induced neurotoxicity. In this sense, it is our interest to study whether Wnt effects on synaptic function are the mediators of neuroprotective events. In this chapter, we describe the analysis of the expression of Wnt receptors, Frizzled (Fzd), on rat neuronal cells. We found that Fzd receptors are expressed in adult rat cortex and hippocampus as well as in cultured hippocampal neurons. An analysis of the Fzd promotor region showed that most Fzd genes present TCF/LEF binding sites, suggesting direct transcriptional regulation by Wnt signaling. In rat brain sections, Fzd-3 was observed in both hippocampal and cortical neurons, and Fzd-1/2 was found to co-localize with PSD-95 in primary cultured hippocampal neurons when they were exposed to the Wnt-5a ligand. All of these data suggest that Wnt signaling may play a role in synaptic function in the adult mammalian central nervous system and that neuroprotective effects against Aβ neurotoxicity can be mediated by these events.

Background

Wnt signaling has been described as an essential pathway in developmental and oncogenic processes [1,2]. Some years ago it was implicated in neurodegenerative disorders such as bipolar disorder [3], schizophrenia [4,5], and Alzheimer's

M.A. Chacón
Centro de Regulación Celular y Patología "Joaquín V. Luco" (CRCP), MIFAB, Facultad de Ciencias Biológicas, Pontificia Universidad Católica de Chile

A. Fisher et al. (eds.), *Advances in Alzheimer's and Parkinson's Disease*,
© Springer 2008

disease (AD) [6,7]. Glycogen synthase kinase-3β (GSK-3β), an evolutionarily conserved kinase, is the central modulator of the Wnt signaling pathway [8]. From the canonical view of the Wnt signaling pathway, the secreted Wnt ligand interacts with a seven-transmembrane frizzled receptor, which transduces its signal by activating disheveled protein, which inactivates GSK-3β through formation of a multiprotein complex. As a consequence of GSK-3β inactivation, intracellular levels of β-catenin increase, allowing its binding to components of a high mobility group family of transcription factors, T-cell factor/lymphoid enhancer-binding factors (TCF/LEF), which regulate the expression of multiple target genes including engrailed and cyclin D1. In contrast, in the absence of Wnt ligand, the activity of GSK-3β is switched on phosphorylating β-catenin for ubiquitin proteosome-mediated degradation [9]. As a consequence, β-catenin levels are decreased in the cytosol, and the expression of Wnt target genes is repressed.

Amyloid-β-peptide neurotoxicity has been associated with neuronal shrinkage and dendritic dystrophy similar to AD neurons that develop neurofibrillar pathology. Primary cultures of fetal rat hippocampal and human cortical neurons exposed to Aβ induced hyperphosphorylation of tau proteins and loss of microtubular network, probably through the activation of GSK-3β [10]. Lithium, a reversible inhibitor of GSK-3β, is able to mimic the Wnt pathway [11] protecting postmitotic neurons from the cytotoxic effects of Aβ, preserving their shape, axonal processes, and neurites [12,13]. Moreover, conditioned media from HEK 293 cells were transfected with an expression vector containing HA-tagged mouse Wnt-3a, and hippocampal cells exposed to 5- to 10-μM fibrils were protected from its cytotoxic effect when co-incubated with HEK-293-conditioned medium containing Wnt-3a protein [14]. This effect was not observed when hippocampal cells were exposed to Aβ fibrils and a control conditioned medium contained the expression vector without the Wnt-3a insert. We have also reported that Wnt signaling blocks the behavioral impairments induced by hippocampal injection of Aβ fibrils [13,15]. Histopathological analysis of hippocampi injected with preformed Aβ fibrils showed abundant neuronal loss in the upper leaf of the dentate gyrus and strong GSK-3β immunoreactivity around deposits of Aβ fibrils, which were also positive for thioflavin S. These data agree with the observation in conditional transgenic mice, which overexpress GSK-3β [16], in which there is a decrease in β-catenin levels and neurodegeneration. Aβ neurotoxicity compromises the stability of cytosolic levels of endogenous β-catenin, likely affecting its translocation to the nucleus, which can be prevented by low doses of lithium.

Although we have demonstrated that activation of the Wnt signaling pathway protects neurons from Aβ-induced toxicity, it is unknown whether upstream components of the cascade are involved in this event. With this in mind, we initiated a study of the Wnt ligand receptors, Frizzled (Fzd). All 10 known mammalian Fzd are seven-transmembrane proteins displaying characteristic motifs that allowed them to be included in the G-protein-coupled receptors superfamily [17,18], although the direct interaction with G-proteins has not been demonstrated. Fzd presents a large cysteine-rich extracellular domain (CRD) responsible for Wnt ligand binding [19,20] and a cytoplasmic tail that significantly differs in length and sequence similarity [21]. Fzd have been shown to signal at least three distinct pathways. In addition to the canonical β-catenin pathway (see below), it has been described in the Rho/ JNK planar cell polarity pathway in *Drosophila*, which regulates developmental processes [22,23], and the Wnt/Ca^{2+} pathway, which signals through hetero-trimeric G-proteins and triggers the release of intracellular Ca^{2+} [24], which activates protein kinase C and calmodulin-dependent protein kinase II [25]. Several studies have demonstrated that the single transmembrane LRP5 and LRP6 proteins are co-receptors for Wnt ligands and that they are required for canonical Wnt signaling [26–28]. A recent article provided a new mechanism whereby the same receptor can be switched between distinct pathways depending on the differential recruitment of a co-receptor by members of the same Wnt ligand family [29]. Mice lacking the fzd3 gene show deficient axonal growth [30], and those lacking the fzd4 gene show altered cerebellar development [31], supporting the idea that Fzd play an important role in central nervous system (CNS) development.

As a first approach, we analyzed the expression of Fzd in rat hippocampal neurons (7 days in culture) and in the rat adult hippocampus and cortex. For this purpose, total RNA was extracted from cultured rat hippocampal neurons or directly from brain tissues. RNA (3 µg) was reverse-transcribed, and the produced cDNA was used for the polymerase chain reaction (PCR) using specific primers for each receptor. As observed in Fig. 1, Fzd are abundantly expressed in rat neuronal cells. Although this is not a quantitative tool, it is interesting to note that both Frizzled-3 and Frizzled-8 seem to be more

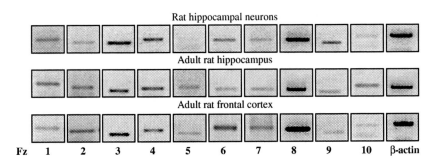

Fig. 1 Frizzled expression pattern in rat central nervous system (CNS). Total RNA was extracted from rat hippocampal neurons and adult rat frontal cortex and hippocampus. Reverse transcription polymerase chain reaction (RT-PCR) assay was carried out using the cDNA obtained and specific primers for each Frizzled. Amplification products were visualized in 1% agarose/TAE gels stained with ethidium bromide

abundant in the three analyzed samples. It has been shown by in situ hybridization that Frizzled-3 is mainly expressed in the hippocampus, specifically in the dentate gyrus [32]; however, in the present study, we detected its expression in both hippocampus and cortex.

The Wnt signaling pathway controls biological processes through the regulation of target gene expression. Several novel target genes have been identified for this signaling pathway. We reported earlier that the activation of this signaling pathway by lithium or Wnt ligands in primary hippocampal neurons induces Wnt target gene expression of the survival gene bcl-2 [33]. It has been already characterized that frizzled 7 [34] and Dfrizzled 2 [35] are Wnt target genes; however, analysis of further frizzled receptors as Wnt target genes remains unexplored. Thus, the promoter regions (4000 bp upstream from ATG) for all fzd human genes were analyzed for TCF/LEF binding sites (A/T A/T CAAAG) through Genomatix software resources (www.genomatix.de). The total numbers and positions of TCF/LEF binding sites as well as the proximal sites are shown in Table 1.

As shown in Table 1 most *fzd* genes present two proximal TCF/LEF binding sites, suggesting that activation of the Wnt signaling pathway might regulate Fzd expression. In the context of synaptic function, Fzd signaling has been shown to regulate the neural potential of progenitors in the developing nervous system [36], and Fzd-9 has been shown to act as a critical determinant of hippocampal development [37], which suggests that the regulation of Fzd by the activation of the Wnt signaling pathway could have important effects on synapse formation.

We have performed immunohistochemical analysis in rat coronal brain sections to study the localization of Fzd proteins. Using an antibody that recognizes Fzd-1 and Fzd-2, we have been unable to detect any signal, probably because of their low expression in tissues. Using an anti-Fzd-3 antibody, we observed positive staining in both the hippocampal region and the cortex as well as in other brain regions (Fig. 2a). On higher magnification micrographs, it is

Table 1 TCF/LEF boxes found in human *Frizzled* genes

No. of TCF boxes (proximal sites)		Basepair upstream of ATG
Frizzled 1	6 (2)	**940, 1087,** 2020, 2553, 3375, 3788
Frizzled 2	6 (4)	**317, 1075, 1742, 1899,** 2247, 3600
Frizzled 3	4 (2)	**434, 1547,** 2570, 3204
Frizzled 4	6 (1)	**1326,** 2534, 2228, 3287, 3404, 3864
Frizzled 5	2 (2)[a]	**1615, 1729**
Frizzled 6	5 (2)	**221, 270,** 2093, 3504, 3904
Frizzled 7	5 (2)[a]	**1057, 1196,** 2382, 1196, 1057
Frizzled 8	2 (1)	**1617,** 2937
Frizzled 9	0[a]	
Frizzled 10	8 (5)[a]	**222, 1027, 1160, 1169, 1286,** 3277, 3720, 3742

Proximal TCF/LEF binding sites (2000 bp upstream ATG) are shown in **bold**.
[a] Sequences with a certain degree of nonidentified nucleotides.

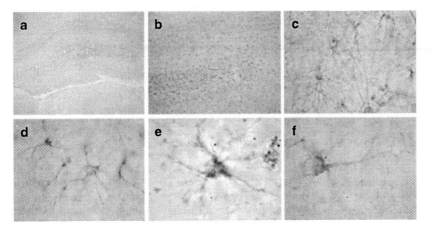

Fig. 2 Immunohistochemical localization of Frizzled-3 in rat CNS. Using a polyclonal antibody against Frizzled-3, free-floating coronal brain sections were permeabilized with 0.1% TX-100, treated with 0.05% H_2O_2 to reduce the endogenous peroxidase activity, blocked with 3% bovine serum albumin (BSA) in phosphate-buffered saline (PBS), and incubated overnight with the primary antibody (1:500). After washes with PBS, brain sections were incubated with the appropriate secondary antibody conjugated with horseradish peroxidase for 1 hour at room temperature. The sections were incubated in a 0.06% diaminobenzidine solution and revealed with 0.01% H_2O_2. **a** ×4. **b** ×10. **c, d** ×40. **e, f** ×100

possible to observe staining of neuronal cells and probably glial cells (Fig. 2b–f). A more exhaustive analysis is necessary to determine the specific distribution of Frizzled-3 in the rat CNS.

It is known that the postsynaptic density protein 95 (PSD-95), a protein crucial for the assembly and localization of multiprotein signaling complexes at specific sites on the plasma membrane [38], interacts with Fzd-1, -2, -4, and -7 through its PDZ domains [39], suggesting that PSD family proteins participate in the recruitment of signaling molecules of the Wnt pathway. As we are interested in studying the Wnt signaling pathway and its relation with the synapse as well as how the activation of this pathway can modulate synaptic processes, we performed immunofluorescence analysis to visualize Fzd-1 and Fzd-2 (red) and PSD-95 (green) in the presence of Wnt-5a in primary cultured hippocampal neurons. The incubation with this ligand induced an increased signal for both Fzd1/2 and PSD-95 with respect to control neurons (data not shown) located in the same regions. (Fig. 3). This suggested that Wnt signaling can regulate the distribution of its receptors through PSD-95.

Thus, we suggest a positive role for Wnt signaling on adult mammalian CNS function, whose impact may be related to the neuroprotective events induced by Wnt ligands.

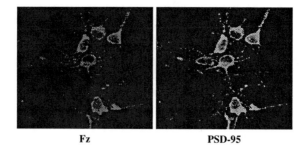

Fz   PSD-95

Fig. 3 Localization of Fzd1/2 and PSD-95 in hippocampal neurons treated with Wnt5a. Hippocampal neurons were treated with conditioned medium Wnt5a, and the distribution of both Fzd1/2 and PSD-95 was studied. As observed, treatment with Wnt5a induced a significant co-localization of both proteins in the neuron soma with respect to control neurons (data not shown), suggesting a role for Wnt ligands in the interaction between Fzd1/2 and the postsynaptic protein and in the localization of Frizzled proteins

References

 1. Nusse R, Varmus HE. Wnt genes. Cell 1992;69(7):1073–1087.
 2. Polakis P. Wnt signaling and cancer. Genes Dev 2000;14(15):1837–1851.
 3. Gould TD, Manji HK. The Wnt signaling pathway in bipolar disorder. Neuroscientist 2002;8:497–511.
 4. Cotter D, Kerwin R, al-Sarraji S, et al. Abnormalities of Wnt signaling in schizophrenia: evidence for neurodevelopmental abnormality. Neuroreport 1998;9:1379–1383.
 5. Miyaoka T, Seno T, Ishino H. Increased expression of Wnt-1 in schizophrenia brains. Schizophr Res 1999;38:1–6.
 6. De Ferrari GV, Inestrosa NC. Wnt signaling function in Alzhemer's disease. Brain Res Brain Res Rev 2000;33(1):1–12.
 7. Anderton BH. Alzheimer's disease: clues from flies and worms. Curr Biol 1999;9(3):R106–R109.
 8. Chen RH, Ding WV, McCormick F. Wnt signaling to beta-catenin involves two interactive components: glycogen synthase kinase-3beta inhibition and activation of protein kinase C J Biol Chem 2000;275(23):17894–17899.
 9. Aberle H, Bauer A, Stappert J, et al. Beta-catenin is a target for the ubiquitin-proteasome pathway. EMBO J 1997;16(13):3797–3804.
10. Busciglio J, Lorenzo A, Yeh F, Yankner BA. β-Amyloid fibrils induce tau phosphorylation and loss of microtubule binding. Neuron 1995;14:879–888.
11. Hedgepeth CM, Conrad LJ, Zhang J, et al. Activation of the Wnt signaling pathway: a molecular mechanism for lithium action. Dev Biol 1997;185(1):82–91.
12. Inestrosa NC, Alvarez A, Godoy J, et al. Acetylcholinesterase-amyloid-beta-peptide interaction and Wnt signaling involvement in Abeta neurotoxicity. Acta Neurol Scand Suppl 2000;176:53–59.
13. Inestrosa NC, De Ferrari GV, Garrido JL, et al. Wnt signaling involvement in beta-amyloid-dependent neurodegeneration. Neurochem Int 2002;41(5):341–344.
14. Alvarez AR, Godoy JA, Mullendorff K, et al. Wnt-3a overcomes beta-amyloid toxicity in rat hippocampal neurons. Exp Cell Res 2004;297:186–196.

15. De Ferrari GV, Chacón MA, Barría MI, et al. Activation of Wnt signaling rescues neurodegeneration and behavioral impairments induced by β-amyloid fibrils. Mol Psychiatry 2003;8:195–208.
16. Lucas JJ, Hernandez F, Gomez-Ramos P, et al. Decreased nuclear beta-catenin, tau hyperphosphorylation and neurodegeneration in GSK-3beta conditional transgenic mice. EMBO J 2001;20(1-2):27–39.
17. Barnes MR, Duckworth DM. Frizzled proteins constitute a novel family of G protein-coupled receptors, most closely related to the secretin family. Trends Pharmacol Sci 1998;19:399–400.
18. Fredriksson R, Lagerström MC, Lundin LG, Schiöth HB. The G-protein coupled receptors in the human genome form five main families: phylogenetic analysis, paralogon groups, and fingerprints. Mol Pharmacol 2003;63:1256–1272.
19. Bhanot P, Brink M, Samos CH, et al. A new member of the frizzled family from Drosophila functions as a Wingless receptor. Nature 1996;382:225–230.
20. Wang Y, Macke JP, Abella BS, et al. A large family of putative transmembrane receptors homologous to the product of the Drosophila tissue polarity gene frizzled. J Biol Chem 1996;271(8):4468–4476.
21. Wodarz A, Nusse R. Mechanisms of Wnt signaling in development. Annu Rev Cell Dev Biol 1998;14:59–88.
22. Moldzik M. Planar polarity in the Drosophila eye: a multifaceted view of signaling specificity and cross-talk. EMBO J 1999;18:6873–6879.
23. Strutt D. Frizzled signalling and cell polarisation in Drosophila and vertebrates. Development 2003;130:4501–4513.
24. Slusarski DC, Corces VG, Moon RT. Interaction of Wnt and a frizzled homologue triggers G-protein-linked phosphatidylinositol signalling. Nature 1997;390:410–413.
25. Kühl M, Sheldahl LC, Park M, et al. The Wnt/Ca^{2+} pathway: a new vertebrate Wnt signaling pathway takes shape. Trends Genet 2000;16:279–283.
26. Pinson KI, Brennan J, Monkley S, et al. An LDL-receptor-related protein mediates Wnt signalling in mice. Nature 2000;407:535–538.
27. Tamai K, Semenov M, Kato Y, et al. LDL-receptor-related proteins in Wnt signal transduction. Nature 2000;407:530–535.
28. Wehrli M, Dougan ST, Caldwell K, et al. Arrow encodes an LDL-receptor-related protein essential for Wingless signaling. Nature 2000;407(6803):527–530.
29. Liu G, Bafico A, Aaronson S. The mechanism of endogenous receptor activation functionally distinguishes prototype canonical and noncanonical Wnts. Mol Cell Biol 2005;25(9):3475–3482.
30. Wang Y, Huso D, Cahill H, et al. Progressive cerebellar, auditory, and esophageal dysfunction caused by targeted disruption of the frizzled-4 gene. J Neurosci 2001;21:4761–4771.
31. Wang Y, Thekdi N, Smallwood PM, et al. Frizzled-3 is required for the development of major fiber tracts in the rostral CNS. J Neurosci 2002;22:8563–8573.
32. Shimogori T, VanSant J, Paik E, Grove EA. Members of the Wnt, Fz, and Frp gene families expressed in postnatal mouse cerebral cortex. J Comp Neurol 2004;473(4):496–510.
33. Fuentealba RA, Farias G, Scheu M, et al. Signal transduction during amyloid-β-peptide neurotoxicity: role in Alzheimer disease. Brain Res Brain Res Rev 2004;47:275–289.
34. Willert J, Epping M, Pollack JR, et al. A transcriptional response to Wnt protein in human embryonic carcinoma cells. BMC Dev Biol 2002;2:8.
35. Cadigan KM, Fish MP, Rulifson EJ, Nusse R. Wingless repression of Drosophila frizzled 2 expression shapes the Wingless morphogen gradient in the wing. Cell 1998;93(5):767–777.
36. Van Raay TJ, Moore KB, Iordanova I, et al. Frizzled 5 signaling governs the neural potential of progenitors in the developing Xenopus retina. Neuron 2005;46:23–36.

37. Zhao C, Aviles C, Abel RA, et al. Hippocampal and visuospatial learning defects in mice with a deletion of frizzled 9, a gene in the Williams syndrome deletion interval. Development 2005;132:2917–2927.
38. Sheng M, Sala C. PDZ domains and the organization of supramolecular complexes. Annu Rev Neurosci 2001;24:1–29.
39. Hering H, Sheng M. Direct interaction of Frizzled-1, -2, -4, and -7 with PDZ domains of PSD-95. FEBS Lett 2002;521(1-3):185–189.

Chapter 34
Cell Models of Tauopathy

J. Biernat, I. Khlistunova, Y-P. Wang, M. Pickhardt, M. von Bergen, Z. Gazova, Eckhart Mandelkow, and Eva-Marie Mandelkow

Introduction

A group of diseases are characterized by the aggregation of the microtubule-associated protein tau in a highly phosphorylated form ("tauopathies"). They include Alzheimer's disease (AD) and frontotemporal dementias (FTDP-17) [1]. These changes lead to loss of synapses and neurodegeneration [2]. Generating reliable models for tau pathology has been difficult because the overproduction of tau by itself does not lead to neurofibrillary aggregation but to axonal transport defects that show up, for example, as motor neuron disease in mice if tau is expressed in the wrong cell types [3]. To generate neurofibrillary pathology, it has been necessary to enhance the toxicity of tau by FTDP-17 mutations or by combining tau mutations with an enhanced β-amyloid (Aβ) load [4–10]. However, there is still a debate on whether tau aggregation is toxic, and if removal of the aggregates is beneficial [11,12]. Furthermore, screening for tau aggregation inhibitors is difficult with mice or neuronal cell cultures derived from them. This problem can be addressed by using inducible cell models where the tau expression can be switched on and off at defined time points.

We have developed several N2a cell lines that allow inducible expression of the repeat domain of tau [13]. They are based on the Tet-On system, where the expression can be induced by the addition of the tetracycline analog doxycycline [14]. Three variants of the repeat domain of tau (tau$_{RD}$) are expressed: the wild-type sequence; a "proaggregation mutant" (ΔK280) that is highly prone to aggregation; and an "antiaggregation mutant" containing additional proline residues that suppress aggregation [15]. These cells display a robust aggregation of tau within a few days, including AD-like paired helical filaments. The cells illustrate that the aggregation of tau$_{RD}$ is toxic and that removal of aggregates is beneficial. Fragmentation by a thrombin-like protease is a prelude to aggregation, whereas phosphorylation in the repeat

J. Biernat
Max-Planck-Unit for Structural Molecular Biology, c/o DESY, Hamburg, Germany

A. Fisher et al. (eds.), *Advances in Alzheimer's and Parkinson's Disease*,
© Springer 2008

domain bears no obvious relation to aggregation. Exposing the cells to inhibitor compounds reduces aggregation and toxicity, and even cells containing aggregates can be rescued. Thus, the cell model is useful in the search for drugs for AD and other tauopathies.

Results

Cell lines

The derivatives of tau used for the inducible N2a cell lines were based on the tau construct K18. It contains the four-repeat domain that is essential for microtubule binding and for the aggregation of tau into paired helical filaments (PHFs) (Fig. 1). The repeat domain forms the core of AD PHFs and can be polymerized into PHFs in vitro.16 The mutation K18-ΔK280 was particularly suitable because it has a high tendency to form β-structure and aggregation [17]. The mutation has been described as one of the tau mutations in a case of FTDP-17 [18,19]. As a contrasting variant, we generated a double proline mutant (PP mutant) where prolines were inserted into each of the two hexapeptide motifs (I277P, I308P) that nucleate β-structure during aggregation. Because the prolines disrupt β-strands, they inhibit aggregation [15,20].

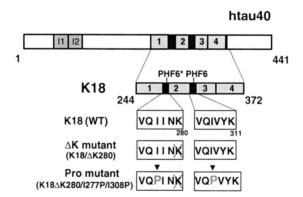

Fig. 1 Tau constructs. *Top:* hTau40, the longest isoform of human CNS (441 residues), containing 4 repeats of ~31 residues each. *Bottom:* The tau$_{RD}$ constructs are based on the four-repeat domain [construct K18, (M)Q244–E372] It contains two hexapeptide motifs that promote PHF aggregation by formation of β-structure ([275]VQIINK[280] in R2 and [306] VQIVYK[311] in R3). The mutant K18-ΔK280 lacks K280 (= ΔK mutant) and therefore aggregates more readily. The double proline mutant K18-ΔK280-PP contains I277P and I308P replacements in the hexapeptide motifs (= PP mutant), which inhibit aggregation because they disrupt a potential β-strand

For establishing the Tet-On inducible N2a cells we generated the host N2a clone with a stably integrated reversed tetracycline transactivator rtTA-S2 under the control of the CMV promoter [21]. Inducible cell lines were then generated by stable transfection of pBI-5 bifunctional vectors encoding the tau_{RD} constructs and a reporter luciferase gene controlled by the P_{bi}-1 promoter. Four cell lines were generated: N2a/Tet-On/pBI-5 (mock), N2a/Tet-On/K18-WT, N2a/Tet-On/K18-ΔK280, and N2a/Tet-On/K18-ΔK280/2P. The cells were selected for clones with the lowest background of luciferase activity and tau_{RD} expression. Doxycycline was chosen at 1 µg/ml for inducing tau_{RD} expression. The yield after 2 days was comparable for the constructs (about 1.2–1.8 µg from 1×10^6 cells).

Aggregation of Tau in Cells

An important question was regarding the aggregation of tau_{RD} and its effects on the cells. One could anticipate several ways of how tau_{RD} might affect cells (e.g., overstabilization of microtubules [22], inhibition of microtubule-based transport) [23,24]. or aggregation. When focusing on the aggregation process, we sought constructs that bind weakly to microtubules (to avoid transport defects) yet can aggregate. The repeat construct K18 met both conditions. Second, we wanted to observe aggregation independently of the expression of soluble tau, which can be achieved by fluorescence microscopy using thioflavin S (ThS), biochemical analysis, and electron microscopy (Fig. 2). ThS is a faithful reporter of the aggregation of tau and can be used to screen for inhibitors of aggregation [25]. Among the three tau_{RD} constructs tested, only the ΔK280 mutant induced a strong reaction with ThS (Fig. 2b), consistent with the observation that this mutant has a strong tendency for spontaneous aggregation in vitro. The wild-type tau_{RD} construct K18 displayed a much weaker aggregation, and aggregation was nearly absent from the inhibitory PP mutant (Fig. 2c). In addition, filaments resembling those of AD PHFs were present in the cells (not shown). Biochemically, the aggregates were demonstrated by sarkosyl extraction [26]. The proteins were separated by sodium dodecyl sulfate (SDS) gel electrophoresis and blotted with antibodies against total tau or phosphorylated tau (Fig. 3). During the first 10 days of K18wt expression, the protein remains mostly in the soluble fraction and is intact (Fig. 3a). The insoluble fraction is minor up to day 11, when the sample reveals both lower cleavage products and higher aggregates.

More details of the aggregation are obtained from the proaggregation mutant K18-ΔK280 (Fig. 3b). Here, the expression leads to a pronounced high-molecular-weight (HMW) smear in the sarkosyl-insoluble pellet, similar to AD aggregates [26,27]. There are also pronounced lower MW breakdown

Fig. 2 Expression and aggregation of tau$_{RD}$. *Left:* Expression of tau in N2a cell models monitored by immunolabeling with antibody K9JA and rhodamine secondary antibody. *Right:* Aggregates monitored by staining with thioflavin S. (a) N2a cells expressing construct K18 tau constructs; (b) K18-ΔK280; (c) K18-ΔK280-PP. Note that appreciable aggregation occurs only in (b)

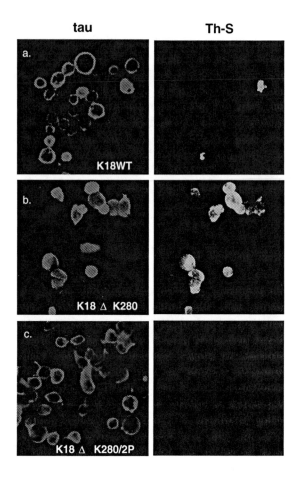

products, mainly bands at relative MW (Mr) 9.9 and 7.3 kDa (fragments F2 and F3, see below). They are smaller than the parent protein (F0 = K18, Mr 12.4 kDa) and fragment F1 (Mr ~11.3 kDa) in soluble tau$_{RD}$ fractions (Fig. 3). The fragmentation becomes visible well before the aggregates appear. Thus, aggregation is strongly correlated with cleavage, and tau fragments accumulate in the pellet. In contrast to the ΔK mutant, the PP mutant shows almost no tendency to aggregate (Fig. 3c), consistent with the insertion of β-breaking prolines into the hexapeptide motifs in R2 and R3. The PP mutant also shows almost no fragments in the pellet, suggesting that fragmentation is related to aggregation.

The phosphorylation of tau in the repeats (at KXGS motifs) was probed with antibody 12E8 (not shown). For all tau$_{RD}$ variants, the phosphorylation remained mainly in the supernatant. This suggests that phosphorylation at KXGS motifs does not enhance the tendency to aggregate [28]. Fragmentation was also less apparent in the phosphorylated protein.

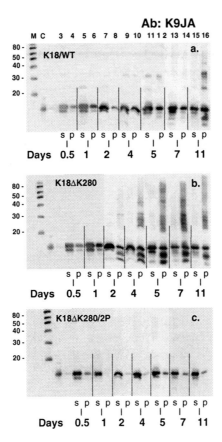

Fig. 3 Kinetics of tau_{RD} expression and aggregation. Lanes labeled P denote sarkosyl-insoluble tau_{RD} species, which are pelletable after sarkosyl extraction and measured by blot analysis and densitometry. S indicates the corresponding soluble proteins. (a)–(c) Blot analysis with antibody K9JA (independent of phosphorylation). (a) Expression of K18; (b) K18-ΔK280 mutant; (c) K18-ΔK280-PP mutant. Note that in (a) a fraction of the K18 protein gradually becomes insoluble (P lanes) but retains its normal apparent molecular weight (MW) until day 11 when a higher MW smear appears. In (c) there is very little aggregation of the PP mutant. By contrast, the ΔK280 mutant (b) shows pronounced aggregation in the pellet, combined with increasing proteolysis (lower bands in the P lanes) and a high-MW smear of aggregates

Fragmentation of Tau$_{RD}$ by a Thrombin-like Protease

The nature of the fragments was determined by sequencing and mass spectrometry. It showed that all fragments started at S258: Fragment F2 represents peptides S258–V363 (106 residues) and S258–P364 (107 residues), F3 is S258–K353 (96 residues) and S258–I360 (103 residues). The cleavage site K257/S258 has been reported as a major site for thrombin [29]. It is consistent with the specificity of thrombin, and the treatment of cells with the thrombin inhibitor PPACK inhibited the fragmentation of tau$_{RD}$ (not shown). Inhibition of other proteases was tested but had no effect. The C-terminal ends of the fragments did not agree with definite protease specificities, possibly owing to some carboxypeptidases or autolytic processes, such as nonenzymatic cleavage behind Asn [27]. The suspected role of thrombin was checked by digestion experiments with K18ΔK280 in vitro. This generated the same major N-terminal cleavage site, which generated a single fragment (F1) extending to the end of the protein (S258–E372).

Aggregation and Toxicity

There is a debate whether the toxicity is due to tau aggregates per se or to some other related events. We approached this question by expressing different tau constructs for different time periods and observing cell degeneration by the lactate dehydrogenase (LDH) assay. Figure 4 shows that the expression of soluble tau$_{RD}$ has no noticeable effect on the viability of the cells, that is, K18wt and the antiaggregation PP mutant have the same level of LDH release as the control cells. The situation is different for the proaggregation ΔK280 mutant, where the toxicity is about twofold higher than in the control. We could show that aggregating species of tau$_{RD}$ cause toxicity, even at a stage where the aggregates are not yet detected by ThS fluorescence (e.g., tau oligomers).

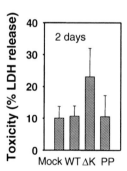

Fig. 4 Cell toxicity of tau$_{RD}$ expression and aggregation. Tau construct K18 or mutants were induced for 2 days. Lactate dehydrogenase (LDH) release was measured as an indicator of cell death, calculated as percent of total LDH (media + lysate). Bar 1 = LDH release of N2a mock cells without doxycycline-induced tau expression (control); bar 2 = N2a cells expressing K18; bar 3 = ΔK mutant; bar 4 = PP mutant. Note that expression of the ΔK mutant strongly increases toxicity, whereas wild-type K18 and the PP mutant remain around control values

Reversibility of Aggregation

Next, we asked whether tau aggregates can be removed. The expression of tau$_{RD}$ was first induced for 5 days by doxycycline until aggregates were clearly present; then the doxycycline was removed, and the cells were assayed several days later (Fig. 5). In all cases, the aggregates disappeared, showing that, in principle, the aggregation is reversible. Again, the aggregation was most pronounced with the ΔK280 mutant. It displayed both the HMW smear and the lower-MW triplet, consisting of the original protein (F0 = K18-Δ280) and the two fragments characteristic of the aggregated pellet (F2, F3). When the expression of tau$_{RD}$ is silenced by removing doxycycline, all these bands and the ThS fluorescence in cells disappear. We conclude that the aggregation of tau$_{RD}$ can be reversed by lowering the concentration, indicating that the protein subunits are exchangeable between the aggregated and soluble state. This is in contrast to the seemingly irreversible aggregation of tau in AD neurofibrillary tangles, which may be the result of secondary modifications.

Inhibition of Aggregation by Small Molecule Inhibitors

Because tau aggregation can be inhibited and reversed, it would be expected that there is a labile equilibrium between soluble and aggregated forms of tau. One should therefore be possible to reverse or suppress the aggregation in cells by inhibitor compounds (Fig. 6). We had previously screened a library for compounds that inhibit the aggregation of tau$_{RD}$ in vitro [25]. Some of these compounds were tested on inducible N2a cell lines (Fig. 7) [50% inhibitory

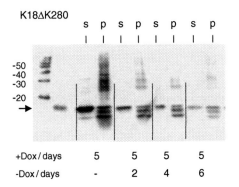

Fig. 5 Reversibility of the aggregation of tau$_{RD}$. Expression of K18 or mutants was first induced for 5 days by doxycycline, then switched off by removal of doxycycline and assayed for expression and aggregation. Shown are blots of supernatants and sarkosyl-insoluble pellets for the case of K18-ΔK280 mutant. The amount of soluble and insoluble material, formed during the 5 days of tau expression, decreases again after silencing the tau gene

N2a/ K18ΔK280 + Dox

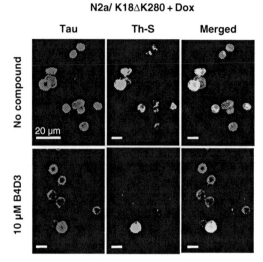

Fig. 6 Tau aggregation inhibition in cells. (a) N2a cells were induced to express K18-ΔK280 for 5 days. *Left column*, staining for tau (antibody K9JA-TRITC); *center*, staining for aggregates (thioflavin S, or ThS); *right*, merged images. Note that the ΔK280 mutant aggregates readily and therefore generates numerous ThS-positive cells (*top row, center*). *Bottom row*: Experiments in the presence of aggregation inhibitor compound B4D3, which strongly reduces the number of ThS-positive cells

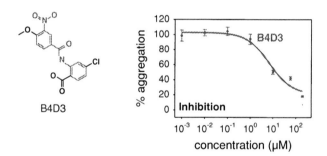

Fig. 7 Tau aggregation inhibition in vitro. Structure of inhibitor compound B4D3 and dose–response curve measured in vitro for the inhibition of aggregation (IC50 values around 10 μM)

concentration (IC50) in the micrometer range in vitro]. Although Tau$_{RD}$ showed robust protein expression and aggregation, the presence of inhibitor compounds strongly reduced aggregation. To test whether aggregates could be disrupted, tau$_{RD}$ was first expressed for 5 days, and then compounds were added for another 2 days. This reduced the level of aggregation to ~50%. This procedure also reduced the toxicity of tau$_{RD}$ to control levels (not shown). We conclude that aggregates of tau$_{RD}$ can be prevented by inhibitor compounds, that even preformed PHFs can be dissolved again, and that removing the aggregates reduced the toxicity of tau to cells.

Discussion

Aggregation and phosphorylation of tau are hallmarks of AD and related tauopathies [1,30]. To study these issues on a cellular and molecular level, we developed several cell models with the following features: (1) The cells are derived from a well characterized neuronal cell line (N2a). (2) The expression of tau$_{RD}$ can be switched on and off. (3) The cell models were made in three variants, expressing the repeat domain of tau in the wild-type sequence or with proaggregation or antiaggregation mutations. (4) The aggregation of tau$_{RD}$ can be monitored in cells by fluorescence, biochemical methods, or electron microscopy. These tools allowed us to test whether the aggregation of tau$_{RD}$ is toxic to cells, if aggregation and toxicity can be reversed, what the relation between the phosphorylation of tau$_{RD}$ and its aggregation is, and if the aggregation can be prevented or reversed by drugs.

One of the problems in the field has been to find indicators of abnormal aggregation in tau-expressing cells. Even strong tau overexpression generally does not cause PHFs, but this can be overcome by using FTDP-17 mutations of tau [4–8,10,31], by adding the cytotoxic Aβ peptide or mutant APP, or by expressing variants of tau. Here we used the four-repeat domain with the ΔK280 mutation because it has a high tendency for aggregation [15,17]. The aggregation can be measured by ThS fluorescence; the aggregates can be visualized by electron microscopy; and incipient aggregates can be revealed by a sarkosyl-insoluble pellet and its accompanying smear at intermediate molecular weights, which is characteristic for AD tau [26] and can be explained by nonenzymatic fragmentation and crosslinking [27].

Although aggregation and phosphorylation occur together in AD, we found no obvious relation in the cell models (not shown). We conclude that phosphorylation of tau's repeat domain does occur in the cell to some extent, but it does not predict aggregation. However, we note that many of the phosphorylation sites of AD tau, notably the SP or TP motifs, lie outside the repeat domain and are not present here. The main phosphorylation sites in tau$_{RD}$ are the KXGS motifs (phosphorylated by the kinase MARK and detected by the antibody 12E8), which promote the dissociation from microtubules [24,32]. Although it is reasonable to assume that breaking tau microtubule bonds is a prerequisite for tau–tau aggregation, it appears that phosphorylation in the repeat domain is not a precursor of PHF–tau, in agreement with earlier observations that phosphorylation of tau$_{RD}$ at the KXGS motifs inhibits aggregation [28].

An unexpected finding was the involvement of specific tau fragments and proteases in the aggregation process. The N-terminal site of fragments F1 to F4 is suggestive of thrombin-like activity, in agreement with Arai et al. [29]. The

C-terminal cleavage site is less well defined, probably due to other proteases. Thus, it is possible that soluble tau_{RD} can be cleaved by thrombin in the first repeat, generating fragment F1 from S258 to the end of the repeat domain, which is later digested from its C-terminal end, generating F2 and F3, which aggregate readily. The HMW smear appears subsequently. The ready aggregation of fragments F2 and F3, but not of F1, may be explained by their being enriched in motifs with a high propensity for β-structure [20,33].

Of particular interest was the question whether the aggregation of tau_{RD} is toxic to the cell. Our results suggest that toxicity is related to aggregation, not merely to the expression of tau_{RD}. We note that the toxicity is already high before abundant aggregation. It is therefore possible that toxicity is caused by species smaller than PHFs (e.g., oligomers). This is reminiscent of the aggregation of Aβ or other peptides where oligomers, rather than polymers, are considered the main toxic species [34–36] Because aggregates of tau_{RD} are toxic, it would be important to find methods to counteract aggregation. The cell's own clearing capacity can be tested by silencing the expression of tau_{RD} (Fig. 5); and indeed soluble tau_{RD} disappears, as does the aggregated tau_{RD}.

Thus, the cell has mechanisms to break down the aggregates. Some proposed catabolic pathways include the ubiquitinproteasome system [37,38], calpain [39,40], or autophagy [41]. As shown elsewhere [42], aggregates of tau_{RD} in vitro are markedly labile and can therefore be disintegrated into subunits that could be digested by various cellular proteases. This explains why toxic aggregates can be removed by the cell by decreasing tau_{RD} production. The consequence is that it should be possible to enhance the cell's self-cleaning capabilities by inhibitory drugs. Such compounds can be identified by in vitro screening [25,43,44], and some of them have been shown to suppress aggregates in cells (Figs. 6, 7). Importantly, these cells also recovered from the toxicity of tau. Thus, the cell models are suitable for developing small-molecule inhibitors of tau aggregation and toxicity.

Acknowledgments We thank Dagmar Drexler and Olga Petrova for excellent technical assistance. We are especially indebted to H. Bujard and K. Schönig (University of Heidelberg) for components and help with the Tet-On system. This work was supported by grants from the Deutsche Forschungsgemeinschaft.

References

1. Lee VM, Goedert M, Trojanowski JQ. Neurodegenerative tauopathies. Annu Rev Neurosci 2001;24:1121–1159.
2. Coleman PD, Yao PJ. Synaptic slaughter in Alzheimer's disease. Neurobiol Aging 2003;24:1023–1027.
3. Terwel D, Lasrado R, Snauwaert J, et al. Changed conformation of mutant tau-P301L underlies the moribund tauopathy, absent in progressive, nonlethal axonopathy of tau-4R/2N transgenic mice. J Biol Chem 2005;280:3963–3973.

4. Lewis J, Dickson DW, Lin WL, et al. Enhanced neurofibrillary degeneration in transgenic mice expressing mutant tau and APP. Science 2001;293:1487–1491.

5. Gotz J, Chen F, van Dorpe J, Nitsch RM. Formation of neurofibrillary tangles in P3011 tau transgenic mice induced by Abeta 42 fibrils. Science 2001;293:1491–1495.

6. Oddo S, Caccamo A, Shepherd JD, et al. Triple-transgenic model of Alzheimer's disease with plaques and tangles: intracellular Abeta and synaptic dysfunction. Neuron 2003;39:409–421.

7. Vogelsberg-Ragaglia V, Bruce J, Richter-Landsberg C, et al. Distinct FTDP-17 mis-sense mutations in tau produce tau aggregates and other pathological phenotypes in transfected CHO cells. Mol Biol Cell 2000;11:4093–4104.

8. DeTure M, Ko LW, Easson C, Yen SH. Tau assembly in inducible transfectants expressing wild-type or FTDP-17 tau. Am J Pathol 2002;161:1711–1722.

9. Ferrari A, Hoerndli F, Baechi T, et al. Beta-amyloid induces paired helical filament-like tau filaments in tissue culture. J Biol Chem 2003;278:40162–40168.

10. Santacruz K, Lewis J, Spires T, et al. Tau suppression in a neurodegenerative mouse model improves memory function. Science 2005;309:476–481.

11. LaFerla FM, Oddo S. Alzheimer's disease: Abeta, tau and synaptic dysfunction. Trends Mol Med 2005;11:170–176.

12. Ashe KH. Mechanisms of memory loss in Abeta and tau mouse models. Biochem Soc Trans 2005;33:591–594.

13. Khlistunova I, Biernat J, Wang Y, et al. Inducible expression of tau repeat domain in cell models of tauopathy: aggregation is toxic to cells but can be reversed by inhibitor drugs. J Biol Chem 2006;281:1205–1214.

14. Gossen M, Bujard H. Studying gene function in eukaryotes by conditional gene inactivation. Annu Rev Genet 2002;36:153–173.

15. Von Bergen M, Barghorn S, Li L, et al. Mutations of tau protein in fronto-temporal dementia promote aggregation of paired helical filaments by enhancing local beta- structure. J Biol Chem 2001;276:48165–48174.

16. Wille H, Drewes G, Biernat J, et al. Alzheimer-like paired helical filaments and antiparallel dimers formed from microtubule-associated protein tau in vitro. J Cell Biol 1992;118:573–584.

17. Barghorn S, Zheng-Fischhofer Q, Ackmann M, et al. Structure, microtubule interactions, and paired helical filament aggregation by tau mutants of fronto-temporal dementias. Biochemistry 2000;39:11714–11721.

18. Rosso SM, van Swieten JC. New developments in frontotemporal dementia and parkinsonism linked to chromosome 17. Curr Opin Neurol 2002;5:423–428.

19. D'Souza I, Schellenberg GD. Regulation of tau isoform expression and dementia. Biochim Biophys Acta 2005;1739:104–115.

20. Von Bergen M, Friedhoff P, Biernat J, et al. Assembly of tau protein into Alzheimer paired helical filaments depends on a local sequence motif ([306]VQIVYK[311]) forming beta structure. Proc Natl Acad Sci U S A 2000;97:5129–5133.

21. Urlinger S, Baron U, Thellmann M, et al. Exploring the sequence space for tetracycline-dependent transcriptional activators: novel mutations yield expanded range and sensitivity. Proc Natl Acad Sci U S A 2000;97:7963–7968.

22. Bunker JM, Wilson L, Jordan MA, Feinstein SC. Modulation of microtubule dynamics by tau in living cells: implications for development and neurodegeneration. Mol Biol Cell 2004;15:2720–2728.

23. Stamer K, Vogel R, Thies E, et al. Tau blocks traffic of organelles, neurofilaments, and APP vesicles in neurons and enhances oxidative stress. J Cell Biol 2002;156:1051–1063.

24. Mandelkow EM, Thies E, Trinczek B, et al. MARK/PAR1 kinase is a regulator of microtubule-dependent transport in axons. J Cell Biol 2004;167:99–110.

25. Pickhardt M, Gazova Z, von Bergen M, et al. Anthraquinones inhibit tau aggregation and dissolve Alzheimer's paired helical filaments in vitro and in cells. J Biol Chem 2005;280:3628–3635.

26. Greenberg SG, Davies P. Preparation of Alzheimer paired helical filaments that display distinct tau proteins by polyacrylamide gel electrophoresis. Proc Natl Acad Sci U S A 1990;87:5827–5831.

27. Watanabe A, Hong WK, Dohmae N, et al. Molecular aging of tau: disulfide-independent aggregation and non-enzymatic degradation in vitro and in vivo. J Neurochem 2004;90:1302–1311.

28. Schneider A, Biernat J, von Bergen M, et al. Phosphorylation that detaches tau protein from microtubules (Ser262, Ser214) also protects it against aggregation into Alzheimer paired helical filaments. Biochemistry 1999;38:3549–3558.

29. Arai T, Guo JP, McGeer PL. Proteolysis of non-phosphorylated and phosphorylated tau by thrombin. J Biol Chem 2005;280:5145–5153.

30. Binder LI, Guillozet-Bongaarts AL, Garcia-Sierra F, Berry RW. Tau, tangles, and Alzheimer's disease. Biochim Biophys Acta 2005;1739:216–223.

31. Perez M, Hernandez F, Gomez-Ramos A, et al. Formation of aberrant phosphotau fibrillar polymers in neural cultured cells. Eur J Biochem 2002;269:1484–1489.

32. Drewes G, Ebneth A, Preuss U, et al. MARK, a novel family of protein kinases that phosphorylate microtubule-associated proteins and trigger microtubule disruption. Cell 1997;89:297–308.

33. Mukrasch MD, Biernat J, von Bergen M, et al. Sites of tau important for aggregation populate beta-structure and bind to microtubules and polyanions. J Biol Chem 2005;280:24978–24986.

34. Lambert MP, Barlow AK, Chromy BA, et al. Diffusible, nonfibrillar ligands derived from Abeta1-42 are potent central nervous system neurotoxins. Proc Natl Acad Sci U S A 1998;95:6448–6453.

35. Walsh DM, Klyubin I, Fadeeva JV, et al. Naturally secreted oligomers of amyloid beta protein potently inhibit hippocampal long-term potentiation in vivo. Nature 2002;416:535–539.

36. Bucciantini M, Giannoni E, Chiti F, et al. Inherent toxicity of aggregates implies a common mechanism for protein misfolding diseases. Nature 2002;416:507–511.

37. David DC, Layfield R, Serpell L, et al. Proteasomal degradation of tau protein. J Neurochem 2002;83:176–185.

38. Kosik KS, Shimura H. Phosphorylated tau and the neurodegenerative foldopathies. Biochim Biophys Acta 2005;1739:298–310.

39. Delobel P, Leroy O, Hamdane M, et al. Proteasome inhibition and tau proteolysis: an unexpected regulation. FEBS Lett 2005;579:1–5.

40. Brown MR, Bondada V, Keller JN, et al. Proteasome or calpain inhibition does not alter cellular tau levels in neuroblastoma cells or primary neurons. J Alzheimers Dis 2005;7:15–24.

41. Nixon RA, Wegiel J, Kumar A, et al. Extensive involvement of autophagy in Alzheimer disease: an immuno-electron microscopy study. Neuropathol Exp Neurol 2005;64:113–122.

42. Li L, von Bergen M, Mandelkow EM, Mandelkow E. Structure, stability, and aggregation of paired helical filaments from tau protein and FTDP-17 mutants probed by tryptophan scanning mutagenesis. J Biol Chem 2002;277:41390–41400.

43. Chirita C, Necula M, Kuret J. Ligand-dependent inhibition and reversal of tau filament formation. Biochemistry 2004;43:2879–2887.

44. Taniguchi S, Suzuki N, Masuda M, et al. Inhibition of heparin-induced tau filament formation by phenothiazines, polyphenols, and porphyrins. J Biol Chem 2005;280:7614–7623.

Chapter 35
Co-expression of FTDP-17 Human Tau and GSK-3β (or APP^{SW}) in Transgenic Mice: Induction of Tau Polymerization and Neurodegeneration

Jesús Avila, Tobias Engel, José J. Lucas, Mar Pérez, Alicia Rubio, and Félix Hernández

Introduction

Alzheimer's disease (AD) is characterized by the presence of two aberrant structures, senile plaques and neurofibrillary tangles (NFTs), present in the brain of the patients [1]. Senile plaques are mainly composed of aggregates of a peptide, β-amyloid peptide (Aβ), that arises from the proteolytic cleavage of a precursor protein (APP) [2,3]. The main component of NFTs is tau protein in hyperphosphorylated form [4]. The phosphorylation of tau protein occurs at two types of residue: proline-directed sites and non-proline-directed sites [5,6].

The presence of NFTs in the brain of AD patients has been correlated with the level of dementia [7,8]. There are other dementias characterized by the presence of aberrant aggregates of phosphorylated tau (tauopathies), where senile plaques are absent [9]. An example is the frontotemporal dementia linked to chromosome 17 (FTDP-17). In some FTDP-17 cases, tau is in mutated form, which facilitates its phosphorylation and aggregation [10,11]. A possible requirement of tau phosphorylation prior to its aberrant aggregation has been suggested [12]. To test that hypothesis, several mouse models expressing tau, bearing various mutations present in FTDP-17 patients, have been developed [13–23]. In addition, a mouse model overexpressing the main tau kinase GSK3β has been developed to increase the level of phosphorylated tau [24]. Phosphorylation of tau by this kinase may play a role in the aberrant aggregation of tau found in a mouse model expressing tau protein with mutations present in FTDP-17 patients [19–25]. In this way, it was found that in the presence of lithium, a specific inhibitor of GSK3, no aberrant tau aggregates were assembled in an FTDP-17 transgenic mouse model [25].

In this chapter, we comment on the consequences of: (1) crossing a mouse bearing three of the mutations present in FTDP-17 (line tau^{VLW}) [19] with a

J. Avila

Centro de Biología Molecular "Severo Ochoa" (CSIC-UAM). Facultad de Ciencias. Campus de Cantoblanco. Universidad Autónoma de Madrid. 28049-Madrid, Spain

A. Fisher et al. (eds.), *Advances in Alzheimer's and Parkinson's Disease*,
© Springer 2008

transgenic mouse overexpressing GSK3β (line GSK-3); or (2) crossing the mouse tau^{VLW} with that one overexpressing the protein APP with the Swedish mutation (mouse APP^{SW}) [26], which promotes the appearance of Aβ, a peptide that could facilitate tau phosphorylation [27].

Tau Pathology in a Double Transgenic Mouse Overexpressing Tau Bearing Three of the Mutations Present in FTDP-17 (tau^{VLW}) and GSK3 β

Double transgenic mice were generated [28], and both tau phosphorylation and tau aggregation were analyzed and compared to those found in wild-type mice or in the single transgenic mice overexpressing tau^{VLW} or GSK3β. Figure 1A shows that in the double transgenic mouse an increase in tau phosphorylation takes place in sarkosyl-insoluble protein from the hippocampus of 18-month-old animals, being that extra band heavily labeled by the antibody 12E8 (which recognizes phosphoSer-266 and 422). As a consequence of that hyperphosphorylation, modified tau showed retarded electrophoretic mobility with an apparent molecular weight of 64 kDa. When the aberrant tau assembly was analyzed by immunoelectron microscopy (Fig. 1B), a lack of aggregates was found in wild-type mice and in the transgenic mouse overexpressing GSK3. Thin filaments were found in the tau^{VLW} mouse [19], whereas the double transgenic mouse contained thick tau filaments in a relatively high amount [28]. We have explored the possibility that overexpression of GSK3β could result in an increase in the formation of Aβ, as

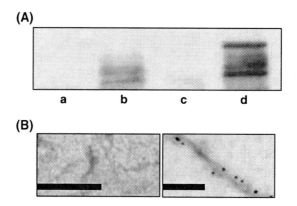

(A)

| a | b | c | d |

(B)

Fig. 1 A Tau phosphorylation in tau^{VLW}/GSK3 transgenic mice. Sarkosyl-insoluble tau protein from (a) wild type, (b) transgenic GSK3, (c) transgenic tau^{VLW}, (d) and double transgenic tau^{VLW}/GSK3 was isolated from hippocampal extracts and the reaction with Aβ 12E8 was determined. **B** Tau filaments found in sarkosyl-insoluble extracts from transgenic tau^{VLW} (*left*) and from double transgenic tau^{VLW}/GSK3 (*right*) mice. Bars = 100 nm

previously reported [30]. However, we were unable to detect such an increase in our animal model.

Tau Pathology in a Double Transgenic Mouse Overexpressing Tau Bearing Three of the Mutations Present in FTDP-17 (Tau^{VLW}), and APP Bearing the Swedish Mutation (APP^{SW})

Double transgenic mice were isolated [29], and tau phosphorylation and aggregation were analyzed and compared to those found in wild-type mouse or in the single transgenic mice overexpressing tau^{VLW} or APP^{SW}. Figure 2A shows that in the double transgenic mouse increased tau phosphorylation takes place, as shown with the antibody 12E8. Again, it was found that, as a consequence of that hyperphosphorylation, modified tau showed retarded electrophoretic mobility, similar to that found in tau^{VLW}/GSK3 mice. The residues Ser-262 and 422 were again phosphorylated in this 64-kDa tau isoform. When the aberrant tau assembly was analyzed (Fig. 2B), the lack of tau aggregates was again found in wild-type mice and the transgenic mice overexpressing APP^{SW}. Thin filaments were found in the tau^{VLW} mouse, whereas the double transgenic mouse contained thick filaments. We analyzed the possibility that APP could facilitate an increase in GSK3 activity, as previously suggested [31,32]. Thus, we cultured cortical neurons, and a decrease in the levels of GSK-3β

(A)

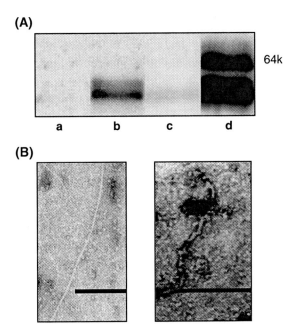

Fig. 2 **A** Tau phosphorylation in Tau^{VLW}/APP^{SW} mice. Sarkosyl-insoluble tau protein from wild-type transgenic APP^{SW}, transgenic tau^{VLW}, and double transgenic tau^{VLW}/APP^{SW} was isolated from hippocampal extracts; and the reaction with Aβ 12E8, which binds to phospho-tau, was determined. **B** Tau filaments found in transgenic tau^{VLW} (*left*) and in double transgenic Tau^{VLW}/APP^{SW} (*right*) mice. Bars = 30 nm and 200 nm, respectively

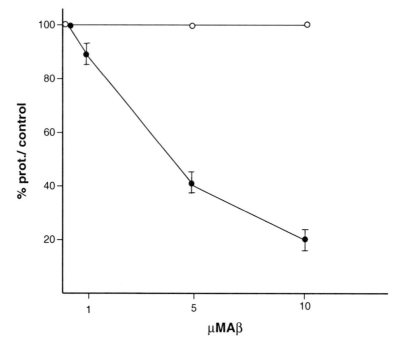

Fig. 3 Aβ addition results in a decrease in GSK3β phosphorylated in serine 9 in cultured cortical neurons. Please note that the addition of increasing amounts of Aβ 1-40 produces a decrease in the phospho-GSK3 (closed symbols) but not in total GSK3 (open symbols)

phosphorylation in serine 9 was observed in the presence of increasing amyloid peptide concentrations (Fig. 3). That decrease results in an increase in GSK3 activity, which promotes tau phosphorylation by the endogenous GSK-3 kinase and likely other protein kinases.

Conclusion

A similar picture seems to emerge from both our double transgenic models (tau^{VLW}/GSK3 and tau^{VLW}/APP^{SW}). Thus, our data suggest that tau isoform p64 may play a role in the formation of thick tau filamentous polymers, which show wide similarity to paired helical filaments found in the brains of AD patients. In fact, in samples from AD patients, hyperphosphorylated tau isoforms that could correspond to the 64 kDa mouse isoforms have been described [33]. Thus, it will be of interest to understand the role of these human or mouse hyperphosphorylated tau isoforms in the formation of tau fibrillar polymers. A possibility that we are currently exploring is that the p64 tau isoform may play a role in the nucleation of tau fibrillar polymers.

References

1. Alzheimer A. Über eine eigenartige Erkrankung der Hirninde. Z Psychiatr Psych Gericht Med 1907;64:146–148.
2. Glenner GG, Wong CW. Alzheimer's disease: initial report of the purification and characterization of a novel cerebrovascular amyloid protein. Biochem Biophys Res Commun 1984;120:885–890.
3. Beyreuther K, Masters CL. Amyloid precursor protein (APP) and beta A4 amyloid in the etiology of Alzheimer's disease: precursor-product relationships in the derangement of neuronal function. Brain Pathol 1991;1:241–251.
4. Grundke-Iqbal I, Iqbal K, George L, et al. Amyloid protein and neurofibrillary tangles coexist in the same neuron in Alzheimer disease. Proc Natl Acad Sci U S A 1989;86:2853–2857.
5. Morishima-Kawashima M, Hasegawa M, Takio K, et al. Proline-directed and non-proline-directed phosphorylation of PHF-tau. J Biol Chem 1995;270:823–829.
6. Xie L, Helmerhorst E, Taddei K, et al. Alzheimer's beta-amyloid peptides compete for insulin binding to the insulin receptor. J Neurosci 2002;22:RC221.
7. Arriagada PV, Marzloff K, Hyman BT. Distribution of Alzheimer-type pathologic changes in nondemented elderly individuals matches the pattern in Alzheimer's disease. Neurology 1992;42:1681–1688.
8. Arriagada PV, Growdon JH, Hedley-Whyte ET, Hyman BT. Neurofibrillary tangles but not senile plaques parallel duration and severity of Alzheimer's disease. Neurology 1992;42:631–639.
9. Lee VM, Goedert M, Trojanowski JQ. Neurodegenerative tauopathies. Annu Rev Neurosci 2001;24:1121–1159.
10. Spillantini MG, Goedert M. Tau protein pathology in neurodegenerative diseases. Trends Neurosci 1998;21:428–433.
11. Spillantini MG, Murrell JR, Goedert M, et al. Mutation in the tau gene in familial multiple system tauopathy with presenile dementia. Proc Natl Acad Sci U S A 1998;95:7737–7741.
12. Kosik KS. Alzheimer's disease: a cell biological perspective. Science 1992;256:780–783.
13. Brion JP, Tremp G, Octave JN. Transgenic expression of the shortest human tau affects its compartmentalization and its phosphorylation as in the pretangle stage of Alzheimer's disease. Am J Pathol 1999;154:255–270.
14. Ishihara T, Hong M, Zhang B, et al. Age-dependent emergence and progression of a tauopathy in transgenic mice overexpressing the shortest human tau isoform. Neuron 1999;24:751–762.
15. Spittaels K, Van den Haute C, Van Dorpe J, et al. Prominent axonopathy in the brain and spinal cord of transgenic mice overexpressing four-repeat human tau protein. Am J Pathol 1999;155:2153–2165.
16. Probst A, Gotz, J, Wiederhold KH, et al. Axonopathy and amyotrophy in mice transgenic for human four-repeat tau protein. Acta Neuropathol (Berl) 2000;99:469–481.
17. Gotz J, Tolnay M, Barmettler R, et al. Human tau transgenic mice: towards an animal model for neuro- and glial-fibrillary lesion formation. Adv Exp Med Biol 2001;487:71–83.
18. Lewis J, McGowan E, Rockwood J, et al. Neurofibrillary tangles, amyotrophy and progressive motor disturbance in mice expressing mutant (P301L) tau protein. Nat Genet 2000;25:402–405.
19. Lim F, Hernandez F, Lucas JJ, et al. FTDP-17 mutations in tau transgenic mice provoke lysosomal abnormalities and tau filaments in forebrain. Mol Cell Neurosci 2001;18:702–714.
20. Allen B, Ingram E, Takao M, et al. Abundant tau filaments and nonapoptotic neurodegeneration in transgenic mice expressing human P301S tau protein. J Neurosci 2002;22:9340–9351.
21. Tanemura K, Murayama M, Akagi,T, et al. Neurodegeneration with tau accumulation in a transgenic mouse expressing V337M human tau. J Neurosci 2002;22:133–141.

22. Tatebayashi T, Miyasaka T, Chui DH, et al. Tau filament formation and associative memory deficit in aged mice expressing mutant (R406W) human tau. Proc Natl Acad Sci U S A 2002;99:13896–13901.

23. Boutajangout A, Authelet M, Blanchard V, et al. Characterization of cytoskeletal abnormalities in mice transgenic for wild-type human tau and familial Alzheimer's disease mutants of APP and presenilin-1. Neurobiol Dis 2004;15:47–60.

24. Lucas JJ, Hernandez F, Gomez-Ramos P, et al. Decreased nuclear beta-catenin, tau hyperphosphorylation and neurodegeneration in GSK-3beta conditional transgenic mice. EMBO J 2001;20:27–39.

25. Perez M, Hernandez F, Lim F, et al. Chronic lithium treatment decreases mutant tau protein aggregation in a transgenic mouse model. J Alzheimers Dis 2003;5:301–308.

26. Mullan M, Crawford F, Axelman K, et al. A pathogenic mutation for probable Alzheimer's disease in the APP gene at the N-terminus of beta-amyloid. Nat Genet 1992;1:345–347.

27. Hardy J, Selkoe DJ. The amyloid hypothesis of Alzheimer's disease: progress and problems on the road to therapeutics. Science 2002;297:353–356.

28. Engel T, Lucas JJ, Gomez-Ramos P, et al. Co-expression of FTDP-17 tau and GSK-3β in transgenic mice induce tau polymerization and neurodegeneration. Neurobiol Aging 2006;27:1258–1268.).

29. Perez M, Ribe E, Rubio A, et al. Characterization of a double (amyloid precursor protein-tau) transgenic: tau phosphorylation and aggregation. Neuroscience 2005;130:339–347.

30. Phiel CJ, Wilson CA, Lee VM, Klein PS. GSK-3alpha regulates production of Alzheimer's disease amyloid-beta peptides. Nature 2003;423:435–439.

31. Hoeflich KP, Luo J, Rubie EA, et al. Requirement for glycogen synthase kinase-3beta in cell survival and NF- kappaB activation. Nature 2000;406:86–90.

32. Farias GG, Godoy JA, Vazquez MC, et al. The anti-inflammatory and cholinesterase inhibitor bifunctional compound IBU-PO protects from beta-amyloid neurotoxicity by acting on Wnt signaling components. Neurobiol Dis 2005;18:176–183.

33. Goedert M. Filamentous nerve cell inclusions in neurodegenerative diseases: tauopathies and alpha-synucleinopathies. Philos Trans R Soc Lond B Biol Sci 1999;354:1101–1118.

34. Soto C, Kascsak RJ, Saborio GP, et al. Reversion of prion protein conformational changes by synthetic β sheet breaker peptides. Lancet 2000;355:192–197.

Chapter 36
Enhanced Activation of NF-κB Signaling by Apolipoprotein E4

Gal Ophir, Liza Mizrahi, and Daniel M. Michaelson

Introduction

Alzheimer's disease (AD) is associated with brain inflammation. This includes the appearance of activated microglia and astrocytes near senile plaques and elevated levels of inflammatory molecules such as cytokines, acute-phase proteins, and complement proteins [1–3]. Genetic and epidemiological studies revealed that the E4 allele of apolipoprotein E (ApoE4) is a major risk factor of AD (for review see ref. 4). Furthermore, brain inflammation is more robust in ApoE4 AD patients than in AD patients who do not carry this allele [5]. The increased brain inflammation in ApoE4 AD patients may be due to direct and isoform-specific effects of ApoE4 on brain inflammatory mechanisms, or it may be a consequence of the increased brain pathology in these patients.

We have recently examined this issue at the animal model level by genome-wide gene expression profiling of the effects of the ApoE genotype on hippocampal gene expression in lipopolysaccharide (LPS)-treated mice transgenic for either ApoE4 or the AD benign allele ApoE3. It revealed that the expression of inflammation-related genes following intracerebroventricular (i.c.v.) injection of LPS was significantly higher and more prolonged in ApoE4 than in ApoE3 transgenic mice [6]. Further clustering analysis and direct measurements of gene expression revealed that the extent of activation of NF-kB regulated genes is greater in the ApoE4 than in the ApoE3 mice [6]. Taken together, these findings suggest that the increased inflammation in ApoE4 mice following brain insults and during aging [7,8] is due, at least in part, to overactivation of the inflammatory response by ApoE4 and that this effect is mediated by dysregulation of the NF-kB signaling pathway.

In the present study we further characterized the isoform-specific effects of ApoE4 on brain inflammation and NF-kB signaling. This was done with immunohistochemical measurements of the effects of the ApoE genotype on

G. Ophir
Department of Neurobiochemistry, George S. Wise Faculty of Life Sciences, Tel Aviv University, Tel Aviv 69978, Israel

A. Fisher et al. (eds.), *Advances in Alzheimer's and Parkinson's Disease*,
© Springer 2008

the kinetics and degree of activation of NF-kB following stimulation with LPS and assessing the extent to which it correlates with distinct markers of microglial activation.

Materials and Methods

Transgenic Mice

Human ApoE3 and ApoE4 transgenic mice were generated on an ApoE-deficient C57BL/6J background, utilizing human ApoE3 and ApoE4 transgenic constructs as previously reported [9]. Briefly, cosmid libraries were constructed from lymphoblasts of humans known to be homozygous carriers for ApoE3 or ApoE4, after which fragments containing human regulatory sequences and the coding sequences for human ApoE were used to produce the transgenic mice. The experiments were performed with ApoE3-453 and ApoE4-81 lineages that were back-bred with genetically homogeneous ApoE-deficient mice (cat. no. N10 JAX; Jackson Labs, Bar Harbor, ME, USA) for more than 10 generations and were heterozygous for the human ApoE transgene and homozygous for mouse ApoE deficiency. The ApoE genotype of the mice was confirmed by polymerase chain reaction (PCR) analysis as previously described [10]. Wild-type C57BL/6J mice were obtained from Jackson laboratories.

Intracerebroventricular Injection of LPS

Four-month-old male human ApoE3 and ApoE4 transgenic mice were used. The mice were anesthetized with ketamine 120 mg/kg (Fort Dodge, Madison, NJ, USA) and injected i.c.v. with either 500 ng LPS (Difco Laboratories, Detroit, MI, USA) (5 µl of 100 µg/ml) or 5 µl of sterile phosphate-buffered saline (PBS) (sham operation) into the left lateral ventricle (1 mm posterior to Bregma; 1.5 mm lateral; 2 mm depth) at a rate of 1 µl/min with the help of a syringe pump (Baby Bee; Bioanalytical Systems, West Lafayette, IN, USA) and a brain infusion cannula (Alzet brain infusion kit; Alza, Mountain View, CA, USA). After each injection, the cannula was left in situ for an additional minute to avoid reflux. The mice were euthanized 24 or 72 hours later ($n = 4$ per experimental group) and perfused transcardially with PBS followed by a fixative solution (4% formaldehyde in PBS, pH 7.4). The brains were removed and postfixed overnight with the same fixative and then transferred for cryoprotection to a 30% sucrose solution for 24 hours at 4 °C. Finally, the brains were embedded in a tissue-freezing medium (Jung; Leica Instruments, Bannockburn, IL, USA) and frozen in Dry Ice-cooled hexane.

Immunohistochemistry

Frozen coronal sections (12 μm thick) were cut at the level of the anterior hippocampus (2.0–2.5 mm posterior to Bregma) and mounted on gelatin-coated slides. Immunohistochemistry was performed as described elsewhere [10] using the Histostain SP kit (Zymed Laboratories, San Francisco, CA, USA) and the following primary antibodies: polyclonal rat anti-mac-2 (1:500) (a gift from Dr. Witz, Tel Aviv University, Tel Aviv, Israel) and rat anti-F4/80 (1:50) (Serotec, Raleigh, NC, USA) for the detection of activated microglia/macrophages, polyclonal rabbit anti-NF-kB p65 subunit (1:200) (Santa Cruz Biotechnology, Santa Cruz, CA, USA), and polyclonal goat anti-mouse interleukin-1β (IL-1β) (1:50) (R&D Systems, Minneapolis, MN, USA).

The NF-kB-positive and IL-1β-positive cells and mac-2- and F4/80-positive activated microglia were determined in the hippocampus by imaging at $100\times$ magnification, followed by counting the number of such cells in the hippocampus on two sections from each animal ($n = 4$). The sections were counted blindly with regard to the genotype and treatment of the mice.

The differences between the numbers of immunohistochemically stained cells were analyzed by two-way analysis of variance (ANOVA) with treatment and genotype as two independent factors. Post hoc comparisons of the results were performed when ANOVA showed a significant difference, utilizing Fisher's LSD test for multiple comparisons.

Results

The levels of NF-kB-positive cells in the hippocampi of ApoE4 and ApoE3 transgenic mice and of wild-type and ApoE-deficient mice were determined immunohistochemically at 24 and 72 hours following the i.c.v. injection of LPS. Representative sections thus obtained at 24 hours are depicted in Fig. 1. As shown, the levels of NF-kB-positive cells were markedly higher at this time point in the LPS-treated ApoE4 mice than in the other mice groups, whose levels of NF-kB-containing cells were low and similar. Quantitation of these results in four mice per group × treatment revealed that the isoform-specific stimulatory effect of ApoE4 on NF-kB was highly significant (Fig. 1B).

Control experiments revealed that the levels of NF-kB-positive cells in LPS-treated mice were markedly higher in all the groups than those of the corresponding sham-treated mice (Fig. 1B,C). The density of NF-kB-containing cells of the ApoE4 mice was unchanged for up to 72 hours following injection of LPS.

In contrast, the number of NF-kB- positive cells in the ApoE3 mice, which was low at 24 hours, increased with time and at 72 hours was similar to that of the ApoE4 mice. The levels of NF-kB-positive cells of the wild-type and ApoE-deficient mice increased progressively with time, but their levels were

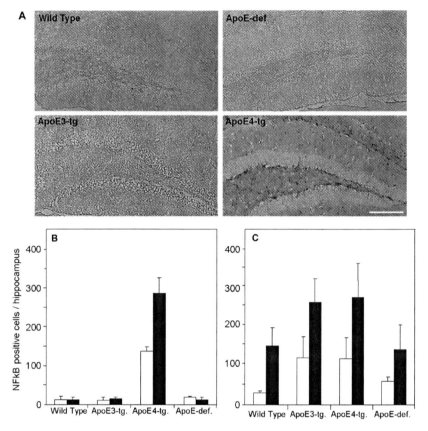

Fig. 1 Effects of apolipoprotein E (ApoE) on the levels of NF-kB-positive cells in the hippocampus following intracerebroventricular (i.c.v.) injection of lipopolysaccharide (LPS). **A** Representative hippocampal sections from wild-type, ApoE3 transgenic (ApoE3-tg.), ApoE4 transgenic (ApoE4-tg.), and ApoE-deficient mice (ApoE-def.) 24 hours after the LPS injection. Quantitation of the immunohistochemical results after 24 hours (**B**) and 72 hours (**C**) was performed as described in Materials and Methods. The results shown represent the mean ± SD of four mice per group. A significant effect of group × treatment at 24 hours ($p < 0.001$) was revealed using two-way analysis of variance (ANOVA) that was associated with a significant difference ($p < 0.001$) between the LPS-treated ApoE4 mice and the other groups. A similar analysis of the data at 72 hours revealed that the treatment had a significant effect ($p < 0.001$)

lower than those of the ApoE4 and ApoE3 transgenic mice (Fig. 1C). Next, we measured the effects of the ApoE genotype and LPS on the hippocampal levels of the cytokine IL-1β, whose expression is known to be regulated by NF-kB11. This revealed that the levels of IL-1β-positive cells of the ApoE4 mice, but not those of the other mouse groups, were markedly elevated 24 hours following the LPS injection; however, this effect was transient and subsided at 72 hours (Fig. 2).

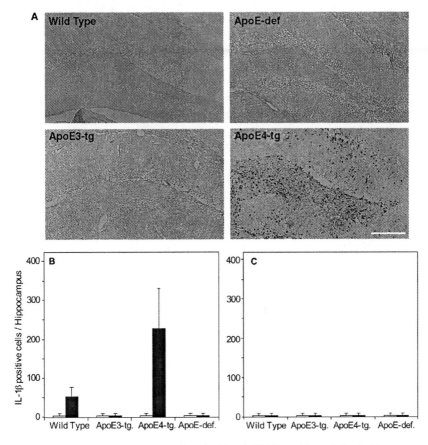

Fig. 2 Effects of ApoE on the levels of interleukin-1β (IL1β)-positive cells in the hippocampus following i.c.v. injection of LPS. **A** Representative hippocampal sections from wild-type, ApoE3 transgenic (ApoE3-tg.), ApoE4 transgenic (ApoE4-tg.), and ApoE-deficient (ApoE-def.) mice 24 hours after the i.c.v. injection of LPS. Quantitation of the immunohistochemical results after 24 hours (**B**) and 72 hours (**C**) was performed as described in Materials and Methods. The results shown represent the mean ± SD of four mice per group. A significant effect of group × treatment at 24 hours ($p < 0.001$) was revealed using two-way ANOVA that was associated with a significant difference between the LPS-treated ApoE4 mice and the other groups ($p < 0.001$)

The ApoE4-dependent elevation of the level of NF-kB-positive cells was associated with translocation of NF-kB from the cytoplasm into the nucleus at 24 hours. However, at 72 hours after the injection of LPS, most of the NF-kB was cytoplasmic (data not shown). Taken together, these observations suggest that, at 24 hours following the injection of LPS, ApoE4 has an isoform-specific stimulatory effect on the levels of NF-kB and triggers its translocation into the nuclei. Consequently, this results in overactivation of NF-kB signaling, which triggers the synthesis of downstream products such as IL-1β. The picture is more complex by 72 hours, at which point the number of NF-kB-positive cells was similar in the ApoE3 and ApoE4 mice (Fig. 1C) and was not associated with either IL-1β immunostaining (Fig. 2C) or translocation of NF-kB into the nuclei (not shown).

The activation of glial cells following stimulation with LPS was determined immunohistochemically utilizing the microglial activation markers mac-2 and F4/80 [11] and the astrocytic marker GFAP. The results obtained with mac-2 and F4/80 are presented in Figs. 3 and 4. As shown, the levels of mac-2-positive cells in the hippocampi of ApoE4 mice were markedly elevated 24 hours after the injection of LPS, whereas the corresponding numbers of cells in the similarly

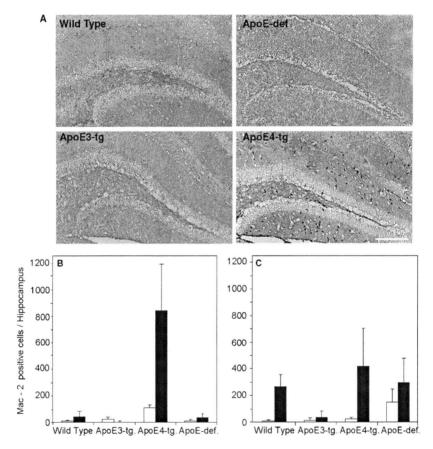

Fig. 3 Effects of ApoE on the levels of mac-2-positive cells in the hippocampus following i.c.v. injection of LPS. **A** Representative hippocampal sections from wild-type, ApoE3 transgenic (ApoE3-tg.), ApoE4 transgenic (ApoE4-tg.), and ApoE-deficient (ApoE-def.) mice 72 hours after the i.c.v. injection of LPS. Quantitation of the immunohistochemical results after 24 hours (**B**) and 72 hours (**C**) was performed as described in Materials and Methods. The results shown represent the mean ± SD of four mice per group. A significant effect of group × treatment at 24 hours ($p < 0.001$) was revealed using two-way ANOVA that was associated with a significant difference ($p < 0.001$) between the LPS-treated ApoE4 mice and the other mouse groups. There were no significant differences at 72 hours

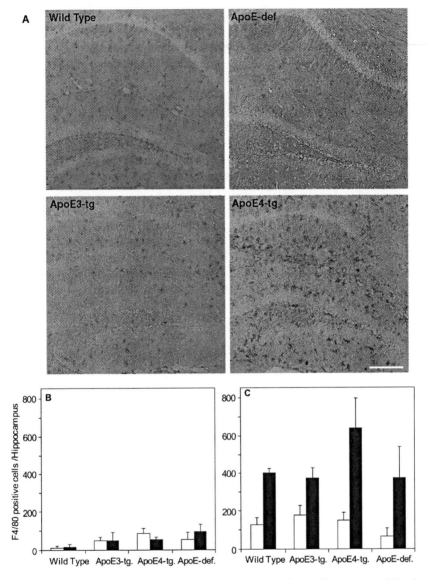

Fig. 4 Effects of ApoE on the levels of F4/80-positive cells in the hippocampus following i.c.v. injection of LPS. **A** Representative hippocampal sections from wild-type, ApoE3 transgenic (ApoE3-tg.), ApoE4 transgenic (ApoE4-tg.), and ApoE-deficient (ApoE-def.) mice 24 hours after the i.c.v. injection of LPS. Quantitation of the immunohistochemical results after 24 hours (**B**) and 72 hours (**C**) was performed as described in Materials and Methods. The results shown represent the mean ± SD of four mice per group. A significant effect of group × treatment at 24 hours ($p < 0.001$) was revealed using two-way ANOVA

treated ApoE3, wild-type, and ApoE-deficient mice were unaffected (Fig. 3A,B). In contrast, the levels of F4/80-positive microglia (Fig. 4B) and of activated astrocytes (data not shown) of all the mice groups were not affected appreciably by LPS at 24 hours. Double labeling immunohistochemical experiments revealed that at 24 hours NF-kB is co-localized with mac-2, suggesting that NF-kB resides in activated microglia (data not shown). At 72 hours following LPS application, F4/80-positive cell levels were elevated in all the mice groups, which was most pronounced in the ApoE4 mice (Fig. 4A,C). The levels of mac-2-positive cells of the wild-type and ApoE-deficient mice also increased by 72 hours, whereas the corresponding mac-2 levels of the ApoE4 mice were lower and those of the ApoE3 mice did not change under these conditions (Fig. 3C). Double labeling experiments revealed that, unlike at 24 hours, at 72 hours following LPS injection NF-kB was mostly astrocytic (data not shown), which is in accordance with previous observations that astrogliosis is detectable 72 hours after the LPS treatment [10]. These findings suggest that the early isoform-specific effect of ApoE on the inflammatory response to LPS is related to enhanced and transient activation of mac-2-positive microglia by ApoE4 and to the concurrent hyperactivation of the NF-kB signaling cascade in these cells. This is followed by activation at 72 hours and, in all the mice groups, of F4/80-positive microglia, astrocytes, and the synthesis of astrocyte-bound NF-kB.

Discussion

The present study showed that the inflammatory response following i.c.v. injection of LPS is activated isoform-specifically by ApoE4 and is associated with enhanced activation of NF-kB signaling. Accordingly, at 24 hours after the injection of LPS, the levels of microglial NF-kB and Mac-2-positive microglia were elevated to a greater extent in the ApoE4 mice than in the corresponding ApoE3, wild-type, or ApoE-deficient mice. Furthermore, this effect was associated with translocation of NF-kB into the nucleus and with enhanced synthesis of IL-1β in the ApoE4 mice. These isoform-specific effects of ApoE4 were transient and were followed by the elevation of F4/80-positive microglia levels and the accumulation of NF-kB in astrocytes. These effects were apparent at 72 hours and were not significantly affected by the ApoE genotype.

This study focused on the 24- to 72-hour period following the injection of LPS, during which time the difference in NF-kB signaling and microglial activation between the ApoE4 and ApoE3 transgenic mice was most pronounced. It should be noted, however, that microglial activation and the level of NF-kB of the wild type and ApoE3 groups, whose responses to LPS at 24 and 72 hours were very low, had a faster response that peaked at 8 hours (not shown). This suggests that the presently observed effects of ApoE4 are mainly due to amplification and prolongation of the microglial and NF-kB responses to LPS. This is in accordance with our recent genome-wide gene expression profiling experiments that revealed that the expression of inflammation-related genes following i.c.v. injection of LPS was significantly higher and more prolonged in ApoE4 mice [6]. Both the basal and LPS-driven elevations in hippocampal ApoE levels were similar in the ApoE4 and ApoE3 mice [10], suggesting that the observed isoform-specific effects of ApoE are due to intrinsic differences between the effects of ApoE4 and ApoE3 on microglial activation and NF-kB signaling.

The initial phase of microglial activation in the ApoE4 mice, which is characterized by transient activation of NF-kB signaling and IL-1β production and by elevated levels of mac-2, is followed by a second phase of microglial activation that is associated with elevated levels of F4/80 (Fig. 4). A similar pattern was observed in the other mouse groups, except that the temporal separation between the expression of mac-2 and F4/80 was less pronounced and the levels of F4/80 were not markedly affected by the ApoE genotype. This is in accordance with previous reports that the expression of mac-2, F4/80, and other microglial molecules is up-regulated differentially following neuroinflammatory insults [12]. Furthermore, because mac-2 plays an important role in phagocytosis [13], it is possible that LPS-triggered phagocytosis occurs earlier in the ApoE4 mice than in the ApoE3 mice.

The present in vivo findings are in accordance with previous in vitro reports that revealed that microglia are activated preferentially by ApoE4 [14–19]. Activation of NF-kB signaling by LPS is affected by Ca^{2+}, [20–23] whereas ApoE4 can stimulate isoform-specific Ca^{2+} influx into neurons and glia [24–27]. Accordingly, it is possible that the effects of ApoE on microglial activation are mediated via intracellular Ca^{2+} and that overactivation of the microglial NF-kB pathway in the ApoE4 mice is due to an increased influx of Ca^{2+} and consequent dysregulation of Ca^{2+} homeostasis. Alternatively, because IL-1β can interact directly with ApoE receptors [28], it is possible that the effects of ApoE on NF-kB signaling are indirect and are due to a downstream modulatory mechanism.

It is important to note that there are genes whose mRNA expression is affected isoform-specifically by ApoE at earlier times after the injection of LPS (e.g., 5 hours) than the presently observed isoform-specific effects of ApoE4 on microglial activation and NF-kB levels [6]. Further in vivo and in vitro kinetic studies are required to decipher the molecular mechanisms underlying the effects of ApoE4 on microglial activation and NF-kB signaling and the possibility that they are mediated indirectly by additional systems.

Conclusion

The present study shows that microglial activation and NF-kB signaling following LPS stimulation are more pronounced in ApoE4 than in ApoE3 transgenic mice. Similar mechanisms may mediate the enhanced brain inflammation that is associated in AD with the ApoE4 genotype.

Acknowledgments We thank Dr. Allen Roses (Duke University, Durham, NC, USA) and Glaxo Wellcome for providing the transgenic mice and Steven Manch for his editorial assistance. This work was supported in part by grants to D.M.M. from the Harry Stern National Center for Alzheimer's disease, the U.S. Binational Science Foundation, and the Eichenbaum Foundation. D.M.M. is the incumbent of the Myriam Lebach Chair in Molecular Neurodegeneration.

References

1. McGeer EG, McGeer PL. Inflammatory processes in Alzheimer's disease. Prog Neuropsychopharmacol Biol Psychiatry 2003;27:741–749.
2. Eikelenboom P, van Gool WA. Neuroinflammatory perspectives on the two faces of Alzheimer's disease. J Neural Transm 2004;111:281–294.
3. Rogers JT, Lahiri DK. Metal and inflammatory targets for Alzheimer's disease. Curr Drug Targets 2004;5:535–551.
4. Roses D. Apolipoprotein E alleles as risk factors in Alzheimer's disease. Annu Rev Med 1996;47:387–400.
5. Egensperger R, Kosel S, von Eitzen U, Graeber MB. Microglial activation in Alzheimer disease: association with APOE genotype. Brain Pathol 1998;8:439–447.
6. Ophir G, Amarglio N, Jacob-Hirsch J, et al. Apolipoprotein E4 enhances brain inflammation by modulation of the NF-kB signaling cascade. Neurobiol Dis 2005;20:709–718.
7. Sabo T, Lomnitski L, Nyska A, et al. Susceptibility of transgenic mice expressing human apolipoprotein E to closed head injury, the allele E3 is neuroprotective whereas E4 increases fatalities. Neuroscience 2000;101:879–884.
8. Murphy MM, Mirtasinovic OM, Sullivan PM. Microarray comparison of gene expression in the hippocampus of aged apoE3 and E4 transgenic mice. Neurobiol Aging 2004;25:316–320.
9. Xu PT, Schmechel D, Rothrock-Christian T, et al. Human apolipoprotein E2, E3, and E4 isoform-specific transgenic mice, human-like pattern of glial and neuronal immunoreactivity in central nervous system not observed in wild-type mice. Neurobiol Dis 1996;3:229–245.
10. Ophir G, Meilin S, Efrati M, et al. Human apoE3 but not apoE4 rescues impaired astrocyte activation in apoE null mice. Neurobiol Dis 2003;12:56–64.
11. Hanada T, Yoshimura A. Regulation of cytokine signaling and inflammation. Cytokine Growth Factor Rev 2002;13:413–421.
12. Reichart F, Rotshenker S. Deficient activation of microglia during optic nerve degeneration. J Neuroimmunol 1996;70:153–161.
13. Rotshenker S. Microglia and macrophage activation and the regulation of complement-receptor-3 (CR3/MAC-1)-mediated phagocytosis in injury and disease. J Mol Neurosci 2003;21:65–72.
14. Laskowitz DT, Goel S, Bennett ER, Matthew WD. Apolipoprotein E suppresses glial cell secretion of TNF alpha. J Neuroimmunol 1997;76:70–74.
15. Barger SW, Harmon AD. Microglial activation by Alzheimer amyloid precursor protein and modulation by apolipoprotein E. Nature 1997;388:878–881.
16. Lynch JR, Morgan D, Mance J, et al. Apolipoprotein E modulates glial activation and the endogenous central system inflammatory response. J Neuroimmunol 2001;114:107–113.
17. Colton CA, Brown CM, Czapiga M, Vitek MP. Apolipoprotein-E allele-specific regulation of nitric oxide production. Ann N Y Acad Sci 2002;962:212–225.
18. Guo L, LaDu MJ, Van Eldik LJ. A dual role for apolipoprotein E in neuroinflammation, anti- and pro-inflammatory activity. J Mol Neurosci 2004;23:205–212.
19. Chen S, Averett WT, Manelli A, et al. Isoform-specific effects of apolipoprotein E on secretion of inflammatory mediators in adult rat microglia. J Alzheimers Dis 2005;7:25–35.

20. Mirzoeva S, Koppal T, Petrova TV, et al. Screening in a cell-based assay for inhibitors of microglial nitric oxide production reveals calmodulin-regulated protein kinases as potential drug discovery targets. Brain Res 1999;844:126–134.
21. Trushin SA, Pennington KN, Algeciras-Schimnich A, Paya CV. Protein kinase C and calcineurin synergize to activate IkappaB kinase and NF-kappaB in T lymphocytes. J Biol Chem 1999;74:22923–22931.
22. Suo Z, Wu M, Citron BA, et al. Persistent protease-activated receptor-4 signaling mediates thrombin-induced microglial activation. J Biol Chem 2003;278:31177–31183.
23. Kim Y, Moon JS, Lee KS, et al. Ca^{2+}/calmodulin-dependent protein phosphatase calcineurin mediates the expression of iNOS through IKK and NF-kappaB activity in LPS-stimulated mouse peritoneal macrophages and RAW 264.7 cells. Biochem Biophys Res Commun 2004;314:695–703.
24. Muller W, Meske V, Berlin K, et al. Apolipoprotein E isoforms increase intracellular Ca^{2+} differentially through an omega-agatoxin IVa-sensitive Ca^{2+}-channel. Brain Pathol 1998;8:641–653.
25. Bacskai BJ, Xia MQ, Strickland DK, et al. The endocytic receptor protein LRP also mediates neuronal calcium signaling via N-methyl-D-aspartate receptors. Proc Natl Acad Sci U S A 2000;97:11551–11556.
26. Veinbergs I, Everson A, Sagara Y, Masliah E. Neurotoxic effects of apolipoprotein E4 are mediated via dysregulation of calcium homeostasis. J Neurosci Res 2002;67:379–387.
27. Qiu Z, Crutcher KA, Hyman BT, Rebeck GW. ApoE isoforms affect neuronal N-methyl-D-aspartate calcium responses and toxicity via receptor-mediated processes. Neuroscience 2003;122:291–303.
28. Noguchi T, Noguchi M, Masubuchi H, et al. IL-1β down-regulates tissue type plasminogen activation by up regulating low density lipoprotein receptor-related protein in AML 12 cells. Biochem Biophys Res Commun 2001;288:42–48.

Chapter 37
Pleiotropic Effects of Apolipoprotein E in Dementia: Influence on Functional Genomics and Pharmacogenetics

Ramón Cacabelos

Introduction

The genetic defects identified in Alzheimer's disease (AD) during the past quarter century can be classified into three main categories: (1) Mendelian or mutational defects in genes directly linked to AD, including 18 mutations in the β-amyloid (Aβ) precursor protein (APP) gene (21q21); 142 mutations in the presenilin 1 (PS1) gene (14q24.3); and 10 mutations in the presenilin 2 (PS2) gene (1q31-q42) [1]. (2) Multiple polymorphic variants of risk characterized in more than 100 genes distributed across the human genome can increase neuronal vulnerability to premature death [1]. Among these genes of susceptibility, the apolipoprotein E (ApoE) gene (19q13.2) is the most prevalent as a risk factor for AD, especially in subjects harboring the ApoE-4 allele, whereas carriers of the APOE-2 allele might be protected against dementia [1–3]. ApoE-related pathogenic mechanisms are also associated with brain aging and with the neuropathological hallmarks of AD [4]. (3) Diverse mutations located in mitochondrial DNA (mtDNA) through heteroplasmic transmission can influence aging and oxidative stress conditions, conferring phenotypic heterogeneity [5]. It is also likely that defective functions of genes associated with longevity may influence neuronal survival, as neurons are potential pacemakers defining life-span in mammals [1]. All these genetic factors may interact in still unknown genetic networks, leading to a cascade of pathogenic events characterized by abnormal protein processing and misfolding with subsequent accumulation of abnormal proteins (conformational changes), ubiquitin-proteasome system dysfunction, excitotoxic reactions, oxidative and nitrosative stress, mitochondrial injury, synaptic failure, altered metal homeostasis, dysfunction of axonal and dendritic transport, and chaperone misoperation [1]. These pathogenic events may exert an additive effect, converging in final pathways leading to premature neuronal death. Some of these mechanisms are

R. Cacabelos
EuroEspes Biomedical Research Center, Institute for CNS Disorders, 15166-Bergondo, Coruña, Spain

A. Fisher et al. (eds.), *Advances in Alzheimer's and Parkinson's Disease*, © Springer 2008

common to several neurodegenerative disorders that differ depending on the gene(s) affected and the involvement of specific genetic networks, together with cerebrovascular factors, epigenetic factors (DNA methylation), and environmental conditions (e.g., nutrition, toxicity, social factors) [1,6,7]. The higher the number of genes involved in AD pathogenesis, the earlier is the onset of the disease, the faster its clinical course, and the poorer its therapeutic outcome [1].

Functional Genomics Studies

Although the amyloid hypothesis is recognized as the *primum movens* of AD pathogenesis [1,8], mutational genetics associated with APP and PSs genes alone (< 10% of AD cases) does not fully explain the neuropathological findings present in AD. Such findings include amyloid deposition in senile plaques and vessels (amyloid angiopathy), neurofibrillary tangle (NFT) formation due to hyperphosphorylation of tau protein, synaptic and dendritic desarborization, and neuronal loss. These changes are accompanied by neuroinflammatory reactions, oxidative stress, and free radical formation probably associated with mitochondrial dysfunction, excitotoxic reactions, alterations in cholesterol metabolism and lipid rafts, deficiencies in neurotransmitter and neurotrophic factor function, defective activity of the ubiquitin-proteasome and chaperone systems, and cerebrovascular dysregulation [1]. All of these neurochemical events are potential targets for treatment [9–11].

The molecular mechanisms underlying Aβ deposition in brain tissue and blood vessels have been elegantly elucidated during the past two decades by many groups all over the world1 [8,9,12], defining the fundamentals for promising therapeutic strategies oriented to inhibit the formation of amyloid deposits and oligomeric Aβ forms or to reduce senile plaque burden [8,9,11,12]. Notwithstanding, the complexity of the pathogenic cascade in AD invites one to predict that many other genetic factors may be involved in the etiology of AD, together with epigenetic phenomena, nutritional factors, and environmental circumstances [1].

Functional genomics studies have demonstrated the influence of many genes on AD pathogenesis and phenotype expression [13,14]. Mutations in the APP, PS1, PS2, and MAPT genes give rise to well characterized differential neuropathological and clinical phenotypes of dementia [1].

Pleiotropic Effects of ApoE

The ApoE gene is highly pleiotropic, participating in multiple metabolic pathways that influence the phenotype profile of dementia (Fig. 1). The analysis of genotype-phenotype correlations has revealed that the presence of the ApoE-4 allele in AD, in conjunction with other factors (genetic or nongenetic), influences disease onset, brain atrophy, cerebrovascular perfusion, blood pressure,

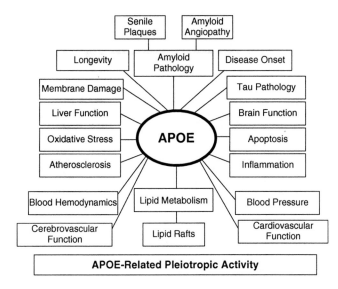

Fig. 1 Pleiotropic effects of apolipoprotein E (ApoE) in dementia and related disorders

Aβ deposition, ApoE secretion, lipid metabolism, brain bioelectrical activity, cognition, apoptosis, and treatment outcome [1–3,13–16] (Fig. 1). The characterization of phenotypic profiles according to age, cognitive performance (MMSE and ADAS-Cog score), serum ApoE levels, serum lipid levels including cholesterol (CHO)—high density (HDL-CHO), low density (LDL-CHO), and very low density (VLDL-CHO) cholesterol—and triglyceride (TG) levels, as well as serum nitric oxide (NO), Aβ, and histamine levels, reveals sex-related differences in 25% of the biological parameters and almost no differences (0.24%) when patients are classified as ApoE-4(−) and ApoE-4(+) carriers (irrespective of the possible six ApoE genotypes), probably indicating that sex-related factors influence these parametric variables more strongly than the presence or absence of the ApoE-4 allele; in contrast, when patients are classified according to their ApoE genotype, dramatic differences emerge among ApoE genotypes (> 45%), with a clear biological disadvantage in ApoE-4/4 carriers who exhibit (1) earlier age of onset, (2) low ApoE levels, (3) high CHO (Fig. 2) and LDL-CHO levels, and (4) low TG (Fig. 3), NO, Aβ, and histamine levels in blood [2,3,13–15].

These phenotypic differences are less pronounced when AD patients are classified according to their PS1 (15.6%) or ACE genotypes (23.52%), reflecting a weak impact of PS1- and ACE-related polymorphic variants on the phenotypic expression of biological markers in AD. However, when ApoE and PS1 genotypes are integrated in bigenic clusters and the resulting bigenic genotypes are differentiated according to their corresponding phenotypes, an almost logarithmic increased expression of differential phenotypes is observed (61.46% variation), indicating the existence of a synergistic effect of the bigenic (ApoE + PS1) cluster on the expression of biological markers associated with

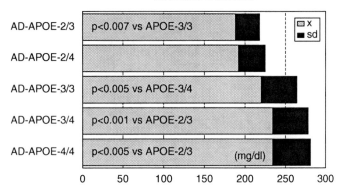

Fig. 2 ApoE-related plasma cholesterol levels in Alzheimer's disease

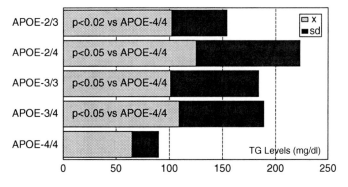

Fig. 3 ApoE-related plasma triglyceride levels in Alzheimer's disease

PS1-related polymorphic variants (PS1-1/1, PS1-1/2, PS1-2/2), apparently unrelated to conventional APP/PS1 mutations, as none of the patients included in the sample were carriers of either APP or PS1 mutations [13]. These examples illustrate the potential additive effects of polymorphic variants of AD-related genes on the phenotypic expression of biological markers.

It has been demonstrated that brain activity slowing is correlated with progressive Global Deterioration Scale (GDS) staging in dementia. In the general population, subjects harboring the ApoE-4/4 genotype exhibit premature slowing in brain mapping activity, represented by increased slow delta and theta activities compared to that in other ApoE genotypes [14]. In patients with AD, slow activity predominates in ApoE-4 carriers with a similar GDS stage, demonstrating the deleterious effect of the ApoE-4 allele on brain function.

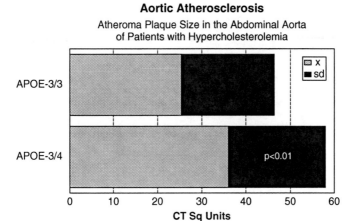

Fig. 4 ApoE-related atheroma plaque size in the abdominal aorta of patients with hypercholesterolemia

All these examples of genotype-phenotype correlations, as a gross approach to functional genomics, illustrate the importance of genotype-related differences in AD and their impact on phenotype expression [13–15]. Most biological parameters, potentially modifiable by monogenic genotypes and/or polygenic

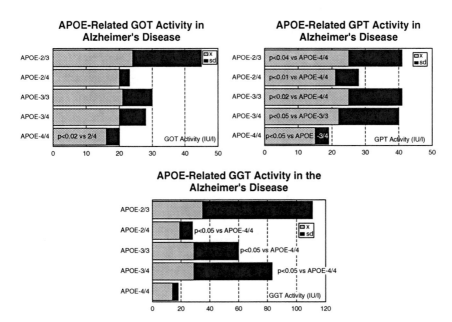

Fig. 5 ApoE-related glutamate oxaloacetate transaminase (GOT), glutamate-pyruvate transaminase (GPT), and γ-glutamyl transaminase (GGT) activities in Alzheimer's disease

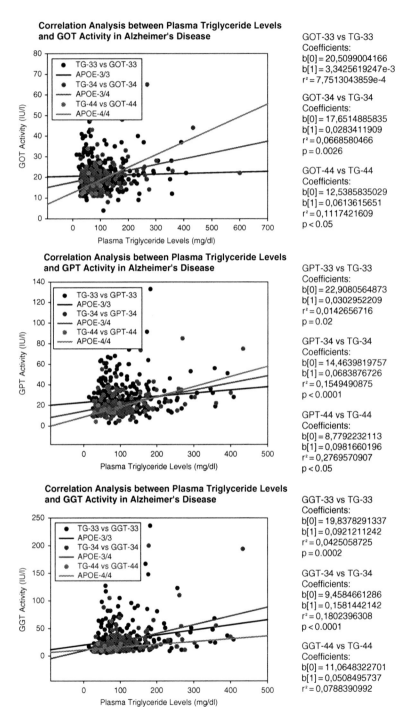

Correlation Analysis between Plasma Triglyceride Levels and GOT Activity in Alzheimer's Disease

GOT-33 vs TG-33
Coefficients:
b[0] = 20,5099004166
b[1] = 3,3425619247e-3
r² = 7,7513043859e-4

GOT-34 vs TG-34
Coefficients:
b[0] = 17,6514885835
b[1] = 0,0283411909
r² = 0,0668580466
p = 0.0026

GOT-44 vs TG-44
Coefficients:
b[0] = 12,5385835029
b[1] = 0,0613615651
r² = 0,1117421609
p < 0.05

Correlation Analysis between Plasma Triglyceride Levels and GPT Activity in Alzheimer's Disease

GPT-33 vs TG-33
Coefficients:
b[0] = 22,9080564873
b[1] = 0,0302952209
r² = 0,0142656716
p = 0.02

GPT-34 vs TG-34
Coefficients:
b[0] = 14,4639819757
b[1] = 0,0683876726
r² = 0,1549490875
p < 0.0001

GPT-44 vs TG-44
Coefficients:
b[0] = 8,7792232113
b[1] = 0,0981660196
r² = 0,2769570907
p < 0.05

Correlation Analysis between Plasma Triglyceride Levels and GGT Activity in Alzheimer's Disease

GGT-33 vs TG-33
Coefficients:
b[0] = 19,8378291337
b[1] = 0,0921211242
r² = 0,0425058725
p = 0.0002

GGT-34 vs TG-34
Coefficients:
b[0] = 9,4584661286
b[1] = 0,1581442142
r² = 0,1802396308
p < 0.0001

GGT-44 vs TG-44
Coefficients:
b[0] = 11,0648322701
b[1] = 0,0508495737
r² = 0,0788390992

Fig. 6 Correlation analysis between plasma triglyceride levels and GOT, GPT, and GGT in Alzheimer's disease

cluster profiles, can be used in clinical trials for monitoring efficacy outcomes. These parametric variables also show a genotype-dependent profile in different types of dementia (e.g., AD versus vascular dementia). For instance, striking differences have been found between AD and vascular dementia in structural and functional genomics studies [2,3,15].

ApoE-Related Lipid Metabolism and Atherosclerosis

ApoE genotypes directly influence lipid metabolism [2,3,15,17] and athero-sclerosis (Fig. 4). The size of atheroma plaques in the abdominal and thoracic aorta of patients with dementia and/or dyslipidemia is significantly larger in ApoE-4 carriers than in ApoE-3 carriers [18] (Fig. 4). In addition, the effect of lipid-lowering agents on atheroma plaques is ApoE-related, with a more effec-tive response in ApoE-3 carriers [18].

Potential Interactions of ApoE with Liver Function and Drug Metabolism

It has been observed that ApoE may influence liver function and drug metabolism by modifying hepatic steatosis and transaminase activity. There is a clear correlation between ApoE-related TG levels (Fig. 3) and glutamate oxaloacetate transaminase (GOT), glutamate-pyruvate transaminase (GPT), and γ-glutamyl transaminase (GGT) activities in AD (Figs. 5, 6). Both plasma TG levels and transaminase activity are significantly lower in AD patients who harbor the ApoE-4/4 genotype, probably indicating that(1) low TG levels protect against liver steatosis and (2) the presence of the ApoE-4 allele influences TG levels, liver steatosis, and transaminase activity. Consequently, it is likely that ApoE influences drug metabolism in the liver through different mechanisms, including interactions with enzymes such as transaminases and/or cytochrome P450-related enzymes encoded in genes of the CYP superfamily.

Pharmacogenetics

Alzheimer's disease patients are currently treated with cholinesterase inhibitors, neuroprotetive drugs, antidepressants, anxiolytics, antiparkinsonian drugs, anticonvulsants, and neuroleptics at a given time of the disease clinical course to palliate memory dysfunction, behavioral changes, sleep disorders, agitation, depression, parkinsonism, myoclonus and seizures, or psychotic symptoms. Many of these substances are metabolized by enzymes known to be genetically variable, including: (1) esterases: butyrylcholinesterase, paraoxonase/arylesterase; (2) transferases: *N*-acetyltransferase, sulfotransferase, thiol methyltransferase, thiopurine methyltransferase, catechol-*O*-methyltransferase, glutathione-*S*-transferases, UDP-glucuronosyltransferases, glucosyltransferase, histamine methyltransferase; (3) reductases: NADPH, quinine oxidoreductase, glucose-6-phosphate dehydrogenase; (4) oxidases: alcohol dehydrogenase, aldehydehydrogenase, monoamine oxidase B, catalase, superoxide dismutase, trimethylamine *N*-oxidase, dihydropyrimidine dehydrogenase; and (5) cytochrome P450 enzymes, such as CYP1A1, CYP2A6, CYP2C8, CYP2C9, CYP2C19, CYP2D6, CYP2E1, CYP3A5, and many others [19]. Polymorphic variants in these genes can induce alterations in drug metabolism, altering the efficacy and safety of the prescribed drugs [19–22].

Drug metabolism includes phase I reactions (i.e., oxidation, reduction, hydrolysis) and phase II conjugation reactions (i.e., acetylation, glucuronidation, sulfation, methylation) [19–22]. The typical paradigm for the pharmacogenetics of phase I drug metabolism is represented by the cytochrome P450 enzymes, a superfamily of microsomal drug-metabolizing enzymes. P450 enzymes represent a superfamily of heme-thiolate proteins widely distributed in bacteria, fungi, plants, and animals, with more than 200 P450 genes identified in various species. The P450 enzymes are encoded in genes of the CYP superfamily and act as terminal oxidases in multicomponent electron transfer chains called P450-containing monooxigenase systems. Some of the enzymatic products of the CYP gene superfamily can share substrates, inhibitors, and inducers, whereas others are quite specific for their substrates and interacting drugs [19–22].

The principal enzymes with polymorphic variants involved in phase I reactions are the following: CYP3A4/5/7, CYP2E1, CYP2D6, CYP2C19, CYP2C9, CYP2C8, CYP2B6, CYP2A6, CYP1B1, CYP1A1/2, epoxide hydrolase, esterases, NADPH-quinone oxidoreductase (NQO1), dihydropyrimidine dehydrogenase (DPD), alcohol dehydrogenase (ADH), and aldehyde dehydrogenase (ALDH). Major enzymes involved in phase II reactions include the following: uridine 5'-triphosphate glucuronosyl transferases (UGTs), thiopurine methyltransferase (TPMT), catechol-*O*-methyltransferase (COMT), histamine methyltransferase (HMT), sulfotransferases (STs), glutathione *S*-transferase A (GST-A), GST-P, GST-T, GST-M, *N*-acetyltransferase NAT2), NAT1, and others [19,23]. The most important enzymes of the P450 cytochrome family in drug metabolism, in decreasing order, are CYP3A4, CYP2D6, CYP2C9, CYP2C19, and CYP2A6 [23–27]. The predominant allelic variants in the CYP2A6 gene are CYP2A6*2 (Leu160His) and CYP2A6del. The CYP2A6*2 mutation inactivates the enzyme and is present in 1% to 3% of Caucasians. The CYP2A6del mutation results in no enzyme activity and is present in 1% of Caucasians and 15% of Asians [23]. The most frequent mutations in the CYP2C9 gene are CYP2C9*2 (Arg144Cys), with reduced affinity for P450 in 8% to 13% of Caucasian, and CYP2C9*3 (Ile359Leu), with alterations in the specificity for the substrate in 6% to 9% of Caucasians and 2% to 3% of Asians.23 The most prevalent polymorphic variants in the CYP2C19 gene are

CYP2C19*2, with an aberrant splicing site resulting in enzyme inactivation in 13% of Caucasians, 23% to 32% of Asians, 13% of Africans, and 14% to 15% of Ethiopians and Saoudians, and CYP2C19*3, a premature stop codon resulting in an inactive enzyme present in 6% to 10% of Asians and almost absent in Caucasians. The most important mutations in the CYP2D6 gene are the following: CYP2D6*2xN, CYP2D6*4, CYP2D6*5, CYP2D6*10, and CYP2D6*17 [24]. The CYP2D6*2xN mutation gives rise to gene duplication or multiplication, resulting in increased enzyme activity, which appears in 1% to 5% of the Caucasian population, 0% to 2% of Asians, 2% of Africans, and 10% to 16% of Ethiopians. The defective splicing caused by the CYP2D6*4 mutation inactivates the enzyme and is present in 12% to 21% of Caucasians. The deletion in CYP2D6*5 abolishes enzyme activity and shows a frequency of 2% to 7% in Caucasians, 1% in Asians, 2% in Africans, and 1% to 3% in Ethiopians. The polymorphism CYP2D6*10 causes Pro34Ser and Ser486Thr mutations with unstable enzyme activity in 1% to 2% of Caucasians, 6% of Asians, 4% of Africans, and 1% to 3% of Ethiopians. The CYP2D6*17 variant causes Thr107Ile and Arg296Cys substitutions, which produce a reduced affinity for substrates in 51% of Asians, 6% of Africans, and 3% to 9% of Ehtiopians; it is practically absent in Caucasians [28].

CYP2D6 Polymorphisms in Alzheimer's Disease

Although initial studies postulated involvement of the CYP2D6B mutant allele in Lewy body formation in both Parkinson's disease and the Lewy body variant of AD, as well as in the synaptic pathology of pure AD without Lewy bodies, subsequent studies in various ethnic groups did not find an association between AD and CYP2D6 variants [28]. However, some conventional antidementia drugs (tacrine, donepezil, galantamine) are metabolized via CYP-related enzymes, especially CYP2D6, CYP3A4, and CYP1A2 [29,30]; and polymorphic variants of the CYP2D6 gene can affect liver metabolism and the safety and efficacy of some cholinesterase inhibitors.

In European countries 6% to 8% of the population are poor metabolizers. In a preliminary study in Spain, 83.8% of the Spanish AD patients were found to be extensive metabolizers (EMs), whereas 8.6% were poor metabolizers (PMs), and 7.4% were ultrarapid metabolizers (UMs) [14]. In a more recent study with a larger sample of AD patients, the distribution of CYP2D6 genotypes in the Spanish AD population was the following: (1) EM (85.30%): *1/*1 (56.90%), *1/*3 (2.59%), *1/*4 (23.28%), *1/*5 (5.17%), *1/*6 (2.59%), *1/*7 (0.86%), *1/*10 (5.17%), *6/*10 (0.86%), *7/*10 (0.86%), and *10/*10 (1.72%); (2) PM (7.35%): *4/*4 (100%); and (3) UM (7.35%): *1xN/*1 (70%), and *1xN/*4 (30%)28 (Fig. 7). Based on these data, it might be inferred that at least 15% of the AD population in Spain have abnormal metabolism of cholinesterase inhibitors. Approximately 50% of this population cluster would show

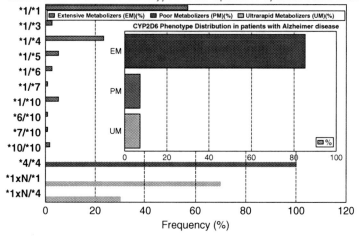

Fig. 7 Distribution of CYP2D6 alleles and genotypes in patients with Alzheimer's disease

ultrarapid metabolism, requiring higher doses of cholinesterase inhibitors to reach a therapeutic threshold, and the other 50% of the cluster would exhibit poor metabolism, displaying potential adverse events at low doses. If we take into account that approximately 60% to 70% of therapeutic outcomes depend on pharmacogenomic criteria (e.g., pathogenic mechanisms associated with AD-related genes), it can be postulated that pharmacogenetic and pharmacogenomic factors are responsible for 75% to 85% of the therapeutic response in AD patients treated with conventional drugs [28]. In addition, some AD-related polymorphisms of risk accumulate in PMs and UMs, contributing to alterations of treatment outcomes in dementia [31–33] (Fig. 8).

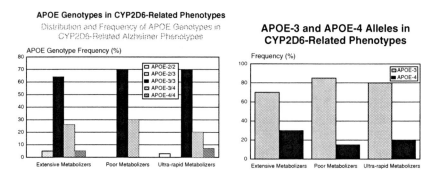

Fig. 8 Distribution of ApoE alleles (right panel) and genotypes (left panel) in CYP2D6-related extensive metabolizers (EMs), poor metabolizers (PMs), and ultrarapid metabolizers (UMs) in Alzheimer's disease

Association of ApoE Genotypes with CYP2D6-Related Phenotypes

The distribution of ApoE alleles and genotypes in CYP2D6-related EM, PM, and UM phenotypes shows a characteristic pattern (Fig. 8). The ApoE-4/4 genotype accumulates in EMs and UMs and is practically absent in PMs; in contrast, ApoE-3/3 and ApoE-3/4 genotypes exhibit an almost identical pattern in EMs, PMs, and UMs. The ApoE-2/2 genotype, which is rare in the normal population ($< 1\%$) [1], is present only in UMs (3%). The typical distribution of ApoE genotypes in AD is characteristically represented in EMs, whereas there is overrepresentation of ApoE-2 and ApoE-4 homozygotes in UMs. This peculiar distribution of ApoE genotypes among CYP2D6-related phenotypes appears to indicate that the absence of homozygous ApoE-4 might influence enzyme activity in PMs and that an overdose of ApoE-2 might account for excess activity in UMs.

Conclusions

Data reported in the literature [2,3,13–18,31–33] and the present results indicate that the pleiotropic activity of ApoE influences the expression of various phenotypic features in AD, including age at onset, amyloid burden in senile plaques and vessels, tau pathology, brain function, apoptosis, inflammation, blood pressure, cardiovascular function, lipid metabolism, cerebrovascular function, brain hemodynamics, atherosclerosis, liver metabolism, transaminase activity, and therapeutic outcomes. The ApoE-related therapeutic response in AD [16,18,31–33] may be mediated via specific mechanisms directly associated with AD pathology and/or an indirect influence on drug metabolism through interaction with cytochrome P450 enzymes encoded in genes of the CYP family.

References

1. Cacabelos R. Molecular genetics of Alzheimer's disease and aging. Genomic Medicine Series. Part 1. Methods Find Exp Clin Pharmacol 200527(suppl A):1–573.
2. Cacabelos R, Fernández-Novoa L, Corzo L, et al. Phenotypic profiles and functional genomics in dementia with a vascular component. Neurol Res 2004;26:459–480.
3. Cacabelos R, Fernández-Novoa L, Corzo L, et al. Genomics and phenotypic profiles in dementia: implications for pharmacological treatment. Methods Find Exp Clin Pharmacol 2004;26(6):421–444.
4. Teter B, Finch CE. Caliban's heritance and the genetics of neuronal aging. Trends Neurosci 2004;27(10):627–632.
5. Wright AD, Jaconson SG, Cideciyan AV, et al. Lifespan and mitochondrial control of neurodegeneration. Nat Genet 2004;36:1153–1158.
6. Jaenisch R, Bird A. Epigenetic regulation of gene expression: how the genome integrates intrinsic and environmental signals. Nat Genet 2003;33(suppl):245–254.

7. Hunter DJ. Gene-environment interactions in human diseases. Nat Rev Genet 2005;6:287–298.
8. Selkoe DJ, Podlisny MB. Deciphering the genetic basis of Alzheimer's disease. Annu Rev Genomics Hum Genet 2002;3:67–99.
9. Selkoe DJ, Schenk D. Alzheimer's disease: molecular understanding predicts amyloid-based therapeutics. Annu Rev Pharmacol Toxicol 2003;43:545–584.
10. Roses AD, Pangalos MN. Drug development and Alzheimer's disease. Am J Geriatr Psychiatry 2003;11:123–130.
11. Cacabelos R, Alvarez XA, Lombardi V, et al Pharmacological treatment of Alzheimer disease: from psychotropic drugs and cholinesterase inhibitors to pharmacogenomics. Drugs Today 2000;6:415–499.
12. Suh YH, Checler F. Amyloid precursor protein, presenilins, and a-synuclein: molecular pathogenesis and pharmacological applications in Alzheimer's disease. Phamacol Rev 2002;54:469–525.
13. Cacabelos R, Lombardi V, Fernández-Novoa L, et al. A functional genomics approach to the analysis of biological markers in Alzheimer disease. In: Takeda M, Tanaka T, Cacabelos R (eds) Molecular Neurobiology of Alzheimer Disease and Related Disorders. Basel: Karger, 2004, pp 236–285.
14. Cacabelos R. The application of functional genomics to Alzheimer's disease. Pharmacogenomics 2003;4(5):597–621.
15. Cacabelos R. Genomic characterization of Alzheimer's disease and genotype-related phenotypic analysis of biological markers in dementia. Pharmacogenomics 2004;5(8):1049–1105.
16. Cacabelos R, Fernández-Novoa L, Pichel V, et al. Pharmacogenomic studies with a combination therapy in Alzheimer's disease. In: Takeda M, Tanaka T, Cacabelos R (eds) Molecular Neurobiology of Alzheimer Disease and Related Disorders. Basel: Karger, 2004, pp 94–107.
17. Cacabelos R, Fernández-Novoa L, Lombardi V, et al. Cerebrovascular risk factors in Alzheimer's disease: brain hemodynamics and pharmacogenomic implications. Neurol Res 2003;25:567–580.
18. Cacauelos R. Pharmacogenomics and Nutrigenomics. Scientific Progress and Pharmaceutical Development. Coruña: Ebiotec Foundation, 2004.
19. Kalow W, Grant DM. Pharmacogenetics. In: Scriver CR, Beaudet AL, Sly WS, et al (eds) The Metabolic Š Molecular Bases of Inherited Disease. New York: McGraw-Hill, 2001, pp 225–255.
20. Weinshilboum R. Inheritance and drug response. N Engl J Med 2003;348:529–537.
21. Evans WE, McLeod HL. Pharmacogenomics—drug disposition, drug targets, and side effects. N Engl J Med 2003;348:538–549.
22. Nebert DW, Jorge-Nebert LF. Pharmacogenetics and pharmacogenomics. In: Rimoin DL, Connor JM, Pyeritz R, Korf BR (eds) Emery and Rimoin's Principles and Practice of Medical Genetics, 4th ed. Edinburgh: Churchill-Livingstone, 2002, pp 590–631.
23. Tribut O, Lessard Y, Reymann JM, et al. Pharmacogenomics. Med Sci Monit 2002;8:152–163.
24. Isaza CA, Henao J, López AM, Cacabelos R. Isolation, sequence and genotyping of the drug metabolizer CYP2D6 gene in the Colombian population. Methods Find Exp Clin Pharmacol 2000;22:695–705.
25. Saito S, Ishida A, Sekine A, et al. Catalog of 680 variants among eight cytochrome P450 (CYP) genes: nine esterase genes, and two other genes in the Japanese population. J Hum Genet 2003;48:249–270.
26. Wooding SP, Watkins SW, Bamshad MJ, et al. DNA sequence variations in a 3.7-kb noncoding sequence 5-prime of the CYP1A2 gene: implications for human population history and natural selection. Am J Hum Genet 2002;71:528–542.
27. Xie HG, Kim RB, Wood AJ, Stein CM. Molecular basis of ethnic differences in drug disposition and response. Annu Rev Pharmacol Toxicol 2001;41:815–850.

28. Cacabelos R. Pharmacogenomics and therapeutic prospects in Alzheimer's disease. Exp Opin Pharmacother 2005;6:1967–1987.
29. Farlow MR. Clinical pharmacokinetics of galantamine. Clin Pharmacokinet 2003;42(15):1383–1392.
30. Jann MW, Shirley KL, Small GW. Clinical pharmacokinetics and pharmacodynamics of cholinesterase inhibitors. Clin Pharmacokinet 2002;41(10):719–739.
31. Cacabelos R, Alvarez A, Fernández-Novoa L, Lombardi VRM. A pharmacogenomic approach to Alzheimer's disease. Acta Neurol Scand 2000;176(suppl):12–19.
32. Cacauelos R. Pharmacogenomics in Alzheimer's disease. Mini Rev Med Chem 2002; 2:59–84.
33. Cacabelos R. Pharmacogenomics for the treatment of dementia. Ann Med 2002; 34:357–379.

Chapter 38
Up-regulation of the α-Secretase Pathway

Falk Fahrenholz, Claudia Prinzen, Rolf Postina, and Elżbieta Kojro

Introduction

In the nonamyloidogenic pathway, the amyloid precursor protein (APP) is cleaved during its transport from the Golgi apparatus to the plasma membrane and at the plasma membrane by the α-secretase within the sequence of the β-amyloid (Aβ) peptides, thereby preventing their formation. Because β- and α-secretases compete for the same substrate, the production of Aβ can be reduced by either inhibition of β- and γ-secretases or activation of the α-secretase. The latter seems to be particularly rewarding, as a soluble fragment of APP is generated, APPsα, with neurotrophic and neuroprotective properties. During the last decade and a half, an increasing number of reports have shown that this fragment has effects antagonistic to those of Aβ: APPsα exogenously applied to cell cultures, hippocampal slices, or the brains of amnestic mice evoked neuroprotective and antiapoptotic action and a positive effect on synaptic plasticity and learning [1]. Two new reports demonstrated that it increases expression of neuroprotective genes in vivo [22] and promotes the proliferation of neural stem cells in the adult subventricular zone [2].

Disintegrin Metalloproteinase ADAM10 as α-Secretase In Vivo

In 1999, we reported that purified ADAM10 and ADAM10 in cellular systems has α-secretase activity [3]. ADAM10 belongs to the membrane-bound disintegrins and metalloproteinases, which are kept in a latently inactive form unless their prodomain is removed by the proteolytic action of proprotein convertases [4]. The mature form is localized mainly on the cell surface [3].

F. Fahrenholz
Institute of Biochemistry, University of Mainz, 55099 Mainz, Germany

A. Fisher et al. (eds.), *Advances in Alzheimer's and Parkinson's Disease*,
© Springer 2008

To demonstrate in vivo α-secretase activity, we overexpressed ADAM10 in APP[V717I] transgenic mice, a mouse model for the amyloid pathology of Alzheimer disease. These animals showed an increase in soluble Aβ peptides and impaired long-term potentiation (LTP) and learning early in life, whereas amyloid plaque formation starts later, after 12 months [5]. The plaque formation seen in the cortex and hippocampus of APP[V717I] mice was almost completely prevented by even moderate overexpression of ADAM10. In contrast, double-transgenic mice expressing a dominant-negative mutant of ADAM10 clearly developed severalfold higher amounts of larger amyloid plaques than APP London transgenic mice. Cognitive characteristics of the mice were analyzed in several ways, including the Morris-Water-Maze test. While transgenic mice carrying the London mutation had impaired spatial learning, overexpression of ADAM10 resulted in improved spatial learning [6].

These results show that ADAM10 acts as an α-secretase in vivo and prevents amyloid plaques and early cognitive deficits in a mouse model. By overexpressing α-secretase there are several beneficial effects that might act synergistically: We found a large increase of APPsα and at the same time APPsβ; and production of the soluble Aβ peptides was reduced. This demonstrates direct competition of ADAM10 with BACE for its substrate in vivo. ADAM10 monotransgenic mice showed normal viability and fertility as well as normal behavioral abilities in several tests.

Up-regulation of the α-Secretase Pathway

How can an increase in α-secretase in the brains of Alzheimer disease (AD) patients be achieved? One possibility is to stimulate signaling pathways that increase the activity or the expression level of ADAM10.

Activation of G Protein-Coupled Receptors and Downstream Signaling Systems

Stimulation of G protein-coupled receptors and the subsequent signal transduction pathways localized in brain areas affected by AD is a valuable approach. This principle has been validated by in vivo studies with a new class of M1 agonists in AD patients [7] and in a transgenic mouse model [8]. It has been shown recently that acetylcholinesterase inhibitors, in addition to their classic role, support the nonamyloidogenic pathway. One mechanism that has been suggested is the enhanced transport of ADAM10 to the cell surface after stimulation of muscarinic receptors by acetylcholine [9]. Activators of the protein kinase C (PKC) stimulated the α-secretase pathway and attenuated symptoms of AD pathology in transgenic mouse models [10]. Considering the beneficial effect of α-secretase overexpression, it is certainly necessary to further explore this approach.

Our aim was to identify new α-secretase activators and to elucidate the signal mechanisms involved. We concentrated on endogenous hormones, in particular on neuropeptides and their G protein-coupled receptors. Such candidates are the pituitary adenylate cyclase-activating polypeptides PACAP27 and PACAP38 and their receptor, the PAC1 receptor. They occur in high concentrations in the cerebral cortex and in the hippocampal formation. Several studies demonstrated the neurotrophic and antiapoptotic actions of PACAP peptides [11] and their involvement in learning processes [12]. These are properties they share with secreted APPsα. This led us to formulate the hypothesis that these properties might be at least partially due to a stimulatory effect of PACAP on the nonamyloidogenic pathway.

Treatment of human neuroblastoma cells endogenously expressing PAC1 receptors with PACAP led to a significant increase of secreted APPsα. To study in more detail the signal mechanisms involved, experiments with various inhibitors were performed on HEK cells overexpressing the PAC1 receptor. By applying various protein kinase inhibitors, we found that there was a major contribution of the MAP kinase pathway and phosphatidyl inositol-3 (PI-3) kinase as well as involvement of protein kinase C in PACAP-induced α-secretase activation (Fig. 1). In a similar series of experiments we excluded the adenylate cyclase protein kinase A (PKA) system as being responsible for PACAP-induced α-secretase activation. The involvement of the PI-3 kinase system suggests that transport of ADAM10 or its substrate to the cell surface might play a role in α-secretase activation. PACAP stimulation did not influence expression of APP or of ADAM10 [17], or 9. Because a peptide transport system has been identified that allows bidirectional transport of PACAP across the blood-brain barrier [13], PACAP receptor agonists may therefore prove to be useful for the treatment of AD.

Characterization of the Human ADAM10 Promoter and Up-regulation of Gene Expression by Retinoic Acid

To elucidate the gene expression of ADAM10, we characterized its promoter. Both human and mouse ADAM10 genes comprise approximately 160 kbp, are composed of 16 exons, and are highly conserved within 500 bp upstream of either translation initiation site. Using luciferase reporter assays, we demonstrated that nucleotides −2179 to −1 upstream of the human ADAM10 translation initiation site represent a functional TATA-less promoter [14].

We found the highest ADAM10 promoter activity in the neural SH-SY5Y cell line followed by neuronal IMR32 cells, with reduced activities in kidney HEK 293 and liver HepG2 cells (Fig. 2). This finding is in agreement with

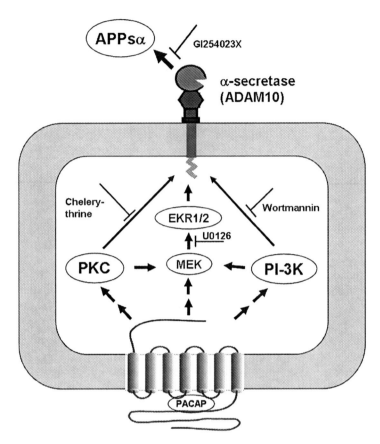

Fig. 1 Model for pituitary adenylate cyclase-activating polypeptide (PACAP)-induced activation of α-secretase. PACAP binds with high affinity to the PAC1 receptor and affects several signaling cascades. The PACAP-induced increase of α-secretase activity was mediated by MAP, PI3, and protein kinase C (PKC) kinases. These led to the phosphorylation of extracellular signal-regulated kinase (ERK). Although PACAP-stimulated ERK phosphorylation was completely inhibited by MEK1,2-specific inhibitors, the effect of PACAP on α-secretase activation was only partly reduced. Therefore, pathways not involved in the MAP kinase cascade contribute to α-secretase activation by PACAP

reports showing that ADAM10 is ubiquitously expressed [15,16] but in addition indicates a favored transcription of ADAM10 in the brain.

By deletion analysis, site-directed mutagenesis, transcription factor overexpression, and electrophoretic mobility shift assays, we identified nucleotides –08 to –300 as the core promoter and found Sp1, USF, and retinoic acid (RA)-responsive elements to modulate its activity. ADAM10 promoter activity was enhanced upon RA treatment, and transcription of endogenous ADAM10 was induced by RA.

Plasma levels of antioxidants including retinol (vitamin A) are depleted in AD patients and subjects with mild cognitive impairment (MCI), a precursor of

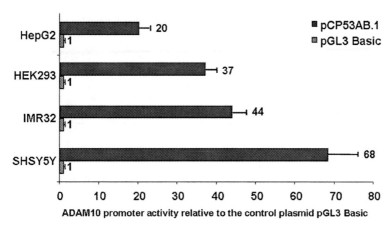

Fig. 2 ADAM10 promoter activity in different cell lines

AD [17]. Furthermore, vitamin A acid is essential for hippocampal long-term plasticity [18,19] and neurogenesis [20]. In addition, genetic and dietary data suggest that defective retinoid transport and function may contribute to a late onset of AD [21]. Our findings suggest that pharmacological targeting of RA receptors may increase expression of the α-secretase ADAM10 with beneficial effects on AD pathology.

References

1. Kojro E, Fahrenholz F. The non-amyloidogenic pathway: structure and function of α-secretases. Subcell Biochem 2005;38:105–127.
2. Caille I, Allinquant B, Dupont E, et al. Soluble form of amyloid precursor protein regulates proliferation of progenitors in the adult subventricular zone. Development 2004;131:2173–2181.
3. Lammich S, Kojro E, Postina R, Gilbert S, et al. Constitutive and regulated α-secretase cleavage of Alzheimer's amyloid precursor protein by a disintegrin metalloprotease. Proc Natl Acad Sci U S A 1999;96:3922–3927.
4. Anders A, Gilbert S, Garten W, et al. Regulation of the α-secretase ADAM10 by its prodomain and proprotein convertases. FASEB J 2001;15:1837–1839.
5. Moechars D, Dewachter I, Lorent K, et al. Early phenotypic changes in transgenic mice that overexpress different mutants of amyloid precursor protein in brain. J Biol Chem 1999;274:6483–6492.
6. Postina R, Schroeder A, Dewachter I, et al. A disintegrin-metalloproteinase prevents amyloid plaque formation and hippocampal defects in an Alzheimer disease mouse model. J Clin Invest 2004;113:1456–1464.
7. Nitsch RM, Deng M, Tennis M, et al. The selective muscarinic M1 agonist AF102B decreases levels of total Abeta in cerebrospinal fluid of patients with Alzheimer's disease. Ann Neurol 2000;48:913–918.
8. Fisher A, Caccamo A, Oddo S, et al. M1 muscarinic agonists attenuate the pathology and restore cognition in animal models for Alzheimer's disease (abstract). AD/PD Conferente, 2005.

9. Zimmermann M, Gardoni F, Marcello E, et al. Acetylcholinesterase inhibitors increase ADAM10 activity by promoting its trafficking in neuroblastoma cell lines. J Neurochem 2004;90:1489–1499.

10. Etcheberrigaray R, Tan M, Dewachter I, et al. Therapeutic effects of PKC activators in Alzheimer's disease transgenic mice. Proc Natl Acad Sci U S A 2004;101:11141–11146.

11. Vaudry D, Gonzalez BJ, Basille M, et al. Pituitary adenylate cyclase-activating polypeptide and its receptors: from structure to functions. Pharmacol Rev 2000;52:269–324.

12. Sacchetti B, Lorenzini CA, Baldi E, et al. Pituitary adenylate cyclase-activating polypeptide hormone (PACAP) at very low dosages improves memory in the rat. Neurobiol Learn Mem 2001;76:1–6.

13. Dogrukol-Ak D, Tore F, Tuncel N. Passage of VIP/PACAP/secretin family across the blood-brain barrier: therapeutic effects. Curr Pharm Des 2004;10:1325–1340.

14. Prinzen C, Muller U, Endres K, et al. Genomic structure and functional characterization of the human ADAM10 promoter. FASEB J 2005;19:1522–1524.

15. Howard L, Lu X, Mitchell S, et al. Molecular cloning of MADM: a catalytically active mammalian disintegrin-metalloprotease expressed in various cell types. Biochem J 1996;317(Pt 1):45–50.

16. Yavari R, Adida C, Bray-Ward P, et al. Human metalloprotease-disintegrin Kuzbanian regulates sympathoadrenal cell fate in development and neoplasia. Hum Mol Genet 1998;7:1161–1167.

17. Rinaldi P, Polidori MC, Metastasio A, et al. Plasma antioxidants are similarly depleted in mild cognitive impairment and in Alzheimer's disease. Neurobiol Aging 2003;24:915–919.

18. Chiang MY, Misner D, Kempermann G, et al. An essential role for retinoid receptors RARbeta and RXRgamma in long-term potentiation and depression. Neuron 1998;21:1353–1361.

19. Misner D, Jacobs S, Shimizu Y, et al. Vitamin A deprivation results in reversible loss of hippocampal long-term synaptic plasticitá. Proc Natl Acad Sci U S A 2001;98:11714–11719.

20. Takahashi J, Palmer TD, Gage FH. Retinoic acid and neurotrophins collaborate to regulate neurogenesis in adult-derived neural stem cell cultures. J Neurobiol 1999;38:65–81.

21. Goodman AB, Pardee AB. Evidence for defective retinoid transport and function in late onset Alzheimer's disease. Proc Natl Acad Sci U S A 2003;100:2901–2905.

22. Stein TD, Anders NJ, DeCarli C, et al. Neutralization of transthyretin reverses the neuroprotective effects of secreted amyloid precursor protein (APP) in APPSW mice resulting in tau phosphorylation and loss of hippocampal neurons: support for the amyloid hypothesis. J Neurosci 2004; 24:7707–7717.

Chapter 39
Frequency and Relation of Argyrophilic Grain Disease and Thorn-Shaped Astrocytes in Alzheimer's Disease

Hirotake Uchikado, Yasuhiro Fujino, Wenlang Lin, and Dennis Dickson

Introduction

Alzheimer's disease (AD) is characterized neuropathologically by neurofibrillary tangles (NFTs), neuropil threads (NTs), and senile plaques (SPs). SPs are complicated lesions that have reactive glia, dystrophic neuritis, and extracellular deposits of β-amyloid. NFTs and NTs are intraneuronal lesions composed of hyperphosphorylated microtubule-associated protein tau. The tau protein that accumulates in these lesions is composed of equal amounts of two major splice variants (due to alternative spicing of exon 10) with three or four repeats (3R and 4R) in the microtubule-binding domain of tau. AD is thus a 3R+4R tauopathy.

Argyrophilic grain disease (AGD) is a disorder associated with accumulation of filamentous tau in small neuronal processes that are argyrophilic with silver stains (e.g., Gallyas stain) resemble grains, so-called "argyrophilic grains" (AGs). Tau immunoreactivity is also present in oligodendroglial coiled bodies and in neurons as pretangles in AGD. AGD preferentially affects neurons and glia in the medial temporal lobe. Thorn-shaped astrocytes (TSAs) are tau-immunoreactive astrocytes found most often in subpial and perivascular spaces of the medial temporal lobe. The frequency and relation of these two tauopathies in AD remain unknown and have been difficult to study until recently with the advent of antibodies specific to 4R tau. Both AGD [1] and TSAs are composed of 4R tau [2]. In addition to the 3R+4R NFTs and NTs, some cases of AD have concurrent 4R tauopathies, including AGD, TSA, or both [3–6]. In the present study, we examined medial temporal lobe sections of AD brain with 4R tau-specific antibody to evaluate the relation between AGD, TSA, and the intensity of 4R tauopathy.

H. Uchikado
Mayo Clinic College of Medicine, Jacksonville, FL 32224, USA

A. Fisher et al. (eds.), *Advances in Alzheimer's and Parkinson's Disease*,
© Springer 2008

Materials and Methods

Paraffin-embedded tissues of 239 cases of pathologically confirmed AD (109 men, 130 women; age range 55–102 years, mean 81.6 years) were obtained from the Mayo Clinic Jacksonville brain bank. The neuropathological diagnosis of AD was established after a standardized dissection, sampling, and staining protocol. Quantitative measurements of SPs ($\times 100$) and NFTs ($\times 400$) in multiple brain regions were performed with thioflavin-S fluorescence microscopy. AGs and TSAs were detected with CP13 and 4R tau-specific monoclonal antibody (ET3).

Coronal sections of the basal forebrain region, which included the lentiform nucleus, amygdala, ambient gyrus, temporal stem, claustrum, and insular cortex, were used. Paraffin-embedded tissue sections were immunostained using phos-phorylation-dependent anti-tau monoclonal antibody (CP13, 1:500) [7] and 4R-tau specific monoclonal antibody (ET3, 1:1000) [8]. To examine the relation between TSA and 4R tauopathy, double immunostaining was performed with antibodies to glial fibrillary acidic protein (anti-GFAP, 1:500; Dako, Glostrup, Denmark) and 4R tau (ET3). To confirm their argyrophilia, sections were also processed with the Gallyas-Braak silver impregnation method.

Ultrastructural characterization of lesions was performed on samples obtained from temporal periventricular white matter. Small blocks of formalin-fixed tissue were postfixed and then processed and embedded in LR white resin. Ultrathin sections were cut and double-immunolabeled using two sizes of immunogold particles with CP13 and GFAP antibodies, according to previously reported methods [9].

Results

Immunohistochemistry

Immunostaining for ET3 or CP13 revealed AGs of the basal forebrain as minute, spindle- or comma-shaped structures (Fig. 1). Most AGs were found in the ambient gyrus and cortical nucleus of the amygdala. These methods also showed TSAs in the subpial and periventricular regions near the temporal horn of the amygdala. TSAs had cytoplasm with a few short, thick processes. TSAs were also detected by the Gallyas silver stain; however, the density of TSAs by

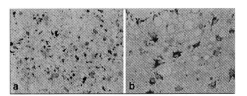

Fig. 1 Argyrophilic grains (AGDs) (**a**) and thorn-shaped astrocytes (TSAs) (**b**), observed using immunohistochemistry

ET3 immunohistochemistry was higher than that of Gallyas-positive TSAs in adjacent sections. No apparent spatial relation was noted between TSA and AGD pathology.

In AD tissues with TSA, double-immunostaining with ET3 and anti-GFAP revealed co-localization of 4R tau and GFAP in some astrocytes. Electron microscopy revealed coexistence of tau filaments and glial fibrils in the cytoplasm of astrocytes (Fig. 2). The two types of filament appeared to aggregate separately.

Semiquantitative Evaluation of 4R Tauopathy and TSA

We first investigated the degree of 4R tauopathy in AD with ET3 immunohisto-chemistry. Some AD tissues had little or no immunoreactivity with ET3, whereas others showed many ET3-immunoreactive NFTs and NTs. Accordingly, the presence and severity of 4R tauopathy (NFTs, NTs, AGD, TSAs) were scored in the basal forebrain on ET3-immunostained sections using a semiquantitative method with scores of 1 to 4 (1, a few; 2, mild; 3, moderate; 4, severe). Likewise, TSAs were assessed semiquantitatively (0, absent; 1, a few; 2, moderate; 3, many).

Frequency of AGD and TSA in AD

In this study, AGDs were confirmed in 67 of 239 cases (28.0%), with the frequency of AGDs in AD tissues increasing with age. Likewise, TSAs were

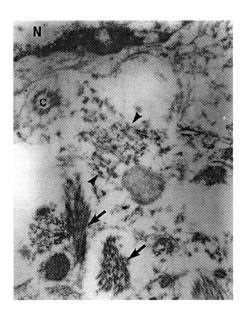

Fig. 2 Immunoelectron micrograph of a TSA shows two types of filaments: glial intermediate filaments (arrows) and immunogold-labeled tau filaments (arrowheads). N, nucleus; c, centriole

seen in 70 of 239 AD cases (29.3%). The frequency of TSA increased with age; it
was 17% in patients ≤ 69 years of age (1/6), 20% in those 70 to 79 years (16/81),
30% in those 80 to 89 years (35/117), and 50% in patients ≥ 90 years (16/32)
(Fig. 3). The percentage of patients affected by some degree of TSA increased
significantly with advancing age. The frequency of AGDs and TSAs in AD
tissues tended to increase with age.

Relation of AGD and TSA with AD Type Pathology

The male/female ratio and average Braak stage, NFTs, and SPs in the limbic
lobe did not differ between AD cases with and without TSA or between AD
cases with and without AGDs [10]. AD pathology did not correlate with AGDs
or TSAs as assessed by the Spearman rank order correlation analysis.

Relation of AGD and TSA with 4R Tau Score in AD

Spearman rank order correlation demonstrated that 4R tau scores correlated
positively with both the presence of AGDs ($r = 0.44$, $p < 0.001$) and the severity
of TSAs ($r = 0.36$, $p < 0.001$). There was a weak correlation between the
presence of AGDs and the severity of TSAs ($r = 0.14$, $p < 0.05$).

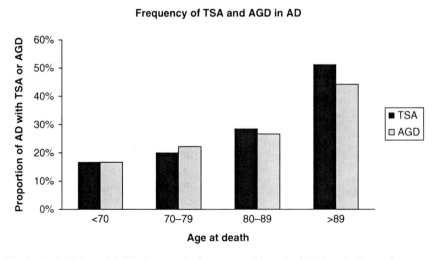

Fig. 3 Both TSAs and AGDs increase in frequency with age in Alzheimer's disease, but cases
with TSAs do not necessarily also have AGDs

Conclusions

AGDs and TSAs are distinctive for medial temporal 4R tauopathies that increase in frequency with age. They are common in AD but are independent of AD pathology. Although there is a weak correlation between AGD and TSA, some cases with AGD do not have significant TSA and vice versa. That AGDs and TSAs are both 4R tauopathies affecting similar parts of the brain, albeit in neurons for AGDs and astrocytes for TSAs, suggests that AGDs and TSAs may have a common pathogenesis. Further neuropathological studies are necessary to resolve this question.

References

1. Togo T, Sahara N, Yen SH, et al. Argyrophilic grain disease is a sporadic 4-repeat tauopathy. J Neuropathol Exp Neurol 2002;61:547–556.
2. Iseki E, Togo T, Suzuki K, et al. Dementia with Lewy bodies from the perspective of tauopathy. Acta Neuropathol (Berl) 2002;105:265–270.
3. Braak H, Braak E. Argyrophilic grain disease: frequency of occurrence in different age categories and neuropathological diagnostic criteria. J Neural Transm 1998;105:801–819.
4. Ikeda K, Akiyama H, Kondo H, et al. Thorn-shaped astrocytes: possibly secondarily induced tau-positive glial fibrillary tangles. Acta Neuropathol (Berl) 1995;90:620–625.
5. Jellinger KA. Dementia with grains (argyrophilic grain disease). Brain Pathol 1998;8:377–386.
6. Schultz C, Ghebremedhin E, Del Tredici K, et al. High prevalence of thorn-shaped astrocytes in the aged human medial temporal lobe. Neurobiol Aging 2004;25:397–405.
7. Jicha GA, Bowser R, Kazam IG, Davies P. Alz-50 and MC-1, a new monoclonal antibody raised to paired helical filaments, recognize conformational epitopes on recombinant tau. J Neurosci Res 1997;48:128–132.
8. Ishizawa T, Ko LW, Cookson N, et al. Selective neurofibrillary degeneration of the hippocampal CA2 sector is associated with four-repeat tauopathies. J Neuropathol Exp Neurol 2002;61:1040–1047.
9. Lin WL, Lewis J, Yen SH, et al. Filamentous tau in oligodendrocytes and astrocytes of transgenic mice expressing the human tau isoform with the P301L mutation. Am J Pathol 2003;162:213–218.
10. Fujino Y, Wang DS, Thomas N, et al. Increased frequency of argyrophilic grain disease in Alzheimer disease with 4R tau-specific immunohistochemistry. J Neuropathol Exp Neurol 2005;64:09–214.

Chapter 40
Cellular Membranes as Targets in Amyloid Oligomer Disease Pathogenesis

Erene W. Mina and Charles G. Glabe

Introduction

Age-related degenerative diseases, including Alzheimer's, Parkinson's, and Huntington's diseases and type II diabetes, manifest uniquely, although a growing body of evidence indicates that they have several striking features in common. Most importantly, they share what is widely recognized as a key pathogenic event: the accumulation of insoluble, misfolded proteins. Regardless of their protein sequences, the end-products of their accumulation, or amyloid fibrils, are assembled via an aggregation pathway involving soluble intermediates known as oligomers. Increasing evidence implicates these oligomers as the primary toxic species in disease, suggesting that their toxicity is intrinsically related to their aggregation state, independent of sequence. Thus, these otherwise unrelated amyloid oligomers may share a common primary pathogenetic mechanism.

This notion of "conformational disease," whereby the formation of a common structural motif leads to a common mechanism of pathogenesis, predicts a common primary target. The peptides and proteins that share this generic propensity for forming amyloid oligomers do not all reside in the same cellular compartments. Some arise from cytosolic proteins, and others are derived from secretory or extracellular proteins. Therefore, the primary target of this common mechanism of pathogenesis should be accessible to both extracellular and cytosolic compartments, pointing to the plasma membrane as a probable candidate. The purpose of this chapter is to review the evidence indicating that amyloid oligomers permeabilize cell membranes and that it represents a key pathological event shared by amyloid-related degenerative diseases.

E.W. Mina
Department of Molecular Biology and Biochemistry, University of California, Irvine, CA 92696-3900, USA

A. Fisher et al. (eds.), *Advances in Alzheimer's and Parkinson's Disease*,   381
© Springer 2008

The Case for Oligomers

Amyloid fibrils are approximately 6 to 10 nm in diameter and are characterized by a cross-β structure, indicating that the backbone hydrogen bonding is parallel to the fibril axis [1–3]. Therefore, aggregation of amyloids requires misfolding of the native protein structure, resulting in a structure that is capable of intermolecular hydrogen bonding to form extended polypeptide strands. This results in aggregates acquiring the ability to propagate indefinitely, forming fibrils. Amyloid from a variety of diseases may also share a common pathway for fibril formation. Recently, the focus has shifted to soluble oligomeric intermediates that have been reported as 3- to 10-nm spherical particles that appear at early incubation times and then disappear as mature fibrils appear [4–6]. These soluble intermediates have been referred to as amorphous aggregates, micelles, protofibrils, prefibrillar aggregates, and β-amyloid-derived diffusible ligands (ADDLs) [4,7–10].

Although there is compelling evidence for a causal role of amyloid-forming proteins in disease, there is additional evidence that the accumulation and deposition of fibrillar material does not correlate well with disease pathogenesis and in some cases may serve a protective function. For example, a significant number of nondemented individuals with Alzheimer's disease (AD) have remarkable amounts of amyloid plaques [11]. In some transgenic mouse models of AD, signs of pathology, including synaptic dysfunction and memory loss, precede the appearance of amyloid deposits [12,13]. Similarly, models of Parkinson's and Huntington's diseases demonstrate that the presence of soluble oligomeric aggregates correlates much better with toxicity than do larger insoluble aggregates or inclusions that may actually have protective properties.

Amyloid oligomers from a wide variety of disease-related and non-disease-related peptides and proteins have been shown to be toxic to cells in culture [14,15], indicating that oligomers are intrinsically toxic, regardless of the protein sequence from which they are derived. Amyloid oligomers also display a common structural motif that is distinct from fibrils as discerned by a conformation-specific antibody that specifically recognizes a common epitope on amyloid oligomers but not fibrils or monomers for a wide spectrum of proteins, including non-disease-related ones [16]. The anti-oligomer antibody blocks the toxicity of several types of amyloid oligomer, indicating that this common structural epitope is related to their generic toxicity.

Plasma Membrane as the Primary Target

That different amyloid oligomers share a common structure and are intrinsically toxic to cells predicts that they have the same primary mechanism of toxicity in their respective degenerative diseases. Indeed, even oligomers formed by proteins that are not disease-related have proven to be as toxic as disease-related oligomers [15]. If there is a common mechanism of pathogenesis set in motion by these oligomers, it would follow that they would act on the same primary target. To be considered a potential target, it would have to be accessible to all types of oligomer—those residing intracellularly as well as extracellularly. Therefore, the most obvious target is the plasma membrane, the interface between the two compartments.

A growing body of evidence points to membrane permeabilization by amyloid oligomers as a common mechanism of pathogenesis in amyloid-related degenerative diseases. An increase in membrane permeability and intracellular calcium concentration has long been associated with amyloid pathogenesis, although questions remain as to the mechanism underlying these observations [17,18]. A gamut of amyloidogenic proteins and peptides, including β-amyloid (Aβ), α-synuclein, islet amyloid polypeptide (IAPP), and polyglutamine, have been reported to form discrete pores or ion-specific channels in membranes in their prefibrillar conformations [5,19–23]. These data culminated in what came to be known as the "channel hypothesis," implicating amyloid peptide channels in the pathogenic ion dysregulation observed in degenerative disease [24]. However, the characteristic ion-specific, stepwise incremental increase in lipid bilayer conductance typically associated with channels is not observable when homogeneous and pure populations of amyloid oligomers are added to lipid bilayers. In fact, we found that the increase in membrane conductance is not ion-selective; rather, it is a generalized increase in lipid bilayer conductance that is not elicited by the application of either fibrils or monomers of the same amyloid proteins, regardless of sequence [25]. Oligomer-elicited permeability is not blocked by Congo red as reported for amyloid channels [25]. Images from atomic force microscopy also reveal that oligomers do indeed disrupt membrane integrity without forming discrete pores [26]. Moreover, the permeabilizing activity of homogeneous oligomers can be reversed, or inhibited, by preincubation with an anti-oligomer antibody [25]. The reasons for the discrepancy between reports of amyloid peptide single-channel insertion and our failure to observe such discrete pores or channels by amyloid oligomers are not clear at this point. Still, we concur on two important points: that amyloids from a variety of peptides permeabilize membranes and that the aggregation state of the peptide is key for this activity.

Growing evidence points to disruption of intracellular Ca^{2+} homeostasis in AD and other amyloidogenic diseases [27,28], and elevated intracellular Ca^{2+} levels are known to trigger apoptosis and/or excessive phosphorylation of key proteins that ultimately lead to cell death [17,29,30]. Prefibrillar amyloid aggregates have been shown to elevate cytosolic Ca^{2+} in neurons [28,31], but such actions have been proposed: (1) to be secondary to the generation of reactive oxygen species [27,32]; (2) to arise from activation of cell surface receptors coupled to Ca^{2+} influx [33,34]; or (3) to result from Ca^{2+} influx across the plasma membrane as a result of either cation-selective channels formed by Aβ itself [28,35] or through general disruption of lipid integrity [25]. However, we found that amyloid oligomers permeabilize cell membranes in a manner similar to synthetic lipid bilayers. Extracellular applications of oligomeric amyloid proteins and peptides induce rapid rises (within seconds) in cytosolic free Ca^{2+}, whereas equivalent amounts of monomers and fibrils show no detectable change in intracellular Ca^{2+} levels [36]. Oligomer-induced Ca^{2+} signals were unaltered in the presence of cobalt, a nonspecific blocker of Ca^{2+}-permeable channels. Moreover, all amyloid oligomers tested induced rapid leakage of fluorescent dyes from cells loaded with fluo-3 and calcein, both polyanionic and relatively cell-impermeant, in agreement with reports showing release of fluorescent dyes from phospholipid vesicles [6,37,38].

In the absence of extracellular Ca^{2+}, oligomer treatment of cells still results in an increase in cytosolic Ca^{2+}, likely owing to liberation of Ca^{2+} from intracellular stores, such as the endoplasmic reticulum and/or mitochondria. Thus it is probable that oligomers exert immediate action to permeabilize the plasma membrane and subsequently penetrate into cells where they similarly disrupt intracellular membranes to cause leakage of sequestered Ca^{2+}, as has been suggested previously [31].

Conclusions

Taken together, these recent data suggest that the membrane-permeabilizing effects of amyloid oligomers may represent the primary common mechanism of amyloid pathogenesis. The bidirectional flux of ions and/or various other molecules across a disrupted membrane may be sufficient to disrupt normal neuronal function and may also be a source of chronic stress in attempting to maintain normal membrane potential. In fact, the observed membrane permeabilization by amyloid oligomers and the simultaneous increase in cytosolic free calcium may be the initiators of several pathogenic pathways, including the production of reactive oxygen species [32], distorted signaling pathways [39,40], and mitochondrial dysfunction [41]. As a result, the energy demands to restore ion homeostasis and membrane potential may overwhelm the cell's machinery, sending it into a downward spiral toward dysfunction and programmed cell death.

References

1. Kirschner DA, Abraham C, Selkoe DJ. X-ray diffraction from intraneuronal paired helical filaments and extraneuronal amyloid fibers in Alzheimer disease indicates cross-beta conformation [erratum]. Proc Natl Acad Sci U S A 1986;83:503–507.
2. LeVine HD. Thioflavine T interaction with synthetic Alzheimer's disease beta-amyloid peptides: detection of amyloid aggregation in solution. Protein Sci 1993;2(3):404–410.
3. Eanes ED, Glenner GG. X-ray diffraction studies on amyloid filaments. J Histochem Cytochem 1968;16:673–677.
4. Harper JD, Wong SS, Lieber CM, Lansbury PT. Observation of metastable Abeta amyloid protofibrils by atomic force microscopy. Chem Biol 1997;4(2):119–125.

5. Lashuel HA, Hartley D, Petre BM, et al. Neurodegenerative disease: amyloid pores from pathogenic mutations. Nature 2002;418(6895):291.
6. Anguiano M, Nowak RJ, Lansbury PT Jr. Protofibrillar islet amyloid polypeptide permeabilizes synthetic vesicles by a pore-like mechanism that may be relevant to type II diabetes. Biochemistry 2002;41(38):11338–11343.
7. Walsh DM, Lomakin A, Benedek GB, et al. Amyloid beta-protein fibrillogenesis: detection of a protofibrillar intermediate. J Biol Chem 1997;272(35):22364–22372.
8. Lambert MP, Barlow AK, Chromy BA, et al. Diffusible, nonfibrillar ligands derived from Abeta1-42 are potent central nervous system neurotoxins. Proc Natl Acad Sci U S A 1998;95(11):6448–6453.
9. Lomakin A, Teplow DB, Kirschner DA, Benedek GB. Kinetic theory of fibrillogenesis of amyloid beta-protein. Proc Natl Acad Sci U S A 1997;94(15):7942–7947.
10. Soreghan B, Kosmoski J, Glabe C. Surfactant properties of Alzheimer's A beta peptides and the mechanism of amyloid aggregation. J Biol Chem 1994;269(46):28551–28554.
11. Terry R. The pathogenesis of Alzheimer disease: an alternative to the amyloid hypothesis. J Neuropathol Exp Neurol 1996;55(10):1023–1025.
12. Hsia AY, Masliah E, McConlogue L, et al. Plaque-independent disruption of neural circuits in Alzheimer's disease mouse models. Proc Natl Acad Sci U S A 1999;96(6):3228–3233.
13. Westerman MA, Cooper-Blacketer D, Mariash A, et al. The relationship between Abeta and memory in the Tg2576 mouse model of Alzheimer's disease. J Neurosci 2002;22(5):1858–1867.
14. Caughey B, Lansbury PT. Protofibrils, pores, fibrils, and neurodegeneration: separating the responsible protein aggregates from the innocent bystanders. Annu Rev Neurosci 2003;26(1):267–298.
15. Bucciantini M, Giannoni E, Chiti F, et al. Inherent toxicity of aggregates implies a common mechanism for protein misfolding diseases. Nature 2002;416(6880):507–511.
16. Kayed R, Head E, Thompson JL, et al. Common structure of soluble amyloid oligomers implies common mechanism of pathogenesis. Science 2003;300(5618):486–489.
17. Mattson MP, Cheng B, Davis D, et al. Beta-amyloid peptides destabilize calcium homeostasis and render human cortical neurons vulnerable to excitotoxicity. J Neurosci 1992;12:376–389.
18. Mattson MP. Calcium and neuronal injury in Alzheimer's disease: contributions of beta-amyloid precursor protein mismetabolism, free radicals, and metabolic compromise. Ann N Y Acad Sci 1994;747:50–76.
19. Arispe N, Pollard HB, Rojas E. Beta-amyloid Ca(2+)-channel hypothesis for neuronal death in Alzheimer disease. Mol Cell Biochem 1994;140:119–125.
20. Arispe N, Rojas E, Pollard HB. Alzheimer disease amyloid beta protein forms calcium channels in bilayer membranes: blockade by tromethamine and aluminum. Proc Natl Acad Sci U S A 1993;90:567–571.
21. Mirzabekov T, Lin MC, Yuan WL, et al. Channel formation in planar lipid bilayers by a neurotoxic fragment of the beta-amyloid peptide. Biochem Biophys Res Commun 1994;202:1142–1148.
22. Hirakura Y, Yiu WW, Yamamoto A, Kagan BL. Amyloid peptide channels: blockade by zinc and inhibition by Congo red (amyloid channel block). Amyloid 2000;7(3):194–199.
23. Mirzabekov TA, Lin M-C, Kagan BL. Pore formation by the cytotoxic islet amyloid peptide amylin. J Biol Chem 1996;271(4):1988–1992.
24. Kagan BL, Azimov RH, Azimova R. Amyloid peptide channels. J Membr Biol 2004;202:1–10.
25. Kayed R, Sokolov Y, Edmonds B, et al. Permeabilization of lipid bilayers is a common conformation-dependent activity of soluble amyloid oligomers in protein misfolding diseases. J Biol Chem 2004;279(45):46363–46366.

26. Green JD, Kreplak L, Goldsbury C, et al. Atomic force microscopy reveals defects within mica supported lipid bilayers induced by the amyloidogenic human amylin peptide. J Mol Biol 2004;342(3):877–887.

27. Mattson MP. Pathways towards and away from Alzheimer's disease. 2004;430(7000):631–639.

28. Kawahara M, Kuroda Y, Arispe N, Rojas E. Alzheimer's beta-amyloid, human islet amylin, and prion protein fragment evoke intracellular free calcium elevations by a common mechanism in a hypothalamic GnRH neuronal cell line. J Biol Chem 2000;275(19):14077–14083.

29. Pierrot N, Ghisdal P, Caumont A-S, Octave J-N. Intraneuronal amyloid-1-42 production triggered by sustained increase of cytosolic calcium concentration induces neuronal death. J Neurochem 2004;88(5):1140–1150.

30. LaFerla FM. Calcium dyshomeostasis and intracellular signalling in Alzheimer's disease. Nat Rev Neurosci 2002;3(11):862–872.

31. Bucciantini M, Calloni G, Chiti F, et al. Prefibrillar amyloid protein aggregates share common features of cytotoxicity. J Biol Chem 2004;279(30):31374–31382.

32. Schubert D, Behl C, Lesley R, et al. Amyloid peptides are toxic via a common oxidative mechanism. Proc Natl Acad Sci U S A 1995;92(6):1989–1993.

33. Blanchard BJ, Chen A, Rozeboom LM, et al. Inaugural article: efficient reversal of Alzheimer's disease fibril formation and elimination of neurotoxicity by a small molecule. Proc Natl Acad Sci U S A 2004;101(40):14326–14332.

34. Guo Q, Furukawa K, Sopher BL, et al. Alzheimer's PS-1 mutation perturbs calcium homeostasis and sensitizes PC12 cells to death induced by amyloid beta-peptide. Neuroreport 1996;8(1):379–383.

35. Kagan BL, Hirakura Y, Azimov R, et al. The channel hypothesis of Alzheimer's disease: current status. Peptides 2002;23(7):1311–1315.

36. Demuro A, Mina E, Kayed R, et al. Calcium dysregulation and membrane disruption as a ubiquitous neurotoxic mechanism of soluble amyloid oligomers. J Biol Chem 2005;280(17):17294–17300.

37. Relini A, Torrassa S, Rolandi R, et al. Monitoring the process of HypF fibrillization and liposome permeabilization by protofibrils. J Mol Biol 2004;338(5):943–957.

38. Janson J, Ashley RH, Harrison D, et al. The mechanism of islet amyloid polypeptide toxicity is membrane disruption by intermediate-sized toxic amyloid particles. Diabetes 1999;48(3):491–498.

39. Mattson MP. Degenerative and protective signaling mechanisms in the neurofibrillary pathology of AD. Neurobiol Aging 1995;16:447–457; discussion 458–463.

40. Saitoh T, Horsburgh K, Masliah E. Hyperactivation of signal transduction systems in Alzheimer's disease. Ann N Y Acad Sci 1993;695:34–41.

41. Shoffner JM. Oxidative phosphorylation defects and Alzheimer's disease. Neurogenetics 1997;1(1):13–19.

Chapter 41
Ganglioside-Dependent Generation of a Seed for Alzheimer's Disease Amyloid

Katsuhiko Yanagisawa

Introduction

A fundamental question regarding the pathogenesis of Alzheimer's disease (AD) is how the nontoxic, monomeric ß-amyloid protein (Aß) is converted into its toxic, oligomeric or polymeric forms. We previously identified a unique Aß species characterized by its binding to GM1 ganglioside (GM1) [1]. Importantly, GM1-bound Aß (GAß) shows a potency to act as a seed for Aß assembly. Interestingly, GAß also shows altered immunoreactivity. On the basis of its molecular characteristics, we hypothesize that Aß adopts an altered conformation through binding to GM1 and acts as a seed for Aß assembly in the brain. We have recently generated a novel monoclonal antibody [2] and validated GAß generation in the brain using this monoclonal antibody. Regarding the mechanism underlying GAß generation, we previously reported that an increased cholesterol level in the host membrane accelerates Aß binding to GM1 through enhanced formation of GM1 clusters in vitro [3]. Because GM1 and cholesterol are major lipid components of microdomains, such as rafts, we extended our study to determine how the GM1 level is altered by risk factors for the development of AD using human apolipoprotein E (ApoE) knock-in mice. In this study, we found that the GM1 level in the microdomain increases in synaptosomes of aged ApoE4-knock-in mice compared with that in the synaptosomes of aged ApoE3-knock-in mice. We also found that the age-dependent increase in GM1 level is sufficient for accelerating the assembly of soluble Aß in vitro [4]. Furthermore, we have recently shown that gangliosides play a pivotal role in the brain area-specific assembly and deposition of Aß.

K. Yanagisawa
National Institute for Longevity Sciences, 36-3, Gengo, Morioka, Obu, Aichi 474-8522, Japan

A. Fisher et al. (eds.), *Advances in Alzheimer's and Parkinson's Disease*,
© Springer 2008

GAß as Seed for Alzheimer Amyloid

Monoclonal Antibody Specific to GAß

To characterize GAß at the molecular level directly and to validate the GAß generation in brain, we generated monoclonal antibodies specific to the altered conformation of Aß. We previously generated an IgM monoclonal antibody using purified GAß from AD brains [5]. The novel immunoglobulin G (IgG) monoclonal antibody 4396C was generated by a genetic class-switch technique from IgM hybridomas. The 4396C specifically recognized GAß generated on the surface of liposomes. On the blot, 4396C reacted with GAß; however, neither monomeric Aß nor authentic GM1 was recognized by 4396C (Fig. 1). Importantly, 4396C also bound to the Aß at the ends of growing fibrils. These results suggest that 4396C specifically recognizes the altered Aß conformation, which is crucial for the initial and consecutive assembly of Aß.

Immunological Detection of GAß in Brain

Having established the binding specificity of 4396C, we attempted to detect GAß in the brain by immunohistochemistry. No immunoreactivity of 4396C was detected in the sections of cerebral cortices of AD brains fixed in formaldehyde. However, neuronal staining by 4396C was observed in the sections of AD brains fixed in Kryofix following sodium dodecyl sulfate (SDS)

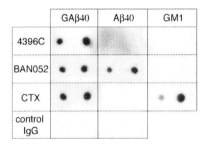 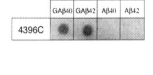

Fig. 1 Characterization of the binding specificity of 4396C by dot blot analysis. *Left*: Liposomes carrying GM1-bound Aß (GAß: GAß40), Aß40, and GM1 ganglioside (GM1) in amounts equal to those contained in blotted liposomes (300 and 600 ng of Aß40; 2 and 4 mg of GM1) were blotted. The blots were incubated with 4396C, BAN052, horseradish peroxidase (HRP)-conjugated cholera toxin subunit B (CTX), or isotype-matched control immunoglobulin G (IgG). *Right*: liposomes carrying GAß (GAß40 and GAß42), prepared using Aß40 or Aß42, in amounts equal to those contained in GAß40 and GAß42 (600 ng of each peptide) were blotted. The blots were incubated with 4396C. (Copyright 2004 by the Society for Neuroscience.)

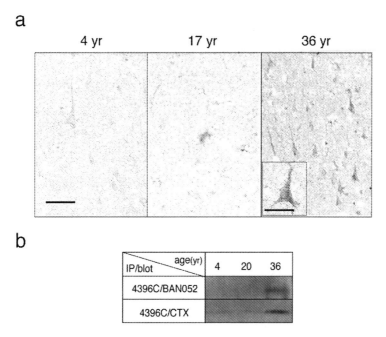

Fig. 2 Immunohistochemistry and immunoprecipitation of GAß in sections of nonhuman primate brains. **a** Immunostaining by 4396C of sections of the cerebral cortices of primate brains, which were fixed in paraformaldehyde, from animals of different ages. Bar = 50 mm. **Inset** Higher magnification. Bar = 20 mm. **b** Immunoprecipitation of GAß by 4396C from cerebral cortices of primates at different ages. Immunoprecipitates were blotted and reacted with BAN052 or HRP-conjugated CTX. (Copyright 2004 by the Society for Neuroscience.)

treatment [2]. Neuronal immunostaining with a granular pattern was also observed in the cerebral cortex of the brains of aged nonhuman primates (Fig. 2a). Furthermore, GAß was immunoprecipitated from the homogenate of the cerebral cortex of the aged animals (Fig. 2b). Based on these results, we conclude that GAß is generated in the brain.

Background to GAß Generation in Brain

Regarding the mechanism underlying the binding of Aß to GM1, Matsuzaki and coworkers previously reported that soluble Aß preferentially binds to GM1 in clusters. They also reported that GM1 clustering is facilitated in a cholesterol-rich environment. These results suggest that GAß generation depends on the GM1 and cholesterol levels in the membranes. Interestingly, we previously reported that local cholesterol content of neuronal membranes, such as synaptic plasma membranes (SPMs), increases with age and ApoE4 expression [4]. Moreover, the GM1 content of detergent-resistant microdomains (DRMs) of SPMs significantly increases with age, and this increase is more pronounced in ApoE4 knock-in mice [4]. Although further studies are needed, we may conclude that alterations in the lipid content of local neuronal membranes facilitate Aß assembly and deposition through GAß generation.

GAß as Determinant for Area-Specific Deposition of Aß

Several mutations of the amyloid precursor protein (APP) are responsible for the development of familial AD and hereditary cerebral amyloid angiopathy (CAA) [6–9]. Among them, several mutations in the Aß sequence induce a distinct phenotype despite the replacement of an amino acid at the same position. That is, the Arctic-type mutation (E22G) induces Aß deposition predominantly in the brain parenchyma, leading to AD development. In contrast, the Dutch-type Aß induces Aß deposition predominantly in the cerebral vessel wall, leading to CAA. These findings led us to hypothesize that the assembly of these hereditary Aß variants is accelerated by as yet unclarified local environmental factors, such as the presence of a specific ganglioside.

Because the phenotype induced by the Arctic-type mutation is essentially the same as that of conventional AD, we expected, on the basis of our previous findings, that the assembly of the Arctic-type Aß is facilitated in the presence of GM1. Van Nostrand and co-investigators previously reported that assembly of the Dutch-type Aß is accelerated in the presence of cultured human cerebro-vascular smooth muscle cells (HCSMs) [10,11]. Interestingly, HCSMs predominantly express GM3 [12]. To examine the possibility that GM1 and GM3 accelerate the assembly of the Arctic- and Dutch-type Aßs, respectively, we performed a kinetic study using liposomes containing GM1 or GM3. Indeed, the assemblies of the Arctic- and Dutch-type Aßs were markedly accelerated in the presence of GM1 and GM3, respectively. Moreover, assembly of the Flemish-type Aß was facilitated in the presence of GD3. It was previously reported that GD3 preferentially accumulates outside the vessel wall; therefore, the GD3-induced assembly of Flemish-type Aß may explain the preferential vessel wall-associated deposition of the Aß.

In contrast to the area-specific expressions of GM3 and GD3, expression of GM1 is extensive and beyond the preferential depositions of the wild- and Arctic-type Aßs. However, it remains to be clarified how GM1 is involved in the pathological assembly and deposition of these various Aß species in the brain. We have recently analyzed GM1 distribution in the brains of three groups of ApoE3 or ApoE4 knock-in mice of different ages [4]. We found that the GM1 level in the DRMs of SPMs increases with age. The age-dependent increase was observed in both the ApoE3 and ApoE4 knock-in mice, although it was more pronounced in the ApoE4 knock-in mice. We also found that the increased level of GM1 in the SPMs is sufficient to induce the assembly of soluble Aß incubated with the SPMs. However, it remains to be clarified how GM1 accumulates in specific membrane microdomains, such as DRMs of SPMs, in an age-dependent and ApoE isoform expression-dependent manner. One possible explanation is that aging and ApoE4 expression induce an increase in the cholesterol content of local membrane microdomains of SPMs, leading to the accumulation and stabilization of GM1 in these domains. Fredman and coworkers have reported that DRMs isolated from the frontal cortex of AD brains, which likely show early pathological changes of AD, have significantly high contents of GM1 and GM2 [1]. Therefore, an alteration in the lipid composition of neuronal membranes may initiate Aß assembly and deposition in the brain through the generation of a seed Aß, such as GAß.

Conclusion

Our studies suggest that Aß assembly and deposition in AD brains are likely induced by the generation of an endogenous seed, such as GAß. However, GM1 is expressed rather extensively in the brain beyond the preferential sites of Aß deposition. Thus, the next challenge is to determine how expression and distribution of GM1 change in AD brains with age and ApoE4 expression, leading to GAß generation.

References

1. Yanagisawa K, Odaka A, Suzuki N, Ihara Y. GM1 ganglioside-bound amyloid ß-protein (Aß): a possible form of pre-amyloid in Alzheimer's disease. Nat Med 1995;1:1062–1066.
2. Hayashi H, Kimura N, Yamaguchi H, et al. A seed for Alzheimer amyloid in the brain. J Neurosci 2004;24:4894–4902.
3. Kakio A, Nishimoto SI, Yanagisawa K, et al. Cholesterol-dependent formation of GM1 ganglioside-bound amyloid ß-protein, an endogenous seed for Alzheimer amyloid. J Biol Chem 2001;276:24985–24990.
4. Yamamoto N, Igbabvoa U, Shimada Y, et al. Accelerated Aß aggregation in the presence of GM1-ganglioside-accumulated synaptosomes of aged apoE4-knock-in mouse brain. FEBS Lett 2004;569:135–139.
5. Yanagisawa K, McLaurin J, Michikawa M, et al. Amyloid ß-protein (Aß) associated with lipid molecules: immunoreactivity distinct from that of soluble Aß. FEBS Lett 1997;420:43–46.
6. Levy E, Carman MD, Fernandez-Madrid IJ, et al. Mutation of the Alzheimer's disease amyloid gene in hereditary cerebral hemorrhage, Dutch type. Science 1990;248:1124–1126.
7. Hendriks L, van Duijn CM, Cras P, et al. Presenile dementia and cerebral haemorrhage linked to a mutation at codon 692 of the ß-amyloid precursor protein gene. Nat Genet 1992;1:218–221.
8. Nilsberth C, Westlind-Danielsson A, Eckman CB, et al. The 'Arctic' APP mutation (E693G) causes Alzheimer's disease by enhanced Aß protofibril formation. Nat Neurosci 2001;4:887–893.

9. Melchor JP, McVoy L, Van Nostrand WE. Charge alterations of E22 enhance the pathogenic properties of the amyloid ß-protein. J Neurochem 2000;74:2209–2212.
10. Davis J, Van Nostrand WE. Enhanced pathologic properties of Dutch-type mutant amyloid b-protein. Proc Natl Acad Sci U S A 1996;93:2996–3000.
11. Van Nostrand WE, Melchor JP, Ruffini L. Pathologic amyloid b-protein cell surface fibril assembly on cultured human cerebrovascular smooth muscle cells. J Neurochem 1998;70:216–223.
12. Yamamoto N, Hirabayashi Y, Amari M, et al. Assembly of hereditary amyloid ß-protein variants in the presence of favorable gangliosides. FEBS Lett 2005;579:2185–2190.
13. Molander-Melin M, Blennow K, Bogdanovic N, et al. Structural membrane alterations in Alzheimer brains found to be associated with regional disease development; increased density of gangliosides GM1 and GM2 and loss of cholesterol in detergent-resistant membrane domains. J Neurochem 2005;92:171–182.

Chapter 42
Evidence That Amyloid Pathology Progresses in a Neurotransmitter-Specific Manner

Karen F. S. Bell and A. Claudio Cuello

Introduction

Alzheimer's Disease (AD), which is the most common cause of cognitive decline in the elderly, progressively worsens with age leading to a specific loss of higher-order mental function (for review see ref. 1). One of the main hallmark lesions of AD pathology is amyloid plaque deposition, which tends to be heavily localized to the cerebral and hippocampal cortices (e.g., refs. 2–8), thus explaining the visible deficits in memory and cognitive function. Although the exact relation between amyloid burden and compromised brain function remains unknown, neurochemical investigations have revealed pronounced neurotransmitter deficits in these particular brain regions, most notably cholinergic in nature [9,10]. Not surprisingly, the neurochemical deficits appear to occur in parallel with structural losses of cortical synaptic profiles [11–13]. Indeed, Terry and coworkers [13] effectively demonstrated that synaptic loss in the mid-frontal brain regions of AD patients is, in fact, what correlates most closely with cognitive deficit compared to other pathological lesions such as the neurofibrillary tangle load or amyloid plaque deposition. Because of this important link between synaptic plasticity and cognitive function, a continued focus in our laboratory is the selective vulnerability of various cortical neurotransmitter-specific presynaptic bouton populations to the amyloid burden.

Neurotransmitter-Specific Structural Vulnerabilities to Extracellular Aβ Burden

Because of the long-standing characterization of the cholinergic deficit in AD, our initial studies focused primarily on the plasticity of the cortical cholinergic presynaptic terminals. Various transgenic animals were investigated, each

K.F.S. Bell
Department of Pharmacology and Therapeutics, McGill University, Montreal, QC, Canada

A. Fisher et al. (eds.), *Advances in Alzheimer's and Parkinson's Disease*,
© Springer 2008

representative of a particular stage of the amyloid pathology. An initial study, investigating an 8-month time point in the singly transgenic mutant mouse models $APP_{K670N, M671L}$ (tg2576 [14] and PSI_{M146L} [15]) as well as the doubly transgenic ($APP_{K670N, M671L}$ + PSI_{M146L} [16]) mouse model revealed striking plasticity of the cortical cholinergic presynaptic boutons. No visible changes were observable (compared to controls) in the PSI_{M146L} model, whereas the cholinergic presynaptic bouton number was up-regulated in the $APP_{K670N, M671L}$ (tg2576) model and significantly lower than controls in the doubly-transgenic ($APP_{K670N, M671L}$ + PSI_{M146L}) model. Interestingly, this initial up-regulation in cholinergic presynaptic bouton density—visible in the $APP_{K670N, M671L}$ (tg2576) model-appeared prior to plaque formation [17], whereas the visible decline in cortical cholinergic number— visible in the doubly transgenic animals-was concomitant with plaque formation [17]. These observations clearly demonstrate a direct link between extracellular β-amyloid (Aβ) aggregation and cholinergic forebrain compromise (as previously seen in our own laboratory (17) and that of Sturchler-Pierrat et al. [18] and later demonstrated by others [19–23] and furthermore serve to demonstrate the ability of Aβ to induce cholinergic dysfunction in a manner similar to that observed in post-mortem AD brains [9,10] (for review see ref. 24).

On the same note, other nontransgenic studies have yielded results that further concur with our observations in the animal models of AD amyloid pathology. For example, Mufson, DeKosky, and their colleagues identified similar up-regulation of the cholinergic terminals, in their case through the cholinergic marker choline acetyltransferase, in patients exhibiting mild cognitive impairment (MCI) [25,26], a condition now widely accepted as prodromic to AD [27]. The explanation for this cholinergic up-regulation remains unclear. However, it is plausible to hypothesize that it may arise as a result of a compensatory mechanism designed to combat increasing Aβ levels as well as Aβ-induced inhibition of acetylcholine release [28]. Alternatively, the up-regulation may result from altered APP processing leading to increased sAPPα secretion. Soluble APPα, which arises from cleavage by both γ- and α-secretase, precludes the formation of Aβ as α-secretase cleaves APP within the Aβ domain. In vitro findings have demonstrated that sAPPα displays both neurotrophic and neuroprotective properties [29–32]. Fahrenholz and collaborators recently created several transgenic mouse models that overexpress either ADAM-10 or a catalytically inactive ADAM-10 mutant [33], where ADAM-10 is currently accepted as the most likely α-secretase candidate [34–41] (for review see ref. 42). Both transgenic models were then crossed with a transgenic mouse model bearing the human APP_{V717I} mutation [43]. The doubly transgenic animals overexpressing ADAM-10 displayed increased sAPPα secretion, reduced Aα peptide formation, prevention of plaque formation, and most strikingly alleviation of the long-term potentiation and cognitive deficits associated with the singly transgenic APP_{V717I} animal [33]. Conversely, in the doubly transgenic animals expressing the catalytically inactive form of ADAM-10, increased plaque size and number were observed [33]. Whether increased levels of sAPPα are in fact responsible for the cholinergic terminal up-regulation, visible in both transgenic mouse models as well as individuals with MCI, remains to be determined. However, it is interesting to note that individuals with AD display significantly lower levels of sAPPα in their cerebrospinal fluid (CSF) than do non-AD patients [44,45].

Assuming that sAPPα elicits a neurotrophic and neuroprotective effect in vivo as well as in vitro, as Fahrenholz and collaborators' results appear to suggest, the beneficial effects of this protein would eventually become counterbalanced by the increasing toxicity of accumulating Aα. Increased concentrations of toxic Aα fragments could then explain the subsequent decline in cholinergic terminal number so classically associated with the AD pathology. To investigate further the possibility that proximal sources of A might elicit greater toxicity on cortical terminals than more distal sources, our group compared cortical cholinergic presynaptic bouton density (via VAChT immunolabeling and subsequent computer-assisted quantification) and overall presynaptic bouton density (via synaptophysin immuno-labeling) in the "plaque adjacent" and "random" neuropil of 8-month-old doubly transgenic $APP_{K670N, M671L} + PSI_{M146L}$ animals, representative of an early stage of amyloid pathology. Plaque proximity ("plaque adjacent" neuropil) was found to reduce cholinergic presynaptic bouton density by 40% and increase synaptophysin immunoreactivity by 9.5% [46]. Furthermore, plaque size was found to correlate negatively with cholinergic terminal density in the "plaque adjacent neuropile" [46].

Because of the visible up-regulation in synaptophysin immunoreactivity, as well as the desire to determine whether other neurotransmitter-specific terminals follow the same fate as the cholinergic system, additional studies were initiated to investigate the status of the glutamatergic and γ-aminobutyric acid (GABA)ergic presynaptic terminals across progressive stages of the amyloid pathology. The tgCRND8 transgenic model of early-onset amyloid pathology [47] was utilized to investigate the "early stage" of amyloid pathology, where "early" is defined as the predominant presence of immature diffuse plaques, in contrast to compact mature amyloid plaques. A similar approach to that previously described [17,46] was utilized to determine the densities of the glutamatergic (vesicular glutamate transporter 1, VGluT-1) and GABAergic (glutamic acid decarboxylase 65, GAD_{65}) presynaptic bouton populations in the "plaque-absent" and "plaque-adjacent" neuropil areas. This investigation revealed novel and significant up-regulation of both the glutamatergic and GABAergic presynaptic boutons compared to controls [48], a situation comparable to that previously observed for the cholinergic terminals [46]. To determine whether the involvement of the glutamatergic and GABAergic populations continued to parallel that of the cholinergic system, we next investigated the status of the two terminal systems in 18-month-old doubly transgenic $APP_{K670N, M671L} + PSI_{M146L}$ animals, representative of "late stage" amyloid pathology. Again, the observed findings clearly mimicked the fate of the cholinergic terminals in that all three neurotransmitter-specific terminal densities were significantly lower than in age-matched controls [49–51].

Amyloid Dependent Neurotransmitter-Specific Dystrophic Neurite Generation

In addition to alterations in presynaptic bouton density, our laboratory was interested in investigating the presence of dystrophic neurites, another pathological lesion clearly associated with AD pathology. Dystrophic neurites were almost always located in the vicinity of plaques and were only rarely found in the "plaque-absent" neuropil. Initial investigations in 8-month-old $APP_{K670N, M671L}$ + $PS1_{M146L}$ doubly transgenic mice found the number of cholinergic dystrophic neurites to be disproportionately large with respect to the incidence of cholinergic presynaptic boutons in the overall cortical presynaptic bouton population. Further investigations using double confocal microscopy labeling for both cholinergic and synaptophysin-immunoreactive dystrophic neurites confirmed preferential recruitment of cholinergic dystrophic neurites into fibrillar amyloid aggregates [46]. This observation, combined with the initial and specific up-regulation of cholinergic presynaptic boutons, led us to hypothesize that the amyloid pathology progresses in a neurotransmitter-specific manner, where the cholinergic terminals appear most vulnerable. Because of the presence of synaptophysin-immunoreactive dystrophic neurites, we extended our investigation to include the glutamatergic and GABAergic terminals. This initiative revealed the novel presence of both glutamatergic and GABAergic dystrophic neurites, once again predominantly surrounding amyloid plaques [48]. We then hypothesized that, like the presynaptic bouton densities, the dystrophic neurites might display neurotransmitter-specific vulnerability to the amyloid burden. To study this possibility further, we investigated the incidence and localization of glutamatergic and GABAergic dystrophic neurites at both an "early" (4-month-old tgCRND8 model) and "late" (18-month-old $APP_{K670N, M671L}$ + $PS1_{M146L}$) stage of amyloid progression. At the early time point, the involvement of cholinergic dystrophic neurites appeared saturated in that the same incidence was observed regardless of plaque size. Interestingly, the incidence was approximately 7% [48], which correlates closely with the incidence of the cholinergic terminals in the overall presynaptic bouton population [52], suggesting that the entire totality or near totality of all cholinergic terminals in the plaque adjacent neuropil become dystrophic. The incidence of glutamatergic dystrophic neurites was found to positively relate to plaque size [48], as might be expected given that the cortical glutamatergic presynaptic terminals account for approximately 60% of all presynaptic boutons in the cortex. Interestingly, little to no evidence of GABAergic dystrophic neurite formation was visible [48] despite the relatively large percent makeup in the overall cortical presynaptic bouton population. The minor participation of the GABAergic terminals remained consistent at late stages of the pathology [49,53,54], suggesting perhaps that this neurotransmitter terminal type may have some type of resistance to amyloid pathology (fibrillar Aβ). The cholinergic and glutamatergic dystrophic neurites both showed a negative relation with decreasing plaque size, suggesting that dystrophic neurites may also go through initial up-regulation (in response to plaque formation), subsequent leveling off (as seen in the cholinergic system), followed by a decreased incidence, likely occurring when the soma becomes unable to support the projection. To determine further whether the cholinergic system is indeed most vulnerable to the amyloid pathology, we quantified the localization of transmitter-specific dystrophic neurites to determine whether certain types were located in closer or farther proximity to the plaque center. This investigation—as expected based on our qualitative observations and our findings involving the presynaptic bouton densities—confirmed our hypothesis that the cholinergic system appears most vulnerable. This was seen by the significantly predominant localization of cholinergic dystrophic neurites to the inner periplaque neuropil compared to glutamatergic dystrophic neuritis, which were consistently located farther away [49,53,54]. This finding supports preferential recruitment of the cholinergic terminals, as the most vulnerable terminals would be expected to be recruited first and would thus be located in closest proximity to the plaque center.

Whether this neurotransmitter-specific progression also exists in human AD pathology remains unknown. However, preliminary findings from our laboratory have identified the novel presence of glutamatergic dystrophic neurites in the human AD brain [49.54] (Fig. 1). Furthermore, and most excitingly, additional preliminary findings demonstrate modest up-regulation of glutamatergic presynaptic boutons in the brains of MCI patients compared to individuals with no

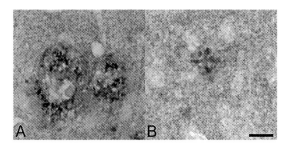

Fig. 1 Preliminary staining of human Alzheimer's disease midfrontal gyrus lamina III tissue. **A, B** Double-labeled light microscopy micrographs of brain tissue immunostained for both Aβ (shown in blue) and glutamatergic presynaptic terminals and dystrophic neurites (shown in brown). Note the grossly distorted dystrophic neurites in contact with fibrillar Aβ material, providing direct evidence for a structural synaptic glutamatergic involvement in the AD pathology. Bar = 20 μm. (The AD midfrontal gyrus lamina III issue was obtained from the Religious Orders Study, RUSH Institute, Chicago, IL, USA.)

cognitive impairment or early AD [54]. A full-length study to confirm these preliminary results is currently underway in our laboratory.

Conclusion

Although the exact mechanism leading to cholinergic basal forebrain attrition remains unknown, the results obtained from past and ongoing studies by our group clearly demonstrate that cortical amyloid burden, in the absence of neurofibrillary pathology, appears sufficient to induce major synaptic alterations with regard to presynaptic bouton density and dystrophic neurite formation. We have identified a novel structural involvement of both the glutamatergic and GABAergic presynaptic populations. We thus hypothesize that amyloid pathology progresses in a neurotransmitter-specific manner, where the cholinergic system appears most vulnerable followed by the glutamatergic system and finally by the somewhat more resilient GABAergic system (unpublished observations). Most importantly, we have been able to validate some of our findings on amyloid pathology in transgenic animal models with similar results obtained from human AD brain tissue. To our knowledge, our group is the first to demonstrate the presence of glutamatergic dystrophic neurites in the human AD brain, clearly demonstrating that neurotransmitter-specific structural deficits extend beyond the classically characterized cholinergic system.

Acknowledgments This research was supported by funds from the Canadian Institutes of Health Research to A.C.C. (grant MOP-37996). The authors thank Professors Karen Duff, Karen Hsiao, Peter St. George-Hyslop, and Don Westaway for their generous donation of transgenic mouse lines and Drs. Shigemoto and Edwards for their kind donation of the anti-VGluT1 and anti-VAChT antibodies, respectively. The authors also express their sincere gratitude to all of the participants of the RUSH Institute's Religious Orders Study, Chicago, IL as well as to Dr. David Bennett for a most valuable collaboration. K.F.S.B. is the recipient of a CIHR Doctoral Research Award and A.C.C. is the holder of the Charles E. Frosst Merck endowed Chair.

References

1. Morris JC, Cole M, Banker BQ, Wright D. Hereditary dysphasic dementia and the Pick-Alzheimer spectrum. Ann Neurol 1984;16:455–466.
2. Geula C, Mesulam MM, Saroff DM, Wu CK. Relationship between plaques, tangles, and loss of cortical cholinergic fibers in Alzheimer disease. J Neuropathol Exp Neurol 1998;57:63–75.
3. Henderson Z. Responses of basal forebrain cholinergic neurons to damage in the adult brain. Prog Neurobiol 1996;48:219–254.
4. Hyman BT, Van Horsen GW, Damasio AR, Barnes CL. Alzheimer's disease: cell-specific pathology isolates the hippocampal formation. Science 1984;225:1168–1170.

5. Kromer Vogt LJ, Hyman BT, Van Hoesen GW, Damasio AR. Pathological alterations in the amygdala in Alzheimer's disease. Neuroscience 1990;37:377–385.
6. Pearson RC, Esiri MM, Hiorns RW, et al. Anatomical correlates of the distribution of the pathological changes in the neocortex in Alzheimer disease. Proc Natl Acad Sci U S A 1985;82:4531–4534.
7. Rogers J, Morrison JH. Quantitative morphology and regional and laminar distributions of senile plaques in Alzheimer's disease. J Neurosci 1985;5:2801–2808.
8. Saper C, German D, White C. Neuronal pathology in the nucleus basalis and associated cell groups in senile dementia of the Alzheimer type: possible role in cell loss. Neurology 1985;35:1089–1095.
9. Bowen DM, Smith CD. Neurotransmitter related enzymes and indices of hypoxia in senile dementia and other abiotrophies. Brain 1976;99:459–496.
10. Davies P, Maloney AJF. Selective loss of central cholinergic neurons in Alzheimer's disease. Lancet 1976;2(8000):1403.
11. DeKosky ST, Scheff SW. Synapse loss in frontal cortex biopsies in Alzheimer's disease: correlation with cognitive severity. Ann Neurol 1990;27:457–464.
12. Masliah E, Terry RD, Mallory M, et al. Diffuse plaques do not accentuate synapse loss in Alzheimer's disease. Am J Pathol 1990;137:1293–1297.
13. Terry RD, Masliah E, Salmon DP, et al. Physical basis of cognitive alterations in Alzheimer's disease: synapse loss is the major correlate of cognitive impairment. Ann Neurol 1991;30:572–580.
14. Hsiao K, Chapman P, Nilsen S, et al. Correlative memory deficits, Abeta elevation, and amyloid plaques in transgenic mice. Science 1996;274:99–102.
15. Duff K, Eckman C, Zehr C, et al. Increased amyloid-beta42(43) in brains of mice expressing mutant presenilin 1. Nature 1996;383:710–713.
16. Holcomb L, Gordon MN, McGowan E, et al. Accelerated Alzheimer-type phenotype in transgenic mice carrying both mutant amyloid precursor protein and presenilin 1 transgenes. Nat Med 1998;4:97–100.
17. Wong TP, Debeir D, Duff, Cuello AC. Reorganization of cholinergic terminals in the cerebral cortex and hippocampus in transgenic mice carrying mutated presenilin-1 and amyloid precursor protein transgenes. J Neurosci 1999;19:2706–2716.
18. Sturchler-Pierrat C, Abramowski D, Duke M, et al. Two amyloid precursor protein transgenic mouse models with Alzheimer disease-like pathology. Proc Natl Acad Sci U S A 1997;94:13287–13292.
19. Boncristiano S, Calhoun ME, Kelly PH, et al. Cholinergic changes in the APP23 transgenic mouse model of cerebral amyloidosis. J Neurosci 2002;22:3234–3243.
20. Gau JT, Steinhilb ML, Kao TC, et al. Stable beta-secretase activity and presynaptic cholinergic markers during progressive central nervous system amyloidogenesis in Tg2576 mice. Am J Pathol 2002;160:731–738.
21. Hernandez S, Sugaya K, Qu T, et al. Survival and plasticity of basal forebrain cholinergic systems in mice transgenic for presenilin-1 and amyloid precursor protein mutant genes. Neuroreport 2001;12:1377–1384.
22. Jaffar S, Counts SE, Ma SY, et al. Neuropathology of mice carrying mutant APP(swe) and/or PS1(M146L) transgenes: alterations in the p75(NTR) cholinergic basal forebrain septohippocampal pathway. Exp Neurol 2001;170:227–243.
23. Masliah E, Rockenstein E, Veinbergs I, et al. beta-amyloid peptides enhance alpha-synuclein accumulation and neuronal deficits in a transgenic mouse model linking Alzheimer's disease and Parkinson's disease. Proc Natl Acad Sci U S A 2001;98:12245–12250.
24. Price DL, Koliatsos VE, Clatterbuck RC. Cholinergic systems: human diseases, animal models and prospects for therapy. Prog Brain Res 1993;98:51–60.
25. DeKosky ST, Ikonomovic MD, Styren SD, et al. Upregulation of choline acetyltransferase activity in hippocampus and frontal cortex of elderly subjects with mild cognitive impairment. Ann Neurol 2002;51:145–155.

26. Ikonomovic MD, Mufson EJ, Wuu J, et al. Cholinergic plasticity in hippocampus of individuals with mild cognitive impairment: correlation with Alzheimer's neuropathology. J Alzheimers Dis 200;35:39–48.

27. Morris JC, Storandt M, Miller JP, et al. Mild cognitive impairment represents early-stage Alzheimer disease. Arch Neurol 2001;58:397–405.

28. Kar S, Seto D, Gaudreau P, Quirion R. Beta-amyloid-related peptides inhibit potassium-evoked acetylcholine release from rat hippocampal slices. J Neurosci 1996;16:1034–1040.

29. Mattson MP, Cheng B, Culwell AR, et al. Evidence for excitoprotective and intraneuronal calcium-regulating roles for secreted forms of the beta-amyloid precursor protein. Neuron 1993;10:243–254.

30. Mattson MP. Secreted forms of beta-amyloid precursor protein modulate dendrite outgrowth and calcium responses to glutamate in cultured embryonic hippocampal neurons. J Neurobiol 1994;25:439–450.

31. Furukawa K, Sopher BL, Rydel RE, et al. Increased activity-regulating and neuroprotective efficacy of alpha-secretase-derived secreted amyloid precursor protein conferred by a C-terminal heparin-binding domain. J Neurochem 1996;67:1882–1896.

32. Small DH, Nurcombe V, Reed G, et al. A heparin-binding domain in the amyloid protein precursor of Alzheimer's disease is involved in the regulation of neurite outgrowth. J Neurosci 1994;14:2117–2127.

33. Postina R, Schroeder A, Dewachter I, et al. A disintegrin-metalloproteinase prevents amyloid plaque formation and hippocampal defects in an Alzheimer disease mouse model. J Clin Invest 2004;113:1456–1464.

34. Buxbaum JD, Liu KN, Luo Y, et al. Evidence that tumor necrosis factor alpha converting enzyme is involved in regulated alpha-secretase cleavage of the Alzheimer amyloid protein precursor. J Biol Chem 1998;273:27765–27767.

35. Esch FS, Keim PS, Beattie EC, et al. Cleavage of amyloid beta peptide during constitutive processing of its precursor. Science 1990;248:1122–1124.

36. Haass C, Koo EH, Mellon A, et al. Targeting of cell-surface beta-amyloid precursor protein to lysosomes: alternative processing into amyloid-bearing fragments. Nature 1992;357:500–503.

37. Koike H, Tomioka S, Sorimachi H, et al. Membrane-anchored metalloprotease MDC9 has an alpha-secretase activity responsible for processing the amyloid precursor protein. Biochem J 1999;343(Pt 2):371–375.

38. Kojro E, Gimpl G, Lammich S, et al. Low cholesterol stimulates the nonamyloidogenic pathway by its effect on the alpha -secretase ADAM 10. Proc Natl Acad Sci U S A 2001;98:5815–5820.

39. Lammich S, Kojro E, Postina R, et al. Constitutive and regulated alpha-secretase cleavage of Alzheimer's amyloid precursor protein by a disintegrin metalloprotease. Proc Natl Acad Sci U S A 1999;96:3922–3927.

40. Marcinkiewicz M, Seidah NG. Coordinated expression of beta-amyloid precursor protein and the putative beta-secretase BACE and alpha-secretase ADAM10 in mouse and human brain. J Neurochem 2000;75:2133–2143.

41. Sisodia SS. Beta-amyloid precursor protein cleavage by a membrane-bound protease. Proc Natl Acad Sci U S A 1992;89:6075–6079.

42. Kojro E, Fahrenholz F. The non-amyloidogenic pathway: structure and function of alpha-secretases. Subcell Biochem 2005;38:105–127.

43. Moechars D, Dewachter I, Lorent K, et al. Early phenotypic changes in transgenic mice that overexpress different mutants of amyloid precursor protein in brain. J Biol Chem 1999;274:6483–6492.

44. Sennvik K, Fastbom J, Blomberg M, et al. Levels of alpha- and beta-secretase cleaved amyloid precursor protein in the cerebrospinal fluid of Alzheimer's disease patients. Neurosci Lett 2000;278:169–172.

45. Lannfelt L, Basun H, Wahlund LO, et al. Decreased alpha-secretase-cleaved amyloid precursor protein as a diagnostic marker for Alzheimer's disease. Nat Med 1995;1:829–832.
46. Hu L, Wong TP, Cote SL, et al. The impact of Abeta-plaques on cortical cholinergic and non-cholinergic presynaptic boutons in Alzheimer's disease-like transgenic mice. Neuroscience 2003;121:421–432.
47. Chishti MA, Yang DS, Janus C, et al. Early-onset amyloid deposition and cognitive deficits in transgenic mice expressing a double mutant form of amyloid precursor protein 695. J Biol Chem 2001;276:21562–21570.
48. Bell KF, de Kort GJ, Steggerda S, et al. Structural involvement of the glutamatergic presynaptic boutons in a transgenic mouse model expressing early onset amyloid pathology. Neurosci Lett 2003;353:143–147.
49. Bell KFS, Ducatenzeiler A, Ribeiro-da-Silva A, et al. The amyloid pathology progresses in a neurotransmitter-specific manner. Neurobiol Aging 2006;27:1644–1657.
50. Bell KFS, Ribeiro-da-Silva A, Cuello AC. Evidence of a neurotransmitter-specific attrition in the Alzheimer's disease like pathology, 8th International Montreal/Springfield Symposium on Advances in Alzheimer Therapy, 8L, 40 (2004).
51. Cuello AC, Bell KFS, Ribeiro-da-Silva A. Amyloid formation provokes transmitter-specific changes and dystrophic neurite formation in the absence of neurofibrillary pathology. Neurobiol Aging 2004;25:S1-S30.
52. Casu MA, Wong TP, De Koninck Y, et al. Aging causes a preferential loss of cholinergic innervation of characterized neocortical pyramidal neurons. Cereb Cortex 2002;12:329–337.
53. Bell KFS,Ribeiro-da-Silva A, Duff K, Cuello AC. A sequential and neurotransmitter-specific involvement of dystrophic neurites in the amyloid pathology. Neurobiol Aging 2004;25(2):S169..
54. Bell KFS, Bennett DA, Cuello AC Amyloid plaque dependent glutamatergic neuritic dystrophy in the Alzheimer's disease brain. Presented at the 7th International AD/PD Conference, 2005.

Chapter 43
Rationale for Glutamatergic and Cholinergic Approaches for the Treatment of Alzheimer's Disease

Paul T. Francis and Sara L. Kirvell

Introduction

Great strides have been made over the last quarter century in identifying and describing the neurochemical pathology of Alzheimer's disease (AD), and they have ultimately led to successful symptomatic treatments for this previously untreatable condition. The mainstay of current treatment for cognitive symptoms is cholinesterase inhibition, and clinical experience suggests that these drugs improve other aspects of the disease as well, such as behavioral symptoms and activities of daily living. Furthermore, with the advent of the N-methyl-D-aspartate (NMDA) antagonist memantine it is time to examine the cliniconeurochemical correlates of the cholinergic system in AD and the interrelation of this system with glutamatergic neurons.

Cholinergic Changes in AD

The cholinergic system arising from the basal forebrain innervates all areas of the cerebral cortex and intrinsic cells such as glutamatergic pyramidal neurons. Activation of cholinergic receptors is generally precognitive, especially by improving attention [1]. Biochemical investigations of biopsy tissue taken from patients who had had AD for 3.5 years (on average) after the onset of symptoms indicate that cholinergic neurotransmitter pathology occurs early in the course of the disease (Table 1).

Presynaptic markers of the cholinergic system are uniformly reduced with choline acetyltransferase (ChAT) activity, high-affinity choline uptake, and acetylcholine (ACh) synthesis all present at about 50% of control values. Furthermore, these changes are correlated with the degree of cognitive

P.T. Francis
Wolfson Centre for Age-Related Diseases, King's College London, Guy's Campus, St Thomas Street, London SE11UL, UK

A. Fisher et al. (eds.), *Advances in Alzheimer's and Parkinson's Disease*,
© Springer 2008

Table 1 Cholinergic markers in AD[a]

Cholinergic marker	Percent of control
ChAT activity	35–50
ACh synthesis	40–50
Choline uptake	60
AChE activity	40–60
Nicotinic binding	60–70
Muscarinic binding	80–100

ChAT, choline acetyltransferase; ACh, acetylcholine; AChE, acetylcholinesterase.
[a] Data summarized from Francis et al. [2,9] and P.T. Francis, unpublished observations.

impairment (dementia rating summary score) in patients with AD [2]. Post-mortem studies have also shown a correlation between ChAT activity and cognitive decline, but there is evidence that there may be initial stabilization or possibly up-regulation in activity early in the disease state [3,4]. Based on our evidence, it appears that neocortical cholinergic innervation is probably dysfunctional at an early stage of the disease, a conclusion substantiated by evidence of similar changes in patients who have displayed clinical symptoms for less than 1 year. It is somewhat difficult to interpret the ChAT data, however, as this enzyme is not the rate-limiting step of ACh synthesis. Other studies have demonstrated a reduction in the number of nicotinic [5] and muscarinic (M2) [6] ACh receptors in AD brains (most of which are considered to be located on presynaptic cholinergic terminals) but relative preservation of postsynaptic muscarinic (M1, M3) receptors. Despite this relative preservation, there is some evidence for disruption of the coupling between the muscarinic M1 receptors, their G-proteins, and second messenger systems [7]. In many studies, nicotinic receptors have been found in lower numbers in the brains of patients with AD [5,8], and this reduction is mainly the $\alpha4\beta2$ subtype with some evidence of relative preservation of $\alpha7$ subtype. We have observed a positive correlation between binding to the $\alpha4\beta2$ subtype and the Mini-Mental State Examination score (P.T. Francis, unpublished observations), suggesting an important role for these receptors.

On the basis of this cholinergic dysfunction, acetylcholinesterase inhibitors (AChEIs) were developed for the treatment of mild-to-moderate AD to increase synaptic concentrations of ACh, thereby permitting the diminished supply of ACh to remain in the synaptic cleft for a longer period of time and, in turn, enhancing cholinergic neurotransmission. Synaptic ACh is then available to activate cholinergic receptors, which are present on glutamatergic pyramidal neurons (as well as other cells) to enhance glutamatergic function [9,10]. Of the four AChEIs licensed by the U.S. Food and Drug Administration (FDA), only donepezil, rivastigmine, and galantamine are in widespread use today. The use of tacrine has been largely discontinued because of its side effects profile—specifically, substantially increased incidence of hepatotoxicity. Although they share a similar mode of action, the three agents have a number of differences that are manifested in their dosing and administration [11]. For example, donepezil, with a half-life of about 70 hours, can be administered once a day, whereas rivastigmine and galantamine, with shorter half-lives (2 hours and 5–7 hours, respectively), are administered twice daily [11]. Again, whereas donepezil and galantamine are principally AChEIs, rivastigmine is equipotent against butyrylcholinesterase, and galantamine is reported to be an allosteric activator of nicotinic inhibitors [11]. However, from a clinical perspective, these drugs have similar efficacy showing modest benefit in most patients. These findings suggest that although cholinergic dysfunction in AD is important and clinically relevant other factors must also be driving cognitive decline.

An interesting finding of the clinical trials with cholinesterase inhibitors is an unexpected reduction in behavioral symptoms of dementia such as apathy, irritability, and aberrant motor activity [12]. A rational biochemical basis for such findings has emerged, with lower ChAT activity being found in brains of AD patients [13] as well as correlations of other cholinergic markers with specific behaviors such as psychosis [6].

Glutamatergic Changes in AD

Glutamate is the primary excitatory neurotransmitter of the central nervous system (CNS) and is used by approximately two-thirds of synapses in the neocortex and hippocampus, whereas ACh is found in perhaps 5% [14]. As such, glutamate is involved in all aspects of cognition and higher mental function. In particular, normal (physiologic) glutamate stimulation of NMDA receptors is essential to learning and memory processes mainly through long-term potentiation (LTP), a form of synaptic strengthening with repeated use. Pyramidal neuron loss is a feature of AD [9,15] and is reflected in loss of the wet weight and total protein of cortical regions (Table 2). Glutamatergic neurons account for many of the neurons lost in the cerebral cortex and hippocampus in AD [15,16]. Similarly, a large decrease in glutamate receptors has been observed in the cortex [17] and hippocampus [18,19] of AD brains, presumably due in part to the accompanying neuronal loss described above. However, because all biochemical data are expressed relative to either wet weight or total protein, unexpected findings can occur later in the disease process in severely affected regions. Markers or extrinsic neurons, such as ChAT activity for cholinergic neurons, are reduced (as they represent only a small proportion of the neuropil) in the temporal and other cortical regions in pathologically severe cases. Markers of pyramidal neurons and their synapses do not always follow this pattern, perhaps because such structures contribute significantly to the total protein. This appears to be the case for β-tubulin and synaptophysin in the present study (Table 2) because relative to total protein neither is reduced in the temporal cortex from AD patients. Similarly, the glial

Table 2 Gravimetric analysis and biochemical markers of pyramidal neurons, synapses, and glia in three cortical regions of brains from patients with AD

Parameter	TCX	PCX	OCX
Wet weight[a]	62	68	100
Total protein[b]	63	65	100
β–Tubulin	100	54	100
GFAP	100	194	100
Synaptophysin	100	60	44
GLT-1	100	100	100
GLAST	100	37	100
D-Aspartate uptake[c]	34	–	–
VGLUT1	100	49	34
VGLUT2	100	100	100
Vesicular glutamate uptake[d]	11	–	–

Values are 100% where value does not differ significantly from appropriate control.
TCX, temporal cortex; PCX, parietal cortex; OCX, occipital cortex; GFAP, glial fibrillary acidic protein; GLT-1, glutamate transporter-1; GLAST, astrocyte-specific glutamate transporter; VGLUT, vesicular glutamate transporter
Data are from Kirvell and Francis, unpublished observations, except where indicated:
[a] Ref. 20. [b] Ref. 21. [c] Ref. 22. [d] Ref. 23.

marker glial fibrillary acidic protein (GFAP) is also unaltered in this region, although it is increased in a region (parietal cortex) that is less severely affected, even at the end-stage of the disease.

In addition to the consequences of glutamatergic cell loss, there is evidence of dysfunction in remaining neurons. For example, although the relative concentration of the major glial glutamate transporter (GLT-1, responsible for 90–95% of glutamate uptake) was unaffected in any of the regions studied (Table 2), the ability of glial cells to remove glutamate from the synaptic cleft (the main method of neurotransmitter inactivation following release) was impaired in several brain regions including the temporal cortex (Table 2) [24]. This impaired functional activity may be due to the action of free radicals on GLT-1 [25]. Similarly, there was a reduction in the vesicular glutamate transporter (VGLUT1) in parietal but not temporal cortex (Table 2), and the activity of this protein was lower in AD temporal cortex than in controls [23]. Again, this suggests an alteration in the activity of the protein by not yet known mechanisms.

As a consequence of these alterations in proteins or protein function, we propose that in AD there is inadequate removal of glutamate in the synaptic cleft between the presynaptic and postsynaptic neurons, creating an excessive level of background "noise" at glutamate receptors in the synapse and adversely affecting the ability of the NMDA receptor to generate LTP [26]. Under normal resting conditions, the NMDA receptor channel is blocked by Mg^{2+} ions in a voltage-dependant manner and becomes responsive to glutamate binding only when the membrane is depolarized. As a consequence of higher background concentrations of glutamate in the synaptic cleft, the membrane tends to be more, or more frequently depolarized, with the result that the Mg^{2+} blockade is less efficient; and the role of NMDA; as a coincidence detector capable of generating LTP, is impaired. It has been proposed that this disruption of glutamatergic neurotransmission contributes to cognitive impairment in AD [27]. Furthermore, when taken to the extreme, excessive (pathological) glutamate acting at all classes of glutamate receptors can cause neuronal death and is likely to be a major factor in stroke [16,26]. It is likely that such excitotoxicity, although not the major cause of cell death, is a contributory factor in AD.

Cholinergic–Glutamatergic Interactions

Cholinergic and glutamatergic systems are often considered in isolation in the context of AD, but this is obviously a gross oversimplification. In the cortex and hippocampus, a major regulatory influence on pyramidal neurons are cholinergic receptors [28–31]. Hence, one of the main actions of cholinesterase inhibitors—to increase the concentration of ACh in the synaptic cleft—leads to increased activity of pyramidal neurons and thus release of glutamate [10]. We have argued that this is one of the main reasons that cholinesterase inhibitors are able to improve cognition [2]. Furthermore, the activity of cholinergic neurons of the nucleus basalis of Meynert are likely to be regulated by glutamatergic input from the amygdala [32] as well as other regions through both NMDA and alpha-amino-3-hydroxy-5-methyl-4-isoxazolepropionic acid (AMPA) receptors [33,34]. It is possible that such receptors are also responsible for the death of some cholinergic neurons if the subunit profile of AMPA receptors is altered to allow additional calcium entry [35,36].

Conclusions

Changes in glutamatergic and cholinergic neurotransmission underlie many of the cognitive changes seen in AD and have formed the basis of the first rational symptomatic treatment for the disease. The considerable interplay between these systems provides an understanding of how drugs targeted at one system (e.g., cholinesterase inhibitors or memantine) may influence another, affecting the neurochemical pathology of AD and hence provide symptomatic benefit.

References

1. Stanhope KJ, Mclenachan AP, Dourish CT. Dissociation between cognitive and motor/motivational deficits in the delayed matching to position test: effects of scopolamine, 8-OH-DPAT and EAA antagonists. Psychopharmacology. 1995;122:268–280.
2. Francis PT, Palmer AM, Snape M, Wilcock GK. The cholinergic hypothesis of Alzheimer's disease: a review of progress. J Neurol Neurosurg Psychiatry 1999;66:137–147.

3. Davis KL, Mohs RC, Marin D, et al. Cholinergic markers in elderly patients with early signs of Alzheimer disease. JAMA 1999;281:1401–1406.

4. DeKosky ST, Ikonomovic MD, Styren SD, et al. Upregulation of choline acetyltransferase activity in hippocampus and frontal cortex of elderly subjects with mild cognitive impairment. Ann Neurol 2002;51:145–155.

5. Court J, Martin-Ruiz C, Piggott M, et al. Nicotinic receptor abnormalities in Alzheimer's disease. Biol Psychiatry 2001;49:175–184.

6. Lai MK, Lai OF, Keene J, et al. Psychosis of Alzheimer's disease is associated with elevated muscarinic M2 binding in the cortex. Neurology 2001;57:805–811.

7. Warpman U, Alafuzoff I, Nordberg A. Coupling of muscarinic receptors to GTP proteins in postmortem human brain: alterations in Alzheimer's disease. Neurosci Lett 1993;150:39–43.

8. Perry EK, Morris CM, Court JA, et al. Alteration in nicotine binding sites in Parkinson's disease, Lewy body dementia and Alzheimer's: possible index of early neuropathology. Neuroscience 1995;64:385–395.

9. Francis PT, Sims NR, Procter AW, Bowen DM. Cortical pyramidal neurone loss may cause glutamatergic hypoactivity and cognitive impairment in Alzheimer's disease: investigative and therapeutic perspectives. J Neurochem 1993;60:1589–1604.

10. Dijk SN, Francis PT, Stratmann GC, Bowen DM. Cholinomimetics increase glutamate outflow by an action on the corticostriatal pathway: implications for Alzheimer's disease. J Neurochem 1995;65:2165–2169.

11. Wilkinson DG, Francis PT, Schwam E, Payne-Parrish J. Cholinesterase inhibitors used in the treatment of Alzheimer's disease: the relationship between pharmacological effects and clinical efficacy. Drugs Aging 2004;21:453–478.

12. Gauthier S, Feldman H, Hecker J, et al. Efficacy of donepezil on behavioral symptoms in patients with moderate to severe Alzheimer's disease. Int Psychogeriatr 2002;14:389–404.

13. Minger SL, Esiri MM, McDonald B, et al. Cholinergic deficits contribute to behavioural disturbance in patients with dementia. Neurology 2000;55:1460–1467.

14. Fonnum F. Glutamate: a neurotransmitter in mammalian brain. J Neurochem 1984;42:1–11.

15. Morrison JH, Hof PR. Life and death of neurons in the aging brain. Science. 1997;278:412–419.

16. Greenamyre JT, Maragos WF, Albin RL, et al. Glutamate transmission and excitotxicity in Alzheimer's disease. Prog Neuropsychopharmacol 1988;12:421–430.

17. Greenamyre JT, Penney JB, Damato CJ, Young AB. Alterations in L-glutamate binding in Alzheimer's and Huntingdon's diseases. Science 1985;227:1496–1499.

18. Greenamyre JT, Penney JB, D'Amato CJ, Young AB. Dementia of the Alzheimer's type: changes in hippocampal L-[^{3}H]glutamate binding. J Neurochem 1987;48:543–551.

19. Procter AW, Wong EH, Stratmann GC, et al. Reduced glycine stimulation of [^{3}H]MK-801 binding in Alzheimer's disease. J Neurochem 1989;53:698–704.

20. Najlerahim A, Bowen DM. Regional weight loss of the cerebral cortex and some subcortical nuclei in senile dementia of the Alzheimer type. Acta Neuropathol (Berl) 1988;75:509–512.

21. Najlerahim A, Bowen DM. Biochemical measurements in Alzheimer's disease reveal a necessity for improved neuroimaging techniques to study metabolism. Biochem J 1988;251:305–308.

22. Procter AW, Francis PT, Holmes C, et al. APP isoforms show correlations with neurones but not with glia in brains of demented subjects. Acta Neuropathol (Berl) 1994;88:545–552.

23. Westphalen RI, Scott HL, Dodd PR. Synaptic vesicle transport and synaptic membrane transporter sites in excitatory amino acid nerve terminals in Alzheimer disease. J Neural Transm 2003;110:1013–1027.

24. Procter AW, Palmer AM, Francis PT, et al. Evidence of glutamatergic denervation and possible abnormal metabolism in Alzheimer's disease. J Neurochem 1988;50:790–802.

25. Keller JN, Mark RJ, Bruce AJ, et al. 4-Hydroxynonenal, an aldehydic product of membrane lipid peroxidation, impairs glutamate transport and mitochondrial function in synaptosomes. Neuroscience 1997;80:685–696.
26. Danysz W, Parsons CG, Quack G. NMDA channel blockers: memantine and amino-aklylcyclohexanes: in vivo characterization. Amino Acids 2000;19:167–172.
27. Francis PT. Glutamatergic systems in Alzheimer's disease. Int J Geriat Psychiatry 2003;18:S15-S21.
28. Chessell IP, Francis PT, Pangalos MN, et al. Localisation of muscarinic (m_1) and other neurotransmitter receptors on corticofugal-projecting pyramidal neurones. Brain Res 1993;632:86–94.
29. Chessell IP, Humphrey PPA. Nicotinic and muscarinic receptor-evoked depolarisations recorded from a novel cortical brain slice preparation. Neuropharmacology 1995;34:1289–1296.
30. Chessell IP, Pearson RCA, Heath PR, et al. Selective loss of cholinergic receptors following unilateral intracortical injection of volkensin. Exp Neurol 1997;147:183–191.
31. Turrini P, Casu MA, Wong TP, et al. Cholinergic nerve terminals establish classical synapses in the rat cerebral cortex: synaptic pattern and age-related atrophy. Neuroscience 2001;105:277–285.
32. Francis PT, Pearson RCA, Lowe SL, et al. The dementia of Alzheimer's disease: an update. J Neurol Neurosurg Psychiatry 1987;50:242–243.
33. Zilles K, Werner L, Qu M, et al. Quantitative autoradiography of 11 different transmitter binding sites in the basal forebrain region of the rat: evidence of heterogeneity in distribution patterns. Neuroscience 1991;42:473–481.
34. Martin LJ, Blackstone CD, Levey AI, et al. Cellular localizations of AMPA glutamate receptors within the basal forebrain magnocellular complex of rat and monkey. J Neurosci 1993;13:2249–2263.
35. Ikonomovic MD, Armstrong DM. Distribution of AMPA receptor subunits in the nucleus basalis of Meynert in aged humans: implications for selective neuronal degeneration. Brain Res 1996;716:229–232.
36. Ikonomovic MD, Nocera R, Mizukami K, Armstrong DM. Age-related loss of the AMPA receptor subunits GluR2/3 in the human nucleus basalis of Meynert. Exp Neurol 2000;166:363–375.

Chapter 44
Cortical Cholinergic Lesion Causes Aβ Deposition: Cholinergic-Amyloid Fusion Hypothesis

Thomas Beach, Pamela Potter, Lucia Sue, Amanda Newell,
Marissa Poston, Raquel Cisneros, Yoga Pandya, Abraham Fisher,
Alex Roher, Lih-Fen Lue, and Douglas Walker

Introduction

Research over the last two decades has established that brain accumulation and histological deposition of a single peptide known as β-amyloid (Aβ) is a critical event in AD. Many investigators believe that Aβ deposition leads to all of the other relevant pathological changes in the disease, including neurofibrillary tangle formation, loss of synapses, neuronal death, and dementia [1–3]. Preventing this accumulation may prevent AD. In simple genetic forms of AD, accumulation of Aβ is caused by inherited gene mutations that result in increased Aβ production [4]. Only a small subset of AD occurs this way, however. In the common form of AD, the reason for Aβ accumulation is not known. If Aβ deposition can be considered to be the essence of AD, the disease transcends human genetics as Aβ deposition is an aging change common to many mammalian species [5–11]. The prevalence of AD increases exponentially with age [12], and the histopathology of AD affects all humans who approach the maximum human life-span [13,14]. The initial pathogenic event of AD therefore must lie within the physiological process of aging.

We have developed an animal model of AD that is initiated by a physiological aging change [15,16]. In primates, cholinergic afferents to the cerebral cortex are depleted during aging, probably owing to the loss or physiological inadequacy of the parent cell bodies in the basal forebrain [17–30]. Evidence from human postmortem studies as well as cell culture and animal experiments has linked cholinergic neurotransmission to Aβ metabolism and deposition. In aging human cerebral cortex, cholinergic deafferentation begins around ages 40 to 50 years [20,25] and is shortly followed by biochemically detectable elevations of cortical Aβ [31] and Aβ deposition [32]. In aging nondemented

T. Beach
Civin Laboratory for Neuropathology, Sun Health Research Institute, Sun City, AZ, USA

A. Fisher et al. (eds.), *Advances in Alzheimer's and Parkinson's Disease*,
© Springer 2008

humans, there is a statistical association between the depletion of cortical cholinergic markers and measures of Aβ accumulation [27,33,34].

In vitro studies have shown that m1 and m3 muscarinic receptor activation leads to changes in the cleavage pattern of Aβ's precursor protein (APPβ) that favor decreased production of Aβ (Processing changes that result in more Aβ formation are termed "amyloidogenic," and changes that favor sAPPα, a different cleavage product, are termed "nonamyloidogenic." [35–50]) This effect has been confirmed by us and others in vivo in animals [51–53] and humans [54–56] Conversely, decreased activation of muscarinic receptors, whether through lesions of cholinergic afferents [57–64] or pharmacological blockade [65], leads to increased expression of APPβ and/or evidence of increased amyloidogenic cleavage. We therefore hypothesized that cortical cholinergic deafferentation would cause increased cortical Aβ production and deposition.

Cortical cholinergic deafferentation, accomplished through lesioning the nucleus basalis magnocellularis (nbm) has long been employed as an animal model of AD. Following the discovery of the cortical cholinergic deficit in AD, many investigators suggested that this could be the initial critical pathogenic event in the disease. Some predicted that degenerating cholinergic fibers might participate in plaque formation [66–68]. Two groups even reported the development of Aβ deposition in rats with nbm lesions [69,70], but this was not replicated by others and these findings were later disputed [71]. One group reported that nbm lesions increased amyloidogenic processing of APPβ [63,64], but they did not measure Aβ directly. Our approach differs from these previous efforts in a number of ways.

We have used rabbits as our experimental animal, rather than rats, as employed by most earlier studies. Rodent Aβ differs from the human peptide by having three different amino acids, and there is some evidence that this difference leads to decreased amyloidogenic processing of APPβ and/or decreased Aβ aggregation [72–74]. It is possible, therefore, that the rodent Aβ sequence may make rodents resistant to Aβ overproduction and/or deposition. Because the rabbit amino acid sequence in and around the Aβ region of APPβ (complete peptides encoded by exons 15, 16, and 17) is identical to that of humans [75,76] this possibility is avoided. The first models of cholinergic deafferentation [77–79] used lesioning methods that were relatively nonspecific in terms of the cell types affected [80]. We have used an immunotoxin that is much more selective [64,81–87]. Finally, previous attempts to link amyloidogenic metabolism and cholinergic deafferentation used surrogate markers for Aβ induction, such as full-length APPβ and sAPPα [57–60,62–64,88] This approach is inadequate. Increased Aβ production can arise from altered processing of APPβ in the absence of concentration changes of the parent molecule, whereas secretion of sAPPα does not always have a reciprocal relation with secretion of Aβ [89–92] We therefore measured Aβ directly, using sensitive enzyme-linked immunosorbent assay (ELISA) methods.

Methods

Cholinergic Therapy of Normal Animals

Three muscarinic agonists, an acetylcholinesterase inhibitor (physostigmine), and nicotine were first administered to normal rabbits to determine the effects of these cholinergic agents on the cerebrospinal fluid (CSF) and cortical Aβ levels [52]. The effects were contrasted with those of a noncholinergic compound, norepinephrine. Three m1-selective muscarinic agonists from the AF series were tested: AF102B, AF150(S), and AF267B [93]. Rabbits received twice-daily subcutaneous injections (total of 2 mg/kg/day). The norepinephrine dose and route was 7.0 mg/kg/day by s.c. osmotic pumps (Durect, Cupertino, CA, USA), which is a dose previously reported to have physiologic central nervous system (CNS) effects in rabbits [94]. For each drug treatment, a control group received sham therapy with a vehicle. In a separate experiment, physostigmine was administered by subcutaneous osmotic pump (Durect) to guinea pigs [51] at a dose of 3 mg/kg/day. All animals were euthanized after 5 days of treatment.

Immunotoxin Lesion of the nbm

Twelve-week-old (2.0–2.5 kg) female New Zealand White rabbits received unilateral intracerebroventricular (i.c.v.) stereotactic injections of 12μl (32.4μg) of an immunotoxin conjugate (Advanced Targeting Systems, Carlsbad, CA, USA) dissolved in sterile saline. The conjugate was composed of saporin, a ribosomal toxin, conjugated to monoclonal antibody ME20.4, which is specific for the p75 neurotrophin receptor (p75NTR, low-affinity nerve growth factor receptor) expressed selectively on cholinergic nbm neurons. The optimal dose of conjugate was determined by pilot experiments. High doses of conjugate, whether delivered by unilateral or bilateral injections, caused increased toxicity (anorexia, lethargy, diarrhea, abnormal posturing, convulsions, death) without an appreciable increase in lesion extent. Coordinates for the injection were, relative to the bregma: AP = 0 mm, L = 2.2 mm, D = 7.5 mm. Additional animals were given a control lesion of the locus ceruleus (lc) to test further the hypothesis that the lesion effects on Aβ are due specifically to loss of cholinergic neurotransmission. The lc noradrenergic neuronal lesions were performed in female New Zealand White rabbits (2.0–2.5 kg) in a manner similar to the nbm lesions, by i.c.v. injection of a specific immunotoxin. The immunotoxin, like that used for the nbm lesion (Advanced Targeting Systems), was composed of a monoclonal antibody bound to the ribosomal toxin saporin. The monoclonal antibody used for the lc lesion was MAB394, which specifically recognizes dopamine β-hydroxylase (DBH), a substance expressed selectively on noradrenergic neurons of the lc. The immunotoxin has previously been shown to effectively lesion the lc in rats, with marked depletions of both noradrenergic neuronal cell bodies in the lc and noradrenergic fibers in the cerebral cortex [95]. The dose of immunotoxin given, 10μg in 12μl, was determined to be optimal in a pilot experiment. Two other groups of rabbits were subjected to alternative treatments, including (1) both nbm and lc lesions; and (2) i.c.v. saline injections.

After survival periods of 3 weeks, 5 weeks, 3 months, and 6 months, the animals were euthanized and the brains removed and cut into 0.4-cm coronal slices. One or two slices containing frontal cortex and hippocampus were fixed in 4% paraformaldehyde for immunohistochemical studies, and the remaining slices were fresh-frozen on slabs of Dry Ice. Immunoperoxidase histochemistry was performed on free-floating sections, as previously described [96]. Cortical cholinergic fibers were visualized using an enhanced acetylcholinesterase (AChE) enzyme histochemical method [97], and cholinergic nbm neuronal cell bodies were demonstrated using the same ME20.4 p75NTR antibody that was used for immunotoxin construction. Tissue localization of Aβ was performed using several antibodies, including 10D5 (amino acids 1–16; Elan Pharmaceuticals, South San Francisco, CA, USA), 4G8 (amino acids 17–24; Signet Laboratories, Dedham, MA, USA), 6E10 (amino acids 1–16, Signet Laboratories), 3D6 (N-terminal-specific for Aβ Elan Pharmaceuticals), 21F12 (end-specific for Aβ42; Elan Pharmaceuticals), and 5C3 (end-specific for Aβ40; Calbiochem, La Jolla, CA, USA). For co-localization experiments, we used monoclonal antibody 10C9 to heparan sulfate proteoglycan (HSPG) (Biogenesis, Brentwood, NH, USA). Control sections were treated by omitting the primary antibody and also (for 10D5 stained sections) by preabsorbing the primary antibody with synthetic Aβ 1-40 (100 μg/ml) (California Peptides, Napa, CA, USA) for 2 hours prior to use. Thioflavin-S staining was used to assess the presence of b-pleated sheet amyloid, and the Campbell-Switzer silver stain [98] was used to demonstrate argyrophilic diffuse and compact amyloid deposits.

Histological vascular Aβ deposition and of cholinergic fiber depletion was quantified by counting the intersections of positively stained blood vessels or fibers within the lines on an ocular grid. The grid position was successively moved through the cortical column and a mean value determined per column for each animal. This method has been described in earlier publications [68,99,100].

Confocal scanning laser fluorescence microscopy was performed on sections stained immunohistochemically for Aβ and HSPG, as described above, except for the substitution of fluorescent signal development in place of the immunoperoxidase method. Sections were first reacted for HSPG with the 10C9 antibody and then for Aβ with the 10D5 antibody. Signal development was achieved with fluorochrome-linked secondary antibodies (Molecular Probes, Eugene, OR, USA), which included goat anti-rat immunoglobulin G (IgG) conjugated to Alexa 488 for HSPG and sheep anti-mouse IgG conjugated to Alexa 568 for Aβ. Control sections were treated identically except for omission of one of the two primary antibodies (antibody to HSPG omitted on one control section, antibody to Aβ omitted on the other control section). Sections were viewed using an Olympus confocal microscope at appropriate excitation wavelengths.

For electron microscopy, sections were initially processed the same as for Aβ immunohistochemistry with the 10D5 antibody, except for omission of nickel ammonium sulfate intensification of the 3,3'- diaminobenzidine reaction product. Following immunostaining, sections were sequentially postfixed for 1 hour at 4°C in 1% glutaraldehyde and 1% osmium tetroxide, both made up in 0.1 M sodium cacodylate buffer, pH 7.4. Following dehydration in a graded ethanol series and propylene oxide, the sections were infiltrated in Polybed 812 resin (Polysciences, Warrington, PA, USA) and polymerized at 65°C. Ultrathin sections were cut at 100 nm and examined without heavy metal staining.

Biochemical analyses of fresh-frozen brain tissue (dorsolateral frontal cerebral cortex ipsilateral to the injection site for all groups, with contralateral cortex and ipsilateral hippocampus also measured in some groups) included choline acetyltransferase (ChAT) enzyme assay [101], ELISA quantification (Biosource) of Aβ and Western blot quantification of secretory APP (sAPPα) and full-length APPβ. Cortical tissue for ELISA was extracted with guanidine as previously described [102]. Additionally, Tris buffer-extracted cortex was prepared for Western blot quantification of APPβ by centrifuging at 100,000 g for 30 minutes, following which the supernatant and pellet were collected. As full-length APPβ is entirely contained in cell membranes, whereas secretory APPβ (mainly sAPPα) is mainly found in extracellular fluid, the supernatant contains predominantly sAPPα, whereas the pelleted membrane fraction, further digested with SDS, contains full-length APPβ.

Cholinergic Therapy of Immunotoxin-Lesioned Animals

The effect of cholinergic therapy on lesioned animals was determined in additional groups of animals. Three sets of New Zealand White rabbits (female, 2.0–2.5 kg) were subjected to unilateral i.c.v. injection of the nbm immunotoxin as described above. One group also received twice-daily s.c. injections of AF267B (2 mg/kg/day), and a second group received physostigmine at a rate of 3 mg/kg/day by s.c. osmotic pumps. The third group received twice-daily s.c. injections of saline. A fourth group received a saline i.c.v. injection (sham nbm lesion) and twice-daily s.c. injections of saline. The animals were euthanized 3 weeks after surgery and processed for immunocytochemistry and biochemistry as described above.

Results

Cholinergic Therapy of Normal Animals

Studies with normal animals indicated that all of the cholinergic agents lowered CNS levels of Aβ in normal animals, whereas norepinephrine did not. The three muscarinic agonists all significantly reduced CSF levels of Aβ40, and two of the three agents significantly reduced CSF Aβ42 [analysis of variance (ANOVA) with post hoc compensated pairwise significance testing using Fisher's least significant difference (LSD) test]. For Aβ40, the CSF concentrations were lowered to 55.5%, 68.5%, and 70.7% of control values for AF267B, AF102B, and AF150S, respectively. For Aβ42, animals had CSF concentrations that were 59.2%, 83.6%, and 77.8%, respectively, of those in the control animals. Immunoblots of CSF confirmed this reduction in Aβ. Concentrations of CSF sAPPα were not significantly affected by drug treatment. All three agonists also lowered cortical Aβ, although this reached the significance level only for Aβ42 and only when the data for all three agents were combined and compared to the control (Wilcoxon rank sum test). Physostigmine did not significantly lower CSF Aβ but lowered cortical Aβ40 and Aβ42 to 86% and 70% of control levels, respectively. Norepinephrine did not significantly alter either CSF or cortical Aβ, and no appreciable trend was discernible (data not shown).

Immunotoxin Lesion of the nbm

Cortical ChAT activity ipsilateral to the toxin injection site was significantly reduced ($p < 0.01$) in all lesioned groups with respect to their control groups. Lesioned animals had cortical and hippocampal ChAT activity that averaged about 50% of that of controls. AChE enzyme histochemistry revealed marked depletion of cortical cholinergic fibers in immunotoxin-injected animals (Fig. 1A,B) and a reduction in the number of cholinergic nbm neurons (Fig. 1D,E). Immunohistochemistry for Aβ with all antibodies revealed frequent positively stained blood vessels throughout the cerebri of immunotoxin-treated animals (Fig. 1C,F). The largest number of vessels were located in the cerebral cortex, although occasional positive vessels were present in diencephalic areas including the thalamus and basal ganglia. Within the cerebral cortex, the superficial layers had the greatest vessel densities. Blood vessels of all sizes were stained, from large arterioles to capillaries; large arterioles were most frequently involved. Staining was localized to the blood vessel walls (Fig. 1F); there was no staining of the vessel luminal contents and no staining of red blood cells. The 5C3 antibody, specific for Ab40, gave results similar to

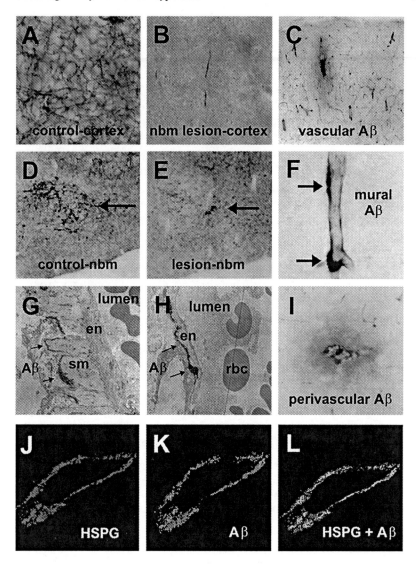

Fig. 1 Comparison of nucleus basalis magnocellularis (nbm)-lesioned and control animals. All panels are from the 6-month survival group. Lesioned animals show a profound reduction of acetylcholinesterase (AChE)-reactive cortical cholinergic fibers (**B**) and nbm neurons (**E**) compared to controls (**A, D**). Lesioned animals display frequent cortical blood vessels that are immunoreactive for β-amyloid (Aβ) (**C**). Deposits of Aβ are found in blood vessel walls (**F**) and perivascular neuropil (**I**). Electron microscopic observation demonstrates that Aβ deposits (black granular material) are found on the abluminal surface of smooth muscle cells (**G**) and endothelium (**H**), consistent with the location of the basal lamina. Confocal fluorescent microscopy confirms this, as immunofluorescence for Aβ co-localizes with that for heparan sulfate proteoglycan (HSPG) (**J–L**), which is a known component of the vascular basal lamina

those seen with antibody 21F12, specific for Aβ42, indicating that both forms of Aβ are present in the vascular deposits. Staining with the 3D6 antibody, specific for the N-terminus of Aβ, was also of similar quality and density, indicating that the complete Aβ peptide is deposited. Intracellular Aβ staining was not present in any brain region. In all immunotoxin-treated animals, the neuropil around occasional blood vessels was also stained (Fig. 1I), resembling the perivascular diffuse plaques sometimes seen in human AD cases. No staining of any type was observed in any control animal or in control sections. Neither the vascular nor perivascular deposits were stained with thioflavin-S, indicating that the deposited Aβ was in a diffuse, nonamyloid form. The vascular deposits were, however, stained with the Campbell-Switzer method for amyloid. Quantification of the density of Aβ-immunoreactive blood vessels showed that histological Ab deposition tended to increase between 3 weeks and 3 months but did not increase appreciably between 3 months and 6 months.

Electron microscopic examination revealed that the immunohistochemical reaction product marking Aβ was localized to the basal lamina covering arteriolar smooth muscle, perivascular cells and endothelial cells (Fig. 1G,H). The reaction product consisted of granular material; and no amyloid fibrils were present. Images obtained using scanning confocal fluorescence laser microscopy were also consistent with a basal lamina localization for Aβ, as the immunofluorescence staining pattern for Aβ matched that of HSPG, an integral component of the vascular basal lamina (Fig. 1J–L).

Immunohistochemical staining for reactive microglia showed the presence of a significant inflammatory reaction of cortical parenchyma (Fig. 2A) in lesioned animals. Staining for ICAM-1, RAGE, and serum amyloid A was positive in lesioned animals only, consistent with vascular cell inflammatory activation in response to Aβ deposition (Figs. 2B-D).

Assays of frontal cerebral cortex for Aβ using ELISA methods showed statistically significant increases ($p < 0.05$) in both Aβ40 and Aβ42 in lesioned animals, relative to controls, in the 3-week, 5-week, and 6-month survival

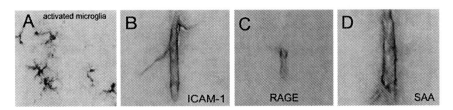

Fig. 2 Evidence of vascular inflammation in nbm-lesioned animals with vascular Aβ deposition: cortical parenchyma with increased numbers of activated microglia, as shown here, stained with an antibody to the rabbit macrophage marker RAM11 (**A**), cortical blood vessels immunohistochemically reactive for intercellular adhesion molecule-1(**B**), and receptors for advanced glycation end-products (RAGE) (**C**) and serum amyloid A (SAA) (**D**)

groups (no measurements were done of the 3-month survival group owing to loss of the frozen tissue). These increases ranged between 1.4- and 2.0-fold for the 3-week and 5-week survival groups, whereas the lesioned animals that survived 6 months had 2.5-fold and 8.0-fold elevations of $A\beta40$ and $A\beta42$, respectively. In the 6-month survival group, $A\beta42$ was increased disproportionately compared to $A\beta40$.

Immunoblot analysis of cortical membrane fractions probed with the 22C11 antibody showed that there was no significant difference between lesioned and control animals regarding the amount of sAPPα or full-length APPβ present (results not shown).

Cholinergic Therapy of Immunotoxin-Lesioned Animals

Both AF267B and physostigmine reduced histological deposition (Fig. 3) and biochemical levels of $A\beta$. Histologic $A\beta$ deposition was reduced to 6.4% and 12.0% of the lesioned untreated group for physostigmine and AF267B, respectively. Analysis of variance showed that the treatment groups differed significantly; furthermore, both AF267B and physostigmine-treated groups differed significantly from the lesioned, untreated group (Fisher's LSD).

Cortical $A\beta42$ concentrations were reduced by physostigmine treatment to 61.8% of that in lesioned rats. $A\beta42$ levels in physostigmine-treated animals were even slightly lower than those of sham-lesioned animals. Treatment with AF267B reduced $A\beta42$ levels to 83.4% of the lesioned, untreated group. The treatment groups were significantly different on ANOVA, and post hoc pairwise comparisons found that physostigmine-treated animals differed significantly from lesioned, untreated animals (Fisher's LSD). For $A\beta40$, only AF267B-treated animals had reduced concentrations (to 85.9%) compared to lesioned, untreated animals.

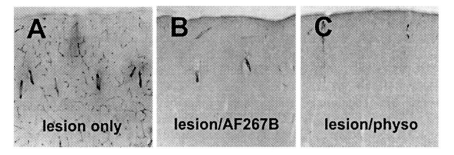

Fig. 3 Reduction of vascular $A\beta$ deposition by cholinergic therapy: nbm-lesioned animal, sham treatment (**A**), and animals treated with AF267B (**B**) or physostigmine (**C**)

Conclusions

We have shown that lesions of the rabbit nbm result in cortical Aβ deposition [15,16]. Lesioned animals have as much as an eightfold increase in cortical Aβ concentration. The Aβ deposits are primarily vascular, with occasional perivascular plaques. We think it likely, based on molecular mechanisms worked out in cell culture studies [35–50], that cholinergic deafferentation leads to increased Aβ production and secretion by deafferented cortical neurons. The increased extracellular fluid concentrations of Aβ lead to precipitation of Aβ on the vascular basal lamina owing to physicochemical affinities between Aβ and the extracellular matrix molecules of the basal lamina [103,104]. The dependence of lesion-induced Aβ deposition on cholinergic processes has been demonstrated by its reduction with cholinergic therapy and by showing that lesioning of the noradrenergic locus ceruleus does not cause Aβ deposition.

The relevance of our animal model for AD depends on the place of the cholinergic deficit in the AD pathogenic cascade. In AD, the cholinergic deficit is accentuated, relative to normal aging, so AD patients, especially those in the younger age ranges, have statistically lower concentrations of cholinergic markers relative to age-matched controls. Controversy exists as to whether the cholinergic deficit in AD occurs at an early stage or a later stage in the disease. Multiple studies were consistent with an early cholinergic deficit [24,105–114], although some recent work has contradicted this, suggesting that it is a late change in the disease [115–118]. It has also been suggested that the cholinergic deficit is merely a secondary change, retrograde degeneration that occurs after cortical degeneration [119]. We believe that the recent studies advocating a late-stage occurrence are flawed in that they used clinical staging of AD and mixed histopathological stages of the disease, resulting in false-negative findings. These studies also ignored earlier investigators who had found that the continuing loss of cortical cholinergic markers with age eventually results in AD-like cholinergic deficits in age-matched controls [17,18,21,22,25,26], Therefore, it will always be difficult, in the "oldest old" to distinguish between AD and controls on the basis of cholinergic markers. We and others have shown that the cortical cholinergic deficit in normal aging humans precedes and is statistically linked to age-related Aβ deposition [20,25,27,31,33,34,120]. Nucleus basalis neuronal cell bodies are also affected at the preclinical stage, as neurofibrillary tangles are universally present in the nbm of nondemented elderly humans (79/80 cases aged 67 and older) [121,122], in many cases in the complete absence of neocortical plaques or tangles. We suggest that this cholinergic deficit of normal aging is the beginning of the AD disease process. Finally, we have causally linked cholinergic deafferentation to Aβ deposition in our animal model, whereas other animal models have shown that neither cortical neuronal destruction [123] nor massive Aβ deposition [124–129] results in a significant and pronounced cortical cholinergic deficit like that seen in AD. The aggregate data strongly indicate that the cholinergic nbm begins to degenerate early in the aging process, leading to cortical cholinergic deafferentation, Aβ deposition, and Alzheimer's disease. Preclinical treatment of at-risk subjects with cholinergic agents may prevent Aβ deposition.

References

1. Neve RL, Robakis NK. Alzheimer's disease: a re-examination of the amyloid hypothesis. Trends Neurosci 1998;21:15–19.
2. Hardy J, Selkoe DJ. The amyloid hypothesis of Alzheimer's disease: progress and problems on the road to therapeutics. Science 2002;297:353–356.
3. Sommer B. Alzheimer's disease and the amyloid cascade hypothesis: ten years on. Curr Opin Pharmacol 2002;2:87–92.
4. Younkin SG. The role of A beta 42 in Alzheimer's disease. J Physiol Paris 1998;92:289–292.
5. Cummings BJ, Satou T, Head E, et al. Diffuse plaques contain C-terminal A beta 42 and not A beta 40: evidence from cats and dogs. Neurobiol Aging 1996;17:653–659.
6. Geula C, Nagykery N, Wu CK. Amyloid-beta deposits in the cerebral cortex of the agedcommon marmoset (Callithrix jacchus): incidence and chemical composition. Acta Neuropathol (Berl) 2002;103:48–58.
7. Maclean CJ, Baker HF, Ridley RM, Mori H. Naturally occurring and experimentally induced beta-amyloid deposits in the brains of marmosets (Callithrix jacchus). J Neural Transm 2000;107:799–814.
8. Head E, McCleary R, Hahn FF, et al. Region-specific age at onset of beta-amyloid in dogs. Neurobiol Aging 2000;21:89–96.
9. Selkoe DJ, Bell DS, Podlisny MB, et al. Conservation of brain amyloid proteins in aged mammals and humans with Alzheimer's disease. Science 1987;235:873–877.
10. Price DL, Martin LJ, Sisodia SS, et al. Aged non-human primates: an animal model of age-associated neurodegenerative disease. Brain Pathol 1991;1:287–296.
11. Gearing M, Rebeck GW, Hyman BT, et al. Neuropathology and apolipoprotein E profile of aged chimpanzees: implications for Alzheimer disease. Proc Natl Acad Sci U S A 1994;91:9382–9386.
12. McDowell I. Alzheimer's disease: insights from epidemiology. Aging (Milano) 2001;13:143–162.
13. Delaere P, He Y, Fayet G, et al. Beta A4 deposits are constant in the brain of the oldest old: an immunocytochemical study of 20 French centenarians. Neurobiol Aging 1993;14:191–194.
14. Bouras C, Hof PR, Giannakopoulos P, et al. Regional distribution of neurofibrillary tangles and senile plaques in the cerebral cortex of elderly patients: a quantitative evaluation of a one-year autopsy population from a geriatric hospital. Cereb Cortex 1994;4:138–150.
15. Beach TG, Potter PE, Kuo YM, et al. Cholinergic deafferentation of the rabbit cortex: a new animal model of Abeta deposition. Neurosci Lett 2000;283:9–12.
16. Roher AE, Kuo YM, Potter PE, et al. Cortical cholinergic denervation elicits vascular A beta deposition. Ann N Y Acad Sci 2000;903:366–373.
17. Mann DM, Yates PO, Marcyniuk B. Monoaminergic neurotransmitter systems in presenile Alzheimer's disease and in senile dementia of Alzheimer type. Clin Neuropathol 1984;3:199–205.
18. Mann DM, Yates PO, Marcyniuk B. Changes in nerve cells of the nucleus basalis of Meynert in Alzheimer's disease and their relationship to ageing and to the accumulation of lipofuscin pigment. Mech Ageing Dev 1984;25:189–204.

19. Mann DM, Yates PO, Marcyniuk B. Alzheimer's presenile dementia, senile dementia of Alzheimer type and Down's syndrome in middle age form an age related continuum of pathological changes. Neuropathol Appl Neurobiol 1984;10:185–207.

20. McGeer PL, McGeer EG, Suzuki J, et al. Aging, Alzheimer's disease, and the cholinergic system of the basal forebrain. Neurology 1984;34:741–745.

21. Mountjoy CQ, Rossor MN, Iversen LL, Roth M. Correlation of cortical cholinergic and GABA deficits with quantitative neuropathological findings in senile dementia. Brain 1984;107(Pt 2):507–518.

22. Rossor MN, Iversen LL, Johnson AJ, et al. Cholinergic deficit in frontal cerebral cortex in Alzheimer's disease is age dependent. Lancet 1981;2:1422.

23. Lowes-Hummel P, Gertz HJ, Ferszt R, Cervos-Navarro J. The basal nucleus of Meynert revised: the nerve cell number decreases with age. Arch Gerontol Geriatr 1989;8:21–27.

24. Perry EK, Blessed G, Tomlinson BE, et al. Neurochemical activities in human temporal lobe related to aging and Alzheimer-type changes. Neurobiol Aging 1981;2:251–256.

25. Perry EK, Johnson M, Kerwin JM, et al. Convergent cholinergic activities in aging and Alzheimer's disease. Neurobiol Aging 1992;13:393–400.

26. Bird TD, Stranahan S, Sumi SM, Raskind M. Alzheimer's disease: choline acetyltransferase activity in brain tissue from clinical and pathological subgroups. Ann Neurol 1983;14:284–293.

27. Beach TG, Honer WG, Hughes LH. Cholinergic fibre loss associated with diffuse plaques in the non-demented elderly: the preclinical stage of Alzheimer's disease? Acta Neuropathol (Berl) 1997;93:146–153.

28. Wenk GL, Pierce DJ, Struble RG, et al. Age-related changes in multiple neurotransmitter systems in the monkey brain. Neurobiol Aging 1989;10:11–19.

29. Beal MF, Walker LC, Storey E, et al. Neurotransmitters in neocortex of aged rhesus monkeys. Neurobiol Aging 1991;12:407–412.

30. Smith DE, Roberts J, Gage FH, Tuszynski MH. Age-associated neuronal atrophy occurs in the primate brain and is reversible by growth factor gene therapy. Proc Natl Acad Sci USA 1999;96:10893–10898.

31. Funato H, Yoshimura M, Kusui K, et al. Quantitation of amyloid beta-protein (A beta) in the cortex during aging and in Alzheimer's disease. Am J Pathol 1998;152:1633–1640.

32. Davies L, Wolska B, Hilbich C, et al. A4 amyloid protein deposition and the diagnosis of Alzheimer's disease: prevalence in aged brains determined by immunocytochemistry compared with conventional neuropathologic techniques. Neurology 1988;38:1688–1693.

33. Beach TG, Kuo YM, Spiegel K, et al. The cholinergic deficit coincides with Abeta deposition at the earliest histopathologic stages of Alzheimer disease. J. Neuropathol. Exp Neurol 2000;59:308–313.

34. Katzman R, Terry R, DeTeresa R, et al. Clinical, pathological, and neurochemical changes in dementia: a subgroup with preserved mental status and numerous neocortical plaques. Ann Neurol 1988;23:138–144.

35. Buxbaum JD, Oishi M, Chen HI, et al. Cholinergic agonists and interleukin 1 regulate processing and secretion of the Alzheimer beta/A4 amyloid protein precursor. Proc Natl Acad Sci U S A 1992;89:10075–10078.

36. Nitsch RM, Slack BE, Wurtman RJ, Growdon JH. Release of Alzheimer amyloid precursor derivatives stimulated by activation of muscarinic acetylcholine receptors. Science 1992;258:304–307.

37. Bymaster FP, Wong DT, Mitch CH, et al. Neurochemical effects of the M1 muscarinic agonist xanomeline (LY246708/NNC11-0232). J Pharmacol Exp Ther 1994;269:282–289.

38. Bymaster FP, Carter PA, Peters SC, et al. Xanomeline compared to other muscarinic agents on stimulation of phosphoinositide hydrolysis in vivo and other cholinomimetic effects. Brain Res 1998;795:179–190.

39. Nitsch RM, Growdon JH. Role of neurotransmission in the regulation of amyloid beta-protein precursor processing, Biochem Pharmacol 1994;47:1275–1284.

40. Nitsch RM. From acetylcholine to amyloid: neurotransmitters and the pathology of Alzheimer's disease. Neurodegeneration 1996;5:477–482.
41. Nitsch RM, Wurtman RJ, Growdon JH. Regulation of APP processing: potential for the therapeutical reduction of brain amyloid burden. Ann N Y Acad Sci 1996;777:175–182.
42. Haring R, Gurwitz D, Barg J, et al. Amyloid precursor protein secretion via muscarinic receptors: reduced desensitization using the M1-selective agonist AF102B. Biochem Biophys Res Commun 1994;203:652–658.
43. Haring R, Gurwitz D, Barg J, et al. NGF promotes amyloid precursor protein secretion via muscarinic receptor activation. Biochem Biophys Res Commun 1995;213:15–23.
44. Eckols K, Bymaster FP, Mitch CH, et al. The muscarinic M1 agonist xanomeline increases soluble amyloid precursor protein release from Chinese hamster ovary-m1 cells. Life Sci 1995;57:1183–1190.
45. Wolf BA, Wertkin AM, Jolly YC, et al. Muscarinic regulation of Alzheimer's disease amyloid precursor protein secretion and amyloid beta-protein production in human neuronal NT2N cells. J Biol Chem 1995;270:4916–4922.
46. Hung AY, Haass C, Nitsch RM, et al. Activation of protein kinase C inhibits cellular production of the amyloid beta-protein. J Biol Chem 1993;268:22959–22962.
47. Pittel Z, Heldman E, Barg J, et al. Muscarinic control of amyloid precursor protein secretion in rat cerebral cortex and cerebellum. Brain Res 1996;742:299–304.
48. Muller D, Wiegmann H, Langer U, et al. Lu 25-109, a combined m1 agonist and m2 antagonist, modulates regulated processing of the amyloid precursor protein of Alzheimer's disease. J Neural Transm 1998;105:1029–1043.
49. Muller DM, Mendla K, Farber SA, Nitsch RM. Muscarinic M1 receptor agonists increase the secretion of the amyloid precursor protein ectodomain. Life Sci 1997;60:985–991.
50. Farber SA, Nitsch RM, Schulz JG, Wurtman RJ. Regulated secretion of beta-amyloid precursor protein in rat brain. J Neurosci 1995;15:7442–7451.
51. Beach TG, Kuo Y, Schwab C, et al. Reduction of cortical amyloid beta levels in guinea pig brain after systemic administration of physostigmine. Neurosci Lett 2001;310:21–24.
52. Beach TG, Walker DG, Potter PE, et al. Reduction of cerebrospinal fluid amyloid beta after systemic administration of M1 muscarinic agonists. Brain Res 2001;905:220–223.
53. Lin L, Georgievska B, Mattsson A, Isacson O. Cognitive changes and modified processing of amyloid precursor protein in the cortical and hippocampal system after cholinergic synapse loss and muscarinic receptor activation. Proc Natl Acad Sci U S A 1999;96:12108–12113.
54. Hock C, Maddalena A, Heuser I, et al. Treatment with the selective muscarinic agonist talsaclidine decreases cerebrospinal fluid levels of total amyloid beta-peptide in patients with Alzheimer's disease. Ann N Y Acad Sci 2000;920:285–291.
55. Nitsch RM, Deng M, Tennis M, et al. The selective muscarinic M1 agonist AF102B decreases levels of total Abeta in cerebrospinal fluid of patients with Alzheimer's disease. Ann Neurol 2000;48:913–918.
56. Basun H, Nilsberth C, Eckman C, et al. Plasma levels of Abeta42 and Abeta40 in Alzheimer patients during treatment with the acetylcholinesterase inhibitor tacrine. Dement Geriatr Cogn Disord 2002;14:156–160.
57. Wallace W, Ahlers ST, Gotlib J, et al. Amyloid precursor protein in the cerebral cortex is rapidly and persistently induced by loss of subcortical innervation. Proc Natl Acad Sci U S A 1993;90:8712–8716.
58. Wallace WC, Bragin V, Robakis NK, et al. Increased biosynthesis of Alzheimer amyloid precursor protein in the cerebral cortex of rats with lesions of the nucleus basalis of Meynert. Brain Res Mol Brain Res 1991;10:173–178.
59. Wallace WC, Lieberburg I, Schenk D, et al. Chronic elevation of secreted amyloid precursor protein in subcortically lesioned rats, and its exacerbation in aged rats. J Neurosci 1995;15:4896–4905.

60. Beeson JG, Shelton ER, Chan HW, Gage FH. Age and damage induced changes in amyloid protein precursor immunohistochemistry in the rat brain. J Comp Neurol 1994;342:69–77.

61. Leanza G. Chronic elevation of amyloid precursor protein expression in the neocortex and hippocampus of rats with selective cholinergic lesions. Neurosci Lett 1998;257:53–56.

62. Lin L, LeBlanc CJ, Deacon TW, Isacson O. Chronic cognitive deficits and amyloid precursor protein elevation after selective immunotoxin lesions of the basal forebrain cholinergic system. Neuroreport 1998;9:547–552.

63. Rossner S, Ueberham U, Yu J, et al. In vivo regulation of amyloid precursor protein secretion in rat neocortex by cholinergic activity. Eur J Neurosci 1997;9:2125–2134.

64. Geula C, Zhan SS. Altered processing of amyloid precursor protein following specific cholinergic denervation of rat cortex. Soc Neurosci Abstr 1997;23:820.

65. Beach TG, Walker DG, Cynader MS, Hughes LH. Increased beta-amyloid precursor protein mRNA in the rat cerebral cortex and hippocampus after chronic systemic atropine treatment. Neurosci Lett 1996;210:13–16.

66. Struble RG, Cork LC, Whitehouse PJ, Price DL. Cholinergic innervation in neuritic plaques. Science 1982;216:413–415.

67. Arendt T, Bigl V, Tennstedt A, Arendt A. Neuronal loss in different parts of the nucleus basalis is related to neuritic plaque formation in cortical target areas in Alzheimer's disease. Neuroscience 1985;14:1–14.

68. Beach TG, McGeer EG. Senile plaques, amyloid beta-protein, and acetylcholinesterase fibres: laminar distributions in Alzheimer's disease striate cortex. Acta Neuropathol (Berl) 1992;83:292–299.

69. Arendash GW, Millard WJ, Dunn AJ, Meyer EM. Long-term neuropathological and neurochemical effects of nucleus basalis lesions in the rat. Science 1987;238:952–956.

70. Fuentes C, Roch G, König N. Light and electron microscopical observations in the nucleus basalis of Meynert and in hippocampus of the rat after injection of a cholinotoxin: degeneration and reorganization. Z Mikrosk Anat Forsch 1987;101:451–460.

71. Thal LJ, Mandel RJ, Terry RD, et al. Nucleus basalis lesions fail to induce senile plaques in the rat. Exp Neurol 1990;108:88–90.

72. DeStrooper B, Simons M, Multhaup G, et al. Production of intracellular amyloid-containing fragments in hippocampal neurons expressing human amyloid precursor protein and protection against amyloidogenesis by subtle amino acid substitutions in the rodent sequence. EMBO J 1995;14:4932–4938.

73. Otvos L Jr, Szendrei GI, Lee VM, Mantsch HH. Human and rodent Alzheimer beta-amyloid peptides acquire distinct conformations in membrane-mimicking solvents. Eur J Biochem 1993;211:249–257.

74. Reaume AG, Howland DS, Trusko SP, et al. Enhanced amyloidogenic processing of the beta-amyloid precursor protein in gene-targeted mice bearing the Swedish familial Alzheimer's disease mutations and a "humanized" Abeta sequence. J Biol Chem 1996;271:23380–23388.

75. Davidson JS, West RL, Kotikalapudi P, Maroun LE. Sequence and methylation in the beta/A4 region of the rabbit amyloid precursor protein gene. Biochem Biophys Res Commun 1992;188:905–911.

76. Johnstone EM, Chaney MO, Norris FH, et al. Conservation of the sequence of the Alzheimer's disease amyloid peptide in dog, polar bear and five other mammals by cross-species polymerase chain reaction analysis. Brain Res Mol Brain Res 1991;10:299–305.

77. Johnston MV, McKinney M, Coyle JT. Evidence for a cholinergic projection to neocortex from neurons in basal forebrain. Proc Natl Acad Sci U S A 1979;76:5392–5396.

78. Mantione CR, Fisher A, Hanin I. The AF64A-treated mouse: possible model for central cholinergic hypofunction. Science 1981;1213:579–580.

79. Muir JL, Page KJ, Sirinathsinghji DJ, et al. Excitotoxic lesions of basal forebrain cholinergic neurons: effects on learning, memory and attention. Behav Brain Res 1993;57:123–131.

80. Lindefors N, Boatell ML, Mahy N, Persson H. Widespread neuronal degeneration after ibotenic acid lesioning of cholinergic neurons in the nucleus basalis revealed by in situ hybridization. Neurosci Lett 1992;135:262–264.

81. Book AA, Wiley RG, Schweitzer JB. Specificity of 192 IgG-saporin for NGF receptor-positive cholinergic basal forebrain neurons in the rat. Brain Res 1992;590:350–355.

82. Book AA, Wiley RG, Schweitzer JB. 192 IgG-saporin. I. Specific lethality for cholinergic neurons in the basal forebrain of the rat. J Neuropathol Exp Neurol., 1994;53:95–102.

83. Heckers S, Ohtake T, Wiley RG, et al. Complete and selective cholinergic denervation of rat neocortex and hippocampus but not amygdala by an immunotoxin against the p75 NGF receptor. J Neurosci 1994;14:1271–1289.

84. Walsh TJ, Kelly RM, Dougherty KD, et al. Behavioral and neurobiological alterations induced by the immunotoxin 192-IgG-saporin: cholinergic and non-cholinergic effects following i.c.v. injection. Brain Res 1995;702:233–245.

85. Wenk GL, Stoehr JD, Quintana G, et al. Behavioral, biochemical, histological, and electrophysiological effects of 192 IgG-saporin injections into the basal forebrain of rats. J Neurosci 1994;14:5986–5995.

86. Wiley RG, Oeltmann TN, Lappi DA. Immunolesioning: selective destruction of neurons using immunotoxin to rat NGF receptor. Brain Res 1991;562:149–153.

87. Fine A, Hoyle C, Maclean CJ, et al. Learning impairments following injection of a selective cholinergic immunotoxin, ME20.4 IgG-saporin, into the basal nucleus of Meynert in monkeys. Neuroscience 1997;81:331–343.

88. Leanza G. Chronic elevation of amyloid precursor protein expression in the neocortex and hippocampus of rats with selective cholinergic lesions. Neurosci Lett 1998;257:53–56.

89. Fuller SJ, Storey E, Li QX, et al. Intracellular production of beta A4 amyloid of Alzheimer's disease: modulation by phosphoramidon and lack of coupling to the secretion of the amyloid precursor protein. Biochemistry 1995;34:8091–8098.

90. Petanceska SS, Nagy V, Frail D, Gandy S. Ovariectomy and 17beta-estradiol modulate the levels of Alzheimer's amyloid beta peptides in brain. Exp Gerontol 2000;35:1317–1325.

91. Savage MJ, Trusko SP, Howland DS, et al. Turnover of amyloid beta-protein in mouse brain and acute reduction of its level by phorbol ester. J Neurosci 1998;18:1743–1752.

92. Vincent B, Smith JD. Effect of estradiol on neuronal Swedish-mutated beta-amyloid precursor protein metabolism: reversal by astrocytic cells. Biochem Biophys Res Commun 2000;271:82–85.

93. Fisher A. Therapeutic strategies in Alzheimer's disease: M1 muscarinic agonists. Jpn J Pharmacol 2000;84:101–112.

94. Czepita D. Influence of alpha and beta-adrenergic stimulators and blockers on the electroretinogram and visually evoked potentials of the rabbit. Biomed Biochim Acta 1990;49:509–513.

95. Wrenn CC, Picklo MJ, Lappi DA, et al. Central noradrenergic lesioning using anti-DBH-saporin: anatomical findings. Brain Res 1996;740:175–184.

96. Beach TG, Tago H, Nagai T, et al. Perfusion-fixation of the human brain for immunohistochemistry: comparison with immersion-fixation. J Neurosci Methods 1987;19:183–192.

97. Tago H, Kimura H, Maeda T. Visualization of detailed acetylcholinesterase fiber and neuron staining in rat brain by a sensitive histochemical procedure. J Histochem Cytochem 1986;34:1431–1438.

98. Braak H, Braak E. Demonstration of amyloid deposits and neurofibrillary changes in whole brain sections. Brain Pathol 1991;1:213–216.

99. Beach TG, McGeer EG. Cholinergic fiber loss occurs in the absence of synaptophysin depletion in Alzheimer's disease primary visual cortex. Neurosci Lett 1992;142:253–256.

100. Geula C, Mesulam MM. Cortical cholinergic fibers in aging and Alzheimer's disease: a morphometric study. Neuroscience 1989;33:469–481.

101. Fonnum, F. Radiochemical micro assays for the determination of choline acetyltransferase and acetylcholinesterase activities. Biochem J 1969;115:465–472.

102. Beach TG, Walker D, Sue L, et al. Immunotoxin lesion of the cholinergic nucleus basalis causes Aβ deposition: towards a physiologic animal model of Alzheimer's disease. Curr Med Chem 2003;3:57–75.

103. Weller RO, Massey A, Newman TA, et al. Cerebral amyloid angiopathy: amyloid beta accumulates in putative interstitial fluid drainage pathways in Alzheimer's disease. Am J Pathol 1998;153:725–733.

104. Weller RO, Massey A, Kuo YM, Roher AE. Cerebral amyloid angiopathy: accumulation of A beta in interstitial fluid drainage pathways in Alzheimer's disease. Ann N Y Acad Sci 2000;903:110–117.

105. Bowen DM, Benton JS, Spillane JA, et al. Choline acetyltransferase activity and histopathology of frontal neocortex from biopsies of demented patients. J Neurol Sci 1982;57:191–202.

106. Bowen DM, Allen SJ, Benton JS, et al. Biochemical assessment of serotonergic and cholinergic dysfunction and cerebral atrophy in Alzheimer's disease. J Neurochem 1983;41:266–272.

107. Francis PT, Palmer AM, Sims NR, et al. Neurochemical studies of early-onset Alzheimer's disease: possible influence on treatment. N Engl J Med 1985;313:7–11.

108. Francis PT, Webster MT, Chessell IP, et al. Neurotransmitters and second messengers in aging and Alzheimer's disease. Ann N Y Acad Sci 1993;695:19–26.

109. Lowe SL, Francis PT, Procter AW, et al. Gamma-aminobutyric acid concentration in brain tissue at two stages of Alzheimer's disease. Brain 1988;111:785–799.

110. Perry EK, Perry RH. A review of neuropathological and neurochemical correlates of Alzheimer's disease. Dan Med Bull 1985;32(suppl 1):27–34.

111. Palmer AM, Gershon S. Is the neuronal basis of Alzheimer's disease cholinergic or glutamatergic? FASEB J 1990;4:2745–2752.

112. Palmer AM. Neurochemical studies of Alzheimer's disease. Neurodegeneration 1996;5:381–391.

113. Procter AW, Lowe SL, Palmer AM, et al. Topographical distribution of neurochemical changes in Alzheimer's disease. J Neurol Sci 1988;84:125–140.

114. Procter AW. Neurochemical correlates of dementia. Neurodegeneration 1996;5:403–407.

115. Davis KL, Mohs RC, Marin D, et al. Cholinergic markers in elderly patients with early signs of Alzheimer disease. JAMA 1999;281:1401–1406.

116. DeKosky ST, Ikonomovic MD, Styren SD, et al. Upregulation of choline acetyltransferase activity in hippocampus and frontal cortex of elderly subjects with mild cognitive impairment. Ann Neurol 2002;51:145–155.

117. Gilmor ML, Erickson JD, Varoqui H, et al. Preservation of nucleus basalis neurons containing choline acetyltransferase and the vesicular acetylcholine transporter in the elderly with mild cognitive impairment and early Alzheimer's disease. J Comp Neurol 1999;411:693–704.

118. Tiraboschi P, Hansen LA, Alford M, et al. The decline in synapses and cholinergic activity is asynchronous in Alzheimer's disease. Neurology 2000;55:1278–1283.

119. Pearson RC, Powell TP. Anterograde vs. retrograde degeneration of the nucleus basalis medialis in Alzheimer's disease. J Neural Transm Suppl 1987;24:139–146.

120. Thomas G Beach, Pamela E Potter, Lucia I Sue PPYPMRPSLBT. Cortical cholinergic deficit is associated with plaque density at preclinical stages of Alzeimer's disease. Presented at the 9th International Conference on Alzheimer's Disease and Related Disorders, 2004.

121. Beach TG, Sue LI, Scott S, Sparks DL. Neurofibrillary tangles are constant in aging human nucleus basalis. Alzheimers Rep 1998;1:375–380.
122. Sassin I, Schultz C, Thal DR, et al. Evolution of Alzheimer's disease-related cytoskeletal changes in the basal nucleus of Meynert. Acta Neuropathol (Berl) 2000;100:259–269.
123. Minger SL, Davies P. Persistent innervation of the rat neocortex by basal forebrain cholinergic neurons despite the massive reduction of cortical target neurons. I. Morphometric analysis. Exp Neurol 1992;117:124–138.
124. Gau JT, Steinhilb ML, Kao TC, et al. Stable beta-secretase activity and presynaptic cholinergic markers during progressive central nervous system amyloidogenesis in Tg2576 mice. Am J Pathol 2002;160:731–738.
125. Jaffar S, Counts SE, Ma SY, et al. Neuropathology of mice carrying mutant APP(swe) and/or PS1(M146L) transgenes: alterations in the p75(NTR) cholinergic basal forebrain septo-hippocampal pathway. Exp Neurol 2001;170:227–243.
126. Bronfman FC, Moechars D, Van Leuven F. Acetylcholinesterase-positive fiber deafferentation and cell shrinkage in the septo-hippocampal pathway of aged amyloid precursor protein London mutant transgenic mice. Neurobiol Dis 2000;7:152–168.
127. Hernandez D, Sugaya K, Qu T, et al. Survival and plasticity of basal forebrain cholinergic systems in mice transgenic for presenilin-1 and amyloid precursor protein mutant genes. Neuroreport 2001;12:1377–1384.
128. Wong TP, Debeir T, Duff K, Cuello AC. Reorganization of cholinergic terminals in the cerebral cortex and hippocampus in transgenic mice carrying mutated presenilin-1 and amyloid precursor protein transgenes. J Neurosci 1999;19:2706–2716.
129. Boncristiano S, Calhoun ME, Kelly PH, et al. Cholinergic changes in the APP23 transgenic mouse model of cerebral amyloidosis. J Neurosci 2002;22:3234–3243.

Chapter 45
Expression of Acetylcholinesterase in Alzheimer's Disease Brain: Role in Neuritic Dystrophy and Synaptic Scaling

David H. Small, Steven Petratos, Sharon Unabia, and Danuta Maksel

Introduction

Although the biochemical and genetic mechanisms that control β-amyloid (Aβ) production and aggregation are now becoming well understood, the mechanism by which Aβ causes neurodegeneration and cognitive decline is still unknown. The buildup of Aβ in the brain may trigger a range of biochemical responses [1]. However, the contribution of each of these responses to the overall pattern of neurodegeneration is unclear.

The neuropathology of Alzheimer's disease (AD) is characterized by amyloid deposition, dystrophic neurites, neurofibrillary tangles (NFTs), gliosis, and cell death. Of these changes, neuritic dystrophy, which is often closely associated with amyloid plaques, is of interest as it is likely that neuritic changes underlie the cognitive deficit [1–4]. Both the amyloid plaques and the NFT-bearing dystrophic neurites that surround the plaques stain strongly for the enzyme acetylcholinesterase (AChE) despite the fact that overall the level of AChE is decreased in the AD brain [5–8]. This suggests that AChE may play some role in the generation of (or response to) amyloid-induced neuritic damage. The observation that levels of AChE change in the AD brain is important because inhibition of AChE activity is associated with improved cognitive performance [9].

The aim of our studies has been to examine the biochemical mechanisms that cause this increase in AChE and to understand the role of AChE and other cholinergic proteins in neuritic dystrophy and repair.

The principal function of AChE is rapid hydrolysis of acetylcholine. There are multiple forms of the enzyme, which differ in molecular weight and hydrodynamic properties (Fig. 1) [10]. All forms of AChE are generated from a single gene on chromosome 7. There are three transcripts (T, H, R) that are produced by mRNA splicing. One transcript (H) encodes a catalytic subunit with a

D.H. Small
Department of Biochemistry and Molecular Biology, Monash University, Victoria 3800, Australia

A. Fisher et al. (eds.), *Advances in Alzheimer's and Parkinson's Disease*, 429
© Springer 2008

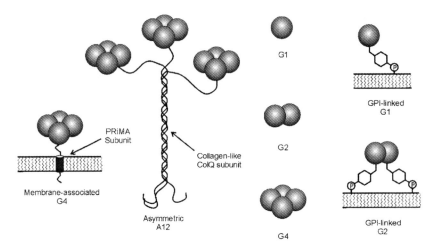

Fig. 1 Multiple isoforms of acetylcholinesterase (AChE) are generated by attachment of a GPI anchor or a noncatalytic subunit [10]

specific C-terminal peptide that directs GPI anchor attachment. Expression is tissue-specific, as GPI-linked forms are found mainly in red blood cells. The major form of AChE in the brain is also membrane-anchored and is encoded by the T transcript. This form is created by attachment of a short membrane anchoring subunit known as PRiMA. Asymmetrical AChE, which is found mostly on basement membranes, is also encoded by the T transcript and is created through attachment of a collagen-like tail subunit (ColQ).

Effect of Aβ on AChE Levels

We have found that the increase in AChE around amyloid plaques and in NFT-bearing neurons is a direct consequence of the accumulation of soluble Aβ in the brain. Aβ can induce AChE production in cultured cells [11,12]. For example, treatment of neuronally differentiated P19 cells with Aβ causes an increase in AChE levels [11]. The effect in P19 cells can be blocked by L-type calcium-channel inhibitors but not by inhibitors of other voltage-gated calcium channels, suggesting that Aβ may directly or indirectly activate L-type channels [11]. Aβ also increases the level of AChE in primary cortical astrocytes in culture [12] indicating that at least some of the AChE in plaques may be derived from astrocytes.

The effect of Aβ on AChE has been observed in vivo. APP CT100 transgenic mice, which express elevated levels of human-sequence Aβ in the brain but do not develop amyloid plaques, have increased levels of AChE [13]. In AD brain, Aβ deposition correlates with an increase in a minor isoform of

AChE that is monomeric and amphiphilic [14]. This minor species has a distinct glycosylation pattern [14] and has been termed Glyc-AChE. Glyc-AChE is also selectively elevated in cerebrospinal fluid (CSF) of patients with AD [15,16]. As the level of Glyc-AChE in CSF increases with increasing severity of AD, we have suggested that Glyc-AChE may have value for monitoring AD progression [17].

The level of Glyc-AChE is also increased in APP (SW) Tg2576 transgenic mice, which develop amyloid plaque pathology [18]. This increase occurs at a very early stage of development (4 months), when soluble Aβ levels are increased but 7 to 8 months prior to amyloid plaque formation. This finding suggests that an increase in soluble Aβ may be responsible for the increase in AChE, consistent with the observation that AChE is also increased in APP CT100 transgenic mice, which do not develop amyloid plaques [13].

Role of α7 Nicotinic Acetylcholine Receptors in AD

Neuronal nicotinic acetylcholine receptors (nAChRs) are distributed through-out the central and peripheral nervous systems [19]. The nAChRs belong to the ligand-gated ion channel family of receptors, all of which have a basic penta-meric structure. There are two major types of nAChR in the brain [19]. One type is heteromeric and contains both α4 and β2 subunits (α4β2 nAChR); it is distributed widely. A second type of receptor is homomeric and contains only α7 subunits (α7 nAChR). The nAChRs are now a major focus of research because of their potential role in memory.

Of the two main types of brain nAChR, the α7 receptor is of particular interest for its role in AD. It is specifically enriched in the hippocampus and neocortex, regions of the brain that are important in memory and cognition [20]. Both the localization of the α7 nAChR on somatic or dendritic spines and the high permeability of the receptor to calcium indicate a role in synaptic plasticity [21].

Recent studies suggest that Aβ binds directly to the α7 nAChRs. However, the exact nature of this interaction remains unclear. Initially, Wang et al. [22] reported that the α7 nAChR and Aβ1-42 can be co-immunoprecipitated from human brain tissue and that neuronal cell lines overexpressing the α7 nAChR can bind Aβ1-42. This binding was reportedly inhibited by the α7 nAChR-specific antagonist α-bungarotoxin. However, the data are conflicting, as some groups have found that Aβ stimulates the receptor, and other groups have described inhibition. For example, Pettit et al. [23], Tozaki et al., [24] and Liu et al. [25] have reported that Aβ blocks the α7 nAChR receptor, whereas Dineley et al. [26,27] found in two separate studies that Aβ stimulates the receptor.

Interestingly, like Glyc-AChE, levels of the α7 nAChR are increased in Tg2576 mice at an early developmental stage [26]. Also like AChE, the level of α7 nAChR is increased in the AD brain [28], and studies by Xiu et al. [29] have shown that Aβ1-42 can up-regulate expression of several nAChR subunits, including the α7 nAChR, in cell culture. Our own studies have shown that the effect of Aβ on AChE in primary cortical neurons is mediated via the α7 nAChR [30]. Agonists of the α7 nAChR specifically increase levels of AChE, whereas antagonists block the effect of Aβ on AChE [30].

Cholinergic Neurites in AD Plaques

There is considerable evidence that many of the dystrophic neurites that decorate amyloid plaques are derived from cholinergic neurons. Struble et al. [31] first reported the presence of AChE-positive neurites in early-stage plaques of aged monkeys. Later choline acetyltransferase (ChAT)-positive neurites were shown to be present around plaques in aged monkeys by Kitt et al. [32] Aberrant sprouting of ChAT-positive cholinergic neurons around amyloid plaques has also been shown to occur in the AD brain [33] as well as in APP transgenic mice [34]. Similar studies have been done by Hu et al. [35] using the vesicular acetylcholine transporter (vAChT) as a marker for cholinergic neurons.

The observation that ChAT immunoreactivity is increased around amyloid plaques is significant because it indicates that cholinergic activity may be increased during early stages of pathogenesis. Like AChE, the ChAT level is decreased overall during the later stages of AD. Most studies on AD brain have been done postmortem on late-stage severely altered AD brains, in which there is considerable neurodegeneration. Thus, the overall decrease in ChAT and AChE seen in AD may be due to a relatively nonspecific neurodegenerative process. However, the observation that Glyc-AChE, ChAT, vAChT, and possibly the α7 nAChR are all increased in association with amyloid plaques raises the possibility that cholinergic activity may be increased during the very early stages of disease pathogenesis.

This possibility is supported by at least two studies. DeKosky et al. [36] have shown that ChAT levels are increased in the hippocampus of individuals with mild cognitive impairment (MCI). Individuals with MCI are a heterogeneous group who can be viewed as being "at risk" of developing AD. It is estimated that 50% to 80% of individuals with MCI eventually develop AD. Therefore, from a clinical standpoint, MCI may represent a continuum between normality and early AD. Evidence that the cholinergic system may be more active than normal during the early stages of AD pathogenesis has also come from longitudinal clinicopathologic studies of Catholic nuns, priests, and brothers (Religious Orders Study), which have shown elevated ChAT activity in MCI patients and a return to normal levels of ChAT in those with early AD [37].

Do AChE Inhibitors Correct Cholinergic Deficiency?

In essence, the cholinergic hypothesis of geriatric memory dysfunction states that at least some of the cognitive deficit in AD is due to a decrease in cholinergic activity. However, it is unclear whether cholinergic activity is decreased in early AD. Therefore, the contribution of the cholinergic system to cognitive dysfunction in early AD may need to be reevaluated. Today, AChE inhibitors are the only drugs that are approved for treatment of the memory deficit in AD [9]. However, the mechanism by which AChE inhibitors provide cognitive benefit is unclear. There is evidence that AChE inhibitors are more effective during the early stages of AD and that at later stages they are become less effective.[9,38] Although it is assumed that AChE inhibitors correct a cholinergic deficiency, if there is little cholinergic loss in early AD this idea seems questionable.

Role of Cholinergic Neurons in Synaptic Scaling

AChE inhibitors may improve cognition by boosting a normal brain mechanism known as synaptic scaling or synaptic compensation [39]. Synaptic scaling is a relative slow mechanism that controls overall neuronal input [40,41]. For example, synaptic scaling has been observed at the neuromuscular junction of *Drosophila*. Altered presynaptic input is compensated for by changes in presynaptic neurotransmitter release or the level of postsynaptic receptors, which helps maintain the excitatory input onto the muscle cell [41].

The significance of synaptic scaling for cognition and memory has been demonstrated by Ruppin and Reggia [42], who have studied the effect of synaptic loss on memory in attractor neural networks. These studies showed that when a critical number of synapses are lost substantial, severe memory loss occurs. However, in AD, amnesia does not appear suddenly but, rather, shows a graded pattern of decline, with recent memories lost first and older, more established memories lost later [43]. Studies using neural network models of associative memory demonstrate that a graded loss of memory can occur when synaptic scaling is present in the network [42].

Cholinergic sprouting may be a compensatory synaptic scaling mechanism that appears in response to synaptic damage caused by the buildup of Aβ. A model of this mechanism is shown in Fig. 2. Aβ-induced neuritic damage at glutamatergic synapses in the cortex or hippocampus, along with an Aβ-induced increase in AChE at cholinergic synapses may cause a decrease in excitatory (glutamatergic and cholinergic) input to cortical or hippocampal neurons. This decreased input may cause a compensatory increase in both ChAT activity in basal forebrain cholinergic neurons and α7 nAChRs on the surface of the cortical or hippocampal neurons.

Based on this model, we propose that AChE inhibitors improve cognition in AD patients by boosting this synaptic scaling mechanism. AChE inhibitors

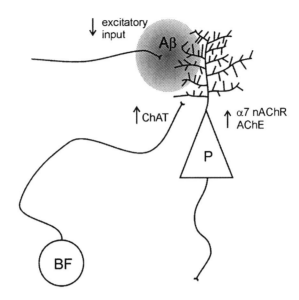

Fig. 2 Model of cholinergic activity may lead to synaptic scaling in the Alzheimer's disease (AD) brain. The buildup of β-amyloid (Aβ) in the cortical or hippocampal neuropil causes neuritic dystrophy and decreases the excitatory input to pyramidal neurons (P) from other cortical or hippocampal neurons. Aβ accumulation also increases AChE, which decreases excitatory cholinergic input from basal forebrain (BF) neurons. This decrease in total neuronal excitation results in a compensatory mechanism involving increased choline acetyltransferase (ChAT) in basal forebrain cholinergic neurons (BF) and increased α7 nicotinic acetylcholine receptors (nAChR) in the pyramidal neurons, which in turn leads to an increase in cholinergic input to the pyramidal neuron

increase excitatory cholinergic input, which would compensate for the decreased excitatory input. By boosting cholinergic activity, AChE inhibitors would increase overall neuronal excitability and thereby assist in preserving memories stored in the neuronal network.

This model of synaptic scaling could explain why some individuals do not respond well to AChE inhibitor therapy. In cases where basal forebrain cholinergic input is decreased (e.g., severe AD), synaptic scaling would not operate effectively, and thus AChE inhibitors would no longer exert their effects efficiently. Thus, decreased responsiveness to AChE inhibitors may be associated with a loss of cholinergic neurons in the basal forebrain.

Conclusion

Our studies and the work of other groups have shown that there is an increase in the levels of AChE, ChAT, and the α7 nAChR in association with Aβ. The increase in ChAT and α7 nAChRs may be due to a compensatory (synaptic scaling) mechanism in which cortical and hippocampal neuronal excitability is maintained by an increase in cholinergic activity. AChE inhibitors may improve cognition by boosting this compensatory mechanism.

References

1. Small DH, Mok SS, Bornstein JC. Alzheimer's disease and Aβ toxicity: from top to bottom. Nat Rev Neurosci 2001;2:595–598.
2. Probst A, Langui D, Ulrich J. Alzheimer's disease: a description of the structural lesions. Brain Pathol 1991;1:229–239.
3. Larner AJ. The cortical neuritic dystrophy of Alzheimer's disease: nature, significance, and possible pathogenesis, Dementia 1995;6:218–224.
4. Knowles RB, Wyart C,. Buldyrev SV, et al. Plaque-induced neurite abnormalities: implications for disruption of neural networks in Alzheimer's disease. Proc Natl Acad Sci U S A 1999;96:5274–5279.
5. Mesulam MM, Geula C. Shifting patterns of cortical cholinesterases in Alzheimer's disease: implications for treatment, diagnosis, and pathogenesis. Adv Neurol 1990;51:235–240.
6. Carson KA, Geula C, Mesulam MM. Electron microscopic localization of cholinesterase activity in Alzheimer brain tissue. Brain Res 1991;540:204–208.
7. Wright CI, Geula C, Mesulam MM. Neurological cholinesterases in the normal brain and in Alzheimer's disease: relationship to plaques, tangles, and patterns of selective vulnerability. Ann Neurol 1993;34:373–384.
8. Geula C, Greenberg BD, Mesulam MM. Cholinesterase activity in the plaques, tangles and angiopathy of Alzheimer's disease does not emanate from amyloid. Brain Res 1994;644:327–330.
9. Small DH. Acetcholinesterase inhibitors for the treatment of dementia in Alzheimer's disease: do we need new inhibitors? Exp Opin Drug Dev 2005;10:817–825.
10. Massoulie J, Sussman J, Bon S, Silman I. Structure and functions of acetylcholinesterase and butyrylcholinesterase. Prog Brain Res 1993;98:139–146.
11. Sberna G, Saez-Valero J, Beyreuther K, et al. The amyloid β-protein of Alzheimer's disease increases acetylcholinesterase expression by increasing intracellular calcium in embryonal carcinoma P19 cells. J. Neurochem. 1997;69:1177–1184.
12. Saez-Valero J, Fodero LR, White AR, et al. Acetylcholinesterase is increased in mouse neuronal and astrocyte cultures after treatment with β-amyloid peptides. Brain Res 2003;965:283–286.
13. Sberna G, Saez-Valero J, Li QX, et al. Acetylcholinesterase is increased in the brains of transgenic mice expressing the C-terminal fragment (CT100) of the β-amyloid protein precursor of Alzheimer's disease. J Neurochem 1998;71:723–731.
14. Saez-Valero J, Sberna G, McLean CA, Small DH. Molecular isoform distribution and glycosylation of acetylcholinesterase are altered in brain and cerebrospinal fluid of patients with Alzheimer's disease. J Neurochem 1999;72:1600–1608.
15. Saez-Valero J, Sberna G, McLean CA, et al. Glycosylation of acetylcholinesterase as diagnostic marker for Alzheimer's disease. Lancet 1997;350:929.
16. Saez-Valero J, Barquero MS, Marcos A, et al. Altered glycosylation of acetylcholinesterase in lumbar cerebrospinal fluid of patients with Alzheimer's disease. J Neurol Neurosurg Psychiatry 2000;69:664–667.
17. Saez-Valero J, Fodero LR, Sjogren M, et al. Glycosylation of acetylcholinesterase and butyrylcholinesterase changes as a function of the duration of Alzheimer's disease. J Neurosci Res 2003;72:520–526.

18. Fodero LR, Saez-Valero J, McLean CA, et al. Altered glycosylation of acetylcholinesterase in APP (SW) Tg2576 transgenic mice occurs prior to amyloid plaque deposition. J Neurochem 2002;81:441–448.

19. Dani JA. Overview of nicotinic receptors and their roles in the central nervous system. Biol Psychiatry 2001;49:166–174.

20. Albuquerque EX, Pereira EF, Castro NG, et al. Nicotinic receptor function in the mammalian central nervous system. Ann N Y Acad Sci 1995;757:48–72.

21. Mann EO, Greenfield SA. Novel modulatory mechanisms revealed by the sustained application of nicotine in the guinea-pig hippocampus in vitro. J Physiol 2003;551:539–550.

22. Wang HY, Lee DH, D'Andrea MR, et al. β-Amyloid(1-42) binds to α7 nicotinic acetylcholine receptor with high affinity. Implications for Alzheimer's disease pathology. J Biol Chem 2000;275:5626–5632.

23. Pettit DL, Shao Z, Yakel, JL. β-Amyloid(1-42) peptide directly modulates nicotinic receptors in the rat hippocampal slice. J Neurosci 2001;21:RC120.

24. Tozaki H, Matsumoto A, Kanno T, et al. The inhibitory and facilitatory actions of amyloid-β peptides on nicotinic ACh receptors and AMPA receptors. Biochem Biophys Res Commun 2002;294:42–45.

25. Liu Q, Kawai H, Berg DK. β -Amyloid peptide blocks the response of α7-containing nicotinic receptors on hippocampal neurons. Proc Natl Acad Sci USA 2001;98:4734–4739.

26. Dineley KT, Westerman M, Bui D, et al. β-Amyloid activates the mitogen-activated protein kinase cascade via hippocampal α7 nicotinic acetylcholine receptors: In vitro and in vivo mechanisms related to Alzheimer's disease. J Neurosci 2001;21:4125–4133.

27. Dineley KT, Bell KA, Bui D, Sweatt JD. β-Amyloid peptide activates α7 nicotinic acetylcholine receptors expressed in Xenopus oocytes. J Biol Chem 2002;277:25056–25061.

28. Yu WF, Guan ZZ, Bogdanovic N, Nordberg A. High selective expression of α7 nicotinic receptors on astrocytes in the brains of patients with sporadic Alzheimer's disease and patients carrying Swedish APP 670/671 mutation: a possible association with neuritic plaques. Exp Neurol 2005;192:215–225.

29. Xiu J, Nordberg A, Zhang JT, Guan ZZ. Expression of nicotinic receptors on primary cultures of rat astrocytes and up-regulation of the α7, α4 and β2 subunits in response to nanomolar concentrations of the β-amyloid peptide(1-42). Neurochem Int 2005;47:281–290.

30. Fodero LR, Mok SS, Losic D, et al. α7-nicotinic acetylcholine receptors mediate an Aβ(1-42)-induced increase in the level of acetylcholinesterase in primary cortical neurones. J Neurochem 2004;88:1186–1193.

31. Struble RG, Cork LC, Whitehouse PJ, Price DL. Cholinergic innervation in neuritic plaques. Science 1982;216:413–415.

32. Kitt CA, Price DL, Struble RG, et al. Evidence for cholinergic neurites in senile plaques. Science 1984;226:1443–1445.

33. Masliah E, Alford M, Adame A, et al. Aβ1-42 promotes cholinergic sprouting in patients with AD and Lewy body variant of AD. Neurology 2003;61:206–211.

34. Hernandez D, Sugaya K, Qu T, et al. Survival and plasticity of basal forebrain cholinergic systems in mice transgenic for presenilin-1 and amyloid precursor protein mutant genes. Neuroreport 2001;12:1377–1384.

35. Hu L, Wong TP, Cote SL, et al. The impact of Aβ-plaques on cortical cholinergic and non-cholinergic presynaptic boutons in alzheimer's disease-like transgenic mice. Neuroscience 2003;121:421–432.

36. DeKosky ST, Ikonomovic MD, Styren SD, et al. Upregulation of choline acetyltransferase activity in hippocampus and frontal cortex of elderly subjects with mild cognitive impairment. Ann Neurol 2002;51:145–155.

37. Ikonomovic MD, Mufson EF, Wuu J, et al. Cholinergic plasticity in hippocampus of individuals with mild cognitive impairment: correlation with Alzheimer's neuropathology. J Alzheimers Dis 2003;5:39–48.
38. Courtney C, Farrell D, Gray R, et al. Long-term donepezil treatment in 565 patients with Alzheimer's disease (AD2000): randomised double-blind trial. Lancet 2004;363:2105–2115.
39. Small DH. Do acetylcholinesterase inhibitors boost synaptic scaling in Alzheimer's disease? Trends Neurosci 2004;27:245–249.
40. Turrigiano GG. Homeostatic plasticity in neuronal networks: the more things change, the more they stay the same. Trends Neurosci 1999;22:221–227.
41. Davis GW, Goodman CS. Synapse-specific control of synaptic efficacy at the terminals of a single neuron. Nature 1998;392:82–86.
42. Ruppin E, Reggia JA. A neural model of memory impairment in diffuse cerebral atrophy. Br J Psychiatry 1995;166:19–28.
43. Storey E, Kinsella GJ, Slavin MJ. The neuropsychological diagnosis of Alzheimer's disease. J Alzheimers Dis 2001;3:261–285.

Chapter 46
Role of β-Amyloid in the Pathophysiology of Alzheimer's Disease and Cholinesterase Inhibition: Facing the Biological Complexity to Treat the Disease

Stefano Govoni, Michela Mazzucchelli, S. Carolina Lenzken, Emanuela Porrello, Cristina Lanni, and Marco Racchi

Introduction

The events leading to neurodegeneration in Alzheimer's disease (AD) brains are largely unknown, challenging the definition of disease-modifying treatment for AD. The current therapeutic option for AD patients is the use of acetylcholinesterase inhibitors (AChEIs), which provide symptomatic relief of some of the clinical manifestations of the disease. Published data suggest that AChEI can also affect putative pathogenetic pathways such as the expression and metabolism of the amyloid precursor protein (AβPP). In this chapter we review the evidence that these drugs, albeit with different quantitative and qualitative effects, can modulate the metabolism and expression of AβPP through mechanisms that involve either indirect cholinergic activation or noncholinergic mechanisms involving the expression of second messenger pathways that modulate AβPP processing.

From a neurochemical point of view, AD is characterized by a consistent deficit in cholinergic neurotransmission, particularly affecting neurons in the basal forebrain [1], a deficit that provided the rationale for the use of AChEIs in the therapy of the disease. Although by their nature AChEIs may offer unique symptomatic short-term intervention, the data emerging from long-term trials (mostly open label) with four such agents (donepezil, metrifonate, rivastigmine, and galantamine) [2] suggest that the clinical effect can be prolonged in some patients for up to 36 months. Stabilization of the cognitive status of the patients suggests an effect of the treatment on pathological features of the disease that go beyond mere substitution of the lacking acetylcholine through inhibition of the degrading enzyme. An action of AChEIs on the expression and/or metabolic processing of the amyloid precursor protein (AβPP), slowing the production of β-amyloid (Aβ), one of the major contributors to the disease

S. Govoni
Department of Experimental and Applied Pharmacology, Centre of Excellence in Applied Biology, Pavia 27100, Italy

A. Fisher et al. (eds.), *Advances in Alzheimer's and Parkinson's Disease*,
© Springer 2008

process [2–4], is the candidate mechanism involved in these putative disease-modifying actions of AChEIs. This hypothesis is reviewed herein.

Effect of Muscarinic Cholinergic Receptor Stimulation on Nonamyloidogenic Metabolism of AβPP

AβPP, an integral membrane protein with a complex proteolytic metabolism [4], generates Aβ by the sequential action of two enzymes: β-secretase and γ-secretase. α-Secretase cleaves AβPP in the Aβ sequence and releases a soluble N-terminal fragment, termed sAβPPα [3], in a process that constitutes the so-called nonamyloidogenic pathway. Cholinergic receptor stimulation promotes the nonamyloidogenic processing of AβPP, which can also lead to reduced formation and release of Aβ [4] suggesting that cholinergic agents may in fact prevent amyloid formation. Indeed, a cholinergic effect on sAβPPα secretion has been also demonstrated for AChEIs, as recently reviewed [3,5] and shown in Fig. 1.

The data reported in Fig. 1 show that the effect exerted by the various AChEIs tested (metrifonate, donepezil, galantamine, ganstigmine) on sAβPP release is concentration-dependent, even if the concentration producing the maximum effect varies among the molecules by at least one order of magnitude.

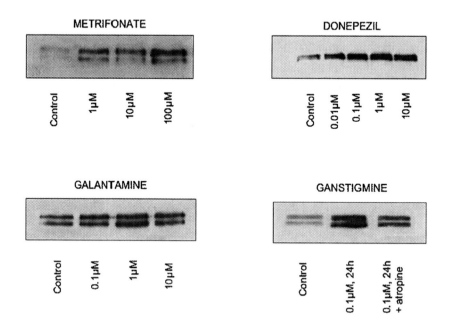

Fig. 1 Effect of three cholinesterase inhibitors on sAβPP secretion from SH-SY5Y neuroblastoma cells and reversal by atropine. See text for further details

In the case of metrifonate, a correlation of the effect on sAβPP release with acetylcholinesterase (AChE) inhibition [7] was shown, but other mechanisms may also participate in promoting the nonamyloidogenic processing (see below). In the case of metrifonate [7] and of ganstigmine (Fig. 1), the effect can be antagonized by atropine, a cholinergic, muscarinic receptor antagonist.

Effect of AChEIs on Expression and Metabolism of AβPP

The results of experimental studies aimed at demonstrating the effect of AChEIs on AβPP metabolism show variability even if the prevailing results suggest stimulation of the secretion of the nonamyloidogenic sAβPPα (Table 1) [5]. Most of the molecules included in the studies cited here and listed in Table 1 are used currently in AD pharmacotherapy, are investigational drugs under clinical evaluation, or were tested in the past as therapeutics for AD (reviewed in Racchi et al. [5]). Notably, among the compounds reported in Table 1 the three AChEIs currently used in AD therapy—donepezil, rivastigmine, galantamine—all stimulate nonamyloidogenic APP processing.

Tacrine was the first drug used to demonstrate modulation of AβPP processing and expression. Somewhat surprisingly, treatment with tacrine inhibited secretion of soluble AβPP derivatives, suggesting that AChEIs may influence negatively the processing of AβPP. However, subsequent experiments with different cholinesterase inhibitors provided evidence that several of them were able to increase the nonamyloidogenic processing of AβPP albeit with relevant differences in the extent and time course of the

Table 1 Effects of AChE inhibitors on AβPP processing and expression

AChE inhibitor	sAβPPα release	Aβrelease	AβPP holoprotein
Metrifonate	↑/↔	↔/↓	↔/↑
Ganstigmine	↑	Nt	↔
Donepezil	↑	Nt	↔
Tacrine	↓	↓	↓/↔
Rivastigmine	↑	Nt	nt
Galantamine	↑	Nt	nt
Physostigmine	↑/↓	↔	↓
Heptylphysostigmine	↑	Nt	nt
Rasagiline analogs	↑	Nt	↔
3,4-Diaminopyridine	↑/↔	↓	↔
Ambenonium	↑	Nt	↑
Phenserine	↓	↓	↓/↔

AChE, acetylcholine esterase; AβPP, β-amyloid precursor protein; sAβPPα, N-terminal soluble fragment; Aβ, β-amyloid.
The symbols indicate an increase (↑), a decrease (↓), or no change (↔) in the parameter indicated. nt, the parameter has not been measured.
See Racchi et al. [5] for further details.

effect (reviewed in Racchi et al. [5]). For example, both metrifonate and ganstigmine increase the levels of soluble AβPP released by the cells in an atropine-sensitive fashion [7,8] (Fig. 1), but metrifonate promotes the release of sAβPP upon short-term treatment [7], and ganstigmine modifies consistently the constitutive release of sAβPP following 24 hours of treatment with only a relatively mild effect in the short term [8]. In both cases, exposure to metrifonate or ganstigmine does not modify APP mRNA levels or isoform ratio, suggesting that the effect of the drug could be limited to a posttranscriptional event [7,8]. The effect of metrifonate and ganstigmine is blocked by atropine and can be ascribed to an indirect cholinergic mechanism involving activation of a protein kinase C (PKC)-dependent signal transduction mechanism [7,8]. Alternatively, some drugs appear to act on AβPP processing via mechanisms that are independent from their activity as AChEIs, such as donepezil, whose effect on AβPP processing appears to be related to modulation of the intracellular trafficking of ADAM 10 [9], or galantamine, which promotes sAβPPα release in cells devoid of AChE activity and cholinergic receptors (Racchi et al., unpublished results).

Signal transduction pathways are particularly interesting when related to the effects of AChEIs on AβPP metabolism. Metrifonate can modulate the levels of PKC [10], and two investigational AChEIs—TV3326 [{N-propargyl-(3R) aminoindan-5-yl}-ethyl methyl carbamate] and TV3279 [{N-propargyl-(3S) aminoindan-5-yl}-ethyl methyl carbamate]—promote the release of sAβPa via the activation of PKC, mitogen-activated protein (MAP) kinase, and tyrosine kinase-dependent pathways [11]. The fact that several molecules can act independently from their activity as an AChEI suggests alternative pathways of modulation of AβPP processing. Another example is the original mechanism of action described for phenserine, which decreases cell-associated AβP levels through interaction with an iron-regulatory element in the 5′-untranslated region of AβP mRNA [12].

Evidence of Gene Expression Modulation by AChEIs

We gained preliminary evidence that AChEIs can modulate the gene expression pattern in neuroblastoma cells. Macroarray analysis (not shown) of mRNA extracted from cells treated with various AChEIs showed that some of them can up-regulate the expression of PKC. In the experiment reported in Fig. 2, SH-SY5Y cells treated with galantamine show increased levels of PKC mRNA and protein. Interestingly, galantamine increased the expression of two kinases, PKCα and PKCε, both of which, upon stimulation, can promote nonamyloidogenic processing of AβPP. However, the increase of PKCε is of particular interest because we recently demonstrated [13] that this isoform is specifically involved in the cholinergic modulation of AβPP processing. These data suggest that another mechanism, through which some AChEIs can favor the

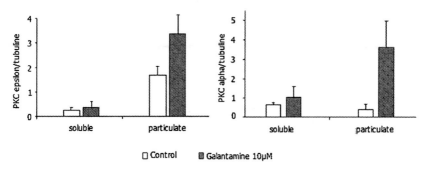

Fig. 2 Galantamine (10 μmol) treatment increases protein kinase C (PKCα, PKCε) expression in SH-SY5Y neuroblastoma cells

nonamyloidogenic processing of AβPP, is through increased expression of enzymes regulating this metabolic pathway.

Conclusions

Substantial evidence has been produced to suggest a significant interaction of AChEIs with the metabolism of AβPP through both cholinergic and non-cholinergic pathways. This may include control of AβPP expression or the expression of proteins participating in signal transduction pathways, which in turn modulate AβPP processing. In almost all cases, altered or modulated AβPP processing reaches the ultimate goal to reduce A production or the production of other amyloidogenic fragments. The major task in future will be to understand whether these mechanisms represent only an interesting aspect of their pharmacological profile or are related also to the clinical action of this class of drugs. In addition, identification of novel specific and unique mechanisms of action of AChEIs related to AβPP processing and Aβ formation may lead to the discovery of new molecules of potential therapeutic application in AD research.

Acknowledgments This work was made possible through grants from the Italian MIUR (prot. #2003057355-2003 and prot. #MM05221899-2000 to S.G.); from the University of Pavia (FAR 2003 to M.R.; Progetto Giovani Ricercatori to M.M., C.L., and E.P.); and from the Italian Ministry of Health (Progetto Finalizzato Alzheimer to S.G. and M.R.). S.C.L. is a recipient of an I.R.E. fellowship from the Italian Embassy in Argentina.

References

1. Perry E, Ziabreva I, Perry R, et al. Absence of cholinergic deficits in "pure" vascular dementia Neurology 2005;64:132–133.

2. Giacobini E. Long-term stabilizing effect of cholinesterase inhibitors in the therapy of Alzheimer' disease. J Neural Transm Suppl 2002;62:181–187.
3. Racchi M, Govoni S. The pharmacology of amyloid precursor protein processing. Exp Gerontol 2003;38(1-2):145–157.
4. Nitsch RM, Slack BE, Wurtman RJ, Growdon JH. Release of Alzheimer amyloid precursor derivatives stimulated by activation of muscarinic acetylcholine receptors. Science 1992;258:304–307.
5. Racchi M, Mazzucchelli M, Porrello E, et al. Acetylcholinesterase inhibitors: novel activities of old molecules. Pharmacol Res 2004;50(4):441–451.
6. Lahiri DK, Farlow MR, Nurnberger JL, Greig NH. Effects of cholinesterase inhibitors on the secretion of beta-amyloid precursor protein in cell cultures. Ann N Y Acad Sci 1997;826:416–421.
7. Racchi M, Sironi M, Caprera A, et al. Short- and long-term effect of acetylcholinesterase inhibition on the expression and metabolism of the amyloid precursor protein. Mol Psychiatry 2001;6(5):520–528.
8. Mazzucchelli M, Porrello E, Villetti G, et al. Characterization of the effect of ganstigmine (CHF2819) on amyloid precursor protein metabolism in SH-SY5Y neuroblastoma cells. J Neural Transm 2003;110(8):935–947
9. Zimmermann M, Gardoni F, Marcello E, et al. Acetylcholinesterase inhibitors increase ADAM10 activity by promoting its trafficking in neuroblastoma cell lines. J Neurochem 2004;90(6):1489–1499.
10. Pakaski M, Rakonczay Z, Fakla I, et al. In vitro effects of metrifonate on neuronal amyloid precursor protein processing and protein kinase C level. Brain Res 2000; 863:266–270.
11. Yogev-Falach M, Amit T, Bar-Am O, et al. Involvement of MAP kinase in the regulation of amyloid precursor protein processing by novel cholinesterase inhibitors derived from rasagiline. FASEB J 2002;16(12):1674–1676.
12. Shaw KT, Utsuki T, Rogers J, et al. Phenserine regulates translation of beta -amyloid precursor protein mRNA by a putative interleukin-1 responsive element, a target for drug development. Proc Natl Acad Sci U S A 2001;98(13):7605–7610.
13. Lanni C, Mazzucchelli M, Porrello E, et al. Differential involvement of protein kinase C alpha and epsilon in the regulated secretion of soluble amyloid precursor protein. Eur J Biochem 2004;271(14):3068–3075.

Chapter 47
Dissociation Between the Potent β-Amyloid Protein Pathway Inhibition and Cholinergic Actions of the Alzheimer Drug Candidates Phenserine and Cymserine

Nigel H. Greig, Tada Utsuki, Qian-sheng Yu, Harold W. Holloway,
Tracyann Perry, David Tweedie, Tony Giordano, George M. Alley,
De-Mao Chen, Mohammad A. Kamal, Jack T. Rogers, Kumar Sambamurti, and
Debomoy K. Lahiri

Introduction

Alzheimer's disease (AD) is characterized by selective neuronal loss, widespread deposition of amyloid fibrils that are comprised of a 39-43 amino acid amyloid-β peptide (Aβ), and the formation of neurofibrillary tangles in the brain of afflicted individuals [1–4]. Currently, cholinesterase inhibitors (ChEIs) are the primary strategy for treating mild to moderate AD subjects in the United States [5–11], to which the recently approved N-methyl-D-aspartate (NMDA) antagonist memantine is sometimes added [8,12]. However, conjoint with the increasing understanding of the complex pathology of AD over recent years, there is an emergence of more novel targets for therapy [3,4,7,9,10,13]. Such targets are based on the actions of Aβ, the inflammatory cascade, and the role of tau proteins in the formation of neurofibrillary tangles, oxidative neuronal damage, and neurotransmitter depletion.

Because both Aβ precursor protein (APP) processing and cholinesterase activity are affected in the AD brain [2–4,7], we examined the effects of several clinically relevant drugs on the APP pathway to elucidate whether common molecular mechanisms linked the two targets. Current research indicates that Aβ plays a key role in the progressive neurodegeneration observed in AD [1–4,7]. The smaller Aβ peptide [molecular weight (MW) ~4.1 kDa] is generated proteolytically from a larger integral membrane protein called APP (about 110-130 kDa). Specifically, APP is cleaved by three enzymes—α-, β-, and γ-secretases—to different protein fragments, including toxic Aβ and several carboxyl-terminal fragments that, additionally, may be involved in the pathogenesis of AD [1–4].

Our goal has been to investigate, design, and synthesize various classes of agents that can potentially lower APP expression and levels, as APP is the

N.H. Greig
Drug Design & Development Section, Laboratory of Neurosciences, Intramural Research
Program, National Institute on Aging, Baltimore, MD 21224, USA

A. Fisher et al. (eds.), *Advances in Alzheimer's and Parkinson's Disease*,
© Springer 2008

originator to all the toxic fragments. Over the last several years, we have examined the effects of a dozen clinically useful drugs/compounds on the amyloid pathway [2,3,7,14–20]. In the present study, our evaluation of the drugs' effects was focused at four levels: (1) APP itself; (2) the toxic cleaved products of APP, explicitly Aβ; (3) anticholinesterase activity; and (4) the chemical structure of the ChEI.

We have developed a family of unique cholinesterase inhibitors, phenserine and analogs, that utilize the optically active tricyclic backbone of the natural product and the classic ChEI, physostigmine (also known as eserine) [14,16] (Fig. 1). They differ from physostigmine, however, on the basis of a common phenylcarbamate structure that, for phenserine, is unsubstituted and for its close analog, cymserine, has an isopropyl in the para (4′) position [21–25]. In contrast to physostigmine, which is short acting and has a brain/plasma concentration ratio of unity [14,15], phenserine and cymserine are reversible and long-acting anticholinesterases that have a brain/plasma concentration ratio of 10:1 and 40:1 and selectivity for acetylcholinesterase (AChE) and butyrylcholinesterase (BuChE), respectively [14,16,22]. Consequently, phenserine has been demonstrated to improve cognitive performance dramatically in rodents over an unusually wide dose range [14,16]. In addition, phenserine administration reduced brain APP production in naive animals and prevented a rise in APP brain levels, which are associated with forebrain cholinergic lesions in rats, a model that mimics the cholinergic loss found in the AD brain [26]. Phenserine is currently being assessed in clinical trials for mild to moderate AD potentially to perturb the cognitive and behavioral symptoms associated with AD and lower brain levels of Aβ [16]. Cymserine analogs are in preclinical development for the potential treatment of severe AD, when BuChE levels are dramatically elevated

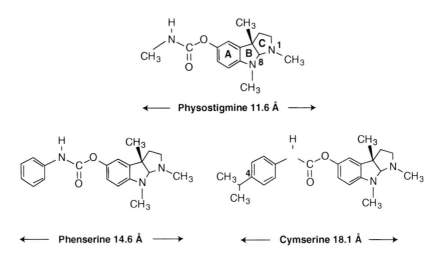

Fig. 1 Chemical structures of physostigmine (top), phenserine (bottom left), and cymserine (bottom right). Hexahydropyrroloindole marked as tricyclic A, B, C rings; N^1 and N^8 nitrogens; and 4′ (para) position are shown

and mismatched to dwindling levels of acetylcholine and AChE [27,28]. In contrast, AD clinical trials with physostigmine, available as a slow-release oral formulation (Synapton; Forest Laboratories, New York, NY, USA) were suspended based on its poor toxicity profile [29].

Consequent to the fact that both APP processing and cholinesterase activity are affected in the AD brain, a focus for our current studies is the molecular changes induced by ChEIs [17–20,30]. In this regard, our present work is a continuation of earlier research on tacrine, a first-generation aminoacridine ChEI with beneficial cognitive action in the treatment of AD that, additionally, showed activity on the APP pathway [19,20]. Specifically, tacrine treatment reduced the levels of secreted APP and Aβ of neural cells in culture; but whether this effect was related to tacrine's chemical structure or anticholinesterase action remained unclear. We therefore have studied the mechanism through which other structurally divergent ChEI drugs, such as phenserine, interact with the cellular processing of APP [17,18,30–33]. Herein, we report that phenserine and cymserine, close structural analogs of physostigmine, reduced levels of APP and Aβ, whereas physostigmine had no effect on neuroblastoma cells. Taken together, the prior tacrine action and the differential effects of three structurally related compounds (phenserine, cymserine, physostigmine) on APP pathways contrast with their similar potency to inhibit either AChE or BuChE, thereby dissociating these two pharmacological actions. Elucidating the mechanism(s) via which selected ChEIs alter APP synthesis/processing and hence lower Aβ deposition may allow optimization of their use and, additionally, aid in the design of better drugs for AD treatment.

Methods

Chemicals

Phenserine ([3aS]- or [–] -phenserine) and cymserine ([3aS]- or [–]-cymserine) were synthesized as their water-soluble tartrate salts, as described previously [21–25] and were of high (> 99.9%) chemical and optical purity. Physostigmine ([3aS]– or [–]-physostigmine) was purchased as eserine salicylate from Sigma (St. Louis, MO, USA), and all other chemicals were of the highest purity and were obtained from Sigma unless stated otherwise.

Lipophilicity Determination

Octanol-water partition coefficients were determined both experimentally and by computation. Octanol solutions (0.5 mM), 5 ml, of phenserine, cymserine, and physostigmine were prepared, and their ultraviolet (UV) absorbences, A1, were determined by spectrophotometer at 254 nm wavelength. The octanol solutions then were vigorously mixed with an equal volume of 0.1 M phosphate buffer (pH 7.4) for 15 minutes. Following separation by centrifugation and appropriate drying, the absorbence of the octanol was again measured, A2. An octanol-water partition coefficient, P, was calculated from the formula: $P = A2/A1 - A2$

Cell Culture

Human neuroblastoma (SK-N-SH) cells were obtained from the American Type Culture Collection (ATCC) and cultured in minimal essential medium (MEM) containing 10% fetal bovine serum (FBS) to 70% confluence, as described elsewhere [18–20]. To initiate the experiment, cells were first fed with fresh medium with low FBS (0.5%) and treated separately with vehicle, phenserine, cymserine, or physostigmine at defined concentrations for specific periods of time (16 and 24 hours).

LDH and MTT Assays

Following drug treatment, cells were allowed to grow, and conditioned medium samples were collected at 16- to 24-hour intervals. The released lactate dehydrogenase (LDH) was measured in the conditioned medium samples. Cells were harvested at the end of the experiment. The cells were resuspended and immediately assayed for MTT [3-(4,5-dimetyl-thiazol-2-yl)-2,5-diphenylte-trazolium bromide]. Both LDH and MTT measurements were undertaken using sensitive and quantitative methods within their linear ranges, as described previously [18,19].

APP Measurement

Levels of secretory APP were assayed in the conditioned medium by Western immunoblotting using 22C11 antibody (Roche), as described previously [18,19].

Aβ Assay

Levels of total Aβx-40 peptide levels were determined in the conditioned medium samples using a sandwich enzyme-linked immunosorbent assay (ELISA) method after the modification of IBL reagents [18–20]. Aβx-40 is the most abundant species among all Aβ isoforms secreted and can be detected in this cell line.

Cholinesterase Measurements

The inhibition of freshly prepared human AChE and BuChE by phenserine and physostigmine was independently measured between the concentrations of 0.3 nM and 10 μM to obtain both an IC50 value (concentration required to inhibit 50% enzyme activity) and the level of enzyme inhibition obtained in the described cell culture studies, as described previously [21–25].

Results

Phenserine, Cymserine, and Physostigmine Assessed at Nontoxic Doses in Human Neuronal (SK-N-SH) Cell Cultures

Human neuroblastoma cells were incubated with vehicle, phenserine, cymserine, or physostigmine as described in the Methods section. When monitoring cells morphologically under a phase contrast microscope, we observed no difference between control and drug-treated cells. This was confirmed by subjecting the cell culture samples to various biochemical viability assays. Cell membrane damage and integrity were measured by assaying the LDH released into the conditioned medium. This assay is both sensitive and quantitative and was performed under linear conditions. Toxicity of the drug is evidenced by a significant increase in LDH activity in drug-treated cells compared to controls, indicative of damage to the cell membrane and loss of cellular integrity. We

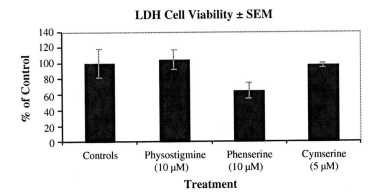

Fig. 2 Comparative effects of physostigmine, phenserine, and cymserine on cell toxicity as measured by lactate dehydrogenase (LDH) assay. All cells were viable and there were no significant increases in LDH levels, associated with a loss of cellular viability. $p > 0.05$, Dunnett's t-test

observed no significant increase in the LDH value between the phenserine-, cymserine-, or physostigmine-treated cells and control samples (Fig. 2).

Effect of Phenserine, Cymserine, and Physostigmine on Cell Viability

In addition to the described cellular toxicity measure, drug effects on cell viability were determined by a MTT test, which measures mitochondrial function, metabolic activity, and thus cellular viability. This assay, likewise, is both sensitive and quantitative; and an increased MTT value from the control is indicative of increased cellular metabolic activity and hence greater viability of the cell under the assay conditions. Based on our MTT results (data not shown), we did not observe any significant change in cell viability with any drug treatment compared to the vehicle control. However, physostigmine treatment resulted in a modest increase in cell viability versus the control. In summary, our MTT results were consistent with the LDH data and demonstrated that all treatments were non-toxic to the cells under the dose and cell culture conditions tested.

Differential Effects of Phenserine, Cymserine, and Physostigmine on Soluble APP Levels

To investigate the effects of these drugs on APP processing pathways, we measured levels of secretory APP, the form predominantly produced in this cell line, in the same conditioned medium samples that were used to assay LDH. Our Western immunoblotting with a monoclonal antibody against total APP revealed several high-molecular-weight bands of 110 to 130 kDa that consisted of different alternatively spliced variants of APP and their posttranslationally modified forms. Notably, phenserine and cymserine treatment reduced levels of APP by 50% and 46% from the control, respectively (Fig. 3). This reduction appeared to occur in all bands of the immunoblot, and action on different isoforms is the focus of current studies.

In contrast, physostigmine treatment did not significantly alter the level of sAPP under the same conditions (Fig. 3). However, as there appeared to be a suggestion of lowered sAPP, although not of statistical significance, a higher concentration of physostigmine (50 μM) was assessed alongside phenserine (50 μM) at 16 hours under similar culture conditions. As illustrated in Fig. 4, the sAPP levels were reduced by phenserine (by 63.9%, $p < 0.05$) but not physostigmine (by 8.4%, $p > 0.05$). In each case the high band was more affected than the low one (phenserine reduction 73.4% and 50.3%, respectively; $p < 0.05$; physostigmine reduction 10.2% and 6.6%, respectively; $p > 0.05$).

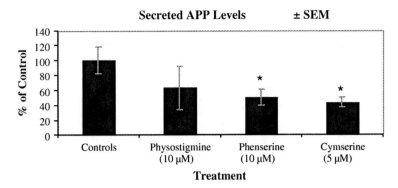

Fig. 3 Comparative actions of physostigmine, phenserine, and cymserine on levels of secreted APP, as quantified by Western immunoblotting (22C11 mAb). The latter two treatments significantly lowered secreted APP levels. *$p < 0.05$ vs. control, Dunnett's t-test

Fig. 4 Comparative actions of 50 μM physostigmine and phenserine on secreted APP levels, quantified by Western immunoblotting (22C11 mAb). *$p < 0.05$ vs. control and physostigmine, Dunnett's t-test

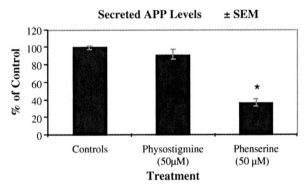

Differential Effects of Phenserine, Cymserine, and Physostigmine on Soluble Aβ

To determine whether the effect of these drugs on sAPP extended to Aβ under the same cell culture conditions, we measured levels of Aβ in the identical conditioned medium samples by a sensitive ELISA. We have previously shown that the human neuroblastoma cells (SK-N-SH) used herein secrete a significant amount of basal Aβ into the conditioned medium, and thus the cell line represents an ideal model to test the effects of potential AD drugs on Aβ levels [17–20,30]. As shown in Fig. 5, phenserine and cymserine treatment reduced levels of Aβx-40 by 36% and 40% from the control value, respectively, whereas physostigmine treatment marginally increased (9%) the level of Aβ.

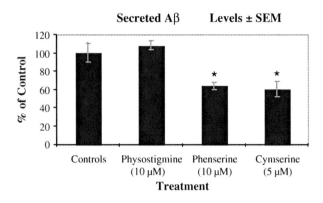

Fig. 5 Effects of physostigmine, phenserine, and cymserine on levels of Aβx-40, measured by a sandwich enzyme-linked immunosorbent assay (ELISA). Phenserine and cymserine significantly lowered secreted Aβ. *$p < 0.05$ vs. controls and physostigmine, Dunnett's t-test

Effects of Phenserine and Physostigmine on Cholinesterase Inhibition

To quantify the level of cholinesterase inhibition associated with the three compounds, we measured their concentration-dependent inhibition of both AChE and BuChE in freshly collected samples of human erythrocytes and plasma, respectively. Inhibition was measured in half-log steps between 1 nM and 10 mM. As shown in Table 1, both phenserine and physostigmine potently inhibited AChE, with similar I C_{50} values. Furthermore, cymserine and physostigmine potently inhibited BuChE. Thus, physostigmine induced approximately similar inhibition of both cholinesterase subtypes, whereas phenserine induced selective inhibition of AChE and cymserine of BuChE. Hence, under the cell culture conditions described, the cholinesterase enzymes were substantially inhibited by phenserine, cymserine, and physostigmine.

Table 1 Effects of phenserine, cymserine, and physostigmine on human AChE and BuChE inhibition

| | AChE inhibition | | BuChE inhibition | |
Treatment	10 μM(%)	AChE IC_{50}[a]	10 μM(%)	BuChE IC_{50}
Physostigmine	100	28 nM	100	14 nM
Phenserine	100	22 nM	85	1.75 μM
Cymserine	100[b]	0.75 μM	△100	50 nM

AChE, acetylcholinesterase; BuChE, butyrocholinesterase.
[a] IC_{50} concentration required to inhibit 50% of enzyme activity.
[b] Similar inhibition at 5 μM.

Pharmacologic Properties of Three Closely Related Carbamate Drugs

Phenserine, a phenylcarbamate of (−)-eseroline, compared to physostigmine, a methylcarbamate (Fig. 1), is a closely related analog with high selectivity for AChE versus BuChE. It is currently being tested in human clinical trials for the treatment of mild to moderate AD [24,26]. Cymserine represents 4′-isopropyl substituted phenserine (Fig. 1), which provides a close structural analog with a preference for BuChE. Compared to physostigmine or tacrine, these phenyl-carbamates are less toxic in animals (LD_{50} = physostigmine 0.6 mg/kg, tacrine 12 mg/kg, phenserine 25 mg/kg and cymserine >30 mg/kg [14–16]). Notably, phenserine robustly enhances cognition in animal models over a broad dose range and is effective at some 2 log concentrations lower than its toxic dosage [16,34]. The replacement of the methylcarbamates of physostigmine with an unsubstituted and substituted phenylcarbamate in phenserine and cymserine, respectively, not only alters the cholinesterase subtype selectivity of the compounds but dramatically impacts their pharmacokinetics [14,16,23]. The lipophilicity of the latter compounds is dramatically greater than that of physostigmine, as reflected in their octanol-water partition coefficients shown in Table 2, where computed and experimentally determined values closely match. Thus, whereas all have sufficient lipophilicilty to pass across biological membranes [35,36], phenserine and cymserine would be expected to do so far more readily.

It has been shown previously that phenserine is absorbed rapidly in rodents and has a bioavailability of 100% compared to 2% for physostigmine after oral administration [15,16]. Like physostigmine, it is rapidly cleared from the body with a pharmacokinetic half-life of 10 to 30 minutes [15,16]. In contrast, phenserine induces an extended duration of AChE inhibition (half-life is 8 hours versus 30 minutes for physostigmine [15,16]) and thereby produces long-lasting stimulation of brain cholinergic function at well tolerated doses. The same properties are shared by cymserine but with a preference for BuChE inhibition. Such analogs have been shown to likewise stimulate central cholinergic activity and readily enter the brain [28]. Indeed, the long duration of action of these phenylcarbamates (numerous hours) coupled with their comparatively short pharmacokinetic half-life (an hour or less) has the potential to reduce dosing frequency, decrease drug exposure, and minimize the dependence

Table 2 Computed and experimentally determined lipophilicity of phenserine, cymserine, and physostigmine

Compound	Clog p	A1	A2	p^*
Physostigmine	0.66	2.418	1.757	2.7
Phenserine	2.22	2.283	2.263	113
Cymserine	3.51	2.312	2.310	1155

of drug action on the individual variations of drug metabolism commonly found in the elderly [15]. These interesting combined properties provide the potential to test the value of either AChE or BuChE inhibition in the treatment of AD, utilizing a pharmacophore that is relatively unencumbered by the dose-limiting adverse actions associated with former ChEIs.

Discussion

As the population of the developed world ages, there is a consequent increasing prevalence of age-related neuropsychiatric disorders, particularly dementias of any cause and AD, for which few treatments are presently available. Current U.S. Food and Drug Administration (FDA)-approved drugs that are based solely on cholinesterase inhibition, together with the NMDA antagonist memantine, have limited value in reducing or halting the progression of AD and primarily provide symptomatic treatment [5,8,37]. Thus, the most important therapeutic effect of current ChEIs in approximately 50% of AD patients is stabilization of their cognitive function, which is maintained at a steady level during a 1-year period of treatment compared to placebo [38]. The true value of this has been a question of recent debate [39,40]. Hence, to gain more than this benefit, other targets for AD drug development are being sought that affect the disease course.

Clearly, one of the targets is based on the known toxic actions of β-amyloid (Aβ) [1–4]. However, based on our expanding basic and clinical knowledge, there are suggestions that the routine use of antiinflammatory agents, antioxidants, free-radical scavengers, cholesterol-lowering statins, certain estrogen preparations, and perhaps cholinomimetics may affect the development of AD by altering pathways centered on Aβ as well as biochemical cascades unassociated with it [4,7,9–11,41]. In addition to antiamyloid strategies, several other noncholinergic agents are currently under development for the treatment and/or prevention of AD, such as transition metal chelators (e.g., clioquinol), herbs (e.g., Ginkgo biloba and curcumin), other neurotransmitter approaches, and vitamin E [42–47]. Whether the specific mechanisms activated by these approaches directly or secondarily stimulate similar pathways initiated by cholinergic therapy or alter Aβ synthesis, deposition, or clearance remains to be elucidated.

The present study was aimed at comparing the effects of the structurally related agents phenserine, cymserine, and physostigmine on amyloid pathways to gain insight into the association between cholinesterase inhibition and the regulation of APP/Aβ levels. These compounds share two important properties: They display anticholinesterase activity, and they are structurally related heterocycle carbamates; specifically, they share the same hexahydroprrolo [2,3b] indole backbone (Fig. 1, ABC tricyclic ring structure) that is connected via the 5-position to closely related carbamate moieties [26,27]. Despite their pharmacokinetic and pharmacodynamic differences, it is their structural similarity together with specific chemical interactions between identical groups within both the carbamate and hexahydropyrroloindole structures of each agent that provides their anticholinesterase action. In this regard, the binding domains and interactions for phenserine and physostigmine with AChE are one and the same. The same is true of cymserine and physostigmine for BuChE [48].

Human AChE and BuChE are large complex proteins comprising one or more catalytic subunits that each contain up to 583 amino acids; they have a relative MW (Mr) of 70 to 80 kDa and are variably glycosylated [49,50]. Even though they have a number of common physiological functions (some cholinergic and others noncholinergically mediated [48,49]) and share a 65% homology in their amino acid sequence, they are products of discrete genes on disparate chromosomes, explicitly, on 3 (3q26) and 7 (7q22) in humans, respectively [48–50]. X-ray crystallography of each enzyme has characterized their three-dimensional architecture and provided structural information regarding the positioning of the catalytically important amino acid residues in them [51–54]. Thus, three major binding domains have been described in AChE and two in BuChE that lie in an internalized and primarily hydrophobic gorge of some 20 Å length that intrudes into each enzyme. Differences in the amino acids that provide the three-dimensional constraints within these gorges account for the contrasting BuChE versus AChE inhibitory activity of 4'-versus unsubstituted phenylcarbamates of eseroline (i.e., cymserine versus phenserine). Deepest in this gorge of both enzymes is a catalytic "acyl" binding domain, which hydrolyzes choline esters through electron transfer in a catalytic triad, termed a "charge relay system." The triad includes a Ser_{200} [Torpedo californica (Tc) AChE amino acid numbering], the imidazole moiety of His_{440}, and the carboxylic acid group of Glu_{327}. A "choline" binding domain resides midway along the gorge, which is centered around Trp_{84}, and a "peripheral" anionic binding site exists at the gorge mouth for AChE but not BuChE [48–54] and appears to have involvement in both ACh regulation and AChE binding with Aβ peptide to form toxic complexes [55].

The choline and esteratic binding sites in both AChE and BuChE interact identically, with common elements of all three compounds studied herein: the carbonyl function ($C=O$) with the esteratic site and the basic N1-nitrogen (Fig. 1, C ring) with the choline site [23,25]. This allows orientation and binding of the three compounds to either AChE, BuChE, or both enzymes to inhibit the interaction and hydrolysis of acetylcholine (ACh), which occurs at the same sites [25]. This defined structure/activity relation (SAR) provides the potential to test preliminary SAR between the agents and effects on the APP secretory pathway.

Several possible outcomes can be predicted from such studies. First, phenserine, cymserine, and physostigmine could lack APP effects. Such a result would dissociate the two phenomena (APP and anticholinesterase activities/mechanisms). Second, the compounds could have similar regulatory effects on APP, either up- or down-regulation. This scenario would support a likelihood of convergence in the SAR between the anticholinesterase and APP-lowering properties of the agents, signifying a possible functional involvement between the two with the potential of shared mechanisms. Third, the agents could differ in their APP regulatory actions. In such an event, the anticholinesterase and APP functions are most probably independent and mediated by disparate mechanisms. Each circumstance would provide SAR insights related to the similarities and differences between the structural analogs. Our results on APP and Aβ for phenserine, cymserine, and physostigmine are indicative of the third state of affairs but for the former two compounds are supportive of the second scenario.

The results reported herein on the action of phenserine on APP and Aβ are consistent with earlier observations [18,30–32]. Previous cell culture studies with human neuroblastoma cell lines, glial cell lines, and PC12 cells have demonstrated that phenserine can consistently reduce both cellular and secreted APP levels. Further detailed studies in human neuroblastoma cells verified that the reduction was both time- and concentration-dependent and occurred without loss in cell viability (similarly measured by quantifying LDH release into the conditioned medium versus untreated controls) [18,30]. Declines in both cellular and secreted APP were associated with a significant reduction in Aβ release, as measured by sandwich ELISA, in both this and prior studies [18,30] and appear to translate to findings in animals [26].

In the current comparative cell culture study, three agents were assessed under similar in vitro conditions, and the divergence of their actions on the cholinesterases and on APP suggests that these defined actions are independent and that a noncholinergic mechanism likely accounts for the action on APP. In this regard, phenserine and cymserine cover the gambit of anticholinesterase actions associated with physostigmine. They potently inhibit AChE and BuChE, as assessed by their IC_{50} values and concentrations used in cell culture, which, by blocking their catabolic action would augment ACh levels and allow activation of both muscarinic and nicotinic receptor-mediated pathways. Whereas the stimulation of these receptors and their triggered biochemical cascades has been shown to alter secreted APP levels [56–59], and would be achieved equally by all three studied compounds, such mechanisms were clearly insufficient alone to affect the changes observed by phenserine and cymserine as assessed by a lack of APP action by physostigmine. Parenthetically, stimulation of m1 and m3 muscarinic ACh receptors increases APP secretion primarily via protein kinase C, as does electrical stimulation [56–59]. Nicotine has been reported to increase secreted APP levels at high doses; and at low and high doses, it lowers secreted APPγ levels, the APP containing the amyloidogenic portion of Aβ [60,61]. It could be that such cholinergically mediated actions are both relevant and evident acutely after stimulation with all three compounds studied but are no longer apparent under relatively steady-state conditions, as achieved by 16 to 24 hours of continuous treatment [2].

The exact mechanism underpinning the differential action of phenserine/cymserine versus physostigmine on APP remains to be fully elucidated and is a focus of our current research. The structures of the compounds assessed herein are closely analogous, varying only in their carbamate moieties [14,16,23,25] (Fig. 1). In this regard, the phenylcarbamates associated with phenserine and cymserine provide greater lipophilicity (Table 2), allowing greater brain delivery [16], and they each provide greater spatial volume versus physostigmine, which has been defined by X-ray crystallography [23,25]. Indeed, it is the phenylcarbamate group and position of substitution (4' for cymserine) that underpins AChE versus BuChE inhibitory action of this class of compounds. Explicitly, the acyl-binding pockets in BuChE and AChE differ from one another in shape and size at their deepest point and can be discriminated by two amino acids positioned at the bottom of the acyl loop: Phe_{288} and Phe_{290} in AChE and Leu_{286} and Val_{288} in BuChE [35–40]. The former amino acids are larger, consequent to their aromatic phenyl residues, whereas those of the latter are smaller and aliphatic. Accordingly, the acyl pocket in BuChE is larger owing to the smaller protruding amino acids; and the 4'-isopropyl substitution of cymserine that elongates the compound (18.1 Å length) compared to unsubstituted phenserine (14.6 Å length) renders cymserine its enzyme subtype selectivity [22,48]. On the contrary, the phenylcarbamate moiety of unsubstituted phenserine can form a π–π interaction with the phenyl groups of Phe_{288} and Phe_{290}, which increases the stability of the transition state associated with its binding, to favor AChE selectivity [23,25]. Physostigmine, with its small methylcarbamate and associated spatial volume is able to potently inhibit either enzyme subtype without such steric restrictions and interactions [25].

Moreover, the presence of such a bulky aromatic residue in phenserine and cymserine allows both hydrophobic and π electron interactions with peptide and protein targets on the basis of potential π–π stacking of the phenyl moiety of the carbamate between closely flanking phenylalanines, tryptophans, or tyrosines. X-ray crystallographic studies of phenserine and cymserine indicate that the tricyclic ring arrangement develops a defined three-dimensional conformation, with the phenylcarbamate rotated orthogonally to it [23,25]. This delimited structure, defined by exact torsion angles that hinder rotation between the carbamate and the tricyclic ring, impede many potential interactions assessable for a less rigid structure and thereby provide the compounds a level of selectivity. Unmistakably, such interactions are not possible for the simple methyl group present in physostigmine. In summary, a rigid space-occupying group in a distinct conformation coupled to a hexahydropyrroloindole moiety emerges as a requirement for APP activity and provides a lead for the design and synthesis of agents with more potency for this target.

In accord with our studies indicating dissociation between the APP/Aβ and ChEI activities of phenserine and cymserine, our comparable analysis of the opposite enantiomer of phenserine, explicitly with (3aR)- or (+)-phenserine that lacks cholinergic activity (IC_{50} AChE and BuChE >10 μM), demonstrated a potent and similar decrease in APP and Aβ levels. This independently substantiates our conclusion and additionally indicates that either chiral orientation is permitted for APP/Aβ activity, a finding confirmed by the APP-lowering activity of both (3aS)-, (–)- and (3aR)-, (+)-N^1-norphenserine [62]. However, such a disassociation of these two functions related to phenserine and cymserine does not take away from other known interactions between AChE and APP and Aβ. As an example among many, the peripheral binding site of AChE has been implicated in neurotrophic activity [63] in addition to AChE-Aβ complex formation with Aβ monomers that have been shown to trigger neurotoxicity via a pathway that involves a loss of Wnt signaling [55]. Such complexes form via electrostatic interactions between a positively charged amino acid cluster present in Aβ, corresponding to residues 12 to 16 (Val_{12}-His_{13}-His_{14}-Gln_{15}-Lys_{16}) and a specific domain localized on the surface of AChE [55].

Recent research from several laboratories indicates the participation of several enzyme activities and proteins in the APP processing pathway [1–4,7]. Whether phenserine and cymserine hinder one or more components of the APP processing machinery to operate efficiently, such as the secretase enzymes and in particular BACE, to thereby decrease APP/Aβ formation is clearly of current interest. Alternatively, a potential interaction may interfere with the binding of APP to a critical cellular protein, such as the iron-regulatory protein (IRP) [31,32]. Of late, an interaction of IRP with the 5′-untranslated region (5′UTR) of APP mRNA has been shown to cause translational enhancement of APP [32]. It is conceivable that phenserine and cymserine may interfere with APP translation consequent to a direct or indirect interaction at the 5′-UTR. Indeed, such an interaction has been described, along with several other factors, such as cytokines, that likewise can regulate APP translation at the level of the 5′-UTR of APP mRNA [17,33]. Thus separately or in harmony, agents such as phenserine and cymserine could interact with BACE, part of the 5′-UTR, or other undefined targets to regulate APP expression directly or indirectly.

Conclusion

We have examined the structure-activity relation underpinning the differential actions of physostigmine versus phenserine and cymserine, three closely related ChEIs, on APP. A hexahydropyrroloindole backbone is a common element to each; although valuable, it is not the sole requirement for APP activity, as the structurally dissimilar ChEI tacrine (9-amino-1,2,3,4-tetrahydroacridine hydrochloride), reduces APP/Aβ levels without bearing structural homology [19,20]. An aryl carbamate with a defined and rigid three-dimensional structure similarly appears to be critical. Further analysis likely will define a number of both discrete and overlapping mechanisms that can influence APP pathways to decrease the amountof Aβ. Close interaction between medicinal chemistry, cell biology, and molecular pharmacology may not only aid in defining the regulatory pathways responsible for controlling APP and Aβ levels in health, aging, and disease but help define the optimal SARs to provide clinically useful agents to normalize APP and Aβ levels. Such research will accelerate the discovery of potential drug targets and therapeutics for AD.

Acknowledgments We are sincerely thankful to Arnold Brossi for intellectual input. This work was supported in part by the Intramural Research Program of the National Institute on Aging (N.G., T.U., Q.Y., H.H.) and grants from Axonyx, Inc, the Alzheimer's Association, and the National Institutes of Health (K.S., J.R., D.L.).

References

1. Hardy J, Selkoe DJ. The amyloid hypothesis of Alzheimer's disease: progress and problems on the road to therapeutics. Science 2002;297:353–356.
2. Sambamurti K, Greig NH, Lahiri DK. Advances in the cellular and molecular biology of the beta-amyloid protein in Alzheimer's disease. Neuromol Med (2002;1:1–31.
3. Lahiri DK, Farlow MR, Greig NH, et al. A critical analysis of new molecular targets and strategies for drug developments in Alzheimer's disease. Curr Drug Targets 2003;4(2):97–112.
4. Sambamurti K, Granholm AC, Kindy MS, et al. Cholesterol and Alzheimer's disease: clinical and experimental models suggest interactions of different genetic, dietary and environmental risk factors. Curr Drug Targets 2004;5(6):517–528.
5. Cummings JL. Use of cholinesterase inhibitors in clinical practice: evidence-based recommendations. Am J Geriatr Psychiatry 2003;11(2):131–145.
6. DeKosky ST. Pathology and pathways of Alzheimer's disease with an update on new developments in treatment. J Am Geriatr Soc 2003;51(5 Suppl Dementia):S314–S320.
7. Lahiri DK, Rogers JT, Greig NH, Sambamurti K. Rationale for the development of cholinesterase inhibitors as anti-Alzheimer agents. Curr Pharm Des 2004;10(25):3111–3119.
8. Farlow MR. Utilizing combination therapy in the treatment of Alzheimer's disease. Expert Rev Neurother 2004;4(5):799–808.
9. Pietrzik C, Behl C. Concepts for the treatment of Alzheimer's disease: molecular mechanisms and clinical application. Int J Exp Pathol 2005;86(3):173–185.
10. Marlatt MW, Webber KM, Moreira PI, et al. Therapeutic opportunities in Alzheimer disease: one for all or all for one? Curr Med Chem 2005;12(10):1137–1147.
11. Standridge JB. Current status and future promise of pharmacotherapeutic strategies for Alzheimer's disease. J Am Med Dir Assoc 2005;6(3):194–199.
12. Rossom R, Adityanjee, Dysken M. Efficacy and tolerability of memantine in the treatment of dementia. Am J Geriatr Pharmacother 2004;2(4):303–312.
13. Doraiswamy PM. Non-cholinergic strategies for treating and preventing Alzheimer's disease. CNS Drugs 2002;16(12):811–824.
14. Greig NH, Pei XF, Soncrant TT, et al. Phenserine and ring C hetero-analogues: drug candidates for the treatment of Alzheimer's disease. Med Res Rev 1995;15(1):3–31.
15. Greig NH, De Micheli E, Holloway HW, et al. The experimental Alzheimer drug phenserine: pharmacokinetics and pharmacodynamics in the rat. Acta Neurol Scand 2000;176:74–84.
16. Greig NH, Sambamurti K, Yu QS, et al. An overview of phenserine tartrate, a novel acetylcholinesterase inhibitor for the treatment of Alzheimer's disease. Curr Alzheimers Res 2005;2:281–291.

17. Lahiri D, Chen D, Vivien D, et al. The role of cytokines in the gene expression of amyloid β-protein precursor: identification of a 5'UTR binding nuclear factor and its implications for Alzheimer's disease. J Alzheimer Dis 2003;5:81–90.

18. Lahiri DK, Farlow MR, Hintz N, et al (2000) Cholinesterase inhibitors, β-amyloid precursor protein and amyloid β-peptides in Alzheimer's disease. Acta Neurol Scand 2000;176:60–67.

19. Lahiri DK, Farlow MR, Sambamurti K. (1998) The secretion of amyloid beta-peptide is inhibited in tacrine-treated human neuroblastoma cells. Mol Brain Res 1998;62:131–140.

20. Lahiri DK, Farlow MR. Differential effect of tacrine and physostigmine on the secretion of the beta-amyloid precursor protein in cell lines. J Mol Neurosci 1996;7(1):41–49.

21. Yu QS, Greig NH, Holloway HW, Brossi A. Syntheses and anticholinesterase activities of (3aS)-N(1), N(8)-bisnorphenserine(3aS)-N(1), N(8)-bisnorphysostigmine, their antipodal isomers, and other potential metabolites of phenserine. J Med Chem 1998;41:2371–2379.

22. Yu QS, Holloway HW, Utsuki T, et al. Phenserine-based synthesis of novel selective inhibitors of butyrylcholinesterase for Alzheimer's disease. J Med Chem 1999;42:1855–1861.

23. Yu QS, Holloway HW, Flippen-Anderson F, et al. Methyl analogues of the experimental Alzheimer drug, phenserine: synthesis and structure/activity relationships for acetyl- and butyrylcholinesterase inhibitory action. J Med Chem 2001;44:4062–4071.

24. Yu QS, Zhu X, Holloway HW, et al. Anticholinesterase activity of compounds related to geneserine tautomers-N-oxides and 1,2-oxazines. J Med Chem 2002;45:3684–3691.

25. Luo X, Yu QS, Zhan M, et al. Novel anticholinesterases based on the molecular skeletons of furobenzofuran and benzodioxepine. J Med Chem 2005;48(4):986–994.

26. Haroutunian V, Greig NH, Utsuki T, et al. Pharmacological modulation of Alzheimer's beta-amyloid precursor protein levels in the CSF of rats with forebrain cholinergic system lesions. Mol Brain Res 1997;46:161–168.

27. Greig NH, Lahiri DK, Sambamurti K. Butyrylcholinesterase: an important new target in Alzheimer's disease therapy. Int Psychogeriatr 2002;14:77–91.

28. Greig NH, Utsuki T, Ingram DK, et al. Selective butyrylcholinesterase inhibition elevates brain acetylcholine, augments learning and lowers Alzheimer β-amyloid peptide in rodent. Proc Natl Acad Sci. USA, 2005;102:17213–8.

29. Coelho F, Birks J. Physostigmine for Alzheimer's disease. Cochrane Database Syst Rev 2001;2:CD001499.

30. Shaw KT, Utsuki T, Rogers J, et al. Phenserine regulates translation of beta-amyloid precursor protein mRNA by a putative interleukin-1 responsive element, a target for drug development. Proc Natl Acad Sci USA 2001;98(13):7605–7610.

31. Rogers JT, Randall JD, Cahill CM, et al. An iron-responsive element type II in the 5'-untranslated region of the Alzheimer's amyloid precursor protein transcript. J Biol Chem 2002;277(47):45518–45528.

32. Venti A, Giordano T, Eder P, et al. The integrated role of desferrioxamine and phenserine targeted to an iron-responsive element in the APP-mRNA 5'-untranslated region. Ann N Y Acad Sci. 2004;1035:34–44.

33. Maloney B, Ge YW, Greig NH, Lahiri DK. Presence of a "CAGA box" in the APP gene unique to amyloid plaque-forming species and absent in all APLP-1/2 genes: implications in Alzheimer's disease. FASEB J 2004;18(11):1288–1290.

34. Patel N, Spangler E, Greig NH, et al. Phenserine, a novel acetylcholinesterase inhibitor, attenuates impaired learning of rats in a 14-unit T-maze induced by blockade of the NMDA receptor. Neuroreport 1998;9:171–176.

35. Greig NH. Drug entry to the brain and its pharmacologic manipulation. In, Handb Exp Pharmacol 1992;103:487–523.

36. Greig NH, Yu QS, Utsuki T, et al. Optimizing drugs for brain action. In: Koliber D, Lustig S, Shapira S (eds) Blood-Brain Barrier Drug Delivery and Brain Pathology. New York: Kluwer Academic/Plenum, 2002, pp 281–309.
37. Thal LJ. Therapeutics and mild cognitive impairment: current status and future directions. Alzheimer Dis Assoc Disord 2003;17(suppl 2):S69–S71.
38. Giacobini E. Cholinesterases: new roles in brain function and in Alzheimer's disease. Neurochem Res 2003;28(3-4):515–522.
39. Courtney C, Farrell D, Gray R, et al. Long-term donepezil treatment in 565 patients with Alzheimer's disease (AD2000): randomised double-blind trial. Lancet 2004;363(9427):2105–2115.
40. Lopez OL, Becker JT, Saxton J, et al. Alteration of a clinically meaningful outcome in the natural history of Alzheimer's disease by cholinesterase inhibition. J Am Geriatr Soc 2005;53:83–87.
41. Greig NH, Mattson MP, Perry T, et al. New therapeutic strategies and drug candidates for neurodegenerative diseases: p53 and TNF-α inhibitors, and GLP-1 receptor agonists. Ann N Y Acad Sci 2004;1035:290–315.
42. Cole GM, Morihara T, Lim GP, et al. NSAID and antioxidant prevention of Alzheimer's disease: lessons from in vitro and animal models. Ann N Y Acad Sci 2004;1035:68–84.
43. Yang F, Lim GP, Begum AN, et al. Curcumin inhibits formation of amyloid beta oligomers and fibrils, binds plaques, and reduces amyloid in vivo. J Biol Chem 2005;280(7):5892–5901.
44. Hoglund K, Thelen KM, Syversen S, et al. The effect of simvastatin treatment on the amyloid precursor protein and brain cholesterol metabolism in patients with Alzheimer's disease. Dement Geriatr Cogn Disord 2005;19(5-6):256–266.
45. Doraiswamy PM, Steffens DC, McQuoid DR. Statin use and hippocampal volumes in elderly subjects at risk for Alzheimer's disease: a pilot observational study. Am J Alzheimers Dis Other Demen 2004;19(5):275–278
46. Ritchie CW, Bush AI, Masters CL. Metal-protein attenuating compounds and Alzheimer's disease. Expert Opin Invest Drugs 2004;13(12):1585–1592.
47. Lahiri DK, Sambamurti K, Bennett DA. Apolipoprotein gene and its interaction with the environmentally driven risk factors: molecular, genetic and epidemiological studies of Alzheimer's disease. Neurobiol Aging 2004;25(5):651–660.
48. Greig NH, Sambamurti K, Yu QS, et al. Butyrylcholinesterase: its selective inhibition and relevance to Alzheimer's disease. In: Giacobini E (ed) Butyrylcholinesterase: Its Function and Inhibition. London: Martin Dunitz, 2003, pp 69–90.
49. Soreq H, Seidman S. Acetylcholinesterase-new roles for an old actor. Nat Rev Neurosci 2001;2:294–302.
50. Massoulie J. Molecular forms and anchoring of acetylcholinesterase. In: Giacobini E (ed) Cholinesterases and Cholinesterase Inhibitors. London: Martin Dunitz, 2000, pp 81–102.
51. Soreq H, Gnatt A, Loewenstein Y, Neville LF. Excavations into the active-site gorge of cholinesterases. Trends Biochem Sci 1992;17:353–358.
52. Dvir H, Wong DM, Harel M, et al. 3D structure of Torpedo californica acetylcholinesterase complexed with huprine X at 2.1 A resolution: kinetic and molecular dynamic correlates. Biochemistry 2002;41:2970–2981.
53. Silman I, Millard CB, Ordentlich A, et al. A preliminary comparison of structural models for catalytic intermediates of acetylcholinesterase. Chem Biol Interact 1999;May 14:119–120.
54. Silman I, Harel M, Axelsen P, et al. Three-dimensional structures of acetylcholinesterase and of its complexes with anticholinesterase agents. Biochem Soc Trans 1994;22:745–749.
55. Inestrosa NC, Sagal JP, Colombres M. Acetylcholinesterase interaction with Alzheimer amyloid beta. Subcell Biochem 2005;38:299–317.

56. Nitsch RM, Farber SA, Growdon JH, Wurtman RJ. Release of amyloid beta-protein precursor derivatives by electrical depolarization of rat hippocampal slices. Proc Natl Acad Sci U S A 1993;90(11):5191–5193.
57. Muller DM, Mendla K, Farber SA, Nitsch RM. Muscarinic M1 receptor agonists increase the secretion of the amyloid precursor protein ectodomain. Life Sci 1997;60(13-14):985–991.
58. Buxbaum JD, Greengard P. Regulation of APP processing by intra- and intercellular signals. Ann N Y Acad Sci 1996;777:327–331.
59. Desdouits-Magnen J, Desdouits F, Takeda S, et al. Regulation of secretion of Alzheimer amyloid precursor protein by the mitogen-activated protein kinase cascade. J Neurochem 1998;70(2):524–530.
60. Utsuki T, Shoaib M, Lahiri DK, et al. Nicotine reduces the secretion of Alzheimer's ß-Amyloid precursor protein containing ß-Amyloid peptide in the rat. J Alzheimers Dis 2002;4:405–415.
61. Lahiri DK, Utsuki T, Chen D, et al. Nicotine reduces the secretion of Alzheimer's beta-amyloid precursor protein containing beta-amyloid peptide in the rat without altering synaptic proteins. Ann N Y Acad Sci 2002;965:364–372.
62. Yu QS, Luo W, Holloway HW, et al. Racemic N^1-norphenserine and its enantiomers: unpredicted inhibition of acetyl- and butyrylcholinesterase and β-amyloid precursor protein in vitro. Heterocycles 2003;61:529–539.
63. Munoz FJ, Aldunate R, Inestrosa NC. Peripheral binding site is involved in the neurotrophic activity of acetylcholinesterase. Neuroreport 1999;10(17):3621–3625.

Chapter 48
Allosteric Potentiators of Neuronal Nicotinic Cholinergic Receptors: Potential Treatments for Neurodegenerative Disorders

Emanuele Sher, Giovanna De Filippi, Tristan Baldwinson, Ruud Zwart, Kathy H. Pearson, Martin Lee, Louise Wallace, Gordon I. McPhie, Martine Keenan, Renee Emkey, Sean P. Hollinshead, Colin P. Dell, S. Richard Baker, Michael J. O'Neil, and Lisa M. Broad

Introduction

Nicotinic acetylcholine (ACh) receptors (nAChRs) are ligand-gated ion channels formed by the assembly of five subunits. Each subunit is composed of a large extracellular N-terminal, four transmembrane regions, and a short extracellular C-terminal [1]. To date, 17 distinct nAChR subunits have been identified. Nicotinic receptors of the neuromuscular junction are composed of $\alpha 1$, $\beta 1$, δ and γ, or ε, with a stoichiometry of $\alpha_2 \beta \delta \gamma / \varepsilon$, whereas neuronal nAChRs are composed of either heteromeric ($\alpha 2$-6 and $\beta 2$-4) or homomeric ($\alpha 7$-10) subunit combinations. The subunit composition and stoichiometry of neuronal nAChRs in their native environment is not fully defined, although peripheral ganglionic nAChRs are known to contain $\alpha 3$ and $\beta 4$ subunits, whereas nAChRs containing $\alpha 4$, $\beta 2$, or $\alpha 7$ subunits are broadly expressed in the central nervous system (CNS) [2].

The orthosteric agonist-binding site is located extracellularly, at the N-terminal. Amino acid loops contributed by two adjacent subunits form a hydrophobic pocket where competitive agonists and antagonists establish high-affinity interactions with specific amino acid residues [3]. Previous pharmacological and biochemical data have been confirmed, and further elaborated, by the new findings derived by the co-crystallization of snail ACh-binding proteins with competitive nicotinic ligands [4].

Noncompetitive modulators (also referred to as allosteric potentiating ligands, or APLs,) of nAChRs that *do not* bind to the same amino acids as ACh or nicotine but modulate the potency and efficacy of ACh have been described [5]. However, highly selective nAChR potentiators have yet to be

E. Sher
Eli Lilly and Company, Lilly Research Centre, Erl Wood Manor, Windlesham, Surrey GU20 6PH, UK

A. Fisher et al. (eds.), *Advances in Alzheimer's and Parkinson's Disease*, 463
© Springer 2008

reported. It was thought that because of the highly conserved site where many of these APLs bind it would have been difficult to find selective ones.

Both inorganic and organic allosteric modulators have been described. Calcium ions bind to amino acids found at the N-terminal, the extracellular side of the protein, and allosterically potentiate various nAChR subtypes [6,7]. Also, galanthamine, physostigmine, and other nonselective nAChR potentiators bind to amino acids at the N-terminus, close to, but apparently distinct from, both the agonist and the calcium-binding site [8]. 5-Hydroxyindole has been shown to potentiate α7 nAChRs selectively (compared to other nAChR subtypes [9]), but it displays cross-reactivity at 5-HT$_3$Rs. [10] Likewise, ivermectin displays selectivity for potentiation of α7 over other nAChR subtypes [11] but also acts on glycine [12] and P2X receptors [13].

Various steroids have been shown to have blocking and/or potentiating effects on nAChRs and other ion channels. Interestingly, 17β-estradiol was recently found to potentiate α4β2 nAChR selectively by binding to extracellular amino acids at the C-terminus of the α4 subunit [14].

We challenged the hypothesis that it would be impossible to find selective APLs by performing a high throughput screen of a large chemical library, searching for potentiators selective for the various nAChR subtypes. We report here on the success in finding allosteric nAChR ligands with various degrees of selectivity.

Materials and Methods

Fluorescent Imaging Plate Reader

Cells stably expressing the human α4β2, α3β4, and α7 nAChRs were grown in 96-well plates [15]. They were loaded with 10 μM Fluo-3-AM/0.05% pluoronic F-127 in Tyrode's buffer (composition, in mM, 137 NaCl, 2.7 KCl, 2.5 CaCl$_2$, 1 MgCl$_2$, 12 NaHCO$_3$, 0.2 NaHPO$_4$, 5.5 glucose; pH 7.3) by incubation at room temperature in the dark. After 1 hour, the medium was replaced with Tyrode's buffer in the absence of Fluo-3, and the plates were transferred to a fluorescent imaging plate reader (FLIPR) (Molecular Devices) for experiments. Fluorescence was recorded every second for 60 seconds; drugs were added after 10 seconds to allow an initial baseline reading. Once added, the drugs were present for the duration of the experiments. Parameters for drug addition to the cell plate were programmed using FLIPR system software, and compound delivery was automated through a 96-tip head pipette device. Responses were measured as the peak fluorescence minus the baseline fluorescence intensity and are expressed as a percentage of a maximal agonist response. Potentiators were preincubated for 10 to 20 minutes before agonist additions.

Data files were saved to the computer and stored for offline analysis using FLIPR system software, Microsoft Excel, and Origin 6.1 (OriginLab, Northampton, MA,

USA). Statistical analysis was performed to determine significant differences ($p < 0.05$) using Student's t-test.

Patch-Clamp Recordings in Cultured Hippocampal Neurons

Primary cultures of hippocampal neurons were prepared from newborn (1- to 2- day-old) Wistar rats (Harlan, Oxon, UK) following a procedure in compliance with the UK Animal (Scientific Procedure) Act 1986. Hippocampi were dissected out in ice-cold phosphate-buffered saline (PBS) containing 10 mM glucose and bovine serum albumin (BSA) 1 mg/ml. The tissue was first enzymatically digested with papain 0.5 mg/ml and deoxyribonuclease 0.25 mg/ml. BSA-containing PBS was replaced with Dulbecco's Modified Eagle Medium supplemented with 10% heat-inactivated fetal bovine serum, penicillin 10,000 U/ml, streptomycin 10,000 µg/ml, and L-glutamine 29.2 mg/ml. The tissue was mechanically triturated by suction using a sterile glass Pasteur pipette; cells were suspended in 10 ml of the above solution and centrifuged; the supernatant was removed, and the cells were resuspended, mechanically dissociated a second time before being counted, and plated at a density of approximately 4×10^5 cells/ml on 12-mm round glass coverslips precoated in poly-D-lysine and mouse laminin. Cells were maintained in an incubator at 37°C in 95% O_2 and 5% CO_2. At 24 hours after plating, the culture medium was replaced with medium deprived of serum to reduce fibroblast proliferation but containing B-27 supplement. Between 5 and 9 days in culture, 5 µM cytosine β-D-arabinofuranoside was added to the medium to reduce the proliferation of nonneuronal cells.

Whole-cell patch-clamp recordings in the voltage-clamp configuration were carried out 18 to 25 days after plating at room temperature. During recordings the cultures were perfused continuously with an extracellular solution containing (in mM): NaCl 145, KCl 2, KH_2PO_4 1.18, Hepes 10, $MgSO_4$ 1.2, $CaCl_2$ 2, D-glucose 11 (pH adjusted to 7.2 with NaOH), kynurenic acid 5 (GluR antagonist), and atropine 0.1 µM (mAChR antagonist). The intracellular solution contained (in mM): $CsSO_4$ 81, NaCl 4, $MgSO_4$ 2, $CaCl_2$ 0.02, BAPTA 0.1, D-glucose 15, Hepes 10, ATP 3, GTP 0.1, and Na phosphocreatine 5 (pH adjusted to 7.2 with CsOH).

Drugs were diluted at their final concentration in the extracellular solution and applied via a multibarreled pipette positioned 150 µm away from the recorded neuron. Recordings were carried out using an Axopatch 200B amplifier (Molecular Devices, Foster City, CA. USA) at a holding potential (V_h) of +20 mV for the detection of spontaneous γ-aminobutyric acid (GABA)-inhibitory postsynaptic currents (IPSCs). The signals were digitized every 50 µsec and analyzed offline with PClamp8 software (Axon Instruments, Molecular Devices). Potentiation of GABA release was quantified by measuring the IPSCs charge over 20 seconds of agonist application alone or in the presence of the potentiators.

Results

CNS Selective nAChRs Potentiators

As a result of high throughput screening, several compounds were found that were able to potentiate the increase in intracellular calcium triggered by a submaximal concentration of a nicotinic agonist in one or more nAChR subtype. Some of these compounds showed an interesting selectivity profile.

One example is shown in Fig. 1, where stable cell lines expressing the most relevant neuronal nAChRs ($\alpha 3\beta 4$, $\alpha 4\beta 2$, $\alpha 7$) were challenged with increasing concentrations of epibatidine, a nonselective nicotinic agonist, in the presence or absence of compound I. The structure of compound I, {[2-(4-fluorophenylamino)-4-methyl-thiazol-5-yl]-thiophen-3-yl-methanone} has recently been disclosed [16].

The compound did not affect the responses of $\alpha 3\beta 4$ receptors but significantly potentiated epibatidine effects on both $\alpha 4\beta 2$ and $\alpha 7$ receptors. The effects of compound I were concentration-dependent, as shown in Fig. 2.

We also found that compound I potentiated the $\alpha 4\beta 4$ neuronal receptor (interesting but not highly relevant based on the low expression of this receptor subtype in mammalian CNS) but, more importantly, had no effect at all on muscle-type nAChRs (not shown). This profile (activating neuronal $\alpha 4\beta 2$ and $\alpha 7$ nAChRs but not ganglionic $\alpha 3\beta 4$ or muscle ones) is unique and offers an alternative to what many groups have been trying to achieve working with competitive agonists.

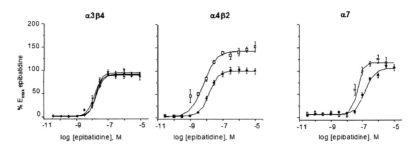

Fig. 1 Stable cell lines expressing $\alpha 3\beta 4$, $\alpha 4\beta 2$, and $\alpha 7$ human nicotinic acetylcholinesterase receptors (nAChRs) were seeded in 96-well plates, loaded with Fluo3, and then challenged with the nonselective nicotinic agonist epibatidine alone (filled circles) or in the presence of 1 μM compound I (preincubated for 10 minutes before agonist addition) (open squares). Although no effects were seen on $\alpha 3\beta 4$, both $\alpha 4\beta 2$ and $\alpha 7$ receptor responses were potentiated. Each point represents the mean ± SE of three experiments (Modified from Broad et al., [16] with permission.)

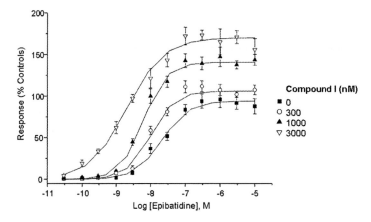

Fig. 2 The new nAChR potentiator dose-dependently potentiates epibatidine responses in human $\alpha4\beta2$ nAChR-expressing cells. Cells were plated and challenged as described in Material and Methods. Each point represents the mean ± SE of three experiments

The lack of effects on recombinant human $\alpha3\beta4$ nAChRs represents a particularly important finding. therefore, we looked carefully also for effects on native human $\alpha3\beta4$ receptors. To this end, we utilized the IMR32 human neuroblastoma cell line [17], which expresses endogenous $\alpha3\beta4$-containing nicotinic receptors. As shown in Fig. 3, compound I did not change the concentration–response curve of epibatidine, confirming the data shown above on recombinant $\alpha3\beta4$ nAChRs receptors. This strongly suggests that this type of molecule could activate central nicotinic mechanisms while being free of the autonomic side effects encountered with many nonselective compounds.

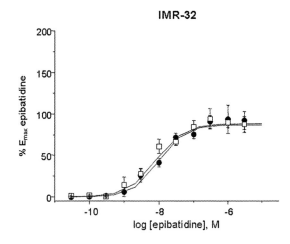

Fig. 3 Potentiator selective for $\alpha4\beta2$ and $\alpha7$ CNS receptors does not potentiate native $\alpha3\beta4$-containing receptors expressed by the human neuroblastoma cell line IMR32. Each point represents the mean ± SE of three experiments

We also evaluated the ability of compound I and its analogs to modulate native central nicotinic receptors. nAChRs are known to stimulate GABA release from hippocampal interneurons (for a review see Sher et al [18]. and the references therein). This presynaptic effect is mediated by various nAChR subtypes including α4β2 and α7. We therefore expected our selective potentiators to be effective in this system. Indeed, ACh-induced GABA release (measured as postsynaptic currents) was significantly and reversibly enhanced by compound I (Fig. 4), confirming that α4β2 and/or α7 receptors native to rat hippocampal neurons are potentiated by this new compound.

Selective α7 nAChR Potentiators

During the course of screening we also identified a series of allosteric potentiators that are even more potent, efficacious, and selective than the one reported above. This class of molecules is highly selective for α7 nAChRs and does not modulate any other nAChR or any other ion channels that we tested.

Figure 5 shows FLIPR-generated curves that summarize the effects of one of the selective α7 potentiators (compound II, structure not yet disclosed) on the agonist–dose response curve: A shift to the left and an increase in the maximal agonist efficacy were found. Figure 5 also shows that the potentiation by compound II is dramatic, irrespective of whether the agonist used is the endogenous neurotransmitter ACh, the second putative endogenous neurotransmitter choline, or exogenous nicotine.

The magnitude of potentiation is marked in comparison to the few known allosteric potentiators for α7 nAChRs, such as ivermectin [11] or 5-hydroxyindole [9]. In particular, compounds such as compound II have dramatic effects on the desensitization of α7 receptors and on their single-channel properties (Sher et al., unpublished data).

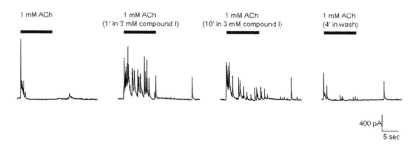

Fig. 4   nAChR-mediated modulation of γ-aminobutyric acid (GABA) release in cultured hippocampal neurones was reversibly potentiated during prolonged application of 3 μM of compound I

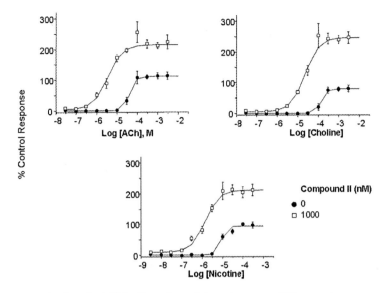

Fig. 5 New selective α7 nAChR allosteric modulator (compound II) potentiates responses in human α7 nAChR-expressing cells. The α7 potentiator causes a significant shift to the left of the agonist dose–response curve and, notably, a significant increase in the maximum response. Note that the potentiating activity is not dependent on the type of agonist used. Cells were plated and challenged as described in Material and Methods. Each point represents the mean ± SE of three experiments

Also we confirmed that the selective α7 potentiators are able to potentiate native α7 nAChRs. With calcium imaging and FLIPR, we found that the very small component of native human α7 nAChRs expressed by IMR32 cells can be potentiated by our compounds, as well as native rat α7 nAChRs, expressed somatically by cultured hippocampal neurons (patch-clamp, not shown).

Similar to compound I described above, the highly selective α7 potentiators also enhanced GABA release from hippocampal neurons. Figure 6 shows results from experiments in which choline was used as a selective agonist of α7 receptors. Under these experimental conditions, compound II dramatically and reversibly potentiated choline-induced GABA release (measured as postsynaptic currents). The effects were also prevented reversibly by the selective α7 antagonist methyllycaconitine citrate (MLA).

Discussion

The α7 and α4β2 nAChRs are normally expressed at high levels in areas involved in learning and memory and are known to play a physiological role in the modulation of neurotransmission in these regions [18,19]. Under

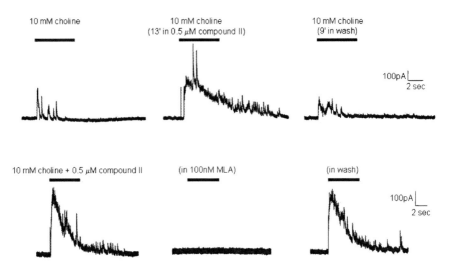

Fig. 6 Choline-induced, α7-mediated, modulation of GABA release in cultured hippocampal neurons was reversibly potentiated during prolonged application of 0.5–1.0 µM of compound II, a selective α7 potentiator. The effect could be blocked by application of methyllycaconitine (MLA) (bottom panels, from a different cell)

pathological conditions, such as Parkinson's and Alzheimer's diseases, reduced cholinergic activity and loss or malfunction of nAChRs is a common feature and can be correlated with cognitive deficits and progressive dementia [20,21]. The involvement of nAChRs in the modulation of cognitive function has been demonstrated in several animal models [22]. In clinical studies, nonselective nAChR agonists are similarly reported to improve attention, memory, and learning in both normal and disease states [22,23]. The therapeutic utility of nonselective nAChR agonists, however, is limited by adverse effects, mainly caused by activation of the peripheral nervous system.

More selective nicotinic agonists (notably for α7 and α4β2 nAChRs) have been recently reported and might circumvent many of the side effects previously found in the clinical situation with compounds also having α3β4 activity [15,24,25].

However, the use of agonists could have other potential limitations: The long-term effects of tonic receptor stimulation and/or receptor desensitization are largely unknown. It is hypothesized (but not proven) that allosteric potentiators might circumvent some of these problems, enhancing receptor activation in a phasic (following endogenous release of the natural neurotransmitter), rather than tonic, manner [26].

It has been suggested that the latter mechanism contributes to the clinical efficacy of galanthamine in AD [27]. Galanthamine is an acetylcholinesterase inhibitor but also has some weak potentiating activity at nAChRs [28]. This dual activity of galanthamine does not help in proving experimentally that selective nAChR potentiation could be beneficial. Furthermore, galanthamine does not discriminate among the various nAChR subtypes, and it has been recently shown to modulate N-methyl-D-aspartate (NMDA) receptors as well [29].

The compounds described here should become useful tools to prove or disprove that this approach can bring therapeutic benefit.

References

1. Lindstrom J. The structures of neuronal nicotinic receptors. In: Clementi F, Fornasari D, Gotti C (eds) Neuronal Nicotinic Receptors, Handbook of Experimental Pharmacology, vol 144. New York: Springer, 2000, pp 101–162.
2. McGehee DS, Role LW. Physiological diversity of nicotinic acetylcholine receptors expressed by vertebrate neurones. Annu Rev Physiol 1995;57:521–546.
3. Karlin A. Emerging structure of the nicotinic acetylcholine receptors. Nat Rev Neurosci 2002;3:102–114.
4. Celie PH, van Rossum-Fikkert SE, van Dijk W, et al. Nicotine and carbamylcholine binding to nicotinic acetylcholine receptors as studied in AChBP crystal structures. Neuron 2004;41:907–914.
5. Pereira EF, Hilmas C, Santos MD, et al. Unconventional ligands and modulators of nicotinic receptors. J Neurobiol 2002;53:479–500.
6. Mulle C, Lena C, Changeux JP. Potentiation of nicotinic receptor response by external calcium in rat central neurons. Neuron 1992;8:937–945.
7. Galzi JL, Bertrand S, Corringer PJ, et al. Identification of calcium binding sites that regulate potentiation of a neuronal nicotinic acetylcholine receptor. EMBO J 1996;15:5824–5832.
8. Pereira EF, Alkondon M, Reinhardt S, et al. Physostigmine and galanthamine: probes for a novel binding site on the alpha 4 beta 2 subtype of neuronal nicotinic acetylcholine receptors stably expressed in fibroblast cells. J Pharmacol Exp Ther 1994;70:768–778.
9. Zwart R, DeFilippi G, Broad LM, et al. 5-Hydroxyindole potentiates human α7 nicotinic receptor-mediated responses and enhances acetylcholine-induced glutamate release in cerebellar slices. Neuropharmacology 2002;43:374–384.
10. Kooyman AR, Van Hooft JA, Vijverberg HPM. 5-Hydroxyindole slows desensitization of 5-HT$_3$ receptor-mediated ion current in N1E-115 neuroblastoma cells. Br J Pharmacol 1993;108:287–289.
11. Krause RM, Buisson B, Bertrand S, et al. Ivermectin: a positive allosteric effector of the α7 neuronal nicotinic acetylcholine receptor. Mol Pharmacol 1998;53:283–294.
12. Shan Q, Haddrill JL, Lynch JW. Ivermectin, an unconventional agonist of the glycine receptor chloride channel. J Biol Chem 2001;276:12556–12564.
13. Khakh BS, Proctor WR, Dunwiddle TV, et al. Allosteric control of gating and kinetics at P2X$_4$ receptor channels. J Neurosci 1999;19:7289–7299.
14. Curtis L, Buisson B, Bertrand S, Bertrand D. Potentiation of human α4β2 neuronal nicotinic acetylcholine receptor by estradiol. Mol Pharmacol 2002;61:127–135.
15. Astles PC, Baker S, Boot JR, et al. Recent progress in the development of subtype selective nicotinic acetylcholine receptor ligands. Curr Drug Targets CNS Neurol Disord 2002;1:337–348.
16. Broad LM, Zwart R, Pearson KH, et al. Identification and pharmacological profile of a new class of selective nicotinic acetylcholine receptor potentiators. J Pharmacol Exp Ther 2006;318:1108–1117.

17. Clementi F, Cabrini D, Gotti C, Sher E. Pharmacological characterization of cholinergic receptors in a human neuroblastoma cell line. J Neurochem 1986;47:291–297.
18. Sher E, Chen Y, Sharples TJW, et al. Physiological roles of neuronal nicotinic receptor subtypes: new insights on the nicotinic modulation of neurotransmitter release, synaptic transmission and plasticity. Curr Top Med Chem 2004;4:283–297.
19. Dani JA. Overview of nicotinic receptors and their roles in the central nervous system. Biol Psychiatry 2001;49:166–174.
20. Court J, Martin-Ruiz C, Piggott M, et al. Nicotinic receptor abnormalities in Alzheimer's disease. Biol Psychiatry 2001;49:175–184.
21. Quik M, Kulak JM. Nicotine and nicotinic receptors; relevance to Parkinson's disease. Neurotoxicology 2002;23:581–594.
22. Levin ED, Rezvani AH. Development of nicotinic drug therapy for cognitive disorders. Eur J Pharmacol 2000;393:141–146.
23. Buccafusco JJ, Letchworth SR, Bencherif M, Lippiello PM. Long-lasting cognitive improvement with nicotinic receptor agonists: mechanisms of pharmacokinetic-pharmacodynamic discordance. Trends Pharmacol Sci 2005;26:352–360.
24. Kem WR. The brain alpha7 nicotinic receptor may be an important therapeutic target for the treatment of Alzheimer's disease: studies with DMXBA (GTS-21). Behav Brain Res 2000;113:169–181.
25. Lloyd GK, Williams M. Neuronal nicotinic acetylcholine receptors as novel drug targets. J Pharmacol Exp Ther 2000;2:461–467.
26. Maelicke A, Schrattenholz A, Samochocki M, et al. Allosterically potentiating ligands of nicotinic receptors as a treatment strategy for Alzheimer's disease. Behav Brain Res 2000;113:199–206.
27. Woodruff-Pak DS, Lander C, Geerts H. Nicotinic cholinergic modulation: galantamine as a prototype. CNS Drug Rev 2002;8:405–426.
28. Samochocki M, Hoffle A, Fehrenbacher A, et al. Galantamine is an allosterically potentiating ligand of neuronal nicotinic but not muscarinic acetylchloline receptors. J Pharmacol Exp Ther 2003;305:1024–1036.
29. Moriguchi S, Marszalec W, Zhao X, et al. Mechanism of action of galantamine on N-methyl-D-aspartate receptors in rat cortical neurons. J Pharmacol Exp Ther 2004;310:933–942.

Index

Aβ Amyloidosis, 124
Aβ deposition
    cholinergic-amyloid fusion hypothesis
        for, 411–421
    cortical cholinergic lesion and, 411–421
Aβ-derived diffusible neurotoxin ligands
        (ADDLs), 293
Aberrant motor behavior, 3
Aβ-induced toxicity, resveratrol
        neuroprotective effects against, 293
Aβ vaccine, 124
    in AD mouse model, 265–271
    gut immune system and advantage of,
        269–270
    immune responses to, 268
    mechanism of, 271
    mediated meningoencephalitis, 268–269
    production of, 266–267
    vaccination and tissue examinations,
        267–268
Acetylcholine, 54
Acetylcholine (ACh) synthesis, 403–404, 407
Acetylcholinesterase enzyme (AChE),
        404–405, 414, 416, 439, 442, 446,
        449, 452–458
    Aβ effect on levels of, 430–431
    expression in AD brain, 429–435
    inhibitors and cholinergic deficiency, 433
    multiple isoforms of, 429–430
Acetylcholinesterase enzyme inhibitors
        (AChEIs)
    effect on AβPP expression and
        metabolism, 441–442
    gene expression modulation, 442–443
Activated microgila in PD, 91–95, 117–119
    damaged dopamine neurons, 92
    neuroprotective and neurotoxic subtypes,
        94–95
    in putamen, 92–93

in substantia nigra and brain regions,
        92–93, 117
Activities of Daily Living Scale
        (ADCS-ADL), 5
Acute akinesia in Parkinson's disease
        patients, 31–34
ADAM10, 442
    promoter activity, 371–373
    as α-secretase, 369–370, 394
Adeno-associated virus vector (AAV)
        vaccine
    in AD mouse model, 265–271
    gut immune system and advantage of,
        269–270
    immune responses to, 268
    production of, 266–267
    vaccination and tissue examinations,
        267–268
Adrenal chromaffin cells, 51
Aging, free radical theory, 192
Akinesia, 25, 31, 35
Alcohol dehydrogenase (ADH), 362
Aldehyde dehydrogenase (ALDH), 362
Aldehyde hydrogenase, 362
Alzheimer Disease Assessment Scale-
        Cognitive Section (ADAS-cog),
        4–5
Alzheimer Disease Collaborative Study-
        Clinical Global Impression of
        Change (ADCS-CGIC), 5
Alzheimer's APP transgenic mice
    immunization with phage-ScFv
        against β-amyloid peptide,
        274–275
    targeting β–amyloid plaques with phage
        anti-AβP ScFv, 275–277
Alzheimer's-associated neurodegeneration,
        decreased ProBDNF as cause,
        279–281

Alzheimer's disease (AD), 1, 41–41, 81, 98
AGD and TSAs relation in, 375–379
amyloid cascade hypothesis, 298–299
amyloid fragments in, 171–173
β-amyloid role in pathophysiology,
439–443
ApoE related
cholesterol levels in, 357–358
GOT, GPT, and GGT activities in, 359
plasma triglyceride levels in, 357–358
Aβ deposition and neuroinflammatory
response in, 114–115
BDNF down-regulation
and functional deficits, 281
loss of synaptic and neuronal
function, 279–281
and pathophysiology, 280
biological complexity in treatment,
439–443
biomarkers for, 169–173
cell cycle reentry in, 138–139
CERAD criteria, 139
cholesterol transport and production in,
211–217
cholinergic changes in, 393–394, 403–405,
407, 420
cholinergic markers in, 404, 420
cholinergic neurites in, 432
CSF biomarkers, 172
CYP2D6 polymorphisms in, 363–364
dysfunction of proteomics-identified
β(1-42)-induced oxidized
proteins, 160
GAβ as seed for amyloid, 387–391
genetic factors, 197–198
glutamatergic changes in, 405–407
inflammation and pathological cascade
in, 116–117
intraneuronal Aβ, 297–301
Khachaturian criteria for, 139
lifestyle-related factors and, 208–209
lipids, 222–223
mechanistic hypothesis of, 191
midlife risk factors for dementia, 207
mitotic abnormalities and signaling,
194–195
α7 nAChRs, 431–432, 434
neuroinflammation and regenerative
pathways in, 115–116
neuronal death in, 133–134
neuropathology of, 429
oxidative insult and neuronal survival
role, 133–142

Aβ, tau, and α-synuclein function
against, 140–142
genetic mutations and risk factors,
135–136
oxidatively modified brain proteins
identification, 151–153
oxidative stress
with genetic mutations and risk
factors, 135–138
and oxidative stress signaling, 192–194
pathology and
cognition, 298
neuronal survival response, 139–140
pathophysiology
β-amyloid role in, 439–443
and BDNF down-regulation, 280
pharmacogenetics, 361–363
plasma triglyceride levels and GOT,
GPT, and GGT in, 360
protein misfolding manifestations,
123–129
redox proteomics identification of
oxidatively modified brain
proteins, 154–160
AD models for Aβ(1-42), 159–160
cholinergic dysfunction, 157–158
CO_2 transport, 159
energy dysfunction, 156–157
excitotoxicity, 157
lipid abnormalities, 157–158
LTP, 158–159
neuritic abnormalities, 158
pH buffering, 159
proteasomal dysfunction, 157
synaptic abnormalities, 158–159
tau hyperphosphorylation, 158
stem cell therapy in, 255–262
tau in, 171–173
tauopathies and, 124–126
two-dimensional maps of brain proteins,
151–152
two-hit hypothesis, 195–198
vascular risk factors in, 206–208
Alzheimer's disease brain
acetylcholinesterase expression in, 429–435
ACh receptors in, 404
glutamatergic dystrophic neurites in,
397–398
Alzheimer's disease mouse model, Aβ
vaccine and AAV in, 265–271
Alzheimer's Disease Neuroimaging Initiative
(ADNI), 183–189
Clinical Coordination Center, 186–187

Clinical Core, 187–188
Neuroimaging Core, 187–188
objectives, 184
structural organization, 185–186
Amantadine sulfate, 33
Ambenonium, 441
γ-Aminobutyric acid (GABA), 38, 61, 465,
468–470
α-Amino-3-hydroxy-5-methyl-4-isoxazole
propionic acid (AMPA)
receptors, 157
Aminoisopropylpropionic acid (AMPA)
receptors, 407
Amnesia, 2
Amplex-Red cholesterol assay, 225
β-Amyloid (βA), 113–119, 123–124, 126,
133, 136–137, 139–142, 154–155,
161, 171–173, 231, 237–239, 265,
291, 298
deposition and neuroinflammatory
response in AD, 114–115
production and cholesterol, 221–228
sGAGs interactions with, 233–235,
247–248
Amyloid-β-peptide, see B-Amyloid
protein
Amyloid-β protein precursor (AβPP) gene,
113–116, 123–124, 128, 135, 137,
139, 154, 194, 197–198, 221–223,
228, 231, 238, 265, 298–300, 305,
337, 355–356, 369–370, 390,
439–442, 445–446
AChE effect on expression and
metabolism of, 441–442
function in stem cell biology, 259–262
nonamyloidogenic metabolism of,
440–442
phenserine and cymserine effect on,
450–451, 456–458
physiological processing, 305–312
ADAM9 and, 305, 311–312
ADAM10 and, 305, 309–313, 394
ADAM17 and, 305, 309–310
up-regulation by PKC agonists,
307–309
α-secretase processing, 305–308
β-Amyloid derived diffusible ligands
(ADDLs), 382
Amyloid oligomer disease pathogenesis,
cellular membranes as targets in,
381–384
Amyloid pathology
dystrophic neurites generation, 396–398

neurotransmitter-specific progression of,
393–398
structural vulnerabilities to extracellular
Aβ burden, 393–396
β-Amyloid peptide, see B-Amyloid protein
Amyloid plaques
targeting with phage anti-AβP ScFv,
275–277
in vivo targeting of, 274
β-Amyloid precursor protein (βAPP),
see Amyloid-β protein precursor
(AβPP) gene
β-Amyloid protein, 293, 305, 318, 325,
334, 337, 369, 375, 382–383,
394, 411, 418, 439, 445, 451–452,
456–458
immunization with phage-ScFv against,
274–275
phenserine and cymserine effect on,
445–458
Amyotrophic lateral sclerosis (ALS), 81,
231, 298
Amyotrophy, 81, 85–86, 128
Anarthria, 32
Annonacine models of PD, 117
Anti-β-amyloid antibodies, intranasal
administration, 273–277
Antithrombotic prophylaxis, 32
Anxiety, 3
A30P α-synuclein mice, 160–161
antibodies and immunohistochemistry, 99
brain tissue extraction, 100
carbonyl detection, 101
DJ-1 co-immunoprecipitates with tau
and phospho-tau in brain
homogenates, 105–106
DJ-1 expression in, 97–109
DJ-1 levels, 102–103
DJ-1 presence in reactive astrocytes in
brain sections from Contursi
kindred, 103–105
DJ-1 siRNA
knockdown confers susceptibility to
oxidative insult in cell culture,
106–107
transfection, 101–102
immunoblot analysis, 100
immunoprecipitation, 101
LDH assay, 102
oxidative stress response, 103, 108
quantification of immunoblots, 100
Apathy, 2–3
Aphagia, 32

APLP-1 proteins, 260
APLP-2 proteins, 260
ApoE-ApoE receptor, 212
Apolipoprotein E3 (ApoE3), 343, 355–356
    alleles, 213, 217
    deficient mice, 344–351
        F4/80-positive microglia, 345,
            348–351
        IL-1β-positive cells, 345–348, 351
        immunohistochemistry, 345
        intracerebroventricular injection of
            LPS, 344
        mac-2-positive microglia, 345,
            348–351
        NF-κB-positive cells, 345–348,
            350–351
    knock-in mice, 387, 390
Apolipoprotein E4 (ApoE4)
    alleles, 135, 137, 213–214, 217, 298
    brain inflammation in AD patients, 343
    isoform-specific effects on brain
        inflammation, 343, 345
    knock-in mice, 387, 390–391
    NF-κB signaling activation by, 343–351
    transgenic mice, 344–351
        F4/80-positive microglia, 345,
            348–351
        IL-1β-positive cells, 345–348, 351
        immunohistochemistry, 345
        intracerebroventricular injection
            of LPS, 344
        mac-2-positive microglia, 345, 348–351
        NF-κB-positive cells, 345–348,
            350–351
Apolipoprotein E2 (ApoE2) alleles, 213, 217
Apolipoprotein E (ApoE) gene, 172, 191,
        206, 208, 211, 213–216, 221
    CYP2D6 polymorphisms association
        with, 365
    functional genomics studies, 356
    interactions with liver function and drug
        metabolism, 361
    pharmacogenetics, 361–363
    pleiotropic effects in dementia, 356–361
    related
        aortic atherosclerosis, 359
        cholesterol levels in AD, 357–358
        GOT, GPT and GGT activity in
            AD, 359
        lipid metabolism and
            atherosclerosis, 361
        plasma triglyceride levels in AD,
            357–358

Apomorphine, 32–33
Apoptosis, 133–134, 247
Apoptosis associated with
        neurodegeneration
    glycosaminoglycans role, 249–253
        animals for, 249
        drugs and chemicals, 249–250
        immunohistochemistry, 250–251
        tissue preparation for histochemical
            assays, 250
APP CT100 transgenic mice, 430–431
APP genes, 114, 171
APP$_{K670N, M671L}$ + PSI$_{M146L}$ transgenic
        mutant mouse models, 394–396
APP$_{K670N, M671L}$ (tg2576) transgenic mutant
        mouse models, 394
APP$_{K670N, M671L\ 03}$ transgenic mutant mouse
        models, 394
APP (SW) Tg2576 transgenic mice, 431
Arctic-type Aβ, 390
Argyrophilic grain disease (AGD), 375–379
    frequency in AD, 377–379
    immunohistochemistry, 376–377
    relation with 4R tau score in AD, 378
Argyrophilic grains (AGs), 375
Arylesterase, 362
A53T α-synuclein mutation, 98, 104
Ateroid, therapeutic effect, 240
ATP synthase-α, 156
Atropine, 440–441
Autologous tissues, 51
Autophagy, 334
Autosomal recessive juvenile parkinsonism
        (ARJP), 127

BACE I, 221, 223
Base excision repair (BER), 17
Basic fibroblast growth factor (bFGF),
        53–54
Basque families with LRRK2 gene, clinical
        and pathological description of, 84
Bcl-2 gene, 320
BDNF mRNA, 280–281
Behavioral-occupational therapy, 24
Behavioral problems, 11
BE-M17 cells, 101–102
Biochemical for PD detection, 41–42
Bioluminescence imaging technique, 175
Bioluminescent imaging
    of brain injury in SBE-luc reporter mice,
        179–181
    excitotoxic and endotoxic brain injury
        in mice, 175–181

method, 176–177
models, 176
results and discussion, 177–181
stereotactic injections, 176–177
tissue preparation, 177
Biomarkers
for AD, 169–173
in body fluids of PD patients, 173
CSF for AD, 172
definitions, 170–171
for PD detection, 35–43, 169–171
biochemical markers, 41–42
definitions, 39
functional imaging, 40
gene expression in blood, 43
genetic markers, 42–43
neuromelanin blood test, 42
tests on olfaction, 41
tests on vision, 41
transcranial ultrasonography, 40
translation of pathological findings,
37–39
qualities, 169–170
uses of, 169–170
Bipolar disorder, 317
Blessed Dementia Scale score, 173
Blood-brain barrier (BBB) in PD, 36–37,
59–60, 240, 273, 275, 277, 371
α3β4 nAChRs receptors, 464, 466–467
α4β2 nAChRs receptors, 466–468, 469–470
Body mass index (BMI), 207
Bone fracture, 32
Bone morphogenetic proteins (BMPs), 175
Boundary shift integral (BSI) methods, 187
Braak stages in PD, 39
Brachydactyly, 13
Bradykinesia, 23, 26, 76, 78–79, 82, 85, 159
Brain
cholesterol homeostasis in, 211–213, 217
inflammation in ApoE4 AD patients, 343
Brain-derived neurotrophic factor (BDNF)
cells, 93–94, 196, 258, 291–292, 294
in AD, 279–281
Brief Smell Identification Test (B-SIT), 41
Bromocriptine, 33
Bronchopneumonia, 32
Butyrylcholinesterase (BuChE), 362, 405,
449, 452–457

Calpains, 309, 334
Campbell-Switzer silver stain, 414, 418
Carbidopa therapy, 76, 78–80, 82–85
β-Carbolines, 36

Carbonic anhydrase II (CA-II), 156, 159–161
Cardiac arrhythmias, 32
Cardinal motor symptoms, 36
Cardiovascular Risk Factors, Aging,
and Dementia (CAIDE) study,
205–209
Caspase-3 activation, 247
Caspase-3 enzyme, 134
Caspase-3 immunoreactivity,
intracerebroventricular AF64A
administration effect on, 251–252
Catalase, 193, 362
Catechins gallate esters, 293
Catechol-O-methyltransferase (COMT), 362
Cathepsin B, 309
CD4 cells, 53
CD8 cells, 53
CDCrel-1 protein, 11
CDNA microarray technology, 19
CDR power of attention tests, 5
CD4+ T cells, 268
CD8+ T cells, 268
Cell line grafting for PD treatment,
51–55
and encapsulation, 52
intracerebral, 54
neurotransmitter and neurotrophic
factor secretion control after,
54–55
neurotransmitter-secreting, 52–53
neurotrophic factor-secreting, 53
simultaneous delivery of
neurotransmitter and
neurotrophic factor, 53–54
Cellular prion protein (PrPᶜ), 305
Central nervous system (CNS), 211, 217, 285,
319, 405, 413, 416, 463
apolipoprotein E and cholesterol
transport in, 213–214
selective nAChRs potentiators, 466–468
Cerebellar ataxia, 11
Cerebellar dysfunction, 76
Cerebral amyloid angiopathy (CAA), 390
Cerebral infarcts, 135, 137
Cerebro spinal fluid (CSF), 171–172, 213,
222, 273–274, 298, 395, 413,
416, 431
ChAT immunoreactivity, 249, 251
intracerebroventricular AF64A
administration effect on, 252–253
ChEI tacrine (9-amino-1,2,3,
4-tetrahydroacridine
hydrochloride), 458

Cholesterol
   biosynthetic pathway, 224
   Aβ production, 221–228
   de novo synthesis assay, 223–226
   homeostasis
      in brain, 211–213, 217
      and cholesterol-lowering drugs,
         214–217, 222
      identifying target enzyme, 226–227
      lowering drugs, 214–217
      synthesis in eukaryotic cells, 212
      transport and production in AD, 211–217
Choline acetyltransferase (ChAT)
   enzyme, 286, 403–404, 415–416, 432–434
   immunohistochemistry, 249
   See also ChAT immunoreactivity
Cholinergic-amyloid fusion hypothesis, for
      Aβ deposition, 411–421
Cholinergic basal forebrain (CBF) neurons,
      285–286
Cholinergic changes in AD, 393–394,
      403–405, 407
Cholinergic neurites, in AD plaques, 432
Cholinergic neurons
   deficits, 4, 6
   role in synaptic scaling, 433–434
Cholinergic therapy
   of immunotoxin-lesioned animals, 415,
      419–420
   of normal animals, 413, 416
Cholinesterase inhibition, β-amyloid role in,
      439–443
Cholinesterase inhibitors (ChEIs), 4, 6,
      445–447, 454
Chorea, 76
Chromaffin cells, 51, 55
Ciliary neurotrophic factor (CNTF), 53, 55
C-Jun N-terminal kinase (JNK), 142, 295
Clioquinol, 454
4396C monoclonal antibody, 388
Cognitive decline, decreased ProBDNF as
      cause of, 279–281
Cognitive impairment symptoms, 23
Cogwheel rigidity, 159
Conformational disease, 381
Continuous administration of MPTP model
      of PD, 62–63
Coomassie Blue staining, 151
Cortical cholinergic lesion, causes of Aβ
      deposition, 411–421
Corticobasal degeneration (CBD), 40, 231
COS-7 cells, 226–227, 311
Creatine kinase BB (CK), 156

Creutzfeldt-Jakob disease (CJD), 172, 231, 307
CSF biomarkers, 172
C3435T polymorphism, 36
Cultured hippocampal neurons, patch-
      clamp recordings in, 465–466
Curcumin, 454
Cu/Zn superoxide dismutase, 193
α-Cyano-4-hydroxycinnamic acid, 153
Cyclooxygenase-2 (COX-2) protein
      expression, 20
Cymserine, 446–447
   analogs, 446
   assessed at nontoxic doses in SK-N-SH
      cells, 449–450
   differential effects on
      soluble Aβ, 451–452, 456–458
      soluble APP levels, 450–451, 456–458
   effect on
      amyloid pathway, 454
      cell viability, 450
      cholinesterase inhibition, 452
      human AChE and BuChE inhibition,
         452, 455–458
   lipophilicity of, 453–454, 457
   pharmacologic properties, 453–454
CYP1A1, 362
CYP1A2, 362–363
CYP2A6, 362
CYP3A4, 362–363
CYP3A5, 362
CYP3A7, 362
CYP1B1, 362
CYP2B6, 362
CYP2C8, 362
CYP2C9, 362
CYP2C19, 362
CYP2D6, 362–365
   ApoE genotypes association with, 365
   genotypes in AD, 364
   polymorphisms in AD, 363–364
CYP2E1, 362
Cysteine-rich extracellular domain
      (CRD), 319
Cytochrome P450 enzymes, 362
Cytokines, from activated microglia
      in putamen in PD, 92
Cytokines gene expression in PD, 91–95
   in hippocampus and putamen, 93–94

Dantrolene, 33
Dardarin gene, 13
Dardarin R1396G mutation, 84
Deep brain stimulation surgery, 51

Delusions, 3
Dementia, 11, 76, 85–86, 128
    ApoE pleiotropic effects in, 355–365
    associated with PD
        behavioral features, 3
        clinical features, 1–3
        cognitive features, 2–3
        treatment, 3–6
Dementia with Lewy bodies (DLB), 2, 9, 34,
        40, 91, 126, 232
    cortical AChE and midfrontal ChAT
        activity, 4
Dementiform cognitive impairments, 38
Depression, 3, 172
Detergent-resistant microdomains (DRMs),
        390–391
Developmental processes, Wnt signaling
        in, 317
Diabetes mellitus, 135, 137
3,4-Diaminopyridine, 441
Dihydropyrimidinase-related protein 2
        (DRP2), 156, 158
Dihydropyrimidine dehydrogenase
        (DPD), 362
2,5-Dihydroxybenzoic acid, 153
2,4-Dinitrophenylhydrazine (DNPH), 149, 151
Disinhibition, 3
Disseminated intravascular coagulation, 32
Distal limb muscle weakness, 81
DJ-1-deficient mice, 65
DJ-1 gene, 13, 15, 76, 97–98, 127, 159, 161
    expression in A30P α-synuclein mice,
        97–109
    mutations in, 171
D-KEFS Test Battery, 5
DNA repair systems
    base excision repair (BER), 17
    mismatch repair system (MMR), 17–18
    mitochondrial dysfunction and
        excitotoxicity sensitivity, 17–20
    nucleotide excision repair (NER), 17
Donepezil, 4, 405, 439–442
Dopa/dopaminequinone, 11
Dopamine β-hydroxylase (DBH), 414
Dopamine (DA), 10, 33, 36–37, 40, 51–55,
        91, 173
    α-hydroxylase immunostaining, 57
    inhibition of UP system in PC12 cells, 67
    metabolism, 57, 67
    producing cells, 51–52
    secreting cell line, 52, 54
    secretion from grafted PC12 cells, 52–54
    secretion from PC12th Tet-Off cells, 54–55

Dopaminergic nigrostriatal system, 37
Dopaminergic system, 38, 54
Dopaminomimetic therapy, 76, 83
Dorsal glossopharyngeus-vagus complex, 117
Down's syndrome (DS), 136, 140, 142, 158,
        192, 194, 300
Downstream signaling systems, α-secretase
        pathway up-regulation, 370–371
Doxycycline, 55
Drosophila, 319, 433
Drosophila melanogaster, 58
DSM-IV criteria, 4, 206
Dual task effect, 26–27
Duodenal ulcer, 32
Dutch-type Aβ, 390
Dysautonomia, 31, 76
Dysexecutive syndrome, 2
Dyskinesia, 25, 78–80
Dystonia (Meige syndrome), 76, 81, 85, 87

EAAT2 (GLT-1), 157
Eeserine, 446
Electrospray ionization (ESI), 152–153
Embryonic stem (ES) cells, 256
Encapsulated cell line grafting, for
        neurological disorders, 52, 55
α-Enolase, 156, 160
Enzyme-linked immunosorbent assay
        (ELISA), 92, 172, 269, 288, 412,
        415, 418, 447–448, 451–452, 456
Epigallocatechin gallate (EGCG), 293–294
Ethylcholine aziridinium (AF64A), 248–251
    intracerebroventricular administration
        effect on
        caspase-3 immunoreactivity,
            251–252
        ChAT immunoreactivity, 252–253
Eukaryotic cells, 17, 212
Excitatory neurotransmitter systems, 38
Excitotoxic and endotoxic brain injury in mice
    bioluminescent imaging, 175–181
        models, 176
        stereotactic injections, 176–177
        tissue preparation, 177
Excitotoxicity sensitivity and impaired DNA
        repair systems, 17–20
Executive functions, abnormalities in, 2

Faber disease, 227
Fabry disease, 227
Familial AD (FAD), 124, 197–198,
        221, 298
Familial PD, 9–15, 35, 43, 76, 97, 127

Family A (German-Canadian) with LRRK2
        gene
    clinical and pathological description of,
        79–81
    neuropathology of, 81
    pedigree of, 80
    PET studies, 82
Family D (Western Nebraska) with LRRK2
        gene
    clinical and pathological description of,
        75–79
    pedigree of, 77
    PET studies, 82
Family 469 with LRRK2 gene, clinical and
        pathological description of, 82
Fasciculations, 81
Fatal familial insomnia, 231
Fetal nigral cells, 51
FINMONICA study, 206
Flemish-type Aβ, 390
Flu, 32
Fluorescent imaging plate reader (FLIPR),
        464–465
Flurbiprofen, 124
FOXO3a, IGF-1 effect involving PI3/AKT
        pathway and, 292
Frizzled genes, 319–320
    expression in rat neuronal cells,
        317–322
    present in proximal TCF/LEF binding
        sites, 320
Frontotemporal dementia (FTDP-17), 325,
        337–338
Frontotemporal dementias (FTDs), 123, 235
    CSF biomarkers for, 172
    protein misfolding manifestations,
        123–129
    tau mutations in, 326, 333
Functional imaging, and biomarkers for PD
        detection, 40

GAβ
    determinant for area-specific deposition
        of Aβ, 390–391
    generation in brain, 389–390
    immunohistochemistry and
        immunoprecipitation of, 389
    immunological detection in brain,
        388–389
    monoclonal antibody specific to, 388
    as seed for Alzheimer amyloid, 388–391
Gait disturbances symptoms, 23
Gait hypokinesia, 25

Galantamine, 4, 405, 439–443
Galanthamine, 464
Ganglioside, 222–223
Ganstigmine, 440–442
Gastric stasis, 32
Gaucher's disease, 128, 227
Gene expression, in blood for PD
        detection, 43
Genetic markers, for PD detection, 42–43
Genetic models, of PD, 63–67
Gerstmann-Straussler-Scheinker
        disease, 232
Ginkgo biloba, 454
Glial cell line-derived neurotrophic factor
        (GDNF), 24, 53–54
Glial cells, 233
Glial fibrillary acidic protein (GFAP), 406
Glial glutamate transporter (GLT-1), 406
Global Cognitive Score (GCS), 281
Glucocerebrosidase gene (GBA), 128
Glucose-6-phosphate dehydrogenase, 362
Glucosyltransferase, 362
Glu46Lys mutation, 10
Glutamate oxaloacetate transaminase
        (GOT), 359–361
Glutamate-pyruvate transaminase (GPT),
        359–361
Glutamatergic changes, in AD, 405–407
Glutamatergic dystrophic neurites, in human
        AD brain, 397–398
Glutamatergic transmission, 38
Glutamate transporter (GLT-1), 155
Glutamine synthase (GS), 156–157
γ-Glutamyl transaminase (GGT), 359–361
Glutathione-S-transferase (GST), 155,
        310, 362
Glutathione-S-transferases A (GST-A), 362
Glutathione-S-transferases M (GST-M), 362
Glutathione-S-transferases P (GST-P), 362
Glutathione-S-transferases T (GST-T), 362
Glycogen synthase kinase-3β (GSK-3β), 318,
        337–338
Glycogen synthase kinase-3 (GSK-3), 126,
        140, 337–338
Glycosaminoglycans (GAGs)
    and analogs in neurodegenerative
        disorders, 231–241
    mimetics, 124
    neuroprotective role of, 247–248
    role in apoptosis associated with
        neurodegeneration, 249–253
    synthesis and physiological functions,
        232–233

therapeutic implications, 237–241
See also Sulfated glycosaminoglycans
(sGAGs)
Glycosylated alpha-synuclein, 11
GM1-bound Aβ (GAβ), *see* GAβ
GM1 ganglioside (GM1), 387–391
GM1-gangliosidosis, 227
GM2-gangliosidosis, 227
G protein-coupled receptors activation, for
α-secretase pathway up-
regulation, 370–371
Gracile axonal dystrophy (gad), 12
G2019S families with LRRK2 gene, clinical
and pathological description of,
83–84
GSHPx, 193
GSSG-R, 193
Gut immune system, 269–270

Hallervorden-Spatz disease, 38
Hallucinations, 3, 38
HEK293 cells, 266–267, 311, 318, 371
Hematoxylin and eosin (H&E) staining, 267
Heme oxygenase-1, 193
Hemiparkinsonism-hemiatrophy, 11
Heparan sulfate proteoglycan (HSPG),
414–415
Heparin, 239, 241
HepG2 cells, 371
Heptylphysostigmine, 441
Hereditary nonpolyposis colorectal cancer
(HNPCC), 17
Hexahydropyrroloindole, 446
High density lipoprotein (HDL), 213, 217
Hippocampal cells, 318
Hippocampal neurons, tea extracts
protection of, 293
Histamine methyltransferase (HMT), 362
HIV-1 Nef gene, 94
Hoehn and Yahr scale, 36
HtrA2/Omi gene, 11–12
Human ADAM10 promoter, and
α-secretase pathway
up-regulation, 371–373
Human amniotic epithelial cell, 54
Human ApoE3 transgenic mice, 344–351
F4/80-positive microglia, 345, 348–351
IL-1β-positive cells, 345–348, 351
immunohistochemistry, 345
intracerebroventricular injection of
LPS, 344
mac-2-positive microglia, 345, 348–351
NF-κB-positive cells, 345–348, 350–351

Human cerebrovascular smooth muscle cells
(HCSMs), 390
Human CNTF protein, 55
Human-derived cell line, intracerebral
grafting for PD, 54
Human embryonic kidney (HEK) cells, 266
Human Frizzled genes,TCF/LEF boxes
found in, 320
Human leukocyte antigen-DR (HLA-DR), 91
Human MeSCs (HMeSCs), 256–262
Human neuroblastoma (SK-N-SH) cells,
61, 448
APP measurement, 448
Aβ assay, 448
cell culture, 448
cholinesterase measurements, 449
LDH and MTT assays, 448
lipophilicity determination, 447–448
phenserine, cymserine, and
physostigmine
assessed at nontoxic doses in,
449–450
differential effects on soluble Aβ,
451–452, 456–458
differential effects on soluble APP
levels, 450–451, 456–458
effect on cell viability, 450
effect on cholinesterase inhibition, 452
effect on human AChE and BuChE
inhibition, 452, 455–458
lipophilicity of, 453–454, 457
pharmacologic properties,
453–454
Human SH-SY5Y neuroblastoma cells, 18
Huntington's disease, 55, 139, 381–382
8-Hydroxy-2'-deoxyguanosine, 41
6-Hydroxydopamine (6-OHDA), 53, 58–59, 91
model of PD, 24, 58–59, 62
8-Hydroxyguanosine (8OHG), 41, 194
3-Hydroxy-3-methylglutaryl-CoA (HMG-
CoA) reductase, 212, 214–216,
222–223, 226
3-Hydroxy-3-methylglutaryl reductase
(HMGR), 211–212, 214, 216,
223, 226
4-Hydroxy-2-trans-nonenal (HNE), 149,
151, 155, 158
Hypercholesterolemia, 135, 137, 221–222
Hyperhomocysteinemia, 135, 137
Hyperhydrosis, 11
Hyperreflexia, 85–86
Hyperthermia, 31–32, 34
Hyposmia, 41

Ideomotor apraxia, 3
Idiopathic Parkinson's syndrome (IPS), 35
βIII-tubulin, 257, 259, 261
Ile93Met mutation, 65
Immunoprecipitation, 150
Immunotoxin lesion of nbm, 413–419
Impaired attention, 2
Impotence, 11
Inducible nitric oxide synthase (iNOS), 149
Insertion/deletion loops (IDLs), 17
Institutional Animal Care and Use
        Committee (IACUC), 176
Insulin-like trophic factor-1 (IGF-1), 291
    neuroprotective effects of, 292
    PI3/Akt pathway and FOXO3a, 292
Interleukin-6 (IL-6), 92–94
Intracerebral cell grafting, 51
    of cell lines, 53
    of human-derived cell line, for PD, 54
Intracerebroventricular AF64A
            administration effect on
    caspase-3 immunoreactivity, 251–252
    ChAT immunoreactivity, 252–253
Intraneuronal Aβ and AD, 297–301
In Vivo Imaging System, 177
Iron-regulatory protein (IRP), 458
Irritability, 3
Islet amyloid polypeptide (IAPP), 383
Isoelectric focusing (IEF), 151

Jejunal volvolus, 32

Kainic acid (KA), 18, 20
    excitotoxin, 176, 179–180
    induced brain injury bioluminescent
            imaging, 179–180
Ki-M1p-positive resting microglia, 92
Krabbe disease, 227
Kufor-Rakeb syndrome, 14
Kuopio MCI study, 206
Kuru, 231
KXGS motifs, 333

Lactate dehydrogenase (LDH), 448–450, 456
Language deficits, 2–3
Laryngitis, 32
L-dopa-responsive parkinsonism, 9–10, 12, 14
Leucine-rich repeat kinase 2 (LRRK2) gene,
    see LRRK2 genes
Leukocyte function associated antigen-I
        (LFA-I), 92
Leu17 peptidyl bond, 310–311
Levodopa therapy, 76, 78–80, 82–85

Lewy bodies disease (LBD), 40–41, 67, 91,
        117, 127
    CSF biomarkers for, 172
    cytokines gene expression in
        hippocampus and putamen,
            93–94
    neuroprotective and neurotoxic subtypes
        of activated microgila in, 94
Lewy bodies (LBs), 9–10, 23, 34, 36, 38, 58,
        60, 63–64, 66, 68, 78, 81, 91, 97,
        107, 117, 123–124, 126–128, 133,
        138–140, 232, 237
Lifestyle-related factors and AD, 208
Lipids and AD, 222–223
Lipophilicity determination, 447–448
Lipopolysaccharide (LPS) endotoxin, 176,
        179–180
Lipopolysaccharide models, of PD, 117
LivingImage software (Xenogen), 177
Low density lipoprotein (LDL), 212,
        215–217
LRP5 proteins, 319
LRP6 proteins, 319
LRRK2 genes, 13–15, 127–128
    Basque families and British Kindred, 84
    clinical and pathological
            characterization, 75–88
    family 469, 82, 86
    family A (German-Canadian), 79–81, 86
    family D (western Nebraska), 75–79, 86
    G2019S families, 83–84
    mutations in, 171
    PET studies, 82
    and ROCO gene, 82–83
    Sagamihara family, 75, 84–86
Lrrk2 protein, 82–83, 85
Lymphoid enhancer-binding factor (LEF), 318
Lysine 63-linked ubiquitylation, 11
Lys16 peptidyl bond, 310–311

Major histocompatibility complex (MHC),
        91–93
Malignant syndrome (MS), 31–33
MAPKKK (mitogen-induced protein kinase
        kinase kinase), 14
MAPT gene, 333, 355–356
Matrix-assisted laser desorption/ionization
        (MALDI), 152–153
MDR1 gene, 36–37
Median forebrain bundle (MFB), 58–59
Memory functions, in PDD, 2
Memory impairment, 2
Mendelian, 355

Meningoencephalitis, 268–269
Mental status, alterations of, 31
Meperidine, 59
Mesenchymal stem cells (MeSCs), 256–258
Metachoromatic leukodystrophy, 227
Methamphetamine (MA), 58
3,4-Methylenedioxymethamphetamine
  (MDMA), 58
1-Methyl-4-phenyl-2,3-dihydropiridinium
  (MPDP+), 60
1-Methyl-4-phenylpyridinium (MPP+), 60
1-Methyl-4-phenyl-1,2,3,
  6-tetrahydropyridine (MPTP)
  intoxication, 36, 52, 58, 63, 91, 97, 108,
    117, 161
  models of PD, 59–60, 62
Metrifonate, 439–442
Michael addition reaction, 149
Microglia in PD, see Activated microgila
  in PD
Mild cognitive impairment (MCI), 136, 172,
    184, 194–195, 205–209, 285, 287,
    297, 372, 394, 432
  in AD, 281, 285
    basocortical cholinergic markers
      in, 286
    cholinotrophic alterations in, 286
    NGF receptor levels in, 286–287
    pro-NGF increases in, 287–288
  vascular risk factors in, 206–208
Mini-Mental Status Examination (MMSE),
  5, 281, 289
Mismatch repair system (MMR), 17–18
Mitochondrial dysfunction, and impaired
  DNA repair systems, 17–20
Mitogen-activated protein kinase (MAPK)
  pathway, 294
Mitogen-activated protein (MAP) kinase, 442
Monoamine oxidase B (MAO-B), 41, 362
Morris-Water-Maze test, 370
Mouse embryonic fibroblast (MEF) cell, 223
Movement, slowness of, 25
MPTP models of PD, 59–60, 62
Msh2 gene, 18–19
Msh2+/- mice, 18–20
MSH2-MSH6 (MutSα), 17
MSH2-MSH3 (MutSβ), 17
Msh2-null mice, 18
MTT [3-(4,5-dimetyl-thiazol-2-yl)-2,
  5-diphenyltetrazolium bromide],
  448, 450
Multidrug resistant protein-1 (MRP-1), 155
Multiple sulfatase deficiency, 227

Multiple system atrophy (MSA), 3, 40
Muscle atrophy, 81
Muscle enzymes, incremental increases of, 31
MutL and Muts homologs, 17

N2a cell lines, 325–328, 330–333
N-acetyltransferase, 362
N-acetyltransferase 1 (NAT1), 362
N-acetyltransferase 2 (NAT2), 362
α–7 nAChRs receptors, 466–470
NADPH oxidase expression, 94, 362
NADPH-quinone oxidoreductase
  (NQO1), 362
N2a/Tet-On/K18-ρK280 cell line, 327
N2a/Tet-On/K18-ρK280/2P cell line, 327
N2a/Tet-On/ K18-WT cell line, 327
N2a/Tet-On/pBI-5 (mock) cell line, 327
Natural products, neuroprotective effects of,
  291–295
Necrosis, 133
Nerve growth factor (NGF), 53, 197, 248,
  285–286, 288–289
N-ethylmaleimide-sensitive factor (NSF), 158
Neural stem cells (NSCs), 255–256,
  260–262
Neuroblastoma cells, 68
Neurodegeneration
  mechanisms of, 149–162
  oxidative insults and cell cycle reentry at,
    135–139
  and tau polymerization, 337–340
Neurodegeneration with brain iron
  accumulation (NBIA), 38
Neurodegenerative disorders
  glycosaminoglycans and analogs in,
    231–241
Neurofibrillary tangles (NFTs), 123, 133,
  138–140, 154, 231, 233, 235,
  237–238, 240, 298, 337, 356,
  375–376, 411, 429
Neuroleptic malignant-like syndrome
  (NMLS), 31–33
Neuroleptic malignant syndrome (NMS), 31,
  33–34
Neurological disorders
  cell line grafting for, 55
  encapsulated cell line grafting for, 55
Neuromelanin blood test, for PD
  detection, 42
Neuronal nicotinic acetylcholine receptors,
  431–432
α-7 Neuronal nicotinic acetylcholine
  receptors, 431–432, 434

Neuronal nicotinic cholinergic receptors, allosteric potentiators of, 463–471

Neuropil threads (NTs), 375, 377

Neuropolypeptide h3, 156–157

Neuropsychiatric Inventory (NPI), 5

Neurotoxins, 36

Neurotransmitter
control after grafting into brain, 54–55
delivery by cell line grafting for PD, 53–54
secreting cell line grafting for PD, 52–53
specific progression of amyloid pathology, 393–398

Neurotrophic factor
control after grafting into brain, 54–55
delivery by cell line grafting for PD, 53–54
secreting cell line grafting for PD, 53

NF-κB signaling, 294
enhanced activation by apolipoprotein E4, 343–351

Nicotinic acetylcholine receptors (nAChRs), 463–464
CNS selective potentiators, 466–468
selective α–7 potentiators, 468–470

Niemann-Pick disease, 227

Nigrostriatal dopaminergic system, 51

NINCDS-ADRDA criteria, 206

3-Nitrotyrosine (3-NT), 149–151

N-methyl-Daspartate (NMDA) antagonist memantine, 157, 403, 405–407, 445, 454, 471

No cognitive impairment (NCI), 285, 287

Nonamyloidogenic metabolism of AβPP, muscarinic cholinergic receptor stimulation effect on, 440–442

Nonsteroidal antiinflammatory drugs (NSAIDs), 124, 136–137

Noradrenergic locus coeruleus, 38

Norepinephrine transporters (NET), 60

North Karelia Project, 206

Nucleotide excision repair (NER), 17

Nucleus basalis magnocellularis, immunotoxin lesion of, 413–419

Nucleus basalis (NB), 285–286

Nucleus basalis of Meynert (nbM), cholinergic neurons deficits in, 4

8-OH-2α-deoxyguanosine, 150

Olfaction, tests for PD detection, 41

Olfactory dysfunction, 41

Oligomers, 382

Omi gene, 11–12

Oncogenic processes, Wnt signaling in, 317

Orthostatic hypotension, 11

Oxidative insults and cell cycle reentry, at neurodegeneration, 135–139

PACAP-induced activation, for α-secretase pathway up-regulation, 371–342

Paclitaxel, 126

PAEL receptor protein, 11

Paired helical filaments (PHFs), 233

Paraoxonase, 362

Paraquat, 36

PARK2, mutations in, 171

PARK8, 13–14, 75–88

Parkin gene, 10, 13, 15, 76, 88, 97, 109, 127, 137, 159
models of PD, 64–65

Parkin-knockout mouse model, 64

Parkin protein, 10–11, 58

Parkinsonian hyperpyrexia (PH), 31–32

Parkinsonism-dementia complex of Guam, 231

Parkinson's disease dementia (PDD), 126
behavioral features, 3
clinical features, 1–3
cognitive features, 2–3
treatment, 3–6

Parkinson's disease (PD), 381–382
activated microgila, 91–95
neuroprotective and neurotoxic subtypes in brain, 94–95
substantia nigra and putamen in brain, 92–93
biochemical pathways mediated by genes implicated in, 97–109
biomarkers for, 35–43, 169–173
blood-brain barrier (BBB) in, 36–37
Braak stages, 39
causes, 35–36
cell cycle reentry in, 138–139
cell grafting for, 51–55
cell replacement therapy for, 256
characteristics, 35
cognitive rehabilitation, 27
cytokines gene expression in, 91–95
hippocampus and putamen, 93–94
definition, 23
dementia associated with
behavioral features, 3
clinical features, 1–3
cognitive features, 2–3
treatment, 3–6
See also Dementia

diagnosis of, 36–37
familial forms, 9–15, 43
  PARK1, 9–10, 76
  PARK2, 10–11, 76, 127
  PARK3, 11–12, 76
  PARK4, 12, 76
  PARK5, 12, 76
  PARK6, 12–13, 76, 127
  PARK7, 13, 65, 76, 127
  PARK8, 13–14, 75–88, 128
  PARK9, 14, 76
  PARK10, 14, 76
  PARK11, 14, 76
genes associated with, 63–67
genetic causes, 36
inflammation and pathological cascade
    in, 117–118
models, 58–67
  annonacine, 117
  continuous administration of MPTP,
    62–63
  genetic models, 63–67
  6-Hydroxydopamine, 58–59, 62
  ideal model, 58–59, 62
  lipopolysaccharide, 117
  MPTP, 59–60, 62, 117
  parkin, 64–65
  rotenone, 60–62, 117, 138
  subcutaneous rotenone exposure, 62
  α-synuclein, 66–67
  Uch-L1, 65
  in vivo environmental models, 62–63
motor
  impairment in, 25–26
  rehabilitation for, 23–28
neurochemical findings, 37–38
neurogenetics in, 9–15
neuroinflammation in early stages,
    113–119
neuronal
  death in, 133–134
  survival response, 139–140
nosology of, 125, 128–129
oxidative insult and neuronal survival
    role, 133–142
  Aβ, tau, and α-synuclein function
    against, 140–142
  genetic mutations and risk factors,
    135–136
oxidative stress, 41, 135–138
  with genetic mutations and risk
    factors, 135–136
paramedical therapies, 24

pathology, 38, 139–140
physical therapy in, 24–25
physiological strategies for motor
    problems in, 26
protein misfolding manifestations,
    123–129
redox proteomics in, 159–161
synucleinopathies, 126–129
therapeutic target in, 68–69
Patients with AD
  acute akinesia in, 31–34
  biomarkers in body fluids of, 173
  cortical AChE activity, 4
  memory impairment in, 2
  midfrontal ChAT activity, 4
  visuospatial function in, 2
  See also Alzheimer's disease (AD)
PC12 cells, 52, 67–68
PC12-GDNF cell line, 54
PC12th Tet-Off cells, 54–55
Peptide mass fingerprinting, search engines
    for, 153
Peptidyl prolyl cis-trans isomerase (Pin 1), 156
Peripheral nervous system (PNS), 214
Peripheral neuropathy, 11
Peroxiredoxins, 193
P-glycoprotein (P-gp) protein, 36
Phenserine, 441–442, 446–447
  assessed at nontoxic doses in SK-N-SH
    cells, 449–450
  differential effects on
    soluble Aβ, 451–452, 456–458
    soluble APP levels, 450–451,
      456–458
  effect on
    amyloid pathway, 454
    cell viability, 450
    cholinesterase inhibition, 452
    human AChE and BuChE inhibition,
      452, 455–458
  lipophilicity of, 453–454, 457
  pharmacologic properties, 453–454
PHFtau proteins, 125
Phosphatidylethanolamine-binding protein
    (PEBP), 157–158
Phosphatidylinositol 3-kinase (PI3K), 291
Phosphoglycerate mutase 1(PGM1), 156
Physiological processing of β-APP
    and TSEs, 305–312
  ADAM9 and, 305, 311–312
  ADAM10 and, 305, 309–313
  ADAM17 and, 305, 309–310
  up-regulation by PKC agonists, 307–309

Physostigmine, 441, 446–447, 464
　assessed at nontoxic doses in SK-N-SH
　　cells, 449–450
　differential effects on
　　soluble Aβ, 451–452, 456–458
　　soluble APP levels, 450–451, 456–458
　effect on
　　amyloid pathway, 454
　　cell viability, 450
　　cholinesterase inhibition, 452
　　human AChE and BuChE inhibition,
　　　452, 455–458
　lipophilicity of, 453–454, 457
　pharmacologic properties, 453–454
Pick's disease, 98, 231
PINK1 (PTEN-induced kinase 1) gene,
　　12–13, 15, 76, 97, 127, 159, 161
　mutations in, 171
Pituitary adenylate cyclase-activating
　　polypeptides (PACAP), 371
Placebo, 4–6, 24, 34
Plasma membrane, as target in amyloid
　　oligomer disease pathogenesis,
　　381–384
Plasmin, 309
PL kindred, British family, 84
P75 neurotrophin receptor (p75NTR),
　　285–289, 413
Positron emission tomography (PET), 36, 40
Postsynaptic density protein 95 (PSD-95),
　　321–322
Postural instability symptoms, 23
Postural reflexes, loss of, 23
Presenilin-1 (PS-1) gene, 114, 124, 135, 137,
　　139, 171, 198, 221–222, 298,
　　355–356
Presenilin-2 (PS-2) gene, 114, 124, 135, 137,
　　139, 171, 198, 221–222, 298,
　　355–356
Presynaptic cholinergic markers, 4
Prion disease, 113, 231, 238
Prion protein, interactions with sGAGs,
　　235–237
ProBDNF level, cause of cognitive decline
　　and neurodegeneration, 279–281
Prodromal AD, TrkA/ProNGF balance in,
　　285–289
Progressive SPS, 41
Progressive supranuclear palsy (PSP), 3, 40, 231
ProNGF, 285–289
Proprioceptive neuromuscular facilitation
　　techniques, 24
Prostate-specific antigen (PSA), 169

Proteasome gelatinase A, 309
Protein kinase A (PKA), 307–308, 371
Protein kinase C (PKC), 291, 305, 307,
　　309–310, 370–371, 442–443
Proteoglycan (PG), 232–233, 237
Proteomics, 150–151
Prototypical dementia syndrome associated
　　with PD (PDD)
　behavioral features, 3
　clinical features, 2
　cognitive features, 2–3
　cortical AChE activity, 4
　memory impairment in, 2
　midfrontal ChAT activity, 4
　treatment, 3–6
　visuospatial function in, 2–3
Psychosis, 2, 11
PTEN (protein tyrosine phosphatase with
　　homology to tensin) gene, 13
Putative neuroprotective therapies, 37
Pyramidal signs, 76

Quinine oxidoreductase, 362

Rapamycin, 292
Rapid-eye-movement (REM) sleep
　　disturbances, 3, 38, 43, 78
Rasagiline analogs, 441
Rat neuronal cells, Wnt receptors and
　　Frizzled expression in, 317–322
Receptor for advanced glycation end-
　　products (RAGE), 193
Rehabilitation for PD
　basic requirements for, 25
　clinical research on, 24
　cognitive rehabilitation, 27
　limitations of, 27–28
　neurophysiological approach to, 25–27
　physiological strategies for motor
　　problems in, 26
　problems in, 25
　symptomatic dopaminergic treatment, 23
Resting tremor, 25, 76, 78, 80, 82, 85, 159
Restless legs syndrome, 41
Resveratrol, neuroprotective effects against
　　Aβ-induced toxicity, 293
Retinoic acid, 18
　α-secretase pathway up-regulation by,
　　371–373
Reverse transcription polymerase chain
　　reaction (RT-PCR), 93
Rho/JNK planar cell polarity
　　pathway, 319

Rivastigmine, 4–6, 405, 439–441
ROCO proteins, 14, 82–83
Rotenone models of PD, 36, 60–62, 138

Sagamihara family with LRRK2 gene,
        clinical and pathological
        description of, 75, 84–85
Sandhoff disease, 227
SAP-2 deficiency, 227
SAP precursor deficiency, 227
SBE-luc reporter mice, 176–178
    bioluminescent imaging of excitotoxic
        brain In, 179
    noninvasive imaging of endotoxic brain
        injury in vivo, 179–181
    specificity of, 177–178
Schizophrenia, 158, 317
α-Secretase, 439
    ADAM10 as, 369–370, 394
    pathway up-regulation, 369–373
        downstream signaling systems,
            370–371
        G protein-coupled receptors
            activation, 370–371
        human ADAM10 promoter and,
            371–373
        PACAP-induced activation, 371–342
        by retinoic acid, 371–373
    processing and β-APP and TSEs,
        305–308, 310, 312
β-Secretase, 124, 305, 311, 369, 439
γ-Secretase, 124, 223, 305, 369, 394, 439
Secreted APP (sAPP), 259–262
Selegiline [(-)-deprenyl], 41
Senile dementia, 191
Senile plaques (SPs), 123, 375
Sepsis, 32
Sequence effect, 26–27
Serin palmitoyltransferase (SPT), inhibition
        of, 222
Serotonin transporters (SERT), 60
Single-chain antibodies (ScFvs),
        273–277
Single photon emission computed
        tomography (SPECT), 36, 40
SK-N-MC cells, see Human neuroblastoma
        (SK-N-SH) cells
Smad-binding elements (SBEs), 175–176,
        179–180
Smad proteins, 175, 179
Smad1 proteins, 175
Smad2 proteins, 175, 179
Smad3 proteins, 175, 179

Smad5 proteins, 175
Smad8 proteins, 175
SNCA, mutations in, 171
γ-Soluble NSF attachment protein (SNAP),
        156, 158
Sphingomyelinase, 227
Sphingomyelin (SM), 223
Spongiform encephalopathies, 231–232
Sporadic frontotemporal dementias, 231
Sporadic PD, 12, 14, 35, 57, 67, 83, 97,
        127–128
Statistical parametric mapping (SPM)
        method, 188
Stem cell biology, APP function in, 259–262
Stem cell therapy in AD, 255–262
    dedifferentiation, 258–259
    selection, 256
    transdifferentiation, 256–258
Stereotactic surface projections (SSP)
        method, 188
Stress-activated protein kinase (SAPK), 295
Subcutaneous rotenone exposure models
        of PD, 62
Substantia nigra (SN), 36–37, 40, 43, 117,
        128, 134, 137–138
Sulfated glycosaminoglycans (sGAGs), 232
    biosynthesis of, 232–233
    interactions with
        β-amyloid peptide, 233–235
        prion protein and peptides, 235–237
        α-synuclein, 235–237
        tau protein and peptides, 235
    therapeutic implications, 237–241
Sulfatidase activator deficiency, 227
Sulfotransferase, 362
Superoxide dismutase, 362
Supranuclear gaze palsy, 85–86
SwissProt, 154
Sympathetic ganglion cells, 51
Synaptic integrity and plasticity, 213–214
Synaptic plasma membranes (SPMs),
        390–391
Synaptic scaling, cholinergic neurons role in,
        433–434
Synaptophysin, 405
Synphilin-1 protein, 11
Synucleinopathies, 3, 38, 126–129, 231
α-Synuclein proteins, 9–10, 35, 40, 58, 61
    Interactions with sGAGs, 235–237
    models of PD, 66–67
α-Synuclein (SNCA) gene, 9–10, 12, 63,
        66–67, 76, 97–109, 123, 126–129,
        133, 137–142, 159–160, 232

Synucleopathies, 34
SYPRO ruby-stained spots, 151

TACE, 309–312
Tacrine, 4, 441, 447
Tau
    phosphorylation, 125–126
        intau^{VLW}/GSK3 transgenic mice,
            338–340
        in tau^{VLW}/APP^{SW} transgenic mice,
            339–340
    polymerization and neurodegeneration,
        337–340
    protein interactions with sGAGs, 235
Tauopathies, 79, 88, 108, 126, 129, 133,
    140–142, 231, 375
    and Alzheimer's disease, 124–126
    cell models, 325–334
        aggregation and toxicity, 330
        aggregation of tau in cells, 327–329
        cell lines, 326–327
        fragmentation of tau$_{RD}$ by thrombin-
            like protease, 329–330
        lihibition of aggregation by small
            molecule inhibitors, 331–333
Tau$_{RD}$
    aggregation and toxicity, 330, 333–334
    expression and aggregation, 327–330
        cell toxicity of, 330, 333–334
        lihibition small molecule inhibitors,
            331–333
        inhibition in vitro, 331
        kinetics of, 328–329
        reversibility, 331
    fragmentation by thrombin-like protease,
        329–330
Tau^{VLW}/APP^{SW} transgenic mice, tau
    phosphorylation in, 339–340
Tau^{VLW}/GSK3 transgenic mice, tau
    phosphorylation in, 338–340
Tay-Sachs disease, 227
T-cell factor (TCF), 318
Tea extracts, hippocampal neurons
    protection by, 293
Ten Point Clock-Drawing test, 5
Tensor-based morphometry (TBM)
    methods, 187
Tet-On system, 325, 327
1,2,3,4-Tetrahydro-isoquinolines, 36
Tg2576 APP transgenic mice, 265, 268
TgCRND8 transgenic model, 395
Tg2576 mice, 431

Thiobarbituric acid reactive substance
    (TBARS), 150
Thioflavin-S staining, 275, 414, 418
Thiol methyltransferase, 362
Thiopurine methyltransferase (TPMT), 362
Thorn-shaped astrocytes (TSAs), 375–379
    evaluation of 4R tauopathy and, 377
    frequency in AD, 377–379
    immunohistochemistry, 376–377
    relation with 4R tau score in AD, 378
Thrombosis, 32
Th1 T cells, 269–270
Th2 T cells, 269–270
Transcranial ultrasonography, for PD
    detection, 40
Transcription actors of FoxO family, trophic
    factors effects on, 292
Transforming growth factor-β1 (TGFβ1)
    protein, 175–179
Transforming growth factor-β2 (TGFβ2)
    protein, 175, 177, 179
Transforming growth factor-β3 (TGFβ3)
    protein, 175, 177, 179
Transforming growth factor-β (TGFβ)
    proteins, 53–54, 175–179
Transgenic Drosophila, 66
Transmissible spongiform encephalopathies
    (TSEs), 305
    physiological processing, 305–312
        ADAM9 and, 305, 311–312
        ADAM10 and, 305, 309–313
        ADAM17 and, 305, 309–310
        up-regulation by PKC agonists,
            307–309
    α-secretase processing, 305–308
Traumatic brain injury, 135, 137
Trimethylamine N-oxidase, 362
Triose phosphate isomerase (TPI), 156
TrkB receptors, 279, 285–289
Trophic factors
    effects on transcription actors of FoxO
        family, 292
    neuroprotective effects of, 291–295
β-Tubulin, 405–406
Tumor necrosis factor-α (TNFα),
    92–94, 309
TV3326 [-ethyl methyl carbamate], 442
TV3279 [-ethyl methyl carbamate], 442
Two-dimensional polyacrylamide gel
    electrophoresis (2D-PAGE), 151
Tyrosine hydroxylase (TH) protein, 61, 92
Tyrosine kinase, 83, 128

Ubiquitin-activating enzyme (E1), 10
Ubiquitin-conjugating enzyme (E2), 10
Ubiquitin C-terminal hydrolase isozyme L1
(Uch-L1), 12, 58, 64–65, 76, 97,
156–157
Ubiquitin proteasome (UP) system, 10, 58,
63, 69, 334
inhibition cause of PD, 67–68
Ubiquitin-protein ligase (E3), 10, 57, 137
UDP-glucuronosyltransferases, 362
Unified Parkinson Disease Rating Scale
(UPDRS), 32–33, 36
Unified Parkinson Disease Rating Scale
(UPDRS) Part III, 5–6
Uridine 5'-triphosphate glucuronosyl
transferases (UGTs), 362
Urinary urgency, 11

Vascular endothelial growth factor
(VEGF), 53
VDAC-1, 156
Vertical gaze palsy, 76
Vesicular acetylcholine transporter
(vAChT), 286, 432
Vesicular glutamate transporter
(VGLUT1), 406

Vision tests, for PD detection, 41
Visuospatial function, 2–3
Vitamin C, 136–137
Vitamin E, 136–137, 155, 454
Volume of interest analyses (VOI) method, 188
Von Economo encephalitis lethargica, 36
Voxel-based morphometry (VBM)
methods, 187

Wild-type C57BL/6J mice, 344–351
F4/80-positive microglia, 345, 348–351
IL-1β-positive cells, 345–348, 351
immunohistochemistry, 345
intracerebroventricular injection of
LPS, 344
mac-2-positive microglia, 345, 348–351
NF-κB-positive cells, 345–348, 350–351
Wnt/Ca^{2+} pathway, 319
Wnt receptors expression, in rat neuronal
cells, 317–322

Xenogeneic monkey brain, PC12 cells
grafting to, 52
3xTg-AD mice, 298, 300–301

Yapsin, 309

Printed in the United Kingdom
by Lightning Source UK Ltd.
128272UK00001B/14/A